AF360790

Challenging the Status Quo to Shape Food Systems Transformation from a Nutritional and Food Security Perspective

Challenging the Status Quo to Shape Food Systems Transformation from a Nutritional and Food Security Perspective

Editors

António Raposo
Renata Puppin Zandonadi
Raquel Braz Assunção Botelho

MDPI • Basel • Beijing • Wuhan • Barcelona • Belgrade • Manchester • Tokyo • Cluj • Tianjin

Editors

António Raposo
CBIOS (Research Center for
Biosciences and Health
Technologies), Universidade
Lusófona de Humanidades e
Tecnologias, Campo Grande 376,
1749-024
Lisboa
Portugal

Renata Puppin Zandonadi
Department of Nutrition,
Faculty of Health Sciences
University of Brasilia
Brasilia
Brazil

Raquel Braz Assunção Botelho
Department of Nutrition,
Faculty of Health Sciences
University of Brasilia
Brasilia
Brazil

Editorial Office
MDPI
St. Alban-Anlage 66
4052 Basel, Switzerland

This is a reprint of articles from the Special Issue published online in the open access journal *Foods* (ISSN 2304-8158) (available at: www.mdpi.com/journal/foods/special_issues/Challenging_Food_Security).

For citation purposes, cite each article independently as indicated on the article page online and as indicated below:

LastName, A.A.; LastName, B.B.; LastName, C.C. Article Title. *Journal Name* **Year**, *Volume Number*, Page Range.

ISBN 978-3-0365-3718-4 (Hbk)
ISBN 978-3-0365-3717-7 (PDF)

Contents

About the Editors

António Raposo

António Raposo is an Assistant Professor at Universidade Lusófona de Humanidades e Tecnologias and he is an Integrated Member of CBIOS (Research Center for Biosciences and Health Technologies). He graduated in Nutritional Sciences by Instituto Superior de Ciências da Saúde Egas Moniz, Portugal, 2009, and obtained his PhD with European Mention in Animal Health and Food Safety from the University Institute of Animal Health and Food Safety, University of Las Palmas de Gran Canaria, Spain, 2013. His main research interests are: studies on the utilization of Catostylus tagi jellyfish in health sciences, particularly as a food ingredient; food habits; food safety evaluation; food innovation; natural food products; food security; and sustainability. He is a member of the editorial boards and an invited reviewer of relevant international peer-reviewed journals in his research field. He has published more than 63 papers in indexed JCR journals, as well as scientific book chapters and a patent model, and co-authored a book in English with a special focus on Nutrition, Food Security and Food Safety Sciences. He has been a Guest Editor of several Special Issues published in high-impact JCR journals such as {Foods}, {Sustainability} and {International Journal of Environmental Research and Public Health}. He has collaborated as a Visiting Professor in Portuguese, Spanish, Chilean, and Vietnamese Universities.

Renata Puppin Zandonadi

Renata Puppin Zandonadi is a researcher and Associate Professor at the University of Brasilia Department of Nutrition and a permanent member of the Postgraduate Program in Human Nutrition (University of Brasília). She graduated in Nutrition from the University of Brasilia and obtained a Master's degree in Human Nutrition (University of Brasilia) and a Ph.D. in Health Sciences (University of Brasilia). She has co-authored and published one book, fifty book chapters, and 115 articles in peer-reviewed journals. She is a reviewer and member of the editorial board of international journals. Her main research interests are focused on nutrition, food, food-related disorders, eating habits, quality of life, sustainability, food security and safety.

Raquel Braz Assunção Botelho

Raquel Botelho graduated in Nutrition from the University of Brasilia and obtained a Master's degree in Food Science from the University of Campinas and Ph.D. in Health Sciences from the University of Brasília. Currently, she is a researcher and Associate Professor of the Nutrition Department at the University of Brasilia and a member of the Postgraduate Program in Human Nutrition. Professor Raquel Botelho acts as a reviewer and member of the editorial board of international journals and has co-authored and published more than 90 articles in scientific journals, seven books, and twenty-one book chapters. Her field of expertise is on nutrition, gastronomy, culinary, food, food-related disorders, eating habits, quality of life, sustainability, food security and safety.

Preface to "Challenging the Status Quo to Shape Food Systems Transformation from a Nutritional and Food Security Perspective"

Food security and nutrition have been emerging as international priorities and have been widely discussed to ensure food security (in all aspects). Efforts are being made to reduce hunger and malnutrition to promote health for individuals, the population, and the planet and ensure strong economies and sustainable development. Unfortunately, what we are experiencing today is that the food system, involving agriculture, livestock, and the food and foodservice industry, as they currently operate, creates challenges to ensuring food security and adequate nutrition. The necessary investment is not always sustainable. Therefore, there is a need to positively change food systems to ensure that the food produced is accessible, sustainable, safe, healthy, and equitable. Attitudes and knowledge of health, food, and nutrition can be key factors in facing food insecurity in several countries. To achieve this goal, attitude changes and improving knowledge about food (involving knowledge of the food chain, health, nutrition, and sustainability) are imperative to face malnutrition and planet degradation by promoting good food choices. In this sense, this book is composed of 23 papers on food systems and food and nutrition security aiming to connect the importance of food systems to change nutrition and food security worldwide. This book involves authors from 19 different countries, including Bangladesh, Brazil, China, Greece, India, Korea, Malaysia, Nigeria, Norway, Poland, Portugal, Korea, Russia, Saudi Arabia, South Africa, Spain, Taiwan, Thailand, The Netherlands, and the USA, reinforcing that attitudes and knowledge of health, food, and nutrition are key factors for facing food insecurity in several countries. The editors are grateful to their families, friends, and colleagues for the support provided. We thank the Brazilian National Council for Scientific and Technological Development (CNPq) for the scientific support of professors Raquel Botelho and Renata Zandonadi. We would also like to thank and congratulate all the authors and the entire MDPI team who allowed the construction of this book to become a reality.

António Raposo, Renata Puppin Zandonadi, and Raquel Braz Assunção Botelho
Editors

 *foods*

Editorial

Challenging the Status Quo to Shape Food Systems Transformation from a Nutritional and Food Security Perspective

António Raposo [1,*], Renata Puppin Zandonadi [2] and Raquel Braz Assunção Botelho [2]

[1] CBIOS (Research Center for Biosciences and Health Technologies), Universidade Lusófona de Humanidades e Tecnologias, Campo Grande 376, 1749-024 Lisboa, Portugal
[2] Department of Nutrition, Faculty of Health Sciences, University of Brasilia, Brasilia 70910-900, Brazil; renatapz@unb.br (R.P.Z.); raquelbotelho@unb.br (R.B.A.B.)
[*] Correspondence: antonio.raposo@ulusofona.pt

Keywords: consumer; environment; food policy; food production; food safety; food security; nutrition; sustainability

Citation: Raposo, A.; Zandonadi, R.P.; Botelho, R.B.A. Challenging the Status Quo to Shape Food Systems Transformation from a Nutritional and Food Security Perspective. *Foods* **2022**, *11*, 604. https://doi.org/10.3390/foods11040604

Received: 16 February 2022
Accepted: 17 February 2022
Published: 20 February 2022

Publisher's Note: MDPI stays neutral with regard to jurisdictional claims in published maps and institutional affiliations.

Food security and nutrition have been prominent elements of the international development agenda. Over time, however, development priorities and challenges have oscillated, and the investment required has not been sustained. A broader consensus has emerged: one that guarantees food security and, in all its aspects, reduces hunger and malnutrition to promote strong economies, human and planetary health, and sustainable development. Our moral imperative is to positively change food systems to ensure that the food we produce is accessible, sustainable, safe, healthy, and equitable for everyone. Therefore, this special issue about Food Systems and Nutritional and Food Security focuses on connecting the importance of food systems to change nutrition and food security around the globe.

In this context, attitude and knowledge of health, food, and nutrition can be one of the keys to facing food insecurity in several countries. Health and nutritional education are imperative to fight malnutrition [1–4]. Knowledge in these aspects is essential in tackling misinformation and promoting good food choices. Since birth, it can be constructed along life using households, schools, the internet, and other channels to allow learning on food, health, and nutrition and stimulate adequate food choices [1–7].

Adequate and healthy diets imply choosing and consuming balanced and adequate foods on nutrients and amount, variety, and sustainable aspects. Therefore, achieving the main social, economic, environmental, cultural, and security goals [1,3,6–16]. Food production must also be alert to the waste produced in the food chain related to environmental pollution and food waste that could be used to feed food-insecure people [10,15,17]. In order to support sustainable and healthy diets and food security, it is also essential to search for alternatives to stimulate local production and regional food consumption. Other priorities are reducing animal suffering and meat consumption, food production in sufficient quantity and quality with the least possible waste production, and understanding the main aspects of food, nutrition, society, and the environment [11–22].

During these past two years of the COVID-19 pandemic, the world has faced new food security and safety challenges. Firstly, food sales worldwide became more complex, and farmers faced many overstocked products leading to food loss [8]. On the other hand, countries' economic crisis conducted to more unemployment, therefore, more food insecurity. Governments need to learn from this period and improve the food supply chain with innovative techniques and logistics to have more organized food systems [8]. To guarantee the four pillars of food security, availability, access, stability, and utilization, Sun and Zhang [9] emphasize the importance of trade openness and sustainable food system strategies. Guiné et al. [7] discuss that the agrifood supply chain should be improved through policies worldwide to promote access to healthy and sustainable food. Authors also call the attention to researchers to rethink the four pillars of food security to include new dimensions like climate change.

Author Contributions: Conceptualization, A.R., R.P.Z. and R.B.A.B.; methodology, A.R., R.P.Z. and R.B.A.B.; resources, A.R., R.P.Z. and R.B.A.B.; data curation, A.R., R.P.Z. and R.B.A.B.; writing—original draft preparation, R.P.Z. and R.B.A.B.; writing—review and editing, A.R., R.P.Z. and R.B.A.B.; visualization, A.R., R.P.Z. and R.B.A.B. All authors have read and agreed to the published version of the manuscript.

Funding: This research received no external funding.

Institutional Review Board Statement: Not applicable.

Informed Consent Statement: Not applicable.

Acknowledgments: We would like to thank and congratulate all the authors who published their manuscripts to this special issue with FOODS/MDPI for this valuable data collection. We also thank the reviewers, the FOODS Editor-in-chief, and the entire MDPI team, without which it would be impossible to construct this successful special issue. We also thank the Brazilian National Council for Scientific and Technological Development (CNPq) for the Support.

Conflicts of Interest: The authors declare no conflict of interest.

References

1. Shapu, R.C.; Ismail, S.; Lim, P.Y.; Ahmad, N.; Garba, H.; Njodi, I.A. Effectiveness of Triple Benefit Health Education Intervention on Knowledge, Attitude and Food Security towards Malnutrition among Adolescent Girls in Borno State, Nigeria. *Foods* **2022**, *11*, 130. [CrossRef] [PubMed]
2. Florença, S.G.; Ferreira, M.; Lacerda, I.; Maia, A. Food Myths or Food Facts? Study about Perceptions and Knowledge in a Portuguese Sample. *Foods* **2021**, *10*, 2746. [CrossRef] [PubMed]
3. dos Santos, E.B.; Maynard, D.D.C.; Zandonadi, R.P.; Raposo, A.; Botelho, R.B.A. Sustainability Recommendations and Practices in School Feeding: A Systematic Review. *Foods* **2022**, *11*, 176. [CrossRef] [PubMed]
4. de Queiroz, F.L.N.; Nakano, E.Y.; Botelho, R.B.A.; Ginani, V.C.; Raposo, A.; Zandonadi, R.P. Eating Competence among Brazilian Adults: A Comparison between before and during the COVID-19 Pandemic. *Foods* **2021**, *10*, 2001. [CrossRef] [PubMed]
5. Cupertino, A.; Maynard, D.; Queiroz, F.; Zandonadi, R.; Ginani, V.; Raposo, A.; Saraiva, A.; Botelho, R. How Are School Menus Evaluated in Different Countries? A Systematic Review. *Foods* **2021**, *10*, 374. [CrossRef] [PubMed]
6. Lu, Y.; Zhang, Y.; Hong, Y.; He, L.; Chen, Y. Experiences and Lessons from Agri-Food System Transformation for Sustainable Food Security: A Review of China's Practices. *Foods* **2022**, *11*, 137. [CrossRef]
7. Guiné, R.D.P.F.; de Jesus Pato, M.L.; da Costa, C.A.; Costa, D.D.V.T.A.D.; da Silva, P.B.C.; Martinho, V.J.P.D. Food Security and Sustainability: Discussing the Four Pillars to Encompass Other Dimensions. *Foods* **2021**, *10*, 2732. [CrossRef]
8. Pu, M.; Chen, X.; Zhong, Y. Overstocked Agricultural Produce and Emergency Supply System in the COVID-19 Pandemic: Responses from China. *Foods* **2021**, *10*, 3027. [CrossRef]
9. Sun, Z.; Zhang, D. Impact of Trade Openness on Food Security: Evidence from Panel Data for Central Asian Countries. *Foods* **2021**, *10*, 3012. [CrossRef]
10. Rao, M.; Bilić, L.; Duwel, J.; Herentrey, C.; Lehtinen, E.; Lee, M.; Calixto, M.A.D.; Bast, A.; de Boer, A. Let Them Eat Fish!—Exploring the Possibility of Utilising Unwanted Catch in Food Bank Parcels in The Netherlands. *Foods* **2021**, *10*, 2775. [CrossRef]
11. Chaosap, C.; Sahatsanon, K.; Sitthigripong, R.; Sawanon, S.; Setakul, J. The Effects of Using Pineapple Stem Starch as an Alternative Starch Source and Ageing Period on Meat Quality, Texture Profile, Ribonucleotide Content, and Fatty Acid Composition of Longissimus Thoracis of Fattening Dairy Steers. *Foods* **2021**, *10*, 2319. [CrossRef] [PubMed]
12. Uyeh, D.D.; Asem-Hiablie, S.; Park, T.; Kim, K.; Mikhaylov, A.; Woo, S.; Ha, Y. Could *Japonica* Rice Be an Alternative Variety for Increased Global Food Security and Climate Change Mitigation? *Foods* **2021**, *10*, 1869. [CrossRef] [PubMed]
13. Soares, P.; Martinelli, S.; Davó-Blanes, M.; Fabri, R.; Clemente-Gómez, V.; Cavalli, S. Government Policy for the Procurement of Food from Local Family Farming in Brazilian Public Institutions. *Foods* **2021**, *10*, 1604. [CrossRef] [PubMed]
14. Njoga, U.; Njoga, E.; Nwobi, O.; Abonyi, F.; Edeh, H.; Ajibo, F.; Azor, N.; Bello, A.; Upadhyay, A.; Okpala, C.; et al. Slaughter Conditions and Slaughtering of Pregnant Cows in Southeast Nigeria: Implications to Meat Quality, Food Safety and Security. *Foods* **2021**, *10*, 1298. [CrossRef]
15. Althumiri, N.A.; Basyouni, M.H.; Duhaim, A.F.; AlMousa, N.; AlJuwaysim, M.F.; BinDhim, N.F. Understanding Food Waste, Food Insecurity, and the Gap between the Two: A Nationwide Cross-Sectional Study in Saudi Arabia. *Foods* **2021**, *10*, 681. [CrossRef]
16. Bvenura, C.; Witbooi, H.; Kambizi, L. Pigmented Potatoes: A Potential Panacea for Food and Nutrition Security and Health? *Foods* **2022**, *11*, 175. [CrossRef]

17. Poto, M.P.; Elvevoll, E.O.; Sundset, M.A.; Eilertsen, K.-E.; Morel, M.; Jensen, I.-J. Suggestions for a Systematic Regulatory Approach to Ocean Plastics. *Foods* **2021**, *10*, 2197. [CrossRef]
18. Anastasiadis, F.; Apostolidou, I.; Michailidis, A. Food Traceability: A Consumer-Centric Supply Chain Approach on Sustainable Tomato. *Foods* **2021**, *10*, 543. [CrossRef]
19. Rahman, M.; Islam, R.; Hasan, S.; Zzaman, W.; Rana, R.; Ahmed, S.; Roy, M.; Sayem, A.; Matin, A.; Raposo, A.; et al. A Comprehensive Review on Bio-Preservation of Bread: An Approach to Adopt Wholesome Strategies. *Foods* **2022**, *11*, 319. [CrossRef]
20. Suhandoko, A.A.; Chen, D.C.-B.; Yang, S.-H. Meat Traceability: Traditional Market Shoppers' Preferences and Willingness-to-Pay for Additional Information in Taiwan. *Foods* **2021**, *10*, 1819. [CrossRef]
21. Iftekhar, A.; Cui, X.; Yang, Y. Blockchain Technology for Trustworthy Operations in the Management of Strategic Grain Reserves. *Foods* **2021**, *10*, 2323. [CrossRef]
22. Chen, L.; Cui, X.; Li, W. Meta-Learning for Few-Shot Plant Disease Detection. *Foods* **2021**, *10*, 2441. [CrossRef]

foods

MDPI

Review

How Are School Menus Evaluated in Different Countries? A Systematic Review

Alessandra Fabrino Cupertino [1], Dayanne da Costa Maynard [1], Fabiana Lopes Nalon de Queiroz [1], Renata Puppin Zandonadi [1], Verônica Cortez Ginani [1], António Raposo [2,*], Ariana Saraiva [3] and Raquel Braz Assunção Botelho [1,*]

[1] Department of Nutrition, University of Brasília, Brasília 70910-900, Brazil; acupertino1006@gmail.com (A.F.C.); day_nut@yahoo.com.br (D.d.C.M.); fabinalon@hotmail.com (F.L.N.d.Q.); renatapz@unb.br (R.P.Z.); vcginani@gmail.com (V.C.G.)

[2] CBIOS (Research Center for Biosciences and Health Technologies), Universidade Lusófona de Humanidades e Tecnologias, Campo Grande 376, 1749-024 Lisboa, Portugal

[3] Department of Animal Pathology and Production, Bromatology and Food Technology, Faculty of Veterinary, Universidad de Las Palmas de Gran Canaria, Trasmontaña s/n, 35413 Arucas, Spain; ariana_23@outlook.pt

* Correspondence: antonio.raposo@ulusofona.pt (A.R.); raquelbabotelho@gmail.com (R.B.A.B.)

Abstract: School meals should focus on quality of life issues, particularly on reducing food shortages, overweight, obesity and its consequences. As an essential tool for quality assurance, creating the menu is an activity of great complexity and requires multidisciplinary knowledge. This activity covers the observation of countless aspects of quality, highlighting nutritional, sensory, cultural, hygienic, and sanitary issues, among others. This study aims to identify and analyze instruments and methods to evaluate school menus in different countries. The authors developed specific search strategies for Scopus, Web of Science, Science Direct, Pubmed, Lilacs, ProQuest Global, and Google Scholar. The included studies' methodological quality was assessed using the statistical analysis and meta-analysis review tool (MASTARI). A total of 16 cross-sectional studies met the inclusion criteria and were analyzed. Brazil and Spain were the countries that presented the highest number of studies ($n = 5$; 31.25% for each). The majority of the studies have a qualitative approach ($n = 12$, 75%), and only 25% ($n = 4$) of the studies present quantitative assessment methods to evaluate school menus. No school menu assessment tools were found to assess all aspects of menu planning. The results show a lack of a methodology or of instruments for evaluating the menus offered for school meals that can contribute to better dietary care offered to students.

Keywords: evaluation; instrument; menu; school feeding

Citation: Cupertino, A.F.; Maynard, D.d.C.; Queiroz, F.L.N.d.; Zandonadi, R.P.; Ginani, V.C.; Raposo, A.; Saraiva, A.; Botelho, R.B.A. How Are School Menus Evaluated in Different Countries? A Systematic Review. *Foods* **2021**, *10*, 374. https://doi.org/10.3390/foods10020374

Academic Editor: Leslie D. Bourquin

Received: 26 December 2020

Accepted: 5 February 2021

Published: 9 February 2021

Publisher's Note: MDPI stays neutral with regard to jurisdictional claims in published maps and institutional affiliations.

1. Introduction

Globally, inadequate nutrition is one of the main challenges representing a threat to individuals' health, including their well-being and productivity. This results in high socioeconomic costs for societies in all regions of the world. Therefore, early healthy eating habits are essential to reduce the risk of immediate and long-term health problems [1].

The school has a strong influence on the students' behavior, and when their environment is adequate, it is considered favorable for the formation and consolidation of healthy habits [2]. According to the Food and Agriculture Organization (FAO), about 40 million children under five years are overweight, and about 120 million children (ages 5 to 10) and adolescents (ages 11 to 19) are obese [1]. Based on the need to change this scenario, the World Health Organization (WHO) defined in 2001 a global strategy for feeding young children, considering respect, protection, and fulfillment of human rights. It recommends that school menus prioritize promoting healthy and adequate food for children and adolescents. Thus, it is recommended to use varied and safe foods, promoting eating habits that contribute to students' growth and development and academic performance [3].

Despite the menu's definition as a list of dishes of a planned meal, the menu goes beyond its initial concept in the scholarly environment. It imposes itself as the vehicle to ensure that school meals meet their proposed aims. The menu may offer foods that contribute to the students' health, considering nutritional, sensory, cultural, and micro-biological aspects [4]. Therefore, the menu also contributes to the formation of healthy eating habits and the preservation of the environment and of food culture, providing nutritional education [5].

Planning and evaluating menus is essential to meet recommendations and contribute to adequate nutrition. School environments must offer healthy and nutritious meals through balanced menus. Families, educators, and governments from different countries should contribute to the formulation, implementation, monitoring, and evaluation of a comprehensive national policy on school feeding [3,6,7].

Despite the existence of criteria and different menu planning methods, there is no consensus on the best protocol for evaluating them, especially concerning school menus [8,9]. In general, in schools, a qualitative assessment of menus is prioritized. It is not clear in the literature whether the assessment instruments meet the recommendations of school feeding programs regarding all their aims, nor how the instruments are applied. It is unknown whether the existing instruments are used to evaluate the executed or planned menus [10,11].

School menu evaluation protocols are essential to ensure their planning. These protocols, defined in this study as a systematic way of evaluating menus, also need to be assessed, disseminated and implemented across different realities. Thus, it is essential to know what protocols are available, what kind of instruments or methods, and how they work, ensuring they comply with the WHO/FAO assumptions about healthy eating. This study is justified by the need to understand how school menus are planned and evaluated worldwide as they directly impact the quality of meals (breakfast, snacks, lunch and/or dinner) served in schools. Therefore, the present systematic review's objective is to identify and analyze protocols for evaluating available school menus, with no restrictions by country for studies included in the review.

2. Materials and Methods

This systematic review focused on protocols for evaluating school menus world-wide. It was prepared according to the report items for systematic reviews and meta-analyses (PRISMA) and its Checklist [12–16]. The protocol was performed according to the following steps.

2.1. Eligibility Criteria

The inclusion criteria were studies of school menu assessment instruments and school menus related to governmental school programs, with no date, language, or publication status limits. The exclusion criteria applied were: (1) comments, letters, conference, review, abstracts, and books; (2) studies in private schools, unrelated to government-subsidized School Feeding Programs; (3) studies that were not performed on school food services; (4) school health programs (Table S1—Supplementary Materials).

2.2. Information Source

Detailed individual search strategies were developed for each database: Pubmed, Scopus, Web of Science, Science Direct, and Lilacs. A search for gray literature was carried out on Google Scholar and for dissertations and theses in ProQuest Global. The last search for all databases was carried out on 7 July 2020. Researchers carefully examined the selected study's reference lists for full reading if any study was not retrieved during the search.

2.3. Search Strategy

The appropriate combination of truncation and keywords were selected and adapted for each database (Table S2—Supplementary Materials). Rayyan software (Qatar Computer

Research Institute, QCRI) was used to select and exclude duplicate articles, and all bibliographic references were included using the Mendeley desktop software. Table S2 shows the search strategy used for the six databases.

2.4. Study Selection

The study screening process was carried out in three phases, with three independent researchers. In phase I, researchers I (A.F.B.C.) and II (D.C.M.) independently selected the articles according to their titles and abstracts. Articles that did not meet the previously established inclusion criteria were excluded. In phase II, a third independent researcher III (F.L.N.Q.) selected the conflicting articles from phase I. In phase III, the selected articles were fully read, and those that met the inclusion criteria were included. Researchers I and II critically evaluated the reference lists of the selected studies and extracted data. Other than titles and abstracts written in English, translation was performed using the Google Translator tool for studies written in other languages.

2.5. Data Collection Process

The selected studies' following characteristics were collected: authors and year of publication, country of research, objective, study design, type of protocol, and school grade focus. The aspects assessed by the protocols were: nutritional, sensory, cultural, sanitary hygiene, and sustainability. Calibration exercises were performed before starting the review to ensure consistency between the two reviewers. Disagreements were solved by discussion, and the third author (F.L.N.Q.) decided on disagreements when these were not solved. These data were synthesized in a standardized table (Table 1).

Table 1. Characteristics of the studies included in the systematic review regarding menu planning evaluation.

Year	Authors	Country	Study Design	Subject	Type of Protocol		Specific Name		Presence of Evaluation Criteria				Type of Data	
					Instrument	Method	Yes	No	Nutritional	Sensorial	Cultural	Sustainability	Qualitative	Quantitative
2015	De Mateo Silleras et al. [17–22]	Spain	Prospective longitudinal study	School menu	X		COMES		YES	NO	NO	NO	X	
2009	Dubuisson et al. [23]	France	Descriptive, cross-sectional study	Elementary school and high school		X		X	YES	NO	NO	NO	X	
2017	Evans and Cade [24]	England	Descriptive, cross-sectional study	Scholars from 08 to 09 years		X		X	YES	NO	NO	NO		X
2018	Llorens-Ivorra et al. [19]	Spain	Observational, cross-sectional study	Primary school (6 to 14 years)	X		EQMES		YES	YES	NO	NO	X	
2013	Patterson et al. [25]	Sweden	Experimental study	Primary school	X			X	YES	NO	NO	YES		X
2020	Martins Rodrigues et al. [2]	Brazil	Descriptive and transversal study	Pre-school	X			X	YES	NO	NO	YES	X	
2013	Longo-Silva et al. [14]	Brazil	Experimental study	Public and private schools		X		X	YES	YES	NO	NO	X	X
2015	Gregoric et al. [26]	Slovenia	Descriptive, cross-sectional study	Primary school (6 to 14 years)		X		X	YES	NO	NO	NO	X	X
2018	González et al. [18]	Spain	Descriptive, cross-sectional study.	Primary school (6 to 14 years)		X		X	YES	YES	NO	NO	X	
2019	Myers et al. [7]	Australia	Descriptive, cross-sectional study.	Public and private schools		X		X	YES	NO	NO	NO	X	
2015	Da Silva Bastos Soares et al. [13]	Brazil	Descriptive, cross-sectional study.	School menus		X		X	YES	YES	NO	NO	X	
2017	Moreira Sampaio et al. [15]	Brazil	Cross-sectional study	School menus		X		X	YES	NO	NO	NO		X
2015	Vidal et al. [16]	Brazil	Cross-sectional study	School menus		X		X	YES	YES	NO	NO	X	
2015	Uriarte et al. [21]	Spain	Descriptive cross-sectional study	School menus		X		X	YES	NO	NO	NO	X	
2019	Soares et al. [20]	Spain	Transversal study	School menus		X		X	YES	YES	NO	NO	X	
2012	Therre et al. [22]	Germany	Transversal study	School menus		X		X	YES	NO	NO	NO	X	

2.6. Risk of Individual Bias in the Included Studies

The quality criteria were synthesized using a statistical review assessment instrument (MASTARI) and the Joanna Briggs Institute protocol to assess the studies' risk of bias. The instrument for assessing the risk of bias included seven questions:

1. Were the analyzed indicators characterized?
2. Were the protocols implemented in the menu evaluation?

3. Did the evaluated indicator respond positively to implementation?
4. Is the study design adequate?
5. Was the statistical analysis adequate to the objective of the study?
6. Did the results answer the main question?
7. In the case of school mentoring services, was a sample of establishments selected to analyze the representative and randomly determined indicators?

After the analysis, the risk of bias was categorized (Table S3) as "High" when the study reached up to 49% of "yes" scores; "Moderate" when the study reached 50–69% of "yes" scores; and "Low" when the study reached more than 70% of "yes" scores.

3. Results

From the 3189 studies initially selected, 52 were selected via their abstracts. After reading full texts, 16 studies met the eligibility criteria and were included in the systematic review. Figure 1 presents a flowchart of the systematic review search process.

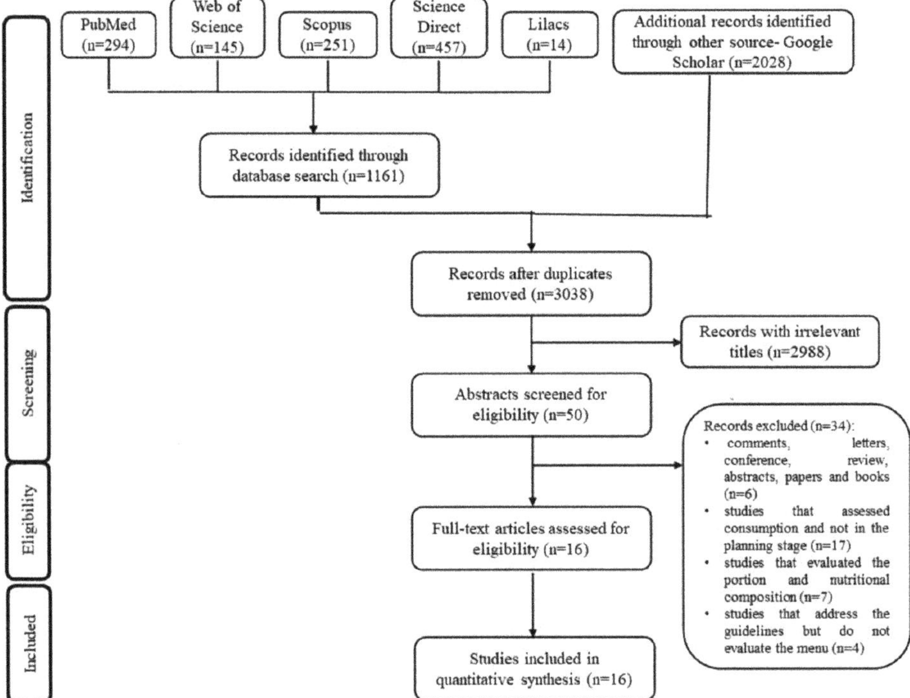

Figure 1. Flowchart of the systematic review search process (adapted from the PRISMA protocol).

3.1. Characterization of Studies

The 16 studies were from Latin America, Europe, and Australia. There is a lack of studies from North America, Central America, Asia, and Africa. The studies presented varied methods for evaluating school menus and challenges in meeting the established recommendations.

Table 1 summarizes the studies included in this review concerning the country and the type of instrument or method for menu evaluation. The studies included in the systematic review were conducted in Brazil ($n = 5$) [2,13–16]; Spain ($n = 5$) [17–21]; Australia ($n = 1$) [7]; Germany ($n = 1$) [22]; France ($n = 1$) [23]; England ($n = 1$) [24]; Sweden ($n = 1$) [25] and Slovenia ($n = 1$) [26], from 2009 to 2018 (Table 1).

Brazil and Spain were the countries that presented the highest number of studies (n = 5; 31.25% for each). In Brazil, the studies mainly discussed a checklist to evaluate the school menu (n = 1) [2], adequacy of serving portions meeting the premises of the national school feeding program (n = 1) [15], and the method of evaluating school menus by the Qualitative Analysis of Menu Preparations (n = 3) [13,14,16]. In Spain, a Brazilian researcher developed a study exploring a method related to the Spanish school feeding program (n = 1) [20]. The study was conducted in a Spanish research center. The number of studies and the public policy presence in studies of school children and adolescents justify Brazil and Spain having the same number of studies (n = 5; 32.25%) [17–21].

In Slovenia, Gregoric [26] proposed an assessment of school menus in three stages. The first was an interview with school managers, a qualitative assessment of menus via the development of a menu quality index defined by food recommendations from the National Dietary Guidelines (NDG). There is also a quantitative assessment of menus in a random sample (considering their weight and nutritional quality). For this last stage, the menus were randomly selected. The served portions were weighed, and the information on recipe, preparation, and techniques was acquired by dietitians from the Institute of Public Health using pre-established methodological instructions and ending with the calculation of nutritional composition.

In Sweden, the study was based on six domains: food groups, nutritional adequacy, food hygiene, eating behavior, sustainability, and political structure. These items were distributed across 110 questions with an organizational, pedagogical, and nutritional approach to be completed by schools. School menus can be classified as probably fulfilled, possibly fulfilled, and unlikely to comply with the recommendations. In the nutritional approach, fat, iron, vitamin D, and fibers were explicitly analyzed by adaptation to national recommendations, which could be assessed as adequate, almost adequate, or inadequate. The instrument used in this study is flexible and can be adapted to other realities, such as other countries seeking this approach model. Access to the instrument is open and available on the online platform [25].

The studies by Myers [7], Dubuisson [23], and Therre et al. [22], conducted respectively in Australia, France, and Germany, similarly approach qualitatively data on school menus in terms of variety, and quality of food, making a comparison with nutritional recommendations. They also approached food and dishes' frequency, looking for information to better evaluate the menu's quality.

3.2. School Menu Assessment Instruments

The instruments used in these studies were: a dietary questionnaire (comedores escolares—COMES) [17], EQ-MEs [19], both developed in Spain, Skolmatsverige [25] from Sweden, and CheckList [2] from Brazil. Of the instruments for evaluating the school menu, only 50% (n = 2) were identified as COMES and EQ-MEs. These are qualitative instruments developed by researchers to evaluate menus regarding the country's national recommendations. Most of the studies have a qualitative approach (n = 13, 81.25%), and only 31.25% (n = 5) of the studies present quantitative data (Table 1).

The studies by Da Silva Bastos Soares et al. [13], Vidal et al. [16], and Longo-Silva [14] presented the Qualitative Evaluation of Menu Preparation (QEMP) as a method of qualitative evaluation of menus using variables related to food and culinary method.

The study by Moreira Sampaio [15] brought a quantitative approach. It sought to assess the portions' nutritional composition, also evaluating the differences between the offered menu and the planned menu. As a result of this study, two school menu proposals were developed based on the nutritional recommendations of the National School Feeding Program (PNAE) [27]. A list of substitutions for handlers to meet possible menu changes was also developed.

Martins Rodrigues [2] proposed and validated a qualitative checklist instrument to evaluate school menus, emphasizing nutritional, sanitary, and sustainability aspects. Based on Brazilian legislation, an instrument with 136 questions was developed, which includes

62 items divided into eight blocks for sanitary hygienic aspects, physical structure, types of equipment, and utensils; 36 items divided into four blocks for nutritional aspects, school menus, and nutritional recommendations for schoolchildren; and 38 items with approaches to sustainability and good environmental practices. There are three possible answers for each item: yes (adequate), no (inadequate), and does not apply. At the end of the checklist's application, the menu and the school food service are evaluated as excellent, good, regular, bad, and terrible. The percentages of adequacy were: excellent—81 to 100; good—61 to 80; regular—41 to 60; bad—21 to 40; and terrible—0 to 20.

Studies in Spain focused on developing researchers' methods for qualitative evaluation of school menus' nutritional balance, meeting national recommendations [17–19,21].

Among the instruments developed in Spain, the COMES questionnaire [17] and the EQ-MEs [19] qualitatively evaluate school menus. COMES includes a table with 15 items. The first eight assess the frequency of food groups. The other seven assess the menus' general characteristics such as fried food, various preparations/ dishes, food preferences, menu rotation, and type of fat used.

The EQ-MEs complements the COMES, with a list of 17 items, specifying food groups and culinary methods, based on recommendations from the Spanish government and a panel of experts. The EQ-MEs assess the quality of school menus and their suitability regarding nutritional aspects in the child population when compared to national recommendations [17,19].

Recently, Soares et al. [20] developed a study in Spain to evaluate the implementation of the Consensus Document on Food in Educational Centers (DCSECE), which includes a minimum set of indicators for the evaluation and follow-up of nutritional recommendations that serve to guide menu planning and how school canteens must respond.

Evans and Cade [24] developed a Quality Index (ranging from 0–21) as a dietary assessment method with nutritional calculation. The portions were weighed, evaluated, and scored using a list of thirteen nutrients and eight foods (five healthy and three restricted/unhealthy). The studies by Myers [7], Dubuisson [23], and Therre et al. [22] emphasize the variety of studies with a qualitative evaluation of school menus, and the difficulty of extending this evaluation due to lack of data, little involvement of the nutrition professionals in planning menus, and different realities in school environments and public policies.

As for the evaluation criteria (nutritional, sensory, cultural, and sustainability), the nutritional criterion was identified in all studies (100%), followed by the sensorial criterion ($n = 6$, 37.5%) and the sustainability criterion (12.5%). No study evaluated the cultural criterion. None of the studies presented three different forms of evaluation criteria.

3.3. Bias Assessment

The studies are heterogeneous, but most, 93.75% ($n = 15$), presented a low risk of bias, and one presented a moderate risk of bias (Table S3—Supplementary File). All studies evaluated specific methods of evaluating school menus and answered the main question (Table S3—Supplementary File).

4. Discussion

This systematic review is the first to analyze how school menus are evaluated in some countries. The results showed that Brazil and Spain were the countries that presented the largest number of studies evaluating menus. In Brazil, adequate and healthy food is a guaranteed right for students enrolled in public, philanthropic, and community organizations, in partnership with the government and guaranteed by the National School Feeding Program. This is characterized as a long-term public policy in Brazil for food and nutritional security and is considered one of the largest, most comprehensive, and long-term programs in the area of school meals globally [28].

In Brazil, the increase in the number of overweight and obese children [29] and the changes in the population's lifestyle and eating habits resulting from growing urbanization reinforce the school's important role in healthy eating habits and health promotion. The school is an environment for health promotion and actions on food and nutritional education [14]. These changes and the consolidation of public policies for schoolchildren may justify the number of studies aimed at evaluating the school menu. However, none of the studies presented instruments or methods for evaluating the four menu planning components.

Spain has one of the highest prevalence of overweight and obesity in Europe. Epidemiological studies show that 26.2% of children aged 6–9 are overweight, and 18.3% obese [21]. In 2005, a consensus on school feeding was published in Spain. This NAOS strategy (Nutrition, Physical Activity, and Prevention of Obesity) focused on healthy eating habits in the school environment and physical activity to fight obesity. Thus, a consensus document on food in educational centers (DCSECE) was defined. This contains a minimum set of indicators for the evaluation and follow-up of nutritional recommendations that guide menu planning and how school canteens should serve these menus [18,20,21,30].

In the NAOS strategy context, Spain consolidates itself with legal guidelines for developing healthier school menus. Several organizations and entities in Spain have developed systems for the supervision and qualitative assessment of school menus based on food groups, type of fat, and fiber and sugar [17]. There is a need to improve evaluation with a quantitative approach.

The other studies, unique among the countries of Europe and Oceania, present various methods of evaluating school menus proposed by researchers to meet national recommendations in the context of school meals. These include a frequency table with comparison to national recommendations, a table of food groups compared to national recommendations, a food group table correlated with a traffic light system, and a quality index of school meals compared to recommendations. The approaches are qualitative and assess the nutritional aspect exclusively, comparing these to nutritional recommendations [7,22–24,26].

Most studies, 81.25% ($n = 13$), presented methods for assessing qualitative data on school menus, i.e., the evaluation of frequency, variety, and quality of food, compared to the nutritional recommendations of the country of the study. These studies show the difficulty of extending evaluation due to lack of data about composition and preparation technique or from the Technical Preparation File (TPF). TPF is an instrument for promoting health, based on specifying culinary preparations with registration of the components and their quantities [31]. It also contains registration of the culinary techniques, the direct and indirect cost, nutrients, and other information for the food and nutrition service. From the TPF, it is possible to estimate the nutrients in the recipe and the portion, to guarantee standardization, and to modify possible misfits. Thus, in the absence of TPF, it is not feasible to accurately assess which nutrients make up a recipe [31]. Added to these difficulties are the scarce involvement of dietitians in menu planning and the various different realities in school environments when dealing with public policies [7,18,22,23,30].

The Qualitative Assessment of Menu Preparations (Avaliação Qualitativa das Preparações do Cardápio—AQPC) [32] is an instrument of menu management to propose balanced meals. It allows the assessment of the nutritional balance and sensory aspects of the menu, considering colors, preparation techniques, repetitions, combinations, the offer of certain foods such as fruits, vegetables, or types of meat, in addition to the sulfur content of the food [32]. The focus is qualitative. The instrument fails to reach all the Brazilian school feeding program recommendations, which presents quantitative, cultural, and sustainability parameters that must be followed in all municipalities in the country [13,14,16].

The checklist instrument proposed by Martins Rodrigues [2] states that school menus need to attend to nutritional, hygienic-sanitary and sustainability aspects. The instrument brings a broad approach to all the inherent aspects of Food and Nutrition security. It is not exclusive to the evaluation of school menus and also addresses aspects of the physical structure of the food production area. The checklist allows for the registration of schools, for the automatic calculation of the percentage of the evaluated blocks' adequacy, and for reassessments and long-term monitoring. With this assessment and its approaches illustrating limitations, the school meal service becomes an appropriate environment focusing on health, encompassing nutritional aspects and sanitary and environmental aspects.

Quantitative data studies [15,24,25] present nutritional calculation for some nutrients such as macronutrients, iron, and vitamin D but do not assess consumption. The served portion is evaluated and not the consumed portion. These studies quantitatively assess the nutritional aspects of school menus and their adequacy given nutritional recommendations.

According to Moreira Sampaio [15], quantitative and qualitative evaluation is essential for students' growth and development and eating habits that will influence adult patterns. Integrating all these aspects is needed to respect the right to healthy food established by the ONU. From this perspective, school feeding seeks to contribute so that food and nutritional security are guaranteed for all.

There is a lack of studies from North America, Central America, Africa, and Asia regarding menu evaluation instruments or methods. Despite this, the USA has two school feeding programs: the NLSP (National Lunch School Program) and the SBP (School Breakfast Program) to ensure that children can access healthy balanced meals within schools. As in other countries, in the USA, 31.8% of children and adolescents aged 2–19 years are overweight or obese [33]. According to Belik [34], America has almost 20 counties with School Meal programs; however, many programs only give small grants to needy schools. The USA presents programs, but no instrument or method to evaluate the menus reported in the literature [34].

In the world scenario during the COVID−19 pandemic, this is even a more evident need. Even before the Pandemic, a FAO [35] publication reveals that approximately 690 million people worldwide, the equivalent of 8.9%, were malnourished. The consequences experienced by the most vulnerable populations in the pandemic, such as the interruption of food supply and family income reduction, make access to adequate food even more distant.

This new scenario reinforces the need to monitor and evaluate school feeding. In this way, it will be possible to stimulate initiatives that respect the recommendations to promote adequate and healthy food. The intended result is to decrease malnutrition and reverse the negative trends of increasing obesity and other chronic diseases related to food. It is crucial to seek this in order to meet what is advocated as a healthy diet, that derives from an adequate nutritional supply and a socially and environmentally sustainable food system [35].

According to Maynard et al. [36], sustainability integrates actions based on three pillars: environmental, social, and economic. Its focus is on the pursuit of quality of life and environmental balance [36]. Besides, Ginani et al. [4] reinforce the importance of using regional foods as a sustainable way of promoting health. Therefore, a greater significance of food is confirmed, which goes beyond the biological and establishes it as a social and cultural act. Food is then an identity marker [4]. However, none of the analyzed methods and instruments covered the cultural aspect, including regionality, and only 12.5% ($n = 2$) observed the menu's sustainability. This fact shows a gap in the field that must be quickly filled, especially facing world events that affect all sustainability dimensions.

In addition to hunger, which causes a great social impact in itself, the environment has suffered from long-term natural and human-made events [37]. The impact of these aspects on the economy is noted in several sectors. Lost natural resources and increased public health costs directly affect the development of areas that are no longer priorities. All these aspects justify the need to value sustainable food systems and include them in planning menus aimed at the community. In the case of children, it is possible to learn about different topics in a transversal way. This also adds to what is recommended as healthy eating.

The studies show the problems that make it impossible for schools to serve their planned menus, failure to deliver goods, the degree of ripeness of fruits and vegetables, and the planned menus' low acceptability. In this context, the need for the dietitian's presence in the school environment is reinforced, playing an essential role in nutritional education and developing TPFs to guarantee adequate nutritional recommendations and minimize the school's negative impacts on the menus.

There are numerous dietary guidelines for proper planning of school menus, but a nutritional assessment of these menus is still a challenge [14,38]. Progress has been made to harmonize the menus provided in schools with standards based on scientific research and nutritional recommendations to achieve adequate nutrition. However, an effort is still needed to plan and evaluate school meals to meet all nutritional requirements, composition, variety, conservation methods, diverse culinary techniques, origins and ethnicities, allergies and intolerances, and hygienic-sanitary standards [39].

The menu stands out as an essential working tool to achieve nutritional recommendations. Planning the menu also contributes to achieving sensory, cultural, and sustainability aspects. Thus, it is relevant to understand the planning of school meals, from food acquisition to final distribution, considering the variables between who buys and distributes. The nutrition professional must follow and participate in the whole process. They should understand the seasonality of products, respect the mapping of farmers' production to ensure the supply of fruits and vegetables systematically, and monitor on-site the stages of meals' preparation and proper distribution [13,18,28].

Few studies present tools for evaluating menus, which, in turn, are mostly based on nutritional criteria. A global assessment of the menu is necessary to offer healthy meals. Thus, the school menu assessment instrument should include qualitative and quantitative aspects, crossing over hygienic and sanitary, cultural and sustainability issues. When planning the menu, using all these tools will allow a careful assessment to meet the proposed objectives.

This review has some limitations since the studies from different countries were collected on the mentioned platforms and, despite the abstract being in English, some studies were written in another languages. Despite the usage of a translation platform, some data could have been missed due to language barriers. There could also be school feeding programs that present instruments or methods to evaluate their menus, but were not published before the moment of this review search on databases.

5. Conclusions

There is a significant opportunity to serve healthy and tasty meals in school food services, to possibly compensate for the user population's daily nutritional deficiencies. Considering that students spend most of their day at school, eating up to three meals, it is imperative to control and evaluate school menus. This is the only way to guarantee healthy eating habits, various food groups, safe food from the hygienic-sanitary point of view, and respect for food and cultural habits. However, the present study did not identify an instrument for evaluating school menus that contemplates a comprehensive approach to aspects that involve adequate nutrition.

National school feeding programs should provide the evaluation stage as a decisive part of planning menus. Countries such as Brazil and Spain, where a more significant number of instruments have been identified, should serve as models to achieve school feeding goals for all students.

Menu planning and evaluation must cover nutritional aspects and cultural, sensory, sanitary, and sustainable principles. School feeding programs should encourage the development of instruments and/or methods for evaluating menus to help dietitians better plan menus for students.

Other search methodologies should be applied to identify school feeding programs and search for instruments not published in the literature. Specific school feeding programs probably present other instruments or methods to evaluate their menus that are not available in the literature. A search of nation's websites regarding school feeding programs could reveal how menus are developed and evaluated, gathering more information on other methods not scientifically described in the literature. Dietitians around the world could be developing or using different methods that could enrich this field of study. Thus, it is planned to gather other experiences in other countries not covered in this research. The current results show a lack of menu evaluation and this could contribute to a population's nutritional inadequacy.

Supplementary Materials: The following are available online at https://www.mdpi.com/2304-8158/10/2/374/s1. Table S1: Full-text excluded articles and reasons, Table S2: Indexers used to select publications that jointly or separately address instruments and methodologies for evaluating school menus, Table S3: The summarized risk of bias assessment of the studies included in this systematic review.

Author Contributions: A.F.C.—Conceptualization; Investigation, Formal analysis, Writing-original draft; D.d.C.M.—Investigation, Formal analysis, writing; F.L.N.d.Q.—Investigation, validation; R.P.Z.—Conceptualization; Supervision, Writing; V.C.G.—Supervision, Review; R.B.A.B.—Conceptualization, Project administration, Supervision, Writing; A.R.—Writing and Visualization; A.S.—Writing and Visualization. All authors have read and agreed to the published version of the manuscript.

Funding: This research received no external funding.

Institutional Review Board Statement: Not applicable.

Informed Consent Statement: Not applicable.

Data Availability Statement: Not applicable.

Acknowledgments: Coordination for the Improvement of Higher Education Personnel (CAPES).

Conflicts of Interest: The authors declare no conflict of interest.

References

1. FAO; WHO. *Sustainable Healthy Diets—Guiding Principles*; FAO: Rome, Italy, 2019; p. 44.
2. Martins Rodrigues, C.; Giordani Bastos, L.; Stangherlin Cantarelli, G.; Stedefeldt, E.; Thimoteo da Cunha, D.; Lúcia de Freitas Saccol, A. Sanitary, nutritional, and sustainable quality in food services of Brazilian early childhood education schools. *Child. Youth Serv. Rev.* **2020**, *113*, 104920. [CrossRef]
3. WHO. Global strategy for infant and young child feeding. In *Fifthy-Fourth World Health Assembly*; WHO: Geneva, Switzerland, 2001; p. 5.
4. Ginani, V.C.; Araújo, W.M.C.; Zandonadi, R.P.; Botelho, R.B.A. Identifier of Regional Food Presence (IRFP): A New Perspective to Evaluate Sustainable Menus. *Sustainability* **2020**, *12*, 3992. [CrossRef]
5. Da Proença, R.P.C. *Qualidade Nutricional e Sensorial na Produção de Refeições*; UFSC: Florianópolis, SC, USA, 2005.
6. Gerritsen, S.; Dean, B.; Morton, S.M.B.; Wall, C.R. Do childcare menus meet nutrition guidelines? Quantity, variety and quality of food provided in New Zealand Early Childhood Education services. *Aust. N. Z. J. Public Health* **2017**, *41*, 345–351. [CrossRef] [PubMed]
7. Myers, G.; Sauzier, M.; Ferguson, A.; Pettigrew, S. Objective assessment of compliance with a state-wide school food-service policy via menu audits. *Public Health Nutr.* **2019**, *22*, 1696–1703. [CrossRef]
8. Ginani, V.C.; Zandonadi, R.P.; Araujo, W.M.C.; Botelho, R.B.A. Methods, Instruments, and Parameters for Analyzing the Menu Nutritionally and Sensorially: A Systematic Review. *J. Culin. Sci. Technol.* **2012**, *10*, 294–310. [CrossRef]
9. Llorens-Ivorra, C.; Quiles-Izquierdo, J.; Richart-Martínez, M.; Arroyo-Bañuls, I. Diseño de un cuestionario para evaluar el equilibrio alimentario de menús escolares. *Rev. Española Nutr. Hum. Dietética* **2015**, *20*, 40. [CrossRef]
10. Camargo, R.G.M.; Caivano, S.; Bandoni, D.H.; Domene, S.M.Á. Healthy eating at school: Consensus among experts TT—Alimentação saudável no ambiente escolar: Consenso entre especialistas. *Rev. Nutr* **2016**, *29*, 809–819. [CrossRef]
11. Gabriel, C.G.; Costa, L.D.C.F.; Calvo, M.C.M.; Vasconcelos, F.D.A.G.D. Planejamento de cardápios para escolas públicas municipais: Reflexão e ilustração desse processo em duas capitais brasileiras. *Rev. Nutr.* **2012**, *25*, 363–372. [CrossRef]
12. Moher, D.; Liberati, A.; Tetzlaff, J.; Altman, D.G.; Altman, D.; Antes, G.; Atkins, D.; Barbour, V.; Barrowman, N.; Berlin, J.A.; et al. Preferred reporting items for systematic reviews and meta-analyses: The PRISMA statement. *PLoS Med.* **2009**, *6*, e1000097. [CrossRef]
13. Da Silva Bastos Soares, D.; Moreira Sampaio Barbosa, R.; Henriques, P.; Camacho Dias, P.; Mendonça Ferreira, D. Análisis de la calidad de los menús del Programa de Alimentación Escolar Nacional en una ciudad de Río de Janeiro—Brasil. *Rev. Chil. Nutr.* **2015**, *42*, 235–240. [CrossRef]
14. Longo-Silva, G.; Toloni, M.; Rodrigues, S.; Rocha, A.; de Taddei, J.A.A.C. Qualitative evaluation of the menu and plate waste in public day care centers in são paulo city, brazil. *Rev. Nutr.* **2013**, *26*, 135–144. [CrossRef]
15. Moreira Sampaio, R.; Cabral Coutinho, M.B.; Mendonça, D.; da Silva Bastos, D.; Henriques, P.; Camacho, P.; Anastácio, A.; Pereira, S. School nutrition program: Assessment of planning and nutritional recommendations of menus. *Rev. Chil. Nutr.* **2017**, *44*, 170–176. [CrossRef]

16. Vidal, G.M.; Veiros, M.B.; Sousa, A.A.D. School menus in Santa Catarina: Evaluation with respect to the national school food program regulations. *Rev. Nutr.* **2015**, *28*, 277–287. [CrossRef]
17. De Mateo Silleras, B.; Martín, M.A.C.; Sainz, B.O.; Enciso, L.C.; De La Cruz Marcos, S.; De Miguelsanz, J.M.M.; Del Río, P.R. Diseño y aplicación de un cuestionario de calidad dietética de los menús escolares. *Nutr. Hosp.* **2015**, *31*, 225–235. [CrossRef]
18. González, S.D.; Bernardo, C.R.; Alonso, O.A.; Cruz, M.; Diez, G.; Díaz, A.; De Epidemiología, U.; Dirección, A.; De Salud, G.; Consejería, P.; et al. Evaluación De La Variedad Y Calidad En Los Menús Escolares De Evaluation of variety and quality in the school menus of Asturias. 2015/2016 La obesidad y el sobrepeso en la infan- cia aumentan las probabilidades de padecer diabetes y enfermedades cardio. *Rev. Esp. Salud Pública* **2018**, *92*, 1–12.
19. Llorens-Ivorra, C.; Arroyo-Bañuls, I.; Quiles-Izquierdo, J.; Richart-Martínez, M. Evaluation of school menu food balance in the Autonomous Community of Valencia (Spain) by means of a questionnaire. *Gac. Sanit.* **2018**, *32*, 533–538. [CrossRef]
20. Soares, P.; Martínez-Milán, M.A.; Comino, I.; Caballero, P.; Davó-Blanes, M.C. Assessment of the consensus document on food in educational centres to evaluate school menus. *Gac. Sanit.* **2019**. [CrossRef]
21. Uriarte, P.S.; Cirarda Larrea, F.B.; Alonso, S.V. Características nutricionales de los menús escolares en Bizkaia (País Vasco, España) durante el curso 2012/2013. *Nutr. Hosp.* **2015**, *31*, 1309–1316. [CrossRef]
22. Therre, P.; Riemer-Hommel, P.; Knoll, M. School meals at secondary schools: An analysis in the district of St Wendel in the province Saarland, Germany. *Gesundheitswesen* **2012**, *74*, 467–475. [CrossRef] [PubMed]
23. Dubuisson, C.; Lioret, S.; Calamassi-Tran, G.; Volatier, J.L.; Lafay, L. School meals in French secondary state schools with regard to the national recommendations. *Br. J. Nutr.* **2009**, *102*, 293–301. [CrossRef]
24. Evans, C.E.L.; Cade, J.E. A cross-sectional assessment of food- and nutrient-based standards applied to British schoolchildren's packed lunches. *Public Health Nutr.* **2017**, *20*, 565–570. [CrossRef] [PubMed]
25. Patterson, E.; Quetel, A.K.; Lilja, K.; Simma, M.; Olsson, L.; Elinder, L.S. Design, testing and validation of an innovative web-based instrument to evaluate school meal quality. *Public Health Nutr.* **2013**, *16*, 1028–1036. [CrossRef] [PubMed]
26. Gregorič, M.; Pograjc, L.; Pavlovec, A.; Simčič, M.; Blenkuš, M.G. School nutrition guidelines: Overview of the implementation and evaluation. *Public Health Nutr.* **2015**, *18*, 1582–1592. [CrossRef]
27. *Ministério da Saúde*; Ministério da Saúde: Brasília, Brasil, 2006.
28. Bicalho Alvarez, D.; Paulo, S.; Slater Villar, B. Universidade de São Paulo Faculdade de Saúde Pública Efeito Da Lei Federal 11.947/09 na Qualidade Nutricional dos Cardápios Propostos Pelo Programa de Alimentação Escolar do Estado de São Paulo. Ph.D. Thesis, Universidade de São Paulo, São Paulo, Brazil, 2017.
29. Instituto Brasileiro de Geografia e Estatística (IBGE). *Pesquisa de Orçamentos Familiares 2017–2018: Análise do Consumo Alimentar no Brasil*; Instituto Brasileiro de Geografia e Estatística: Rio de Janeiro, Brazil, 2020.
30. Seiquer, I.; Haro, A.; Cabrera-Vique, C.; Muñoz-Hoyos, A.; Galdó, G. Evaluación nutricional de los menús servidos en las escuelas infantiles municipales de Granada. *An. Pediatr.* **2016**, *85*, 197–203. [CrossRef]
31. Akutsu, R.D.; Botelho, R.A.; Camargo, E.B.; Savio, K.E.O.; Araujo, W.C. A ficha técnica de preparação como instrumento de qualidade na produção de refeições The technical cards as quality instrument for good manufacturing process. *Rev. Nutr.* **2005**, *18*, 277–279. [CrossRef]
32. Veiros, M.; Martinelli, S. Avaliação Qualitativa das Preparações do Cardápio Escolar—AQPC Escola. *Nutr. Pauta* **2012**, *20*, 3–12.
33. Schober, D.J.; Carpenters, L.; Currie, V.; Yaroch, A.L. Evaluation of the LiveWell@ School Food Initiative Shows Increases in Scratch Cooking. *J. Sch. Health* **2016**, *86*, 604–611. [CrossRef]
34. Belik, W.; de Souza, L.R. Algumas Reflexões Sobre Os Programas De. *Security* **2009**, *1*, 209.
35. FAO. *The State of Food Security and Nutrition in the World*; FAO: Rome, Italy, 2020; ISBN 978-92-5-109888-2.
36. Da Maynard, D.C.; Vidigal, M.D.; Farage, P.; Zandonadi, R.P.; Nakano, E.Y.; Botelho, R.B.A. Environmental, Social and Economic Sustainability Indicators Applied to Food Services: A Systematic Review. *Sustainability* **2020**, *12*, 1804. [CrossRef]
37. Li, Y.; Li, Y.; Kappas, M.; Pavao-Zuckerman, M. Identifying the key catastrophic variables of urban social-environmental resilience and early warning signal. *Environ. Int.* **2018**, *113*, 184–190. [CrossRef]
38. Lavall, M.J.; Blesa, J.; Frigola, A.; Esteve, M.J. Nutritional assessment of the school menus offered in Spain's Mediterranean area. *Nutrition* **2020**, *78*, 110872. [CrossRef] [PubMed]
39. Matić, I.; Jureša, V. Compliance of menus with nutritional standards in state and private kindergartens in Croatia. *Rocz. Państw. Zakł. Hig.* **2015**, *66*, 367–371. [PubMed]

 *foods*

Review

Sustainability Recommendations and Practices in School Feeding: A Systematic Review

Emanuele Batistela dos Santos [1], Dayanne da Costa Maynard [2], Renata Puppin Zandonadi [2], António Raposo [3,*] and Raquel Braz Assunção Botelho [1,*]

1 Department of Food and Nutrition, Federal University of Mato Grosso, Cuiabá 78060-900, Brazil; emanuelebatistela.ufmt@gmail.com
2 Department of Nutrition, University of Brasília, Brasília 70910-900, Brazil; day_nut@yahoo.com.br (D.d.C.M.); renatapz@unb.br (R.P.Z.)
3 CBIOS (Research Center for Biosciences and Health Technologies), Universidade Lusófona de Humanidades e Tecnologias, Campo Grande 376, 1749-024 Lisboa, Portugal
* Correspondence: antonio.raposo@ulusofona.pt (A.R.); raquelbotelho@unb.br (R.B.A.B.)

Abstract: Considering the importance of schools for sustainable food offers and the formation of conscientious citizens on sustainability, this systematic review aimed to verify the recommendations on sustainability in school feeding policies and the sustainability practices adopted in schools. The research question that guided this study is "what are the recommendations on sustainability in school feeding policies and the sustainability practices adopted in schools?". This systematic review was prepared according to PRISMA, and its checklist was registered in PROSPERO. Specific search strategies for Scopus, Web of Science, Pubmed, Lilacs, Google Scholar, and ProQuest Dissertations & Theses Global were developed. The included studies' methodological quality was evaluated using the Meta-Analysis Statistical Assessment and Review Instrument (MASTARI). A total of 134 studies were selected for a full reading. Of these, 50 met the eligibility criteria and were included in the systematic review. Several sustainability practices were described. The most cited are school gardens and education activities for sustainability. However, actions carried out in food services were also mentioned, from the planning of menus and the purchase of raw materials (mainly local and organic foods, vegetarian/vegan menus) to the distribution of meals (reduction of organic and inorganic waste: composting, recycling, donating food, and portion sizes). Recommendations for purchasing sustainable food (organic, local, and seasonal), nutrition education focused on sustainability, and reducing food waste were frequent; this reinforces the need to stimulate managers' view, in their most varied spheres, for the priority that should be given to this theme, so that education for sustainability is universally part of the curricula. The importance of education in enabling individuals to promote sustainable development is reaffirmed in Sustainable Development Goal 4 (SDG 4). The development of assessment instruments can help monitor the evolution of sustainable strategies at schools and the main barriers and potentialities related to their implementation.

Keywords: school feeding; school meals; sustainability

Citation: dos Santos, E.B.; da Costa Maynard, D.; Zandonadi, R.P.; Raposo, A.; Botelho, R.B.A. Sustainability Recommendations and Practices in School Feeding: A Systematic Review. *Foods* **2022**, *11*, 176. https://doi.org/10.3390/foods11020176

Academic Editor: Theodoros Varzakas

Received: 8 December 2021
Accepted: 4 January 2022
Published: 10 January 2022

1. Introduction

School feeding programs, widely spread across the globe, are recognized as an essential strategy for achieving goals in various sectors of society, including education, health, social protection, and agriculture. Recognized as the most prominent social protection network globally, even with the effects suffered by the COVID-19 pandemic, they appear as a robust investment in human capital that will guarantee the economic growth of nations [1]. The relationship between school feeding and educational and nutritional outcomes is widely investigated in the literature [2–5]. However, more recently, its role in conducting actions aimed at sustainability has been studied to mitigate the global challenges that threaten human and planet health in the 21st century [6–8].

A product of the concern with the environmental impacts derived from the world pattern of production and consumption in the second half of the 20th century, the term sustainable development refers to satisfying the needs of the present without compromising the ability of future generations to meet their own needs [9]. Integrating the sustainability pillars is necessary to increase productive potential, guaranteeing equal opportunities for all without putting the environment at risk [10]. The three main pillars (environmental, economic, and social) have been studied for years, and, recently, the cultural and health pillars were also linked to sustainability [11].

Through school meals and educational practices, students become aware of the impacts of individual and collective choices [6,12–14], consequently generating better environmental, economic, and social outcomes. The essential role of education for achieving a more sustainable future in environmental, economic, and social aspects was recognized by the United Nations in the Decade of Education for Sustainable Development, which aimed to integrate the values, principles, and practices of sustainable development in all aspects of education [15]. Therefore, education for sustainability is a powerful tool capable of providing the knowledge, skills, and awareness needed by young people to deal with the various problems that threaten the integrity of the planet and human health and well-being [16]. In this sense, it is essential to conduct a process that considers, in addition to global issues, those that are local and common to the participants' routine and that integrates a holistic perspective, allowing for informed decision-making, individually or collectively [15,17]. The literature also highlights the importance of using school meals as a tool for nutritional education and education for sustainable consumption and practical learning activities, such as school gardens, cooking activities, and field visits to small local farmers [6,12,18–20].

The relationship between sustainability and school feeding also occurs at the level of decisions made at all stages of meal production. It is known that food production is associated with significant environmental impacts. Although elements before or after the meals' preparation are responsible for most of these effects (such as field production, transport, and food waste) [21–23], the choices made by school food services influence this dynamic, determining the degree to which they employ actions to mitigate the environmental impacts generated in this process. Some examples are the purchase of organic and local food, encouraging the consumption of fresh vegetables at meals, controlling the supply of meat, and actions such as adjusting portions, donating food, composting, purchasing products with minimum packaging, recycling, and reducing energy and water consumption [24–30].

The adoption of sustainable practices often generates results that simultaneously reach the different dimensions of sustainability. Reducing energy and water consumption in the production of meals and adjusting the size of food portions, for example, can represent actions of economic and environmental sustainability [29]. Also, the donation of food, raw or prepared, provided that it is in perfect hygienic and sanitary condition, can be observed from social and environmental sustainability perspectives [26,27].

Establishing a close relationship between school meals and small farmers to purchase locally sourced food favors the increase in income and class organization. Consequently, the economic development of the region, as well as contributing to the food and nutritional security of farmers and their families [31,32], goes beyond environmental to social and economic sustainability dimensions. Due to the benefits for both students and farmers widely recognized in the literature, the practice of buying local food is encouraged by public policies for school meals in different parts of the world [33,34].

Therefore, the range of activities involved with school feeding generates unique challenges and opportunities from the point of view of sustainability in the environmental, economic, and social dimensions [7,8,12,35]. In this sense, school feeding programs are part of the strategies used to achieve the Sustainable Development Goals of the 2030 Agenda [36].

Although school feeding policies in some countries already present recommendations on sustainability in their guiding principles [34,37,38], and the literature presents different sustainability practices employed in this context, bringing to light a body of evidence on this topic will be helpful for decision-makers at the government level to create or even revise guidelines for their school feeding policies, incorporating the principles of sustainability. Therefore, the research question that guided this study is "What are the recommendations on sustainability in school feeding policies and the sustainability practices adopted in schools?". In addition, the findings may help policymakers and members of the school community, within their local context, in the development of sustainability practices linked to school and school feeding. Considering the importance of schools to offer sustainable food and in the formation of conscientious citizens who are able to act on the challenges related to sustainability in the contemporary world [6], this systematic review aimed to verify the recommendations on sustainability in school feeding policies and the sustainability practices adopted in schools.

2. Materials and Methods

This systematic review was prepared according to the Preferred Reporting Items for Systematic Reviews and Meta-Analyses (PRISMA), and its checklist [39] was registered in PROSPERO [CRD42021264978]. The protocol was performed according to the following steps.

2.1. Inclusion and Exclusion Criteria

The inclusion criteria were studies that described the recommendations on sustainability in school feeding policies and the sustainability practices adopted in schools, in environmental, social, and/or economic aspects, with no date and language limits. Legislations of school feeding policies and programs found in the studies reference lists had their full text analyzed to identify sustainability recommendations. The exclusion criteria were: (1) comments, letters, conferences, reviews, abstracts, reports, undergraduate works, discussion papers, and books, (2) studies carried out outside schools or in which the school was not responsible for the action, (3) studies in which practices were not performed or studies where activities were punctual, (4) studies focused on the supplier or that only reported purchases, and (5) studies that did not describe sustainability practices (Appendix A).

2.2. Information Source

Detailed individual search strategies were developed for each database: MEDLINE via Pubmed, Embase, Scopus, Web of Science, and Lilacs. A search for gray literature was performed on Google Scholar and for dissertations and theses in ProQuest Global. Additionally, we examined the reference lists of the selected articles as relevant studies could have been missed during the data search. The last search in all databases was carried out on 30 June 2021.

2.3. Search Strategy

The appropriate combinations of truncation and keywords were selected and adapted for the search in each mentioned database (Table S1—Supplementary Materials). We used Rayyan software (Qatar Computer Research Institute (QCRI)) to select and exclude duplicate articles, and all references were managed by Mendeley desktop software.

2.4. Study Selection

Two phases were necessary for the selection. In phase 1, researchers I (EBS) and II (DCM) independently reviewed the titles and abstracts of all references identified from databases. EBS and DCM excluded the articles that did not meet the eligibility criteria. In phase 2, the full texts of the selected articles were fully read by the same reviewers (EBS, DCM), and only those that met the inclusion criteria were included. In both phases, the disagreements were discussed until a consensus was reached between the two reviewers. A third reviewer (RBAB) made the final decision in situations without consensus. EBS, an examiner, critically evaluated the list of references of the selected studies. Additional studies were added by the third examiner (RBAB) and the expert (RPZ).

2.5. Data Collection Process

Two reviewers independently (EBS, DCM) collected the following characteristics from the selected studies by authors and year of publication, country of research, the objective of the study, methods, sustainability dimensions, and main results referring to the identified sustainability practices. Calibration exercises were conducted before starting the review to ensure consistency among reviewers. Disagreements were solved by discussion, and the third reviewer (RBAB) adjudicated unresolved disagreements. These data were synthesized by three reviewers (EBS, DCM, and RBAB) using a standardized table containing the following information: reference, authors, year, country, objectives, type of school management (public, private), teaching stage (according to the teaching stages of each country), participants (individuals, schools, or municipalities), sustainability dimensions (environmental, economic, and social), and main results referring to the identified sustainability practices.

2.6. Risk of Individual Bias in the Included Studies

The quality criteria were synthesized using a statistical review assessment instrument (MASTARI) and the Joanna Briggs Institute protocol to assess the risk of bias in the studies. The instrument for assessing the risk of bias included seven questions:

1. Were the practices identified characterized?
2. Has the practice been implemented in schools?
3. Did the practices present a positive implementation response?
4. Was the study design appropriate?
5. Was the statistical analysis adequate to the objective of the study?
6. Did the results answer the main question?
7. In the case of the schools, was the sample of establishments selected for analysis representative and randomly determined?

The categorization of the risk of bias followed the percentage of "yes" score: "High" for up to 49%, "Moderate" for between 50 and 69%, and "Low" for more than 70%.

3. Results

The researchers retrieved 1763 studies from the electronic databases; 1319 titles and abstracts were evaluated after removing the duplicates, and, after reading the abstracts, 134 studies were selected for a full reading. Of these, 50 met the eligibility criteria and were included in the systematic review. At the same time, recommendations on sustainability were found in 11 governmental school feeding policies or programs and 5 in other available non-governmental school feeding programs/initiatives retrieved from the studies' reference lists (Figure 1). The latest available versions were evaluated for governmental and non-governmental school feeding policies or programs.

3.1. Studies Characteristics

Regarding sustainability practices, the studies included (n = 50) were conducted between 1991 and 2021, in the following countries: United States (n = 22), Brazil (n = 7), Spain (n = 3), Italy (n = 3), South Africa (n = 2), Canada (n = 2), England (n = 2), Denmark (n = 1), Finland (n = 1), Ghana (n = 1), India (n = 1), Japan (n = 1), Wales (n = 1), Kenya (n = 1), and Tanzania (n = 1). A parallel study was carried out in the United States and Cuba. The characteristics of the analyzed studies are presented in Table 1.

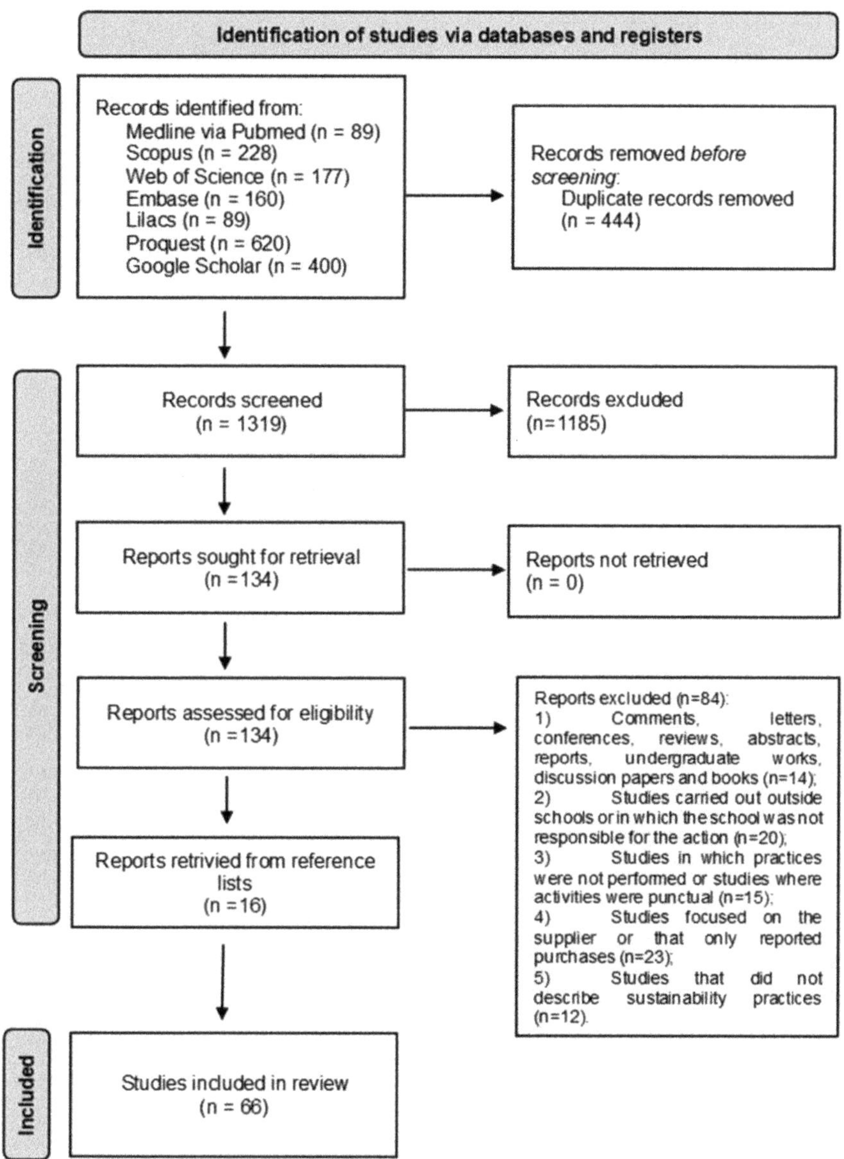

Figure 1. Flowchart of the systematic review search process. Adapted from PRISMA protocol.

Table 1. Main descriptive characteristics and results from the included studies.

Author (Year) Country	Objectives	School Management (SM) Teaching Stage (TS) Participants (P) Sustainability Dimension (SD)	Main Sustainability Practices Identified
Mann (1991) [40] USA	To assess the solid waste management practices in school food, and to develop and assess a decision model for solid waste management in school food services.	SM: Public, private TS: Not informed P: School food service directors ($n = 458$) SD: Environmental, economic	Recycling, purchase of bulk products, and reusable dispensing devices.
Ghiselli (1993) [41] USA	To analyze waste and disposal practices in Indiana's school food service, and the feasibility of reducing it through permanent service and product recycling.	SM: Public, private TS: Elementary, middle, high school P: School food service directors ($n = 237$) SD: Environmental, economic	Recycling.
Hackes; Shanklin (1999) [42] USA	To identify resource allocation decisions, policies, and procedures used by school food service directors that were based on pollution prevention, product stewardship, and sustainable development.	SM: Not informed TS: Not informed P: School food service directors ($n = 168$) SD: Environmental, economic	Recycling; energy policy: solid waste and water.
Albertse, Mancusi-Materi (2000) [43] South Africa	To illustrate how the initiation of school children into innovative technologies has fostered mechanisms of social mobilization towards enhanced food security in South Africa.	SM: Not informed TS: Not informed P: Students, parents (n = not informed) SD: Environmental, economic, social	Irrigation system for water reuse and school garden.
Wadsworth (2002) [44] USA	To conduct a curriculum assessment of an after-school program on food choices that minimize energy, natural resources used, and pollution generated in food processing, packaging, and transportation.	SM: Public TS: Elementary P: Students ($n = 240$) SD: Environmental	Nutritional education focused on the sustainability of the food system; Cooking activities.
Lima (2006) [45] Brazil	To analyze the management of a School food service unit in the State of Santa Catarina, based on the introduction of organic foods.	SM: Public TS: Elementary P: Representatives of the Department of Education, a state school, the School Feeding Council, students ($n = 21$) SD: Environmental	Organic school garden and feeding program; Control of non-organic waste generation.
Vogt (2006) [46] USA	To identify district/community characteristics supporting buying food locally, the perceived benefits and barriers in buying locally, and generate solutions to encountered issues in California.	SM: Public TS: Not informed P: School food service directors, farmers ($n = 37$) SD: Environmental, economic, social	Participation in the "Farm-to-School" program (local foods), school garden, recycling, composting, and vegetarian/vegan meals.

Table 1. *Cont.*

Author (Year) Country	Objectives	School Management (SM) Teaching Stage (TS) Participants (P) Sustainability Dimension (SD)	Main Sustainability Practices Identified
Sonnino (2009) [47] Italy	To examine how city authorities have integrated different (and at times contrasting) quality conventions in school meals in Rome.	SM: Public TS: Not informed P: Representatives of the sectors involved in school feeding (n = not informed) SD: Environmental, economic, social	Purchase of organic food; Adoption of social and environmental criteria for contracting food services.
Izumi, Alaimo, Hamm (2010) [48] USA	To identify why farmers, school food service professionals, and food distributors participate in farm-to-school programs and the opportunities and challenges for purchasing food at local schools.	SM: Public TS: Not informed P: School food service professionals, farmers, food distributors (n = 18) SD: Local, economic, social	Participation in the "Farm-to-School" (local food).
Baca (2011) [49] USA	To investigate the status of food waste management programs, recycling of packaging waste, and cost of waste hauling in school nutrition programs in the USA.	SM: Not informed TS: Not informed P: Child nutrition directors (n = 79) SD: Environmental, economic, social	Food donation, composting, donation of waste for animal feed, recycling.
Bennell (2012) [50] Wales	To explore the development of the Education for Sustainable Development and Global Citizenship through case studies of Welsh primary schools.	SM: Not informed TS: Elementary, middle P: Students, teachers, support staff (n = 46) SD: Environmental, economic	School garden, recycling, energy audit, and sustainability aspects.
Bucher (2012) [51] USA/Cuba	To examine how pedagogies of sustainability are embedded in socio-cultural contexts and policy structures and driven by the localized actions of teachers.	SM: Public TS: Elementary, primary, high school P: Teachers, community members (n = 12) SD: Environmental, social	Environmental education; school garden.
Jones et al. (2012) [12] England	To examine the associations between the promotion of sustainable food and student self-reported fruit and vegetable consumption and associated behaviors.	SM: Not informed TS: Elementary P: Students (n = 1435) SD: Environmental	Participation in a sustainable food program (education for sustainability, use of sustainable food).

Table 1. *Cont.*

Author (Year) Country	Objectives	School Management (SM) Teaching Stage (TS) Participants (P) Sustainability Dimension (SD)	Main Sustainability Practices Identified
Lombardini, Lankoski (2013) [52] Finland	To examine the effects of forced restriction of food choice through a natural field experiment, the Helsinki vegetarian day.	SM: Not informed TS: Elementary, high school, vocational P: Schools (n = 43) SD: Environmental	Vegetarian day.
O'Brien (2013) [53] USA	To explore efforts by some independent schools to develop education and act in ways that promote environmental sustainability and social equity.	SM: * Independent TS: High school P: Schools (n = 5) SD: Environmental, social	Education for sustainability.
Orme et al. (2013) [54] England	To report on an evaluation of the Food for Life Partnership program, a multi-level initiative in England promoting healthier nutrition and food sustainability awareness for students and their families.	SM: Not informed TS: Elementary P: Teachers, students (n = 152) SD: Environmental	Formation of a leadership group in a sustainable food program (tasting new dinners; visits to local farmers to buy and prepare food to share with the school).
Rilla (2013) [55] USA	To examine the design features of schoolyard gardens in the Unified School District of Los Angeles and see how they are a way to encourage community involvement.	SM: Public TS: Elementary, middleP: Schools (n = 5) SD: Environmental	School garden.
Shuttleworth (2013) [56] USA	To investigate the curricular, pedagogical, and assessment strategies of three teachers when they teach the social issues of sustainability education.	SM: Public, private TS: Elementary, middle, high school P: Teachers (n = 4) SD: Environmental, locial	Education for sustainability.
Barnett (2014) [57] USA	To examine the founding and first ten years of operation of a charter school committed to ecological literacy and sustainability.	SM: Public TS: Elementary, middle P: Founders, alumni (n = not informed) SD: Environmental	Ecological literacy.
Galli et al. (2014) [8] Italy	To explore the role of new public-private partnerships for promoting more sustainable school meal services, by drawing on the theory of co-production.	SM: Public TS: Not informed P: Representatives from the food service and education sector, parents (n = not informed) SD: Environmental, social	Short supply chain; organic food; use of food produced on confiscated land; exchange of mineral water for filtered water; single dish menu.

Table 1. *Cont.*

Author (Year) Country	Objectives	School Management (SM) Teaching Stage (TS) Participants (P) Sustainability Dimension (SD)	Main Sustainability Practices Identified
He, Mikkelsen (2014) [58] Denmark	To examine the possible influence of organic food policies on Danish school feeding systems on the development of healthier school food environments.	SM: Public TS: Elementary P: School food service supervisors ($n = 92$) SD: Environmental	Organic food.
Keller (2014) [59] USA	To examine how educators are fostering sustainability through cultivating nature awareness in young children.	SM: Public, independent TS: Elementary P: Scholar, education director, teachers, principal ($n = 6$) SD: Environmental, economic, social	Ecological literacy; school garden; field trips to farmers; local and organic foods.
Bamford (2015) [60] USA	To discover the relationships between educational experience and sustainability attitudes and behaviors, the motivation behind these behaviors, and establish their role in educational programs.	SM: Not informed TS: Elementary P: Students, teachers ($n = 102$) SD: Environmental, social	Sustainability curriculum; school garden; field trips.
Black et al. (2015) [24] Canada	To describe the development of a tool to assess the integration of healthy and environmentally sustainable food initiatives in schools and characterize a sample of schools using this tool.	SM: Public TS: Elementary, secondary P: Food service worker, teachers, school administrators ($n =$ not informed) SD: Environmental, economic, social	School garden; composting; local, organic food with minimal packaging; vegetarian dishes.
Coe (2015) [61] USA	To understand how a school gardening program and its ecology curriculum influences students' environmental perceptions and attitudes.	SM: Public TS: Elementary P: Students ($n = 21$), parents ($n = 3$), staff ($n = 3$) SD: Environmental	Ecology curriculum; organic school garden; rainwater collection cistern; composting.
Fabri et al. (2015) [62] Brazil	To identify and analyze the use of regional foods in the school meals of a Brazilian city.	SM: Public TS: Not informed P: City ($n = 1$) SD: Environmental, economic, social	Regional food.

Table 1. *Cont.*

Author (Year) Country	Objectives	School Management (SM) Teaching Stage (TS) Participants (P) Sustainability Dimension (SD)	Main Sustainability Practices Identified
Strohl (2015) [20] USA	To investigate how science education is structured to develop scientifically literate students.	SM: Not informed TS: Elementary P: Teachers ($n = 2$) SD: Environmental	School garden; food literacy; scientific literacy; cooking activities.
Triches (2015) [14] Brazil	Report the actions taken with schoolchildren in a municipality, combining changes in food consumption and production and linking health and sustainability.	SM: Public TS: Not informed P: City ($n = 1$) SD: Environmental, economic, social	Local foods; organic school garden; teaching for sustainability; cooking activities; use of returnable juice bottles.
Fernandes et al. (2016) [63] Ghana	To describe the adaptation of the School Meals Planner Package to reality in Ghana during the 2014 to 2015 school year.	SM: Public TS: Not informed P: Districts of Ghana ($n = 42$) SD: Environmental, economic, social	Meal package plan (local food).
Bareng-Antolin (2017) [64] USA	To identify practices, perceived benefits, barriers, and resources needed to implement and maintain a gardening program in high schools.	SM: Public; Private TS: High school P: Teachers; School administrators or staff, community volunteers, support organizations ($n = 42$) SD: Environmental, economic, social	School garden; food donation.
Borish, King, Dewey (2017) [65] Kenya	To understand how a school feeding and agroforestry program impacts the surrounding community's human, financial, natural, and social capital.	SM: Public TS: Elementary P: Community members ($n = 64$) SD: Environmental, economic, social	Agroforestry project (teaching on agroforestry practices).
Laurie, Faber, Maduna (2017) [66] South Africa	To evaluate knowledge, perceptions, and practices about food production among students and educators, management, and gardening activities in the National School Feeding Program schools.	SM: Public TS: Elementary P: Garden administrators, garden workers, teachers, students ($n = 3355$) SD: Environmental, economic	School garden.

Table 1. *Cont.*

Author (Year) Country	Objectives	School Management (SM) Teaching Stage (TS) Participants (P) Sustainability Dimension (SD)	Main Sustainability Practices Identified
Soares et al. (2017) [67] Spain	To identify and characterize initiatives that promote the purchase of locally-sourced foods to supply schools and the schools carrying out the initiatives.	SM: Public, private TS: Kindergarten, elementary, high school, special school P: Informants from the Ministries of Education and Agriculture (n = Not informed) SD: Environmental, economic, social	Local foods; organic food.
Garcia (2018) [68] Brazil	To analyze the actions of the National School Feeding Program in the city of Marechal Cândido Rondon-PR.	SM: Public TS: Kindergarten, elementary, high school, special school P: Family farmers, nutritionists, managers, cooks, teachers (n = 125) SD: Environmental, economic, social	Sustainability training; partnerships for environmental preservation, short circuit sales, and certification of organic food; competition and recipe booklet for the use of organic products and valorization of work.
Huston (2018) [69] USA	To highlight how leadership affects the implementation of education forsustainability in two K-6 elementary schools in rural New England.	SM: Public TS: Elementary P: School staff members (n = 23) SD: Environmental, economic, social	Education for sustainability; participation in the "Farm-to-School" program; student participation in the local food pantry.
Lagorio et al. (2018) [27] Italy	To use a case study in Italy to illustrate an effective and reliable strategy to reduce food waste in public school canteens.	SM: Public TS: Elementary, high school P: Municipal Councillors of Social Policies and of Education (n = 2) SD: Environmental, social	Portion adequacy; food donation.
Lehnerd (2018) [70] USA	To investigate the adoption and the potential impacts of the Farmers' Market Nutrition Incentive and Farm to School programs.	SM: Not informed TS: Elementary, middle P: Farmer, food service administrators or principals, students (n = 721) SD: Environmental, economic, social	Participation in the "Farm to School" program (school garden; local foods)

Table 1. *Cont.*

Author (Year) Country	Objectives	School Management (SM) Teaching Stage (TS) Participants (P) Sustainability Dimension (SD)	Main Sustainability Practices Identified
Powell, Wittman (2018) [71] Canada	To investigate the farm-to-school movement in British Columbia, where concerns related to education and health have been the main vectors of farm-to-school mobilization.	SM: Public TS: Not informed P: Farm-to-school actors ($n = 30$) SD: Environmental, economic, social	Participation in the "Farm-to-School" program (local food, food literacy, school garden).
Roy et al. (2018) [72] India	To explore and further explain the phenomena of supplier participation in addressing the sustainability-oriented objectives of a supply chain.	SM: Public TS: Elementary, upper elementary P: Unit President, purchasing, quality end operation managers ($n = 4$) SD: Environmental, economic, social	Sustainable management of supply chains.
Elkin (2019) [73] USA	To explore the three domains of sustainability of the Farm-to-School program (classroom, cafeteria, and community) developed in a California School District.	SM: Public TS: Elementary, middle P: School district ($n = 1$) SD: Environmental, economic, social	Participation in the "Farm-to-School" program (local food, school garden; teaching about food, farming, and agriculture).
Lopes, Basso, Brum (2019) [74] Brazil	To evaluate the functioning of the market generated by the National School Feeding Program in the school network of Ijuí, RS, Brazil, from the standpoint of short agrifood chains.	SM: Public TS: Elementary P: Education Secretary, nutritionist, school director ($n = 3$) SD: Environmental	School Garden; environmental education.
Santos et al. (2019) [75] Brazil	To implement a school vegetable garden using recyclable materials.	SM: Public TS:Elementary P: School ($n = 1$) SD: Environmental, economic	Organic school garden with recycled material (tires).
Blondin et al. (2020) [25] USA	To assess the Meatless Monday campaign's nutritional, environmental, and environmental impacts in the National School Lunch Program in a US school district.	SM: Public TS: Not informed P: School district ($n = 1$) SD: Environmental, economic	Reduced meat supply.
Derqui, Grimaldi, Fernandez (2020) [26] Spain	To understand the level of awareness about food waste generated, of interventions applied to minimize it, and to categorize the schools and prioritize a list of interventions to reduce food waste in school canteens.	SM: Public, private TS: Elementary, high school P: School headteachers ($n = 420$) SD: Environmental, economic, social	Certification and training (sustainability); flexible servings; composting; food donation; noise reduction,; communication (adjustment of the quantity produced); reduced use of paper/water/energy.

Table 1. *Cont.*

Author (Year) Country	Objectives	School Management (SM) Teaching Stage (TS) Participants (P) Sustainability Dimension (SD)	Main Sustainability Practices Identified
Izumi et al. (2020) [35] Japan	To explore factors that minimize lunch waste in Tokyo elementary schools and consider how such factors can be modified and applied in US schools.	SM: Public TS: Elementary P: School dietitians ($n = 5$) SD: Environmental, economic	Social norms (avoid waste); exposure to unknown/unappreciated foods; pedagogical practices; portion adequacy; recycling; composting.
Prescott et al. (2020) [76] USA	To identify potential school meal recovery options, their prevalence, and systems factors influencing school food waste recovery across three Northern Colorado school districts.	SM: Public TS: Not informed P: Individuals engaged in food recovery ($n = 28$) SD: Environmental, social	Composting; sharing table; food donation.
Virta, Love (2020) [77] USA	To identify how fishes are implemented in school programs, their impacts, and the enabling factors to support these programs.	SM: Public TS: Elementary, middle, high school P: Seafood processors, Oregon Seafood Commission leaders, school district food service leaders, school kitchen managers ($n = 6$) SD: Environmental, economic, social	Participation in the "Fish to School" program (offer and education about local seafood).
Perez-Neira et al. (2021) [30] Spain	To assess the greenhouse gas emissions reduction of agroecological policies implemented in public food procurement, specifically for school canteens.	SM: Public TS: Pre-School, lementary P: School canteens ($n = $ not informed) SD: Environmental, economic, social	Purchase of local, organic, and seasonal food (agro-ecology policies).
Rector et al. (2021) [78] Tanzania	To assess the state of adolescent school nutrition interventions in Dodoma, Tanzania.	SM: Public TS: High School P: School administrators, teachers, students, parents ($n = $ not informed) SD: Environmental	School garden.
Toledo (2021) [79] Brazil	To evaluate the "Educational Garden" Program to promote adequate and healthy food in the school environment.	SM: Public TS: Elementary, high school SD: Environmental	School Garden; environmental education.

* Independent schools: Non-profit private schools independent in philosophy, administration, and funding.

In the United States, where the largest number of studies was identified ($n = 22$), sustainability practices mainly involved educational activities for sustainability [20,44,53,56,57,59–61,69], waste reduction [40–42,46,49,61,76], school gardens [20,46,55,59–61,64], and participation in programs that promote closer ties between schools and producers [46,48,69,70,73,77]. Practices such as food donation [49,64,69,76], strategies for the rational use of water and energy [42,61], the offer of vegetarian/vegan menus or with reduced meat supply [25,46], and the use of local and organic foods (not mentioning participation in specific programs for this purpose) [59] were less mentioned. The study was carried out in parallel in the United States and Cuba in the context related to the experiences of urban school gardens in Philadelphia (USA) and Havana (Cuba) [51].

In Canada, the identified sustainability practices ($n = 2$) involved school gardens, purchase of local and organic foods, participation in programs that promote closer ties between schools and local farmers, waste reduction, and the use of vegetarian dishes [24,71]. In Brazil, the only country in Latin America in which studies were identified ($n = 7$), the most cited practices were related to school gardens [14,45,74,75,79] and educational activities for sustainability, including training for those involved in the operationalization of the National School Feeding Program (PNAE) [14,68,74,79]. However, the studies also cited activities to reduce waste [14,45,75], use of local [14] and regional [62] foods, partnerships for the development of sustainability activities (such as environmental preservation), and to encourage both the use of organic food and to value the work of those involved in all food production [68].

Among European countries, three studies were conducted in Spain and Italy, two in England, and one in Denmark, Finland, and Wales. The most common practices involved were buying local and organic food [8,30,47,58,67]. In Italy, socio-environmental criteria in hiring school food services were also mentioned [47]. Among the European studies, practices linked to changes in menus or portions were also identified [8,26,27,52], sustainability certification [26], waste reduction and energy and/or water savings [26,50], participation in a program to encourage sustainable eating [12,54], school garden [50], and practices aimed at social sustainability (such as food donation and the use of food from land confiscated from criminal organizations) [8,26,27].

On the African continent, two studies were identified in South Africa [43,66] and one in Ghana [63], Tanzania [78], and Kenya [65]. The identified practices focused on activities related to school gardens [43,66,78]. However, the teaching of agroforestry practices, water-saving, and the development and adoption of a meal planning package nutritionally balanced meals, with locally sourced ingredients, were also identified in the studies [43,63,65].

Among Asian countries, one study was identified in Japan [35] and one in India [72]. In Japan [35], the study demonstrated that the reinforcement of social norms not to waste and factors related to the planning of menus, pedagogical practices, and recycling and composting activities, contributed to reducing the food waste in schools. In India, the sustainable management of supply chains was studied based on one of the companies responsible for the school feeding program, which considers the integration of economic and non-economic issues in the generation of value in the supply chain [72].

In 20% of the studies ($n = 10$), it was impossible to identify information about the responsibility for managing schools. Most studies were performed in public schools (64%; $n = 32$). Studies conducted in public and private schools corresponded to 14% ($n = 7$), and only one study was conducted in a private school. All stages of the education system, including the earliest (pre-school) and the final years (high school), were mentioned, but the stage referring to primary or elementary education was identified in most studies (68%; $n = 34$). In 27.4% of the studies ($n = 14$), it was impossible to obtain this information. The studies used quantitative (30%; $n = 15$), qualitative (48%; $n = 24$), or mixed methods (22%; $n = 11$).

3.2. Identified Sustainability Practices

The environmental dimension of sustainability was identified in all studies that cited sustainability practices ($n = 50$), alone (26%; $n = 13$), or together with the other considered dimensions. In most studies (44%; $n = 22$), it was possible to identify practices related to the three sustainability dimensions (environmental, economic, and social). The combination of environmental and social dimensions was identified in 14% of the studies ($n = 7$), and the environment with economic dimension in 16% ($n = 8$). It is important to highlight that for identifying the sustainability dimensions, some practices, in isolation, represented the attendance of more than one of the mentioned dimensions. In contrast, in other cases, the identification of different sustainability dimensions in the same study resulted from different practices cited by the authors. Sustainability practices adopted in schools are presented in Table 1. The activities were described according to the dimensions of sustainability (Figure 2).

Figure 2. Identified sustainable practices in schools according to the environmental, economic, and social dimensions.

The involvement with school gardens and education activities for sustainability represented the most commonly reported practices, being identified in 36% ($n = 18$) [14,20,24,43,45,46,50,51,55,59–61,64,66,74,75,78,79] and 28% ($n = 14$) [14,20,35,44,51,53,56,57,59–61,69,74,79] of the studies, respectively. Among the studies that described the use of school gardens, 8% ($n = 4$) [14,45,61,75] described the cultivation of organic foods and, in the context of education practices for sustainability, travel field studies [59,60] and cooking activities [14,20,44,69] were also cited.

Another frequently cited category of sustainability practices concerns schools food supply initiatives. Actions to purchase or encourage the employment of local or short-chain foods, including participation in programs such as "Farm to School" and "Fish to School" were cited by 26% (n = 13) of the studies [8,14,24,30,46,48,59,67,69–71,73,77] and organic foods were observed in 18% (n = 9) [8,24,30,45,47,58,59,67,68]. Participation in a sustainable food consumption promotion program (4%; n = 2) was also identified [12,54]. Although little mentioned, practices that integrated socio-environmental and economic dimensions in the contracts (4%; n = 2), the use of regional (2%; n = 1), seasonal (2%; n = 1), and produced foods in lands confiscated from criminal organizations, were also observed [8,30,47,62,72].

Regarding the adoption of measures to reduce waste, recycling [35,40–42,46,49,50] and composting [24,26,35,46,49,61,76] were the most reported, described in 14% (n = 7) of the analyzed studies. These represent important strategies for the control of organic and inorganic waste. Food donation was a practice identified in 10% (n = 5) of the studies [26,27,49,64,76] and food portion size adjustment in 6% (n = 3) [26,27,35]. Sharing tables [76], single-course menu [8], reinforcement of social norms [35], donation of food waste for animal feeding [49], team communication to adjust the amount produced, and noise reduction in the cafeteria to allow a more comfortable environment [26] were also identified practices for the reduction of organic waste, to a lesser degree than the others previously mentioned. Other less cited practices involving the control of non-organic waste generation were the use of reusable devices [40,45], the purchase of products in bulk or with minimum packaging [24,40], the use of returnable bottles [14], reduced use of paper [26], or the replacement of mineral water by filtered water [8].

The adoption of strategies that involved saving water or energy was cited in 10% (n = 5) of the studies [26,42,43,50,61], represented by activities such as reduced use of energy, energy audits, reduced use of water, and installing a cistern for collection rainwater or irrigation system for water reuse.

As for menu actions, practical studies were identified as offering vegetarian/vegan meals [24,46,52], reducing meat supply [25], planning a menu that included the exposure of students to unfamiliar or unappreciated foods [35], and the adoption of a meal planning package that facilitated the planning of nutritionally balanced meals with locally sourced ingredients [63].

Although mentioned in only one study, partnerships for the development of sustainability activities (such as environmental preservation) and certification were also identified [26,68].

3.3. Sustainability Recommendations in School Feeding Policies/Programs

Among the 11 policies under government responsibility identified in the studies that mentioned sustainability recommendations, 73% (n = 8) were national in scope, 18% (n = 2) state, and 9% (n = 1) municipal. The European continent had the highest number of policies/programs (64%; n = 7), identified in Italy, England, Finland, Spain, Sweden, and Germany [37,80–85]. School feeding policies/programs were also identified in Brazil, Japan, and the United States [34,86–89] (Table 2).

The most mentioned aspects were the origin and type of food used in school meals, such as organic, local or shorter transport distances, seasonal, agroecological, and sustainable [34,37,80–87]. Other examples cited in this category, although less frequently, were reducing meat, increasing consumption of vegetables, reducing carbon emissions, typical foods, and respecting local traditions [34,37,82,84,86].

Table 2. Identified Sustainability recommendations in governmental school food policies documents.

Year (Reference)	Document	Document Type	Responsibility	City/Country	Identified Sustainability Recommendations
Municipality of Barcelona (2020) [82]	This Is Not a Drill. Climate Emergency Declaration, Barcelona.	Declaration	Municipal	Barcelona (Spain)	Implementation and promotion of healthier and low-carbon diets in schools through the use of seasonal, local and organic foods; reduction of animal protein intake (especially red meat) and ultra-processed foods.
The National Food Agency (2021) [84]	Good school meals. Guidelines for primary schools, secondary schools, and youth recreation centers.	Guideline	National	Sweden	Topics on menu planning (including, among others, reducing meat and increasing vegetables, legumes, fruits, and cereals, choosing organic foods, and observing seasonality), measures to prevent food waste, reducing energy consumption, and transport distance.
Brazil (2009)/Brazil (2020) [34,86]	Law $n°$ 11.947, from 16 June 2009/Resolution FNDE $n°$ 06, May 2020.	Law/Resolution	National	Brazil	Support for sustainable development through purchasing local food from family farming, preference for organic and agroecological food, observation of sustainability in menu planning, and nutrition education actions, seasonality; local traditions.
Italian Ministry of Health (2021) [37]	National guidelines for hospital, care, and school catering Decree 28 October 2021.	Guideline	National	Italy	Seasonality; local, short-chain, organic, and typical foods; environmental protection; animal welfare; local traditions; fair trade; food recovery; reduction of food waste and non-organic residues; food education aimed at conscientious and sustainable consumption; social and environmental criteria for contracts.
Cabinet Office Japan (1954) [88]	School Lunch Program Act.	Law	National	Japan	Respect for nature; a positive attitude towards environmental conservation; a sense of valuing the work of those involved in food production; food education; generation of a correct understanding of the production, distribution, and consumption of food.

Table 2. *Cont.*

Year (Reference)	Document	Document Type	Responsibility	City/Country	Identified Sustainability Recommendations
United States Department of Agriculture (2015) [89]	Updated Offer versus Serve Guidance for the National School Lunch Program and School Breakfast Program Effective Beginning School Year 2015–2016.	Guidance	National	USA	The possibility of the student refusing some of the foods offered to reduce food waste in school feeding programs.
Santa Catarina (2018) [87]	Law 17.504, 10 April 2018.	Law/Resolution	State	Brazil	Preference for the purchase of organic vegetables by schools, foreseeing a gradual increase in the percentage of purchases.
National Nutrition Council (2017) [81]	Eating and learning together–recommendations for school meals.	Recommendations	National	Finland	Sustainable development and environmental issues concerning food acquisition, food choices, and waste reduction, citing, among others, seasonality; favoring the consumption of domestic vegetables; assembly of dishes by students; possibility of repetition.
Department for Education (2021) [80]	School food standards practical guide.	Guidance	National	England	It recommended sustainable procurement, including the use of fresh, seasonal, sustainable, and locally sourced ingredients, sustainable fish purchase, waste reduction, and school gardens.
Consejo Interterritorial de Sistema Nacional de Salud (2010) [83]	Consensus document about food in educational centers.	Consensus	National	Spain	It informed that the possible incorporation of organic food in school lunches might have advantages about sustainability and protection of the environment. However, it considered no evidence to affirm that organic foods are nutritionally better or safer.
Senate Administration (2017) [85]	Reorganization of the school lunch at open and affiliated all-day primary schools and for support centers in Berlin.	Handout	State	Berlin (Germany)	The establishment of criteria for quality assessment, considered a priority about the price when hiring school food suppliers (organic food corresponds to one of the quality criteria).

Regarding other non-governmental programs and initiatives (n = 5), two were identified in the United States [90,91], two were global in scope [92,93], and one was identified in England [94]. The set of activities observed in these programs involved encouraging the purchase of local, seasonal, and sustainable food [91,92,94], school gardens [91,94], visits to local farmers [91,94], cooking and nutritional education activities [91,94], waste reduction [90], and specific actions for each school to train people to generate environmental and sustainability awareness [93] (Table 3).

3.4. Risk of Bias

Among the studies analyzed, 49 had a low risk of bias and 1 had a moderate risk. All studies implemented the practices and answered the main research question (Table S2—Supplementary Materials).

Table 3. Other available non-governmental school feeding programs/initiatives retrieved from the studies.

Initiative	Country	Description
Food for Life Partnership (FFLP) [94]	England	A program with a whole-school approach that addresses healthy, tasty, and sustainable eating through four areas of development: food quality, food leadership and food culture, food education, and community and partnerships. "Food quality" includes, among other factors, the use of fresh, seasonal, local, and organic foods, meat that meets animal welfare standards, marine conservation certified fish, and eggs from free range hens. "Nutritional education" includes the development of cooking skills, planting food and visiting or receiving visits from farmers, in addition to ethical and environmental issues around food choices.
Eco-Schools [93]	Global	A global program of sustainable schools that aims to train people with an environmental and sustainability conscience. The program is based on seven steps, the: formation of an Eco Committee (in which students play a main role) to discuss environmental and social actions for the school, conduction of a sustainability audit, preparation, monitoring, and evaluation of the action plan, linking of activities to the curriculum, information and involvement with the community, and production of an ecological code that represents the school's commitment to sustainability.
World Food Programme (WFP)'s Home Grown School Feeding [92]	Global	An initiative in which the World Food Program works with governments to develop school food policies that seek to improve student nutrition and support the local economy through the connection between school food and local farmers.
Smarter Lunchrooms Movement (SLM) [90]	USA	The initiative generated by research in schools is used to create lunchrooms that encourage healthy food choices and reduce waste, using a strategy with little or no cost.
Farm to School (FTS) [91]	USA	It connects schools and local food producers to offer fresh and healthy food to students. It is based on local food purchasing activities, education about food, nutrition, health, agriculture, and hands-on learning activities (school gardens, including visits to local farmers and culinary classes).

4. Discussion

In 2019, the EAT-The Lancet Commission established universal strategies and recommendations to achieve food system transformation, striving for human health and environmental sustainability. The need to improve availability and access to healthy foods from sustainable food systems and educate individuals on these topics using food programs was reinforced [95]. Therefore, the school is an opportune locus for sustainability practices. The environmental, economic, and social effects of the actions carried out in school food services and the education process will echo both in society's present and future.

Vegetable gardens and education activities for sustainability were the most cited practices among the studies. School gardens are essential tools to support community and school feeding programs by using their produce in student meals and training vegetable growing skills [19,43,66]. However, the support of the school administration, the availability of space and resources to purchase tools and supplies, teacher training, the integration of the garden into the school curriculum, sharing activities with community members, and the presence of a coordinator to organize activities are identified as key factors in determining the results of the implementation and continuity of school gardens [66,96,97].

Education plays a central role in enabling students to think and act critically on current and future global challenges, including climate change, environmental degradation, biodiversity loss, poverty, and inequality [16]. Therefore, the literature has strongly recommended and evidenced the association between school gardens and educational processes aimed at health, environment, and sustainability [19,24,61]. Gonsalves et al. [19] emphasize the role of school gardens "in environmental and nature education, in local food biodiversity and conservation, food, eco-literacy, diets, nutrition and health, and agricultural education, contributing to the search for a food system more sustainable."

The importance of education in enabling individuals to promote sustainable development is reaffirmed in Sustainable Development Goal 4 (SDG 4). SDG 4 is integrated with other indivisible goals, comprising the environmental, economic, and social dimensions of Agenda 2030, an action plan for people, the planet, and prosperity [98].

In line with this purpose, schools can observe efforts to integrate environmental education, ecological literacy, or education for sustainability in their curricula [14,56,57,60,69]. However, a recent study [99] that assessed the extent to which environmental issues are integrated into primary and secondary education policies and curricula in 46 member states of the United Nations Educational, Scientific, and Cultural Organization (UNESCO) demonstrated that despite 92% of the documents making some reference to environmental issues, the depth of this inclusion was low. Terms such as "climate change" and "biodiversity" were rarely mentioned. What has been done is not enough to ensure that learning helps in confronting current global challenges. The document recommends, among other actions, that greater emphasis needs to be placed on environmental issues in education, integrating them into curricula and overcoming the focus on cognitive knowledge, and training all teachers and school leaders in education for sustainable development [99]. Also, considering the food system's environmental, economic, and social impacts, the efforts to integrate themes involving sustainable food consumption into educational practices are worthy of note [12,44]. These approaches are fundamental because education can increase adherence to different sustainability practices since, although observed in this study, many had a low frequency of realization.

The use of local or organic foods, often set by government regulations, stems from concerns about students' health, the living conditions of farmers, and the environment [34,47,48,70,100]. Initiatives involving the use of local and organic foods were also frequently cited among the analyzed studies [8,14,24,30,47,58,59,67].

The creative food purchase policy incorporates social and environmental criteria into the contracts, going beyond economic considerations and encouraging the purchase of local food [33]. Examples of buying local and organic food are worldwide, as in the

Brazilian case at the National School Feeding Program, where the purchase of food from local farmers is a compulsory item provided in its legal framework. Organic food is preferred in public purchases, the standardized instruments for this type of purchase in the country's program [34]. Through Home Grown School Feeding, the World Food Program (WFP) works with governments in 46 countries to develop national policies that provide adequate food for students and ensure local development by purchasing food from family farmers [92]. There are also experiences linked to the Farm To School programs, which purchase local food and develop educational activities related to agriculture, food, health, and nutrition [70,71]. All initiatives that connect schools to family farmers are vital because the benefits of school meals go beyond the boundaries of schools and reach family farmers. These involve economic (increase in income, price support, and inclusion in the market), social (food security, living conditions, and social inclusion), and environmental (crop diversification and greater production of organic food) aspects [31].

Reducing food waste and controlling non-organic waste represent initiatives that must be implemented in school food services. Studies in different parts of the world have demonstrated that these places are major food waste generators, causing environmental, economic, and social impacts [101–104]. Concerning non-organic waste, a study carried out in northern Colorado, USA identified that factors such as the speed of the service line, the quality of food, the cost, and the difficulty managers have in understanding the impact of their decisions at a systemic level, affected the ability to reduce or recover these wastes [105].

In our systematic review, composting was the most cited practice for reducing the generation of organic waste and recycling non-organic waste. However, it is important to highlight that even among the studies in which the performance of waste management practices was cited, they were often not reported among the participants or were reported by a minority of them [26,40–42,46,49,76], demonstrating that adopting waste reduction strategies in school meals is not yet routine practice.

Among the strategies to promote the reduction of food waste, the literature discusses the importance of integrating this theme and the sustainability of the food system in pedagogical practices, in addition to actions aimed at improving operations and planning, team communication, and the involvement of students in waste management activities [35,102]. Food donation can represent a successful experience to mitigate the impacts of the production of meals by reducing waste and serving people in vulnerable situations, with relatively few investments [27]. However, the main barriers related to food donation and food recovery in this context involve concerns about responsibility, cost, inconsistent food waste, policy confusion, and the sanitary quality of food [76]. As for non-organic waste, among other recommendations, a study indicated that school food services could incorporate packaging waste in purchasing processes, as they do not always control the packaging used by manufacturers [105].

The adoption of some strategies related to saving water and energy was mentioned among the analyzed studies. In school feeding, studies that reported the environmental impacts of the choices made by food services regarding the origin and types of food purchased (fresh or not, and from different groups, such as meat and vegetables) demonstrated a significant contribution from phases before the production of meals [21,23,106]. However, considering that during the production of meals both water and energy are essential factors for the operation of the service, the training of a school's employees and the monitoring of the intended use of these resources is necessary. Instruments created to evaluate sustainability practices in food services, which include among their analysis categories the rational use of energy and water, are helpful tools in this regard [107,108].

Environmental and health damage to the population generated by how the food system has been operated is already well established [109]. Two of the factors contributing to the harmful effects of this modus operandi are meat production, especially red meat, and food waste, which are responsible, among other factors, for a significant emission of greenhouse gases into the environment and/or consumption of freshwater [23,106,110,111].

Therefore, some practices related to the offer of vegetarian/vegan menus, with a reduction in the meat offering, the adequacy of the portion sizes, or the adoption of the single-course scholar menu have been reported in the literature [8,25–27,35,112]. Some instruments have been proposed to allow the planning of more environmentally sustainable menus based on reducing carbon and/or water footprints while addressing nutritional, economic, and cultural dimensions [7,113–116]. In addition, the definition of criteria for planning sustainable menus in the context of school meals has also been described [117].

Other less mentioned strategies involved using regional foods, environmental certification, and the development of partnerships to carry out environmental preservation activities.

According to Morgan [118], "the creation of a sustainable school foodservice is the litmus test of a country's commitment to sustainable development, as it involves nothing less than the health and well-being of young people and vulnerable people". In this sense, several efforts were made to strengthen the role of school feeding in achieving nutritionally adequate diets for students and meet the principles of sustainability in the three dimensions: environmental, economic, and social. However, it is noteworthy that, despite the general premise established in the literature of the potential effects of sustainability practices in school in mitigating global challenges, the wide scope of school feeding and the variability of characteristics and challenges experienced between different regions of the globe, including different regulations, economic, social, political and cultural conditions, demand specific solutions, adapted to each local context.

5. Limitations

This review has some limitations. First, it was not possible to state that the school feeding policies/programs that mentioned concerns about sustainability were exhausted, since the policies were found in the studies reference lists that had their full text analyzed. In addition, some of these policies, written in a non-English language, were translated through a translation platform. Therefore, some information may have been lost due to language barriers. Despite these limitations, these findings evidenced different recommendations that reinforced the importance of actions, which ranged from the choice of sustainable foods to the strengths of nutrition and sustainable consumption practices education.

6. Conclusions

There is an imminent need to ensure the prosperity of nations, anchored in the priorities of protecting the health of people and the planet and guaranteeing adequate living conditions, reducing social inequalities. It involves offering food in terms of education, enabling students to make conscious choices consistent with this need. In this sense, schools and school feeding programs have all the necessary characteristics for developing practices that aim at sustainability in the environmental, economic, and social dimensions, given their scope and the different perspectives that can be worked.

The present study identified sustainability recommendations in 16 governmental and non-governmental policies/programs. Recommendations for purchasing sustainable food (organic, local, and seasonal), nutrition education focused on sustainability, and reducing food waste were frequent.

Several sustainability practices were described in this systematic review, such as the use of school gardens and education activities for sustainability. Actions carried out in food services were also mentioned, from the planning of menus and the purchase of raw materials (mainly local and organic foods, vegetarian/vegan menus) to the distribution of meals (especially practices to reduce waste organics and inorganics such as composting, recycling, donating food, and adjusting portion sizes).

The findings reinforce the need to stimulate managers' views, in their most varied spheres of power, for the priority that should be given to this theme, so that education for sustainability is universally part of the curricula, and so that food services can equip themselves with the knowledge and tools necessary to carry out sustainability practices in their daily activities.

Lastly, further investigations to evaluate these practices are needed to examine the evolution of their adoption and the main barriers and potentialities related to their implementation. With a specific look at the school field, assessment instruments can help with this monitoring.

Supplementary Materials: The following are available online at https://www.mdpi.com/article/10.3390/foods11020176/s1, Table S1: Databases and terms used to search references on sustainability practices adopted in schools; Table S2: Quality criteria of the studies selected for the systematic review.

Author Contributions: E.B.d.S., conceptualization, investigation, methodology, writing—original draft, writing—review and editing; D.d.C.M., investigation and methodology; R.P.Z., data curation, formal analysis, supervision, validation and writing—review and editing; A.R., writing—review and editing, visualization, project administration. R.B.A.B., data curation, formal analysis, supervision, validation and writing—review and editing. All authors have read and agreed to the published version of the manuscript.

Funding: This research received no external funding.

Institutional Review Board Statement: Not applicable.

Informed Consent Statement: Not applicable.

Data Availability Statement: No data availability.

Acknowledgments: The authors would like to acknowledge the "National Council for Scientific and Technological Development–CNPq".

Conflicts of Interest: The authors declare no conflict of interest.

Appendix A

Table A1. Full-text articles excluded, with reasons.

Author (Year)	Reason for Exclusion
Alexandre et al. (2016) [119]	2
Amarante (2016) [120]	1
Andreatta et al. (2021) [121]	4
Anton-Peset, Fernandez-Zamudio and Pina (2021) [13]	2
Batlle-Bayer et al. (2021) [122]	3
Braun et al. (2018) [123]	4
Brena (2017) [124]	3
Carvalho (2009) [125]	5
Coleman et al. (2011) [126]	1
Colombo et al. (2019) [7]	3
Colombo et al. (2020) [127]	2
Constanty (2014) [128]	4
Constanty and Zonin (2016) [129]	4
Conner et al. (2010) [130]	2
Damapong, Kongnoo and Monarumit (2013) [131]	1
Dirks (2011) [132]	3
Eich (2015) [133]	5
Ellinder et al. (2020) [113]	2
Elnakib et al. (2021) [134]	2
Colombo (2021) [135]	3
Ferderbar (2013) [136]	2
Filippini et al. (2018) [137]	5
Fitzsimmons and O'Hara (2019) [138]	4
Franzoni (2015) [139]	4
Gaddis and Jeon (2020) [38]	1
Ghattas et al. (2020) [140]	2
Granillo-Macías (2021) [141]	3

Table A1. *Cont.*

Author (Year)	Reason for Exclusion
Green (2016) [142]	2
Gregolin et al. (2017) [143]	4
He (2013) [144]	1
Hendler, Ruiz and Oliveira (2021) [145]	3
Henry-Stone (2008) [146]	3
Hodgkinson (2011) [147]	3
Johnston et al. (2009) [148]	5
Jones (2012) [149]	5
Kipfer (2018) [150]	3
Koch (2000) [151]	2
Lalli (2020) [152]	5
Lauffer (2019) [153]	2
Lawless (2013) [154]	3
Lindgren (2020) [112]	3
Løes; Nölting (2011) [28]	1
Løes; Nölting (2009) [155]	1
McCarty (2013) [156]	2
Medina (2009) [157]	5
Melão (2012) [158]	4
Mikkola (2010) [159]	1
Moss Gamblin (2013) [160]	5
Mosiman (2014) [161]	4
Morgan and Morley (2003) [162]	1
Morgan and Sonino (2007) [33]	1
Morgan (2008) [118]	5
Mota, Silva and Pauletto (2021) [163]	5
Muansrichai, Panyasing and Yonvanij (2015) [164]	5
Nunes et al. (2018) [165]	4
Nuutila, Risku-Norja and Arolaakso (2019) [166]	3
Orr (2020) [167]	3
Otsuki (2011) [168]	4
Padilha et al. (2018) [169]	4
Osowski and Fjellström (2018) [170]	1
Polo et al. (2017) [171]	1
Prescott et al. (2019) [172]	2
Rambing et al. (2020) [173]	5
Redman (2013) [174]	2
Resque et al. (2019) [175]	4
Ribeiro, Ceratti and Broch (2013) [176]	4
Rodrigues et al. (2020) [107]	2
Santos et al. (2014) [177]	4
Schachtner-Appel (2019) [178]	2
Scott (2011) [179]	2
Silva and Sousa (2013) [180]	4
Silva and Pedon (2015) [181]	4
Silva, Gehlen and Schultz (2016) [182]	4
Silva, Dias and Amorim (2015) [183]	4
Soares (2011) [184]	4
Soares et al. (2017) [185]	4
Solof (2014) [186]	2
Szinwelski et al. (2015) [187]	4
Trott (2017) [188]	2
Turpin (2009) [189]	4
Vasconcelos, Vieira and Rodrigues (2014) [190]	2
Valadão and Sousa (2018) [191]	1
Wade (2019) [192]	3
Wickramasinghe et al. (2016) [193]	1

Legend—Exclusion criteria: (1) Comments, letters, conferences, reviews, abstracts, reports, undergraduate works, discussion papers, and books, (2) studies carried out outside schools or in which the school was not responsible for the action, (3) studies in which practices were not performed or studies where activities were punctual, (4) studies focused on the supplier or that only reported purchases, and (5) studies that did not describe sustainability practices.

References

1. WFP. *State of School Feeding Worldwide 2020*; World Food Programme: Rome, Italy, 2020; p. 260.
2. Aurino, E.; Gelli, A.; Adamba, C.; Osei-Akoto, I.; Alderman, H. Food for thought? Experimental Evidence on the Learning Impacts of a Large-Scale School Feeding Program. *J. Hum. Resour.* **2020**, *57*. [CrossRef]
3. Jomaa, L.H.; McDonnell, E.; Probart, C. School feeding programs in developing countries: Impacts on children's health and educational outcomes. *Nutr. Rev.* **2011**, *69*, 83–98. [CrossRef] [PubMed]
4. Metwally, A.M.; El-Sonbaty, M.M.; El Etreby, L.A.; Salah El-Din, E.M.; Abdel Hamid, N.; Hussien, H.A.; Hassanin, A.M.; Monir, Z.M. Impact of National Egyptian school feeding program on growth, development, and school achievement of school children. *World J. Pediatr.* **2020**, *16*, 393–400. [CrossRef]
5. Locatelli, N.T.; Canella, D.S.; Bandoni, D.H. Positive influence of school meals on food consumption in Brazil. *Nutrition* **2018**, *53*, 140–144. [CrossRef] [PubMed]
6. Oostindjer, M.; Aschemann-Witzel, J.; Wang, Q.; Skuland, S.E.; Egelandsdal, B.; Amdam, G.V.; Schjøll, A.; Pachucki, M.C.; Rozin, P.; Stein, J.; et al. Are school meals a viable and sustainable tool to improve the healthiness and sustainability of children´s diet and food consumption? A cross-national comparative perspective. *Crit. Rev. Food Sci. Nutr.* **2017**, *57*, 3942–3958. [CrossRef]
7. Colombo, P.E.; Patterson, E.; Elinder, L.S.; Lindroos, A.K.; Sonesson, U.; Darmon, N.; Parlesak, A. Optimizing School Food Supply: Integrating Environmental, Health, Economic, and Cultural Dimensions of Diet Sustainability with Linear Programming. *Int. J. Environ. Res. Public Health* **2019**, *16*, 3019. [CrossRef]
8. Galli, F.; Brunori, G.; Di Iacovo, F.; Innocenti, S. Co-Producing Sustainability: Involving Parents and Civil Society in the Governance of School Meal Services. A Case Study from Pisa, Italy. *Sustainability* **2014**, *6*, 1643–1666. [CrossRef]
9. WCED. *Report of the World Commission on Environment and Development: Our Common Future*; Oxford University Press: Oxford, UK, 1987; p. 400.
10. United Nations. Report of the United Nations Conference on Environment and Development. In Proceedings of the UN Conference on Environment and Development, Rio de Janeiro, Brazil, 3–14 June 1992; p. 72.
11. Von Koerber, K.; Bader, N.; Leitzmann, C. Wholesome Nutrition: An example for a sustainable diet. In Proceedings of the Nutrition Society, Berlin, Germany, 20–23 October 2015; Cambridge University Press: Cambridge, UK, 2017; Volume 76, p. 34.
12. Jones, M.; Dailami, N.; Weitkamp, E.; Salmon, D.; Kimberlee, R.; Morley, A.; Orme, J. Food sustainability education as a route to healthier eating: Evaluation of a multi-component school programme in English primary schools. *Health Educ. Res.* **2012**, *27*, 448–458. [CrossRef]
13. Antón-Peset, A.; Fernandez-Zamudio, M.A.; Pina, T. Promoting food waste reduction at primary schools. A case study. *Sustainability* **2021**, *13*, 600. [CrossRef]
14. Triches, R.M. Promoção do consumo alimentar sustentável no contexto da alimentação escolar. *Trab. Educ. Saúde* **2015**, *13*, 757–771. [CrossRef]
15. UNESCO. UN. Decade of ESD. Available online: https://en.unesco.org/themes/education-sustainable-development/what-is-esd/un-decade-of-esd (accessed on 31 December 2021).
16. UNESCO. Education for Sustainable Development. Available online: https://en.unesco.org/themes/education-sustainable-development (accessed on 25 November 2021).
17. Wamsler, C. Education for sustainability: Fostering a more conscious society and transformation towards sustainability. *Int. J. Sustain. High. Educ.* **2020**, *21*, 112–130. [CrossRef]
18. Benn, J.; Carlsson, M. Learning through school meals? *Appetite* **2014**, *78*, 23–31. [CrossRef]
19. Gonsalves, J.; Hunter, D.; Lauridsen, N. School gardens: Multiple functions and multiple outcomes. In *Agrobiodiversity, School Gardens and Healthy Diets: Promoting Biodiversity, Food and Sustainable Nutrition*; Routledge: London, UK, 2020; p. 345.
20. Strohl, C.A. Scientific Literacy in Food Education: Gardening and Cooking in School. Ph.D. Thesis, University of California, Davis, CA, USA, 2015.
21. Mistretta, M.; Caputo, P.; Cellura, M.; Cusenza, M.A. Energy and environmental life cycle assessment of an institutional catering service: An Italian case study. *Sci. Total Environ.* **2019**, *657*, 1150–1160. [CrossRef] [PubMed]
22. Jungbluth, N.; Keller, R.; König, A. ONE TWO WE—life cycle management in canteens together with suppliers, customers and guests. *Int. J. Life Cycle Assess.* **2015**, *21*, 646–653. [CrossRef]
23. De Laurentiis, V.; Hunt, D.V.L.; Rogers, C.D.F. Contribution of school meals to climate change and water use in England. *Energy Procedia* **2017**, *123*, 204–211. [CrossRef]
24. Black, J.L.; Velazquez, C.E.; Naseam, A.; Chapman, G.E.; Carten, S.; Edward, J.; Shulhan, S.; Stephens, T.; Rojas, A. Sustainability and public health nutrition at school: Assessing the integration of healthy and environmentally sustainable food initiatives in Vancouver schools. *Public Health Nutr.* **2015**, *18*, 2379–2391. [CrossRef] [PubMed]
25. Blondin, S.A.; Cash, S.B.; Griffin, T.S.; Goldberg, J.P.; Economos, C.D. Meatless Monday National School Meal Program Evaluation: Impact on Nutrition, Cost, and Sustainability. *J. Hunger Environ. Nutr.* **2020**, 1–13. [CrossRef]

26. Derqui, B.; Grimaldi, D.; Fernandez, V. Building and managing sustainable schools: The case of food waste. *J. Clean. Prod.* **2020**, *243*, 118533. [CrossRef]

27. Lagorio, A.; Pinto, R.; Golini, R. Food waste reduction in school canteens: Evidence from an Italian case. *J. Clean. Prod.* **2018**, *199*, 77–84. [CrossRef]

28. Løes, A.-K.; Nölting, B. Increasing organic consumption through school meals-lessons learned in the iPOPY project. *Org. Agric.* **2011**, *1*, 91–110. [CrossRef]

29. Maynard, D.d.C.; Vidigal, M.D.; Farage, P.; Zandonadi, R.P.; Nakano, E.Y.; Botelho, R.B.A. Environmental, Social and Economic Sustainability Indicators Applied to Food Services: A Systematic Review. *Sustainability* **2020**, *12*, 1804. [CrossRef]

30. Perez-Neira, D.; Simón, X.; Copena, D. Agroecological public policies to mitigate climate change: Public food procurement for school canteens in the municipality of Ames (Galicia, Spain). *Agroecol. Sustain. Food Syst.* **2021**, *45*, 1528–1553. [CrossRef]

31. Cervantes-Zapana, M.; Yagüe, J.L.; De Nicolás, V.L.; Ramirez, A. Benefits of public procurement from family farming in Latin-AMERICAN countries: Identification and prioritization. *J. Clean. Prod.* **2020**, *277*, 123466. [CrossRef]

32. Da Cunha, W.A.; de Freitas, A.F.; Salgado, R.J.S.F. Efeitos dos programas governamentais de aquisição de alimentos para a agricultura familiar em Espera Feliz, MG. *Rev. Econ. Sociol. Rural* **2017**, *55*, 427–444. [CrossRef]

33. Morgan, K.; Sonnino, R. Empowering consumers: The creative procurement of school meals in Italy and the UK. *Int. J. Consum. Stud.* **2007**, *31*, 19–25. [CrossRef]

34. Brasil. Lei no 11.947, de 16 de Junho de 2009. 2009. Available online: https://www.fnde.gov.br/index.php/legislacoes/institucional-leis/item/3345-lei-n-11947-de-16-de-junho-de-2009 (accessed on 31 December 2021).

35. Izumi, B.T.; Akamatsu, R.; Shanks, C.B.; Fujisaki, K. An ethnographic study exploring factors that minimize lunch waste in Tokyo elementary schools. *Public Health Nutr.* **2020**, *23*, 1142–1151. [CrossRef]

36. WFP. *How School Meals Contribute to the Sustainable Development Goals*; World Food Programme: Rome, Italy, 2017; p. 8.

37. Ministero della Salute. *Linee di Indirizzo Nazionale per la Ristorazione Ospedaliera, Assistenziale e Scolastica*; Ministero della Salute: Rome, Italy, 2021; Volume 1, pp. 1–65.

38. Gaddis, J.E.; Jeon, J. Sustainability transitions in agri-food systems: Insights from South Korea's universal free, eco-friendly school lunch program. *Agric. Human Values* **2020**, *37*, 1055–1071. [CrossRef]

39. Page, M.J.; McKenzie, J.E.; Bossuyt, P.M.; Boutron, I.; Hoffmann, T.C.; Mulrow, C.D.; Shamseer, L.; Tetzlaff, J.M.; Akl, E.A.; Brennan, S.E.; et al. The PRISMA 2020 statement: An updated guideline for reporting systematic reviews. *PLoS Med.* **2021**, *18*, e1003583. [CrossRef] [PubMed]

40. Mann, N.L. A Decision Model for Solid Waste Management in School Food Service. Ph.D. Thesis, Texas Woman's University, Denton, TX, USA, 1991.

41. Ghiselli, R.F. Reusing, Reducing, and Recycling Solid Waste in Indiana School Food Service: A Cost-Effective Approach. Ph.D. Thesis, Purdue University, West Lafayette, IN, USA, 1993.

42. Hackes, B.L.; Shanklin, C.W. Factors Other than Environmental Issues Influence Resource Allocation Decisions of School Foodservice Directors. *J. Am. Diet. Assoc.* **1999**, *99*, 944–949. [CrossRef]

43. Albertse, G.; Mancusi-Materi, E. Children ensuring their own food security in South Africa. *Development* **2000**, *43*, 105–108. Available online: https://doi-org.ez54.periodicos.capes.gov.br/10.1057/palgrave.development.1110126 (accessed on 31 December 2021). [CrossRef]

44. Wadsworth, K.G. A Process and Outcome Evaluation of EarthFriends: A Curriculum Designed to Teach Elementary School-Aged Children to Make Environmentally Sustainable Food Choices. Ph.D. Thesis, Columbia University, New York, NY, USA, 2002.

45. Lima, E.E. Alimentos Orgânicos na Alimentação Escolar Pública Catarinense: Um Estudo de Caso. Master's Thesis, Universidade Federal de Santa Catarina, Florianópolis, Brazil, 2006.

46. Vogt, R.A. Improving School Nutrition with Sustainable Food Systems. Ph.D. Thesis, University of California, Davis, CA, USA, 2006.

47. Sonnino, R. Quality food, public procurement, and sustainable development: The school meal revolution in Rome. *Environ. Plan. A Econ. Space* **2009**, *41*, 425–440. [CrossRef]

48. Izumi, B.T.; Alaimo, K.; Hamm, M.W. Farm-to-School Programs: Perspectives of School Food Service Professionals. *J. Nutr. Educ. Behav.* **2010**, *42*, 83–91. [CrossRef]

49. Baca, J. Investigating Waste Management Programs in School Foodservice Operations. Master's Thesis, Texas Woman's University, Denton, TX, USA, 2011.

50. Bennell, S.J. ESDGC in Primary Schools: Exploring Practice, Development and Influences. Ph.D. Thesis, Bangor University, Bangor, UK, 2012.

51. Bucher, K.A. Sowing City Schools: Teachers and Garden Education in Havana and Philadelphia. Ph.D. Thesis, Indiana University, Bloomington, IN, USA, 2012.

52. Lombardini, C.; Lankoski, L. Forced Choice Restriction in Promoting Sustainable Food Consumption: Intended and Unintended Effects of the Mandatory Vegetarian Day in Helsinki Schools. *J. Consum. Policy* **2013**, *36*, 159–178. [CrossRef]

53. O'Brien, B. Religion, Ethics, Nature in Secondary School Education: Exploring Religion's Role in Sustainability Trends. Ph.D. Thesis, University of Florida, Gainesville, FL, USA, 2013.

54. Orme, J.; Matthew, J.; Salmon, D.; Weitkamp, E.; Kimberlee, R. A process evaluation of student participation in a whole school food programme. *Health Educ.* **2013**, *113*, 168–182. [CrossRef]

55. Rilla, C. School Garden Design as Catalyst for Environmental Education and Community Engagement: Los Angeles Unified School District Case Studies. Master's Thesis, University of Southern California, Los Angeles, CA, USA, 2013.
56. Shuttleworth, J.M. Teaching Sustainability as a Social Issue: Learning from Three Teachers. Ph.D. Thesis, Columbia University, New York, NY, USA, 2013.
57. Barnett, C.M. Paradigm Shift in Public Education: Ridge and Valley Charter School 2000–2014. Ph.D. Thesis, Drew University, Madison, NJ, USA, 2014.
58. He, C.; Mikkelsen, B.E. The association between organic school food policy and school food environment: Results from an observational study in Danish schools. *Perspect. Public Health* **2014**, *134*, 110–116. [CrossRef] [PubMed]
59. Keller, R.A. Fostering sustainability: A qualitative interview study exploring how educators work to cultivate nature awareness in young children. Master's Thesis, Mills College, Oakland, CA, USA, 2014.
60. Bamford, K. The role of motivation and curriculum in shaping prosustainable attitudes and behaviors in students. Master's Thesis, The University of Vermon, Burlington, NJ, USA, 2015.
61. Coe, M.A. Influential Environments: School Gardens Impacting Arizona Children's Environmental Perspectives. Master's Thesis, The University of Arizona, Tucson, AZ, USA, 2015.
62. Fabri, R.K.; Proenca, R.P.D.C.; Martinelli, S.S.; Cavalli, S.B. Regional foods in Brazilian school meals. *Br. Food J.* **2015**, *117*, 1706–1719. [CrossRef]
63. Fernandes, M.; Galloway, R.; Gelli, A.; Mumuni, D.; Hamdani, S.; Kiamba, J.; Quarshie, K.; Bhatia, R.; Aurino, E.; Peel, F.; et al. Enhancing Linkages Between Healthy Diets, Local Agriculture, and Sustainable Food Systems: The School Meals Planner Package in Ghana. *Food Nutr. Bull.* **2016**, *37*, 571–584. [CrossRef] [PubMed]
64. Bareng-Antolin, N. High School Gardens Program across the Nation: Current Practices, Perceived Benefits, Barriers, and Resources. Master's Thesis, University of Nevada, Las Vegas, NV, USA, 2017.
65. Borish, D.; King, N.; Dewey, C. Enhanced community capital from primary school feeding and agroforestry program in Kenya. *Int. J. Educ. Dev.* **2017**, *52*, 10–18. [CrossRef]
66. Laurie, S.M.; Faber, M.; Maduna, M.M. Assessment of food gardens as nutrition tool in primary schools in South Africa. *South Afr. J. Clin. Nutr.* **2017**, *30*, 20–26. [CrossRef]
67. Soares, P.; Martínez-Mián, M.A.; Caballero, P.; Vives-Cases, C.; Davó-Blanes, M.C. Local food production for school feeding programmes in Spain. *Gac. Sanit.* **2017**, *31*, 466–471. [CrossRef]
68. Garcia, J.R.N. O Programa Nacional de Alimentação Escolar como Promotor do Desenvolvimento Rural Sustentável e da Segurança Alimentar e Nutricional em Marechal Cândido Rondon—PR. Master's Thesis, Universidade Estadual do Oeste do Paraná, Marechal Cândido Rondon, Brazil, 2018.
69. Huston, W.G. Implementing Education for Sustainability in Rural Elementary Schools: A Collective Case Study of How Leadership Impacts the Implementation of Education for Sustainability in Rural Elementary Schools in Northern New England. Ph.D. Thesis, New England College, Henniker, NH, USA, 2018.
70. Lehnerd, M.E. Investigating the Adoption and Impact of Nutrition Incentive and Farm to School Programs. Ph.D. Thesis, Tufts University, Medford, MA, USA, 2018.
71. Powell, L.J.; Wittman, H. Farm to school in British Columbia: Mobilizing food literacy for food sovereignty. *Agric. Human Values* **2018**, *35*, 193–206. [CrossRef]
72. Roy, V.; Charan, P.; Schoenherr, T.; Sahay, B.S. Ensuring supplier participation toward addressing sustainability-oriented objectives of the mid-day meal supply chain: Insights from The Akshaya Patra Foundation. *Int. J. Logist. Manag.* **2018**, *29*, 456–475. [CrossRef]
73. Elkin, S. Comprehensive Farm-to-School: A Mixed-Methods Case Study of the Classroom, Cafeteria, and Community. Ph.D. Thesis, The University of Vermont, Burlington, NJ, USA, 2019.
74. Lopes, I.D.; Basso, D.; Brum, A.L. Cadeias agroalimentares curtas e o mercado de alimentação escolar na rede municipal de Ijuí, RS. *Interações* **2019**, *20*, 543–557. [CrossRef]
75. Dos Santos, A.M.; Baracuhy, M.P.; Furtado, D.A.; Leal de Morais, F.T.; da Silva Felix, A. Implementation of a project for the organic agriculture experience in rural schools: Climate studies, vegetable gardens, and free-range poultry production. *J. Anim. Behav. Biometeorol.* **2019**, *7*, 66–72.
76. Prescott, M.P.; Grove, A.; Bunning, M.; Cunningham-Sabo, L. A systems examination of school food recovery in Northern Colorado. *Resour. Conserv. Recycl.* **2020**, *154*, 104529. [CrossRef]
77. Virta, A.; Love, D.C. Assessing fish to school programs at 2 school districts in Oregon. *Health Behav. Policy Rev.* **2020**, *7*, 557–569. [CrossRef]
78. Rector, C.; Afifa, N.N.; Gupta, V.; Ismail, A.; Mosha, D.; Katalambula, L.K.; Vuai, S.; Young, T.; Hemler, E.C.; Wang, D.; et al. School-Based Nutrition Programs for Adolescents in Dodoma, Tanzania: A Situation Analysis. *Food Nutr. Bull.* **2021**, *42*, 378–388. [CrossRef]
79. Toledo, A.D. Promoção da Alimentação Saudável no Ambiente Escolar: Avaliação do Programa Horta Educativa em Escolas Estaduais de São Paulo. Ph.D. Thesis, Universidade de São Paulo, Sao Paulo, Brazil, 2021.
80. Department for Education. School Food Standards Practical Guide. Available online: https://www.gov.uk/government/publications/school-food-standards-resources-for-schools (accessed on 31 December 2021).
81. National Nutrition Concil. *Eaten and Learning Together—School Meal Recommendation*; National Nutrition Concil: Helsinki, Finland, 2017.

82. Ajuntament de Barcelona. *Declaració d' Emergència Climàtica*; Ajuntament de Barcelona: Barcelona, Spain, 2020.
83. Consejo Interterritorial de Sistema Nacional de Salud. *Documento de Consenso Sobre la Alimentacion en los Centros Educativos*; Consejo Interterritorial de Sistema Nacional de Salud: Madrid, Spain, 2010.
84. The National Food Agency. *National Guidelines for Meals at School. Preschool Class, Elementary School, High School and Leisure Center*; The National Food Agency: Uppsala, Sweden, 2021.
85. Senate Department for Education, Youth and Family. *Reorganization of the School Lunch at Open and Affiliated All-Day Elementary Schools as Well as for Support Centers in the State of Berlin*; Senate Department for Education, Youth and Family: Berlin, Germany, 2017; p. 28.
86. Brasil. Fundo Nacional de Desenvolvimento da Educação. Resolução no 06, de 08 de Maio de 2020. Dispõe sobre o Atendimento da Alimentação Escolar aos Alunos da Educação Básica no Âmbito do Programa Nacional de Alimentação Escolar—PNAE. 2020. Available online: https://www.fnde.gov.br/index.php/acesso-a-informacao/institucional/legislacao/item/13511-resolu%C3%A7%C3%A3o-n%C2%BA-6,-de-08-de-maio-de-2020 (accessed on 31 December 2021).
87. Santa Catarina. Lei no 17504, de 10 de abril de 2018. Altera a ementa e o art. 1o, da Lei no 12.282, de 2002, que Dispõe Sobre o Fornecimento de Alimentos Orgânicos na Merenda Escolar nas Unidades Educacionais do Estado de Santa Catarina. 2018. Available online: https://leisestaduais.com.br/sc/lei-ordinaria-n-17504-2018-santa-catarina-altera-a-ementa-e-o-art-1o-da-lei-no-12-282-de-2002-que-dispoe-sobre-o-fornecimento-de-alimentos-organicos-na-merenda-escolar-nas-unidades-educacionais-do-estado-de-santa-catarina (accessed on 31 December 2021).
88. E-Gov. Law Search. School Lunch Law. Available online: https://elaws.e-gov.go.jp/document?lawid=329AC0000000160 (accessed on 31 December 2021).
89. United States Department of Agriculture. Offer versus Serve Guidance for the National School Lunch Program and the School Breakfast Program. 2015. Available online: https://www.fns.usda.gov/cn/updated-offer-vs-serve-guidance-nslp-and-sbp-beginning-sy2015-16 (accessed on 31 December 2021).
90. California School Nutrition Association. 6 Principles of Behavioral Economics Used in Smarter Lunchrooms | Smarter Lunchrooms Movement. Available online: http://www.calsna.org/documents/events/2017Conference/Handouts/B6WasteNot.pdf (accessed on 24 November 2021).
91. National Farm to School Network. About Farm to School. Available online: https://www.farmtoschool.org/about/what-is-farm-to-school (accessed on 24 November 2021).
92. WFP. Home Grown School Feeding. Available online: https://www.wfp.org/home-grown-school-feeding (accessed on 24 November 2021).
93. Foundation for Environmental Education. Eco Schools. Available online: https://www.ecoschools.global/ (accessed on 2 December 2021).
94. Food for Life Partnership. Welcome to Food for Life. Available online: https://www.foodforlife.org.uk/ (accessed on 24 November 2021).
95. Willett, W.; Rockström, J.; Brent, L.; Springmann, M.; Lang, T.; Vermeulen, S.; Garnett, T.; Tilman, D.; DeClerck, F.; Wood, A.; et al. Food in the Anthropocene: The EAT–Lancet Commission on healthy diets from sustainable food systems. *Lancet* **2019**, *393*, 447–492. [CrossRef]
96. Hoover, A.; Vandyousefi, S.; Martin, B.; Nikah, K.; Cooper, M.H.; Muller, A.; Marty, E.; Duswalt-Epstein, M.; Burgermaster, M.; Waugh, L.; et al. Barriers, Strategies, and Resources to Thriving School Gardens. *J. Nutr. Educ. Behav.* **2021**, *53*, 591–601. [CrossRef]
97. Burt, K.G.; Luesse, H.B.; Rakoff, J.; Ventura, A.; Burgermaster, M. School Gardens in the United States: Current Barriers to Integration and Sustainability. *Am. J. Public Health* **2018**, *108*, 1543–1549. [CrossRef] [PubMed]
98. UN General Assembly. Transforming Our World: The 2030 Agenda for Sustainable Development. Available online: https://www.refworld.org/docid/57b6e3e44.html (accessed on 25 November 2021).
99. UNESCO. *Learn for Our Planet: A Global Review of How Environmental Issues Are Integrated in Education*; United Nations Educational, Scientific and Cultural Organization: Paris, France, 2021; p. 50.
100. Soil Association. *Food for Life: Healthy, Local, Organic School Meals*; Soil Association: Bristol, UK, 2003; p. 117.
101. Pancino, B.; Cicatiello, C.; Falasconi, L.; Boschini, M. School canteens and the food waste challenge: Which public initiatives can help? *Waste Manag. Res.* **2021**, *39*, 1090–1100. [CrossRef]
102. Derqui, B.; Fernandez, V.; Fayos, T. Towards more sustainable food systems. Addressing food waste at school canteens. *Appetite* **2018**, *129*, 1–11. [CrossRef] [PubMed]
103. García-Herrero, L.; De Menna, F.; Vittuari, M. Food waste at school. The environmental and cost impact of a canteen meal. *Waste Manag.* **2019**, *100*, 249–258. [CrossRef]
104. Liu, Y.; Cheng, S.; Liu, X.; Cao, X.; Xue, L.; Liu, G. Plate Waste in School Lunch Programs in Beijing, China. *Sustainability* **2016**, *8*, 1288. [CrossRef]
105. Palmer, S.; Herritt, C.; Cunningham-Sabo, L.; Stylianou, K.S.; Prescott, M.P. A Systems Examination of Food Packaging and Other Single-Use Item Waste in School Nutrition Programs. *J. Nutr. Educ. Behav.* **2021**, *53*, 380–388. [CrossRef]
106. González-García, S.; Esteve-Llorens, X.; González-García, R.; González, L.; Feijoo, G.; Moreira, M.T.; Leis, R. Environmental assessment of menus for toddlers serviced at nursery canteen following the Atlantic diet recommendations. *Sci. Total Environ.* **2021**, *770*, 145342. [CrossRef]

107. Rodrigues, C.M.; Bastos, L.G.; Cantarelli, G.S.; Stedefeldt, E.; Cunha, D.T.; de Freitas Saccol, A.L. Sanitary, nutritional, and sustainable quality in food services of Brazilian early childhood education schools. *Child. Youth Serv. Rev.* **2020**, *113*, 104920. [CrossRef]

108. Maynard, D.D.C.; Zandonadi, R.P.; Nakano, E.Y.; Raposo, A.; Botelho, R.B.A. Green Restaurants ASSessment (GRASS): A Tool for Evaluation and Classification of Restaurants Considering Sustainability Indicators. *Sustainability* **2021**, *13*, 10928. [CrossRef]

109. Swinburn, B.A.; Kraak, V.I.; Allender, S.; Atkins, V.J.; Baker, P.I.; Bogard, J.R.; Brinsden, H.; Calvillo, A.; De Schutter, O.; Devarajan, R.; et al. The Global Syndemic of Obesity, Undernutrition, and Climate Change: The Lancet Commission report. *Lancet* **2019**, *393*, 791–846. [CrossRef]

110. Mekonnen, M.M.; Hoekstra, A.Y. A Global Assessment of the Water Footprint of Farm Animal Products. *Ecosystems* **2012**, *15*, 401–415. [CrossRef]

111. Food and Agriculture Organization of the United Nations. *Food Wastage Footprint: Impacts on Natural Resources*; Food and Agriculture Organization of the United Nations: Rome, Italy, 2013.

112. Lindgren, N. The political dimension of consuming animal products in education: An analysis of uppersecondary student responses when school lunch turns green and vegan. *Environ. Educ. Res.* **2020**, *26*, 684–700. [CrossRef]

113. Elinder, L.S.; Colombo, P.E.; Patterson, E.; Parlesak, A.; Lindroos, A.K. Successful Implementation of Climate-Friendly, Nutritious, and Acceptable School Meals in Practice: The OPTIMAT^{TM} Intervention Study. *Sustainability* **2020**, *12*, 8475. [CrossRef]

114. Benvenuti, L.; De Santis, A.; Santesarti, F.; Tocca, L. An optimal plan for food consumption with minimal environmental impact: The case of school lunch menus. *J. Clean. Prod.* **2016**, *129*, 704–713. [CrossRef]

115. Ribal, J.; Fenollosa, M.L.; García-Segovia, P.; Clemente, G.; Escobar, N.; Sanjuán, N. Designing healthy, climate friendly and affordable school lunches. *Int. J. Life Cycle Assess.* **2016**, *21*, 631–645. [CrossRef]

116. Rossi, L.; Ferrari, M.; Martone, D.; Benvenuti, L.; De Santis, A. The Promotions of Sustainable Lunch Meals in School Feeding Programs: The Case of Italy. *Nutrients* **2021**, *13*, 1571. [CrossRef]

117. Bianchini, V.U. Critérios de Sustentabilidade para o Planejamento de Cardápios Escolares no Âmbito do Programa Nacional de Alimentação Escolar. Master's Thesis, Universidade Federal de Santa Catarina, Florianópolis, Brazil, 2017.

118. Morgan, K. Greening the Realm: Sustainable Food Chains and the Public Plate. *Reg. Stud.* **2008**, *42*, 1237–1250. [CrossRef]

119. Alexandre, V.P.; Gomes, L.O.F.; Silva, S.U.; Almeida, G.M.; Martins, K.A.; Monego, E.T.; Sousa, L.M.; Campos, M.R.H. Do campo à escola: Compra de alimentos da agricultura familiar pelo Programa Nacional de Alimentação Escolar em Territórios da Cidadania de Goiás. *Segurança Aliment. Nutr.* **2016**, *23*, 1049–1064. [CrossRef]

120. Amarante, E.A.L. Caracterização da Comercialização de Alimentos Através do Programa Nacional de Alimentação Escolar-PNAE-no Município de Marechal Cândido Rondon-PR Como Estratégia para Promover a Sustentabilidade nos Empreendimentos da Agricultura Familiar. Undergraduate Thesis, Universisdade Estadual do Oeste do Paraná, Marechal Cândido Rondon, Brazil, 2016. Final paper.

121. Andreatta, T.; Martins, R.; Camara, S.B.; Gelatti, E. Efetividade do Programa Nacional de Alimentação Escolar sob a perspectiva dos agricultores familiares do município de Panambi-RS. *Agric. Fam. Pesqui. Formação Desenvolv.* **2021**, *15*, 135–155. [CrossRef]

122. Batlle-Bayer, L.; Bala, A.; Aldaco, R.; Vidal-Monés, B.; Colomé, R.; Fullana-I-Palmer, P. An explorative assessment of environmental and nutritional benefits of introducing low-carbon meals to Barcelona schools. *Sci. Total Environ.* **2021**, *20*, 143879. [CrossRef]

123. Braun, C.L.; Rombach, M.; Häring, A.M.; Bitsch, V. A Local Gap in Sustainable Food Procurement: Organic Vegetables in Berlin's School Meals. *Sustainability* **2018**, *10*, 4245. [CrossRef]

124. Breña, G.M. The Garden Talks: Cultivating Desire for Garden Education at Killip Elementary School. Master's Thesis, Northern Arizona University, Flagstaff, AZ, USA, 2017.

125. Carvalho, D.G. O Programa Nacional de Alimentação Escolar e a sustentabilidade: O caso do Distrito Federal. Master's Thesis, Universidade de Brasília, Brasília, Brazil, 2009.

126. Coleman, K.J.; Shordon, M.; Caparosa, S.L.; Pomichowski, M.E.; Pasadena, C.; Dzewaltowski, D.A.; Manhattan, K. Changing nutrition policies and environments in low-income schools using implementation models: The healthy options for nutrition environments in schools (ONES) intervention. In *Obesity*; Wiley Online Library: Hoboken, NJ, USA, 2012; Volume 19, p. 124.

127. Colombo, P.E.; Patterson, E.; Lindroos, A.K.; Parlesak, A.; Elinder, L.S. Sustainable and acceptable school meals through optimization analysis: An intervention study. *Nutr. J.* **2020**, *19*, 61. [CrossRef]

128. Constanty, H.F.P.-H. Contribuições do PNAE na Sustentabilidade dos Agricultores Familiares: O Caso do Município de Marechal Cândido Rondon. Master's Thesis, Universidade Estadual do Oeste do Paraná, Marechal Cândido Rondon, Brazil, 2014.

129. Constanty, H.F.P.-H.; Zonin, W.J. Programa Nacional de Alimentação Escolar (PNAE) e sustentabilidade: O caso do município de Marechal Cândido Rondon. *Desenvolv. Meio Ambient.* **2016**, *36*, 371–392. [CrossRef]

130. Conner, D.S.; Abate, G.; Liquori, T.; Hamm, M.W.; Peterson, H.C. Prospects for more healthful, local, and sustainably produced food in school meals. *J. Hunger Environ. Nutr.* **2010**, *5*, 416–433. [CrossRef]

131. Damapong, S.N.; Kongnoo, W.; Monarumit, P. Lesson learned: Improvement of food quality in school and child development center for sustainable child nutrition. In Proceedings of the 20th International Congress of Nutrition, Granada, Spain, 15–20 September 2013; S. Karger AG: Basel, Switzerland, 2013; Volume 63, p. 1073.

132. Dirks, K.J. Sowing Seeds for Learning: A Case Study of a School Garden in Pocahontas, Iowa. Master's Thesis, Iowa State University, Ames, IA, USA, 2011.

133. Eich, A.B. Percepções: Agricultura familiar e políticas públicas para alimentação escolar no município de São Paulo das Missões-RS. *Direitos Sociais Políticas Públicas* **2015**, *3*, 22–44.
134. Elnakib, S.A.; Quick, V.; Mendez, M.; Downs, S.; Wackowski, O.A.; Robson, M.G. Food Waste in Schools: A Pre-/Post-Test Study Design Examining the Impact of a Food Service Training Intervention to Reduce Food Waste. *Int. J. Environ. Res. Public Health* **2021**, *18*, 6389. [CrossRef]
135. Colombo, P.E. Optimizing school meals today: A pathway to sustainable dietary habits tomorrow. Ph.D. Thesis, Karolinska Institutet, Stockholm, Sweden, 2021.
136. Ferderbar, C.A. The Effects of In-Nature and Virtual-Nature Field Trip Experiences on Proenvironmental Attitudes and Behaviors, and Environmental Knowledge of Middle School Students. Ph.D. Thesis, Cardinal Stritch University, Glindale, CA, USA, 2013.
137. Filippini, R.; De Nonib, I.; Corsia, S.; Spigarolo, R.; Bocchi, S. Sustainable school food procurement: What factors do affect the introduction and the increase of organic food? *Food Policy* **2018**, *76*, 109–119. [CrossRef]
138. Fitzsimmons, J.; O'Hara, J.K. Market Channel Procurement Strategy and School Meal Costs in Farm-to-School Programs. *Agric. Resour. Econ. Rev.* **2019**, 1–26. [CrossRef]
139. Franzoni, G.B. Inovação social e tecnologia social: O caso da cadeia curta de agricultores familiares e a alimentação escolar em Porto Alegre/RS. Master's Thesis, Universidade Federal do Rio Grande do Sul, Porto Alegre, Brazil, 2015.
140. Ghattas, H.; Choufani, J.; Jamaluddine, Z.; Masterson, A.R.; Sahyoun, N.R. Linking women-led community kitchens to school food programmes: Lessons learned from the Healthy Kitchens, Healthy Children intervention in Palestinian refugees in Lebanon. *Public Health Nutr.* **2020**, *23*, 914–923. [CrossRef]
141. Granillo-Macías, R. Logistics optimization through a social approach for food distribution. *Socioecon. Plann. Sci.* **2021**, *76*, 100972. [CrossRef]
142. Green, S. Using a Pen Pal Program to Assess Student Learning through Culture and School Gardens. Master's Thesis, Iowa State University, Ames, IA, USA, 2016.
143. Gregolin, G.C.; Gregolin, M.R.P.; Triches, R.M.; Zonin, W.J. Política pública e sustentabilidade: Possibilidade de interface no Programa Nacional de Alimentação Escolar-PNAE. *Emancipação* **2017**, *17*, 199–216.
144. He, C. Assessment of the impact of organic school meals to improve the school food environment and children's awareness of healthy eating habits. In Proceedings of the 20th International Congress of Nutrition, Granada, Spain, 15–20 September 2013; S. Karger AG: Basel, Switzerland, 2013; p. 840.
145. Hendler, V.M.; Ruiz, E.N.F.; Oliveira, L.D. de Sociobiodiversidade na escola, promoção da saúde, da sustentabilidade e da cultura: Um movimento em construção no município de Mostardas/RS. *Agric. Fam. Pesqui. Formação Desenvolv.* **2021**, *15*, 115–134.
146. Henry-Stone, L.R. Cultivating Sustainability through Participatory Action Research: Place-Based Education and Community Food Systems in Interior Alaska. Ph.D. Thesis, University of Alaska Fairbanks, Fairbanks, AK, USA, 2008.
147. Hodgkinson, T.M. Translating Sustainability: The Design of a Secondary Charter School. Ph.D. Thesis, The University of Iowa, Iowa City, IA, USA, 2011.
148. Johnston, Y.; Denniston, R.; Morgan, M.; Bordeau, M. Rock on Cafe: Achieving Sustainable Systems Changes in School Lunch Programs. *Health Promot. Pract.* **2009**, *10*, 100S–108S. [CrossRef] [PubMed]
149. Jones, P.K. The City Feeds the Poor: The Struggle for Sustainable Food Systems in San Francisco. Ph.D. Thesis, University of California, Santa Cruz, CA, USA, 2012.
150. Kipfer, H.J. Gaining Consensus on Implementation, Sustainability, and Benefits of School Garden Programming in Washington, D.C. Ph.D. Thesis, West Virginia University, Morgantown, WV, USA, 2018.
151. Koch, P.A. A comparison of two nutrition education curricula: Cookshops and food & environment lessons. Ph.D. Thesis, Columbia University, New York, NY, USA, 2000.
152. Lalli, G.S. School meal time and social learning in England. *Camb. J. Educ.* **2020**, *50*, 57–75. [CrossRef]
153. Lauffer, B. Education for Sustainability: The Effectiveness of a Place Based Environmental Education Unit in Food Waste Reduction on the Environmental Awareness of Fourth Grade Students. Ph.D. Thesis, University of South Carolina, Columbia, SC, USA, 2019.
154. Lawless, T.H. Harnessing the Impact of Schools: New Insights for Sustainable Community Development. Ph.D. Thesis, Arizona State University, Tempe, AZ, USA, 2013.
155. Løes, A.-K.; Nölting, B. Organic school meal systems–towards a more sustainable nutrition. *Agron. Res.* **2009**, *7*, 647–653.
156. McCarty, C.A. Alaskan Adolescent Nutrition Project. Master's Thesis, University of Alaska Anchorage, Anchorage, AK, USA, 2013.
157. Medina, J.L. A Dose-Response Analysis of a School-Based Nutrition Intervention in Middle School Children. Master's Thesis, University of Texas, Austin, TX, USA, 2009.
158. Melão, I.B. Produtos sustentáveis na alimentação escolar: O PNAE no Paraná. *Cad. IPARDES Estud. Pesqui.* **2012**, *2*, 87–105.
159. Mikkola, M. *Role of Public Catering and Use of Organic Food in Educational Contexts: Creating Centres for Sustainable Food Systems*; Bioforsk Organic Food and Farming: Tingvoll, Norway, 2010.
160. Moss Gamblin, M.K. (Kate) Becoming a Sustainability Chef: An Empirical Model of Sustainability Perspectives in Educational Leaders. Ph.D. Thesis, University of Toronto, Toronto, ON, Canada, 2013.
161. Mosimann, E.N. Agricultura familiar e alimentação escolar nas encostas da serra geral de Santa Catarina: Desafios e potencialidades. Master's Thesis, Universidade Federal de Santa Catarina, Florianópolis, Brazil, 2014.

162. Morgan, K.J.; Morley, A. *School Meals: Healthy Eating and Sustainable Food Chains*; Regeneration Institute, Cardiff University: Cardiff, UK, 2003.
163. Mota, J.S.; Silva, D.W.; Pauletto, D. A inserção de produtos da Sociobiodiversidade na alimentação escolar no município de Santarém, PA. *Agric. Fam. Pesqui. Formação Desenvolv.* **2021**, *15*, 92–114. [CrossRef]
164. Muansrichai, S.; Panyasing, S.; Yonvanij, S. Learning and practicing in accordance with the sufficiency economy philosophy of third grade primary students in the local northeast socio-culural context of Thailand. *Int. J. Appl. Bus. Econ. Res.* **2015**, *13*, 4605–4619.
165. Nunes, E.M.; Morais, A.C.; Aquino, J.R.; Gurgel, I.A. O Programa Nacional de Alimentação Escolar (PNAE) como política de inclusão na agricultura familiar do Nordeste do Brasil. *Rev. Grifos* **2018**, *45*, 114–139. [CrossRef]
166. Nuutila, J.; Risku-Norja, H.; Arolaakso, A. Public kitchen menu substitutions increase organic share and school meal sustainability at equal cost. *Org. Agric.* **2019**, *9*, 117–126. [CrossRef]
167. Orr, E. Sustainable Waste Management in Schools. Master's Thesis, University of Rhode Island, Kingston, RI, USA, 2020.
168. Otsuki, K. Sustainable partnerships for a green economy: A case study of public procurement for home-grown school feeding. *Nat. Resour. Forum* **2011**, *35*, 213–222. [CrossRef]
169. Padilha, N.; Corbari, F.; Zanco, A.M.; Canquerino, Y.K.; Alves, A.F. The contribution of PNAE to sustainable rural development in the municipality of Pitanga—PR. *Braz. J. Dev.* **2018**, *4*, 4351–4365.
170. Osowski, C.P.; Fjellström, C. Understanding the ideology of the Swedish tax-paid school meal. *Health Educ. J.* **2019**, *78*, 388–398. [CrossRef]
171. Polo Galante, A.; Mireles, M.; Kiamba, J.; Mamadoultaibou, A.; Sablah, M.; Boitshepo, G.; Sanou, D. Innovative home grown school feeding linked to family farming: FAO school food and nutrition approach in sub-Saharan Africa. In Proceedings of the 21st International Congress of Nutrition, Buenos Aires, Argentina, 15–20 October 2017; Karger: Basel, Switzerland, 2017; p. 747.
172. Prescott, M.P.; Burg, X.; Metcalfe, J.J.; Lipka, A.E.; Herritt, C.; Cunningham-Sabo, L. Healthy Planet, Healthy Youth: A Food Systems Education and Promotion Intervention to Improve Adolescent Diet Quality and Reduce Food Waste. *Nutrients* **2019**, *11*, 1869. [CrossRef] [PubMed]
173. Rambing, D.R.; Wahyuni, C.U.; Isfandiardi, M.A.; Ssekalembe, G. Teachers' interpretation and behavior in conducting a Clean and Healthy Lifestyle Program at elementary schools in Kediri District. *Eur. J. Mol. Clin. Med.* **2020**, *7*, 3966–3976.
174. Redman, E. Development, Implementation, and Evaluation of Sustainability Education through the Integration of Behavioral Science into Pedagogy and Practice. Ph.D. Thesis, Arizona State University, Tempe, AZ, USA, 2013.
175. Resque, A.G.L.; Coudel, E.; Piketty, M.-G.; Cialdella, N.; Sá, T.; Piraux, M.; Assis, W.; Page, C. Le Agrobiodiversity and Public Food Procurement Programs in Brazil: Influence of Local Stakeholders in Configuring Green Mediated Markets. *Sustainability* **2019**, *11*, 1425. [CrossRef]
176. Ribeiro, A.L.d.P.; Ceratti, S.; Broch, D.T. Programa Nacional de Alimentação Escolar (PNAE) e a participação da agricultura familiar em municípios do Rio Grande do Sul. *Rev. Gedecon. Rev. Gestão Desenvolv. Context* **2013**, *1*. Available online: http://revistaeletronica.unicruz.edu.br/index.php/GEDECON/article/view/282 (accessed on 7 December 2021).
177. Santos, F.; Fernandes, P.F.; Rockett, F.C.; de Oliveira, A.B.A. Evaluation of the inclusion of organic food from family-based agriculture in school food in municipalities of rural territories of the state of Rio Grande do Sul, Brazil. *Cien. Saude Colet.* **2014**, *19*, 1429–1436. [CrossRef]
178. Schachtner-Appel, A.E. Design, Implementation, and Evaluation of a Mindfulness Focused Nutrition Promotion Program to Balance USDA School Nutrition Goals with Food Waste Reduction in Elementary Schools. Ph.D. Thesis, University of Maryland, College Park, MD, USA, 2019.
179. Scott, R.A. The Nested Systems of Sustainability Education: Charting a Pathway to Ecological Learning. Master's Thesis, Prescott College, Prescott, AZ, USA, 2011.
180. Silva, A.P.F.; de Sousa, A.A. Organic foods from family farms in the National School Food Program in the State of Santa Catarina, Brazil. *Rev. Nutr.* **2013**, *26*, 701–714. [CrossRef]
181. Silva, A.B.; Pedon, N.R. Reprodução do campesinato através de políticas públicas voltadas para a agricultura familiar: A dinâmica do Programa Nacional de Alimentação Escolar (PNAE) em Ourinhos-SP. *Rev. Nera* **2015**, *18*, 92–109. [CrossRef]
182. Silva, D.W.; Gehlen, I.; Schultz, G. Family farm, public policies and citizenship: Connections built from the operation of the program national school feeding. *Redes* **2016**, *21*, 121–145. [CrossRef]
183. Silva, M.G.; Dias, M.M.; Amorim Junior, P.C.G. Mudanças Organizacionais em Empreendimentos de Agricultura Familiar a partir do Acesso ao Programa Nacional de Alimentação Escolar. *Rev. Econ. Sociol. Rural* **2015**, *53*, 289–304. [CrossRef]
184. Soares, P. Análise do programa de aquisição de alimentos na alimentação escolar em um município de Santa Catarina. Master's Thesis, Universidade Federal de Santa Catarina, Florianópolis, Brazil, 2011.
185. Soares, P.; Martinelli, S.S.; Melgarejo, L.; Cavalli, S.B.; Davó-Blanes, M.C. Using local family farm products for school feeding programmes: Effect on school menus. *Br. Food J.* **2017**, *119*, 1289–1300. [CrossRef]
186. Solof, L.E. Bay Area Student Involvement in the Environmental and Food Justice Movements: A Narrative of Motivations, Experiences, and Community Impact. Ph.D. Thesis, The University of San Francisco, San Francisco, CA, USA, 2014.
187. Sziwelski, N.K.; Teo, C.R.P.A.; Gallina, L.d.S.; Grahl, F.; Filippi, C. Programa Nacional de Alimentação Escolar (PNAE) [National School Feeding Program] implications in the income and organization of family farmers. *Rev. Bras. Políticas Públicas* **2015**, *5*, 220–239.

188. Trott, C.D. Engaging Key Stakeholders in Climate Change: A Community-Based Project for Youth-Led Participatory Climate Action. Ph.D. Thesis, Colorado State University, Fort Collins, CO, USA, 2017.
189. Turpin, M.E. A Alimentação Escolar como Fator de Desenvolvimento Local por meio do Apoio aos Agricultores Familiares. *Segur. Aliment. Nutr.* **2009**, *16*, 20–42. [CrossRef]
190. Vasconcelos, M.G.; Vieira, S.S.; Rodrigues, V.W.B. Utilização de boas práticas de cultivo e manejo de hortaliças para uma alimentação escolar saudável. *Rev. Extensão* **2014**, *13*, 61–69. [CrossRef]
191. Valadão, W.B.; de Sousa, J.M.M. O PNAE em Viçosa-MG: Reflexões sobre a interface entre a produção e comercialização de alimentos advindos da agricultura familiar e a agroecologia. In Proceedings of the VI Congresso Latino-Americano de Agroecologia; X Congresso Brasileiro de Agroecologia; V Seminário de Agroecologia do Distrito Federal e Entorno, Brasília, Brazil, 12–15 September 2017; Cadernos de Agroecologia: Rio de Janeiro, Brazil, 2018.
192. Wade, S.A. Exploring the Influence of Agriculture Lessons on the Agricultural Knowledge and Beliefs of 4th Grade Students. Ph.D. Thesis, Grand Canyon University, Phoenix, AZ, USA, 2019.
193. Wickramasinghe, K.W.; Scarborough, P.; Townsend, N.; Goldacre, M.; Rayner, M. Estimating the changes in nutritional quality and environmental impact of primary school meals if all meals met the new school food plan standards in England. In Proceedings of the World Congress of Cardiology & Cardiovascular Health, Mexico City, Mexico, 4–7 June 2016; Global Heart: Geneva, Switzerland, 2016; Volume 11, p. e128.

Review

A Comprehensive Review on Bio-Preservation of Bread: An Approach to Adopt Wholesome Strategies

Mizanur Rahman [1,*], Raihanul Islam [1], Shariful Hasan [1], Wahidu Zzaman [1], Md Rahmatuzzaman Rana [1], Shafi Ahmed [2], Mukta Roy [1], Asm Sayem [1], Abdul Matin [3], António Raposo [4,*], Renata Puppin Zandonadi [5], Raquel Braz Assunção Botelho [5] and Atiqur Rahman Sunny [6,7,*]

[1] Department of Food Engineering and Tea Technology, Shahjalal University of Science and Technology, Sylhet 3100, Bangladesh; raihanul08@student.sust.edu (R.I.); sharifulh.nahid@gmail.com (S.H.); wahid-ttc@sust.edu (W.Z.); rzaman-fet@sust.edu (M.R.R.); muktaroy-fet@sust.edu (M.R.); asm.sayem-fet@sust.edu (A.S.)

[2] Department of Agro Product Processing Technology, Jashore University of Science and Technology, Jashore 7400, Bangladesh; shafi.fet@gmail.com or shafi@just.edu.bd

[3] Department of Food Processing and Engineering, Chattogram Veterinary and Animal Sciences University, Chittagong 4203, Bangladesh; abmatin@cvasu.ac.bd

[4] CBIOS (Research Center for Biosciences and Health Technologies), Universidade Lusófona de Humanidades e Tecnologias, Campo Grande 376, 1749-024 Lisboa, Portugal

[5] Department of Nutrition, Campus Darcy Ribeiro, University of Brasilia, Asa Norte, Distrito Federal, Brasília 70910-900, Brazil; renatapz@unb.br (R.P.Z.); raquelbotelho@unb.br (R.B.A.B.)

[6] Department of Genetic Engineering and Biotechnology, Shahjalal University of Science and Technology, Sylhet 3100, Bangladesh

[7] Suchana Project, WorldFish, Bangladesh Office, Gulshan, Dhaka 1213, Bangladesh

* Correspondence: mizan-fet@sust.edu (M.R.); antonio.raposo@ulusofona.pt (A.R.); a.rsunny@cgiar.org (A.R.S.)

Citation: Rahman, M.; Islam, R.; Hasan, S.; Zzaman, W.; Rana, M.R.; Ahmed, S.; Roy, M.; Sayem, A.; Matin, A.; Raposo, A.; et al. A Comprehensive Review on Bio-Preservation of Bread: An Approach to Adopt Wholesome Strategies. *Foods* **2022**, *11*, 319. https://doi.org/10.3390/foods11030319

Academic Editor: Maria Grazia Volpe

Received: 8 December 2021
Accepted: 17 January 2022
Published: 24 January 2022

Publisher's Note: MDPI stays neutral with regard to jurisdictional claims in published maps and institutional affiliations.

Abstract: Bread is a food that is commonly recognized as a very convenient type of food, but it is also easily prone to microbial attack. As a result of bread spoilage, a significant economic loss occurs to both consumers and producers. For years, the bakery industry has sought to identify treatments that make bread safe and with an extended shelf-life to address this economic and safety concern, including replacing harmful chemical preservatives. New frontiers, on the other hand, have recently been explored. Alternative methods of bread preservation, such as microbial fermentation, utilization of plant and animal derivatives, nanofibers, and other innovative technologies, have yielded promising results. This review summarizes numerous research findings regarding the bio-preservation of bread and suggests potential applications of these techniques. Among these techniques, microbial fermentation using lactic acid bacteria strains and yeast has drawn significant interest nowadays because of their outstanding antifungal activity and shelf-life extending capacity. For example, bread slices with *Lactobacillus plantarum* LB1 and *Lactobacillus rossiae* LB5 inhibited fungal development for up to 21 days with the lowest contamination score. Moreover, various essential oils and plant extracts, such as lemongrass oil and garlic extracts, demonstrated promising results in reducing fungal growth on bread and other bakery products. In addition, different emerging bio-preservation strategies such as the utilization of whey, nanofibers, active packaging, and modified atmospheric packaging have gained considerable interest in recent days.

Keywords: bakery products; bio-preservatives; shelf-life; mold spoilage; antifungal activity; microbial fermentation; active packaging

1. Introduction

For thousands of years, bread is still one of the dominant food sources of the human diet, with the manufacturing of yeast-based and sourdough bread being one of the earliest biotechnological mechanisms [1]. Amidst its medium growth rate (122,000 t in 2007 to 129 t in 2016), it earned approximately \$358 billion in global revenue in 2016 [2]. It is also a magnificent energy source, including protein, iron, calcium, and various vitamins [3–6]. Commercially available bread and biscuits contain nearly 7.5 and 7.8% protein [7]. Bakery products are ideal for fiber addition, as fiber intake has declined in the European diet partially due to cereal adjustments.

Due to the easily spoiled nature of this food, its quality and palatability degrade during preservation, resulting in changes in physiological, biochemical, sensorial, and microbial properties [8]. Mold and fungal deterioration are the primary causes of significant financial detriments in packed bread items. They may also be considered as mycotoxins sources [9], posing issues to the people's health [10–13] and putting a significant financial strain on bakeries. Penicillium (*Penicillium chrysogenum*, *Penicillium roqueforti*, and *Penicillium brevicompactum*), Aspergillus (previously Eurotium), Wallemia, and other familiar molds such as Rhizopus, *Chrysonilia sitophila*, and Mucor are the main genera involved in the spoiling of bakery items [14–16]. Yeasts, specifically *Saccharomycopsis fibuligera* and *Hyphopichia burtonii*, can also contribute to the "chalk mold" problem [7]. Due to these microbial attacks, a high proportion of food waste is being produced around the globe. For example, in Germany in 2015, 34.7% of total bread was lost [17]. Article [18] estimated losses of 10% in Brazil and presumably other nations with a tropical environment. Additionally, mycotoxin infection in food items is claimed to be a universal issue [19–22] and it occurs in nearly 25% of all grain yields globally [23]. A study conducted in Poland identified nine categories of causes of losses based on a quantitative study within the four considered sectors of the bakery enterprise: (i) raw materials magazine—mechanical damage, magazine pests, spoiling, molding, and impurities; (ii) production section—hygiene and sanitary requirements, technical breakdowns; (iii) final product magazine—damaged packaging, hygiene and sanitary requirements/food safety hazards, technical breakdowns; and (iv) final product transport—errors in placed orders, damaged unit packaging, technical breakdowns, incomplete collective packaging [24]. In all of them, mold and fungal deterioration may cause losses.

Today's food industry faces a tremendous problem in producing goods that are not only productive but also wholesome for customers, as well as much more long-lasting. The use of organic preservatives has the feasibility of meeting both needs [25,26]. Natural antimicrobial preservatives have been the subject of extensive research due to the growing evidence of the harmfulness of chemical preservatives and their impacts on consumer health [27–29]. The restoration of chemical preservatives—such as propionates and sorbates—in bread and other bakery goods is of considerable interest [25].

Chemical preservatives, such as calcium propionate, are commonly applied to expand the microbial lifespan of bread [30,31]. However, prolonged exposure to chemical preservatives may pose a health risk. Thus, using bio-preservation techniques on bread can aid in solving this issue and preventing economic loss caused by fungi, mold, or yeast [32]. Bio-preservatives are organic resources that can be utilized to lower or eliminate microbial populations while improving food quality. It is a revolutionary concept for processing and preserving perishable fresh goods and is generally recognized as safe (GRAS) for food usage. Numerous distinct studies have been conducted on bio-preservation. Thus, the objective of this review was to integrate all studies conducted on the spoilage problem of bread, as well as the preservation techniques, most notably bio-preservation, used to extend its shelf life.

2. Materials and Methods

A search strategy was conducted on key literature databases such as SCOPUS, PubMed, ResearchGate, and Google Scholar. The keywords mentioned in Table 1 were used to find articles, starting with the primary keywords, then joined with the secondary keywords using the set operator AND. Article references were also reviewed to evaluate other relevant publications.

The studies related to the spoilage and bio-preservation of bread were included in this review. Additionally, some supporting facts about the varieties and economic relevance of bread was added. It summarized only previously conducted, scientifically validated findings. Articles published in regional languages have been excluded entirely. Papers that lacked sufficient data on the subject or failed to adequately characterize the impact of bio-preservatives on bread were omitted entirely. This topic does not include articles on chemical preservation of bread, nutritional analysis, bread formulation and innovation, or functional aspects of bread. No papers with inconclusive outcomes were considered for this article. Finally, to study this clean-level alternative, only works focusing on recent advances in this field were cited, with the exception of a few fundamental principles.

Table 1. Keywords used for literature search.

Primary Keywords	Secondary Keywords [a]
Bio-preservation of bread	Nanoparticles incorporated into bread
Bread spoilage	Bakery product storage life
Staling of bread	Spoilage control of bread
Bread preservation by natural anti-microbes	Lactic acid bacteria used in bread preservation
Mold spoilage of bread	Statistics on bakery waste

[a] Literature search was conducted by using primary keywords in combination with secondary keywords.

3. Bread and Its Types

The world's expanding, aging, more urbanized, and nomadic population calls for a more accessible, nutritious, and safe food supply. Consuming wholemeal or multigrain bread may provide the most outstanding contribution to the necessary nutritional intake [33]. It is remarkable that white bread now surpasses pasta, pseudocereals, rice, and tubers in terms of consumption.

There are several varieties of bread (Figure 1), with national and regional variants on fundamental types and an abundance of specialty varieties, such as malt loaves, soda loaves, milk loaves, fruit loaves, and cream loaves [34]. Additionally, there is high protein bread, added malt grain bread, high fiber, multigrain bread, wheat germ bread, soft grain bread, ethnic multigrain bread, slimming and healthy high fiber bread, bread for special dietary needs, and cereal bread other than wheat crispbread. The bakery industry does not use the same terminology to describe bread and its quality characteristics, leading to misunderstandings regarding bread quality among consumers.

Figure 1. Different types of bread around the world.

The most prominent example references French bread, which in the United Kingdom, is often referred to as baguette-style bread. In contrast, in the United States, the word may apply to pan bread with lean (i.e., low-fat) formulas, such as those adopted to make French toast. The term "French" bread is not as well understood in France as in the United

Kingdom. Correspondingly, the conception of "toast bread" has a distinct connotation in France, the United States of America, the United Kingdom, and elsewhere [35]. The description of a specific bread type will always explain its substantial look, most often with its outer structure. Consequently, the length and diameter of baguettes are often used to classify them, whereas the pan form of other bread is used to classify Middle Eastern and conventional Indian bread. Even surface markings may require definition, as the quantity and pattern of cuts on the dough surface may become an essential aspect of the traditional appearance of the product.

4. Economic Importance of Bakery Product

Bread is consumed in a diversity of shapes and forms all over the world, with an average annual consumption of 70 kg per capita; however, Europe consumes less bread, with an average annual consumption of 59 kg [36–38]. Bread products have evolved into various shapes and flavors, each with its own unique set of features. It is more favorable in an urbanized environment because it requires no additional preparation and is immediately consumable. Bread consumption has already increased in areas where it was previously considered a luxury item. For example, between 1980 and 2008, per capita wheat consumption increased by 44% in Africa due to its convenience and ease of preparation [39]. Annually, almost 9 billion kg of bread items are manufactured [40], with an expected annual per capita intake of 41 to 303 kg [41]. Moreover, Asians are increasing their bread consumption. The average volume of bread and bakery products consumed per person is predicted to reach 120.5 kg in 2021 and an annual growth of 4.87% is estimated from 2021 to 2026. Globally, the majority of income is earned in China (US $275,427 million in 2021) [42]. It is worth noting that India produced 257 thousand metric tons of baked goods in fiscal 2020, compared to 275 thousand metric tons in fiscal 2018 [43] in Europe; the consumption of bread and bakery products climbed from 29,641 million kg in 2010 to 30,328 million kg in 2017 and is predicted to reach 30,352 million kg in 2021 [44]. In Chile, people consume about 86 kg per capita/year of bread, the highest per capita among Latin American countries. Argentina presents the second-largest consumption at 50 kg per capita/year [45]. In Brazil, bread consumption is about 34 kg per capita/year.

5. Spoilage Concerns of Bread

Bread quality deteriorates after baking, resulting in significant monetary losses for the bakery sector and the customer [46]. Bread spoilage is a complicated process that undergoes chemical (changes in nutraceutical value, rancidity), physical (moisture redistribution, staling), and microbiological (yeast, bacterial spoilage, and mold) alterations and contributes to the "staling process" of bread [11,47]. Staling shortens the shelf life of bread, which is determined as the period that food remains "acceptable" for consumption under certain storage conditions. Acceptable means retaining the desired sensory, chemical, physical, and biological attributes in addition to being safe [48]. The other significant reason for the reduction of the shelf life of bakery goods through post-baking storage is microbial spoilage, which ultimately results in detectable mold growth and the generation of mycotoxins that are not identifiable. High moisture levels (a_w = 0.94–0.99) stimulate the development of approximately all bacteria, yeasts, and molds [49–52]. We will review some of the most familiar concerns about bakery spoilage in the following sections.

5.1. Physical Spoilage

Moisture absorption and loss can cause physical changes in low and intermediate moisture products, as well as chemical and microbiological deterioration. As well, the staling of bakery items is the most severe issue with physical deterioration [51]. The total staling mechanism involves two distinct phenomena: the intrinsic firming of the cell wall substance related to starch re-crystallization at the time of storage and the firming effect induced by moisture migration from crumb to crust [17,53]. It has an eloquent economic impact on the baking business. Furthermore, bread loses nutritional value and chemical stability after baking, usually linked with rancidity, particularly in bread processed with whole wheat flour or possessing a high-fat content [54]. By breaking down unsaturated fatty acids via malodorous aldehydes, autolytic free radicals, short-chain fatty acids, and ketones are combined, leading to rancidity [55].

5.2. Chemical Spoilage

Rancidity is the most common type of chemical spoilage that occurs after baking high-fat bread products [54]. It is known as lipid degradation, and it results in off-odors and flavors. Rancidity lessens the bread's shelf life and makes it highly unhealthy for consumers. There are two different types of rancidity in general: oxidative and hydrolytic rancidity [51].

Oxidative rancidity induces the degradation of unsaturated fatty acids by oxygen using an autolytic free-radical process. As a result, foul-smelling aldehydes, ketones, and short-chain fatty acids are produced. These free radicals and peroxides developed by lipid oxidation may also negatively impact food composition by bleaching pigments (for example, lycopene in tomato paste in pizza), deteriorating protein, and spoiling specific vitamins (for example, vitamins A and E). In contrast to oxidative rancidity, hydrolytic rancidity develops without oxygen triglyceride hydrolysis and the consequent discharge of malodorous fatty acids and glycerol. Moisture and endogenous enzymes such as lipases and lipoxygenases trigger rancidity problems.

5.3. Microbiological Spoilage

Bread ingredients promote the growth and proliferation of microbes during various phases of bread preparation, processing, packing, and storage. Study [56] found that molds, yeasts, and bacteria were the primary sources of microbiological spoilage in bread, and that by using cluster analysis, the enzymic degradation induced by lipoxygenase may be separated from each other and preserve bread analogs after 48 h, prior to apparent indications of decomposition. They can survive in the baking environment and multiply in various conditions, including those where other bacteria are not competitive [57].

Water activity (a_w) has the most significant impact on the microbial spoilage of bakery items. Microbiological spoilage is not an issue with low moisture bread items (a_w 0.6). The primary spoilage pathogens in moderate moisture (aw 0.6–0.85) are osmophilic yeasts and molds. Mainly all yeasts, molds, and bacteria can thrive in high-moisture conditions (a_w 0.94–0.99) [58,59].

5.3.1. Bacterial Spoilage

Spore-forming bacteria are another source of concern for bread quality and safety. This problem is specific to bakery goods with a high moisture content as most bacteria need high a_w to survive [51]. The key pollutants are ingredients and/or bakery equipment, as spore-forming bacteria are almost certainly produced on the extraneous surface sections of grains and in the surrounding air of the bakery [60].

Bacillus subtilis is the main cause of bacterial spoilage. Its spores form endospores, which possibly sustain baking and sprout and develop inside the bread within 36–48 h, forming the distinctive soft, fibrous, brown mass with a ripe pineapple or melon odor leading to the exoneration of volatile components, for example, diacetyl, acetoin, acetaldehyde, and isovaleric-aldehyde [53]. *Bacillus subtilis* spores are heat-resistant; after 20 min at 65 °C, 55% remain active in amylase [7]. Other Bacillus species have been identified that cause bacterial spoilage of bread, including *Bacillus amyloliquefaciens, Bacillus pumilus, Bacillus licheniformis, Bacillus megaterium,* and *Bacillus cereus* [60] (Table 2). According to [61,62], *B. amyloliquefaciens* could be the primary species responsible for rope spoilage. "Ropey" bread can be distinguished by brown to black discoloration, the release of a rotten fruit odor, and the formation of sticky breadcrumbs as a result of protein and starch degradation during bacterial growth [63,64].

Table 2. Significant microbes responsible for microbial bread spoilage.

Major Spoilage Concerns	Spoilage Agents	Influencing Factor/Species	Properties of The Microbe's Colony	Issues	Reference
Physical	Staling	Starch retrogradation Moisture distribution Protein Baking process Storage condition		Crust softening Crumb firming	[51,55]
Chemical	Rancidity	Lipid Oxygen Short-chain fatty acid aldehyde		Off-flavors Off-odors Loss of vitamin (A,E) Protein degradation	[54]
Microbial	Bacteria	*Bacillus subtilis*	Irregular shape	Rotten fruit odor Sticky breadcrumbs Protein degradation Black discoloration	[7,60,65]
		Bacillus licheniformis	White/dull color		
		Bacillus amyloliquefaciens	Gram-positive		
		Bacillus pumilus	Ellipsoidal		
	Mold	*Rhizopus nigricans*	Grey, fluffy, fast spread	Mycotoxin Product loss Chalk mold	[66–69]
		Penicillium expansum	Blue/green, slow spread		
		Aspergillus niger	Black, fluffy, sporehead		
		Cladosporium	Dark green, flat, spreads slowly		
	Yeast	*S. cerevisiae*	Rapid growth, thermotolerance	Chalky bread White dot on bread Esteric off-odor	[46,70–72]
		H. burtonii *Scopsis fibuligera* *P. anomala*	Slow growth, low, white, spreading colonies		

5.3.2. Fungal Spoilage

Wheat, rice, and maize are the three most significant cereal crops grown globally [63]. In the Western world, wheat is the most considerable cereal for making bread, accompanied by rye. Barley, oats, and the gluten-free crops quinoa, millet, sorghum, buckwheat, and amaranth are all important cereals [46,63]. The contamination of cereal grains and their products occurs after harvest, during storage time, and pre- and post-processing. An estimated 40 species of fungi have been discovered in baked goods, most of which are considered contaminants [64]. Among these, Penicillium, Aspergillus, Rhizopus, and Wallemia are the most prevalent fungi species liable for the fungal deterioration of bread [8,14,15]. An illustration of fungal spoilage is shown in Figure 2. Some of the fungal spoilage types are discussed below:

1. Mold Spoilage: Mold in bread is a serious concern as it creates mycotoxin, which is harmful to public health [65–67]. Mold growth and mycotoxin production in crops and foods are challenges in emerging countries in Africa and Asia, leading to increased humidity and temperature [68,69]. It also shortens the shelf-life of bakery items containing moderate and high moisture content. Most molds can generally grow at a_w values greater than 0.80, except for a few xerophilic molds, which may thrive at a_w values as low as 0.65. While baking, molds and spores may be rendered inactive. However, since the environment within a bakery are not hygienic and might be a source of contamination, the mold contamination of bread can occur during chilling, slicing, wrapping, and storage [53]. Moisture precipitates on the inner area of the package when hot bread is wrapped, promoting mold growth. The most frequent

molds involved in bread spoilage are Rhizopus nigricans, which has fluffy white mycelium and black spots of sporangia, and *Penicillium expansum* or *P. stolonifer*, and *Aspergillus niger*, which has greenish to black conidial heads and are referred to as "bread mold" [53,70,71] (Table 2). Aspergillus, Mucorales, Cladosporium, Penicillium, and Neurospora species have all been found in wheat bread, with *Penicillium* spp. being the most frequently occurring variety [11].

2. Yeast Spoilage: Yeast spoilage is the type of microbial spoilage with the lowest prevalence. It occurs most frequently during cooling, and more significantly, during the slicing stage in the case of industrial bread. Off-odors in bread may indicate yeast spoilage. Yeast spoilage can occur in two different forms of yeast [72]. (a) An alcoholic or esteric off-odor is generally identified when fermentative yeasts cause spoilage. Most seen is *Saccharomices cerevisiae*, which is commonly employed as baker's yeast. (b) The other type of contamination is due to filamentous yeasts. The familiar circumstance described as "chalky bread" develops in this situation, showing white dots in the crumb. Mold growth is sometimes confused with this type of spoilage. However, yeasts create single cells and reproduce through budding; thus, they can be differentiated. *P. anomala, Scopsis fibuligera*, and *H. burtonii* promote the prime spoilage of bread goods by growing in white, low, spreading colonies that symbolize a scattering of chalk dust on the product surface. Visible yeast generation is typically linked with goods with a high a_w and short shelf life.

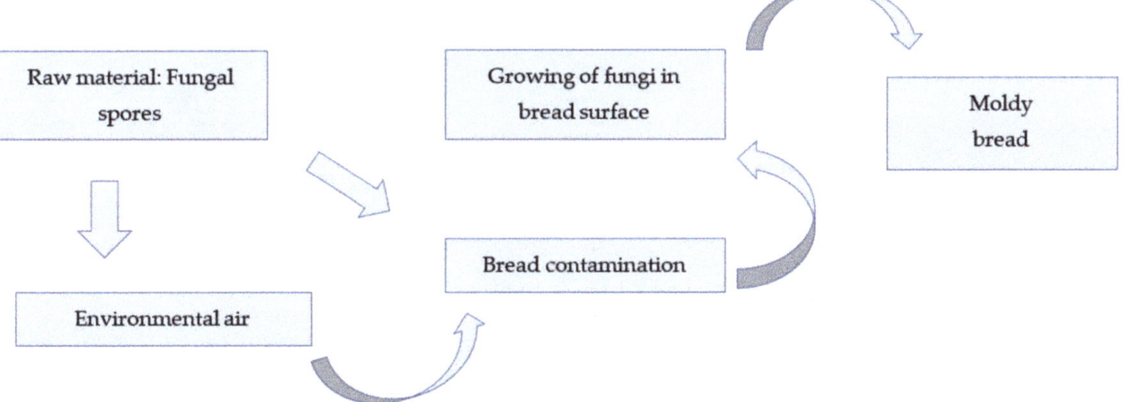

Figure 2. Fungal spoilage procedure. Adapted from ref. [73].

6. Bio-Preservation to Control the Spoilage of Bread

The bakery sector uses product reformulation, which is the most popular, practical, and cost-effective solution to prevent post-baking contamination. This is done by lowering the product's a_w and pH, which are related to the microorganisms' shelf life. They also use chemical preservatives directly in the product or on its surface to prevent bacterial and microbial spoilage [51]. According to [74], chemical preservatives inhibit microbial metabolism by denaturing cell protein or causing physical damage to the cell membrane. Propionic acid, as well as its salt, are the most widely used chemical preservatives in bread [7]. It helps prevent mold deterioration and bread ropiness that occurs due to *B. subtilis*.

However, they are continually investigated due to the possibility of developing chronic non-communicable diseases [75]. As a result, bio-preservatives have emerged as a favored solution to these shortcomings, with the intention to generate "clean label" foods [15,76,77]. Bio-preservatives can also be adopted as natural antifungal substances to prevent fungal degradation and prolong shelf life, reducing public health risks [2,12]. Bio-preservatives such as lactic acid bacteria, essential oils, or natural nanoparticles are becoming more popular because of consumer apprehensions about applying chemicals in food. According to [78–83], a good bio-preservative should have the following characteristics: it should have an expansive antibacterial spectral range, be non-toxic to humans, be suitable for lower doses, have some slight impact on product pH, not impair product odor, color, or flavor at the proposed level of its use, be accessible in a dry state, have a higher water solubility, be non-corrosive, be unreactive, and have no detrimental effects on fermentation or bread character. We will review a few of the bio-preservatives in the following sections.

6.1. Microbial Fermentation

Traditionally, organic and flavorful bread with a long shelf life was achieved naturally through an expanded fermenting operation: sourdough [11,84–89]. The word "sourdough" describes a particular bread recipe in which flour, water, LAB, and yeast organisms are fermented together [89–93]. Because of their remarkable antifungal activity, lactic acid bacteria (LAB) and antagonistic yeasts have received particular attention among natural agents [93–97] and are herein discussed below as well as presented in a tabular form (Table 3).

Table 3. Preservation technique by microbial fermentation to improve bread shelf life.

Microbes	Product	Starter Culture Used, Compounds	Shelf Life/Fungal Inhibition	Reference
Lactic acid bacteria	Pan bread	*Lactobacillus plantarum*	7 days after baking, *A. niger* growth was lower	[98–100]
	Quinoa and rice bread	*Lactobacillus reuteri, Lactobacillus brevis.*	2 days extended shelf life	[93,101–103]
	Bread	*Lactobacillus plantarum*	>14 days extended shelf life	[104–107]
	Gluten-free breads	*Lactobacillus amylovorus*	4 days extended shelf life	[99]
	Bread	*P. acidilactici* KTU05-7, *P. pentosaceus* KTU05-8, and KTU05-10	8 days fungal growth inhibition	[100]
	Bread	*Lactobacillus hammesii*	6 days extended mold-free shelf life.	[108–111]
	Bread	*Lactobacillus plantarum* 1A7	Up to 28 days fungal inhibition	[104]
Yeast	Pan bread	*Penicillium anomala* SKM-T	Overall storage life is 6–8 days, when appearing with fewer fungi count	[105]
	Wheat sourdough	*W. anomalus* LCF1695	Up to 14 days shelf life	[104]

Lactic acid bacteria: LAB metabolic products enhance bread's organoleptic and technological aspects, as well as its textural characteristics [81,82,112–116], along with its shelf life, nutritional value [83,117–121], and beneficial aspect (anticarcinogenic and cholesterol reduction abilities), during the fermentation of the dough [84–86,122–126]. They can also be adopted to replace chemical preservatives, ensuring a clean label and increased consumer acceptance [87,126–129].

Lactic acid bacteria (LAB) have been utilized in fermented foods for over about 4000 years [88]. It is naturally found in foods or introduced as pure cultures [89]. It is also GRAS-certified (generally recognized as safe) and has an extended application history in various cereal fermentations, particularly in the baking industry. LAB's adaptability is remarkable, not just in terms of catabolic and anabolic pathways but also in changing environmental conditions [90,130–134].

It has also been employed as a starter culture in the food business for centuries, which may produce several bioactive compounds, along with fatty acids, bacteriocins, organic acids, hydrogen peroxide, indole lactic acid, and phenyl lactic acid [91]. They also have an anti-aflatoxigenic effect [92,135–138]. Particular lab strains that have gripping bio-preservation action on bread when adopted as starter cultures include *Lactobacillus amylovorous* DSM 19,280, *Lactobacillus fermentum* Te007, *Lactobacillus acidophilus* ATCC 20079, *Lactobacillus paralimentarius* PB127, *Lactobacillus brevis* R2D, *Lactobacillus rossiae* LD108, *Lactobacillus hammesii*, *Lactobacillus paracasi* D5, *Pediococcus pentosaceus* KTU 05-8 and KTU 05-10, *Lactobacillus pentosus* G004, *Lactobacillus plantarum*, *Lactobacillus reuteri* R29, *Lactobacillus rhamnosus*, *Lactococcus* BSN, *Pediococcus acidilactici* KTU05-7, as well as *Leuconostoc citreum* C5 and HO12 [93,139–142]. Additionally, adding 15–20% sourdough significantly (p = 0.0001) increased bread volume and crumb porosity, based on the LAB strain, and reduced acrylamide formation by an average of 23% (for LUHS51 and LUHS206) and 54% (for LUHS71 and LUHS225), respectively, in comparison with regular bread [94]. Also, the most dominating species of the conventional sourdough microbiota, *Lactobacillus sanfranciscensis*, has been found to have a favorable impact on several important quality features of sourdough, notably dough rheological qualities, bread texture and aroma, and shelf-life conservation [95,96,143–145].

According to [97], LAB mix culture-activated bread samples could tolerate fungal deterioration until the fourth day. It was also discovered that the primary products of LAB fermentation, such as lactic and acetic acid, inhibited further fungal growth in *Mucor* sp. and *Rhizopus* sp. by up to 40% and 20%, respectively, when compared to a control bread sample. Also, the development of *A. niger* was observed in a study by [98,146]. They found that its growth in pan bread containing LAB isolate was slower than the control bread without the isolate after 7 days of baking. Moreover, the antifungal efficacy of *Lb. amylovorus* DSM19280 as a sourdough starter culture was evaluated by [99]. The result showed that, when it was applied, the bread's shelf life was increased by 4 days, relative to the control samples, which had mold detectable after only 2 days. Moreover, [100,147] used *P. acidilactici* KTU05-7, *P. pentosaceus* KTU05-8, and KTU05-10 strains on sourdough bread in another experiment. Their findings showed that adding sourdough made with these strains in bread decreased fungal deterioration more than control samples and suppressed fungal growth over an 8-day storage period, whereas control bread had visible fungi colonies. Furthermore, to investigate the antifungal activity of *Lactobacillus rossiae* LB5 and *Lb. plantarum* LB1, bread slices with *Lactobacillus rossiae* LB5 and *Lb. plantarum* LB1 were mixed with *Penicillium roqueforti* DPPMAF1 by [101,148]. Mycelial development was seen in the wheat germ bread sample after 21 days of inoculation with only a 10% contamination score. In comparison to control samples injected with *Cladosporium* spp., *Aspergillus clavatus*, *Penicillium roquefortii*, or *Mortierella* spp., [102] studied the implications of using two active propionate providers, *Lactobacillus dioliyorans* and *Lactobacillus buchneri*, in bread restoration, and discovered that mold development was inhibited for more than 12 days.

- Yeast: Numerous authors have experimented with using incompatible yeasts as bio-control agents. They can be used as bio-preservatives as they retain some of the essential features that enhance their acceptability. They compete for nutrients with fungal pathogens and their higher rate of nutrient utilization significantly contributes to a bio-preservative nature. Yeast produces killer toxins, also called mycotoxins, which showed bioprotective attributes against food spoiling microorganisms and pathogens [149]. Some yeast genera produce extra- and intracellular compounds which possess antibacterial properties. Production of ethanol of high concentration and organic acids which results in the change of pH of the medium also responsible for the effectiveness of yeasts as bio-preservatives [150]. Many of them can sustain residence on dry surfaces due to their low requirements for water and nutrients [103]. With *Lb. plantarum* 1A7 as a starter, the yeast *Wickerhamomyces anomalus* LCF1695 (previously recognized as *Pichia anomala*) was occupied for sourdough fermentation. It was found that when *P. roqueforti* DPPMAF1 was artificially inoculated (102 conidia ml1) with this combination and stored at room temperature, the microbiological shelf life was elongated to at least 14 days [104]. As well as *Penicillium paneum* KACC

44834, outgrowth on white pan bread leavened with *Penicillium anomala* SKM-T was significantly reduced compared to standard baker's yeast, and improved the shelf life [105]. In contrast, bread composed of propionic acid from cultured yeast extract contained less ethanol and had a better shelf life against mold growth than bread formulated with non-fermented yeast extract [106].

6.2. Plant Extracts as Bio-Preservative

Plant extracts have been extensively studied as bio-preservatives, as plants contain many essential antifungal compounds, for example, phenolic compounds, glucosinolates, cyanogenic glycosides, oxylipins, and alkaloids [46]; there is a thriving interest in natural ingredients with multifunctional properties in food as well. Most edible plant parts include trace amounts of plant defense substances (phenolic acids), categorized as hydroxybenzoic and hydroxycinnamic acids [109]. Hydroxybenzoic acids are frequently found in larger phenolic compounds, such as hydrolyzable tannins. Hydroxycinnamic acids occur as esters of glycerol, tartaric, shikimic, and quinic acids, as well as glycosylated derivatives [110]. Plant extracts can inhibit harmful bacteria from adhering to the host cell membrane. As a result, it reduces bacterial attachment to host cell surface membranes, and thus sometimes it becomes a potential anti-adhesive agent [143].

Study [111] investigated the antifungal efficacy of different raisin extracts and by-products in traditional bread. Compared to a sample containing no preservatives, the bread produced with raisin paste and raisin water separate (7.5%) exhibited the best mold-reducing abilities. Leaf extracts of cherry laurel (*Prunus laurocerasus* L.) were recommended as potential bio-preservatives after demonstrating a very low MIC (mg/mL) against a variety of bread spoilage fungi [112]. Further, [113] showed that bread produced with a mix of sourdough and pea flour hydrolysate fermented by the antifungal strain *Lb. plantarum* 1A7 had the most extended shelf life. It is also effective against *P. roqueforti* DPPMAF1. Moreover, bread formulations containing free or liposome-encapsulated garlic extract (0.65 mL/100 g of dough) were found to be more microbiologically stable than controls, inhibiting molds such as *P. herquei*, *F. graminearum*, and *A. flavus* for five days [12].

6.3. Essential Oil as Bio-Preservative

Plant essential oils are receiving much attention in the food sector because of their potential as decontaminating agents, food flavoring agents, and natural food preservation agents [114]. They are also GRAS (Generally Recognized as Safe) [115]. EOs are formed from the largest category 1-phytoanticipins, intended antimicrobial elements found in plants and applied in the food, pharmaceutical, agronomic, and cosmetic industries [116]. The other categories include (cat. 2) inducible preformed compounds and (cat. 3) phytoalexins, which comprise activated restraining substances when the plant is attacked by a pathogen [117]. These substances can be found in the skins, shells, bark, and cereal bran of fruits, vegetables, and plants [110].

Numerous research has been conducted to figure out the efficacy of essential oils in enhancing the shelf life of bread (Table 4). Carvacrol and eugenol, two antifungal components found in essential oils, could be regarded as powerful antifungal agents. The inner mitochondrial membranes of fungal cells can be largely destroyed, while the cell wall is completely destroyed by them [114]. Thus, essential oils work as antifungal agents. The effects of thyme, clove, cinnamon oils, and orange, sage, and rosemary oils on rotting fungi in rye bread have been examined. Oils of thyme, clove, and cinnamon were known to suppress spoilage fungi, but oils of orange, sage, and rosemary had only a minor impact [118]. Among them, the potential of using marjoram and sage essential oils on bread is neglected owing to their low acceptability in terms of flavor and odor, despite their demonstrated mold inhibiting capabilities [119]. Study [120] used a disc diffusion experiment to study the antifungal effect of several EOs and reported that cinnamon and mustard EOs might cause a 100% inhibition in *P. roqueforti* multiplication with only 1 µL of EO introduced into the Petri dish system. In comparison, eugenol in cinnamon EO [121] and allyl isothiocyanate in mustard EO are the primary antifungal active components [120]. Furthermore, [122] investigated the antifungal activity of thyme essential oil in par-baked bread (0, 15, and 30 g sourdough/100 g dough) using the macro-dilution method with changed pH (4.8, 5.0, 5.5, and 6.0), a_w (0.95 and 0.97), and temperature (22 and 30 °C). Despite thyme oil's strong in 1vitro potential, there was no noticeable shelf-life expansion for par-baked bread. In the case of rosemary oil, [123] found that applying 50.0 µL/mL rosemary essential oil inhibited both the *Penicillium* sp. and the *Aspergillus* sp. fungi tested. After 8 days of preservation at 25 °C, the amount of fungal generation in the dough containing pure oil and the dough containing microencapsulated oil decreased by at least 0.7 and 1.5 log cycles, respectively, in comparison to the control. A study by [124] evaluated essential oils of oregano (Origanum vulgar) and clove bud (*Syzygium aromaticum*), which were processed using low-speed mixing and ultrasonication to create coarse emulsions (1.3–1.9 µm) and nanoemulsions (180–250 nm). The results showed that both essential oils significantly reduced yeast and mold counts in sliced bread during 15 days. Apart from this, bread was treated with lemongrass EO to the amount of 125 to 4000 $µL/L_{air}$ where *P. expansum* generation was inhibited for 21 days at 20 °C [125].

Although using essential oils in the bread industry has many benefits, one major downside is that the consumer does not always appreciate the flavor and aroma they impart. Article [126] documented color changes in food because of essential oils.

Table 4. Preservation of bread by plant essential oils.

Essential Oils	Targeted Molds	Results	Reference
Thyme	*Aspergillus niger* *P. paneum*	No noticeable shelf-life extension	[122]
Lemongrass	*P. expansum*	Mold growth was inhibited for 21 days	[125]
Rosemary	*Penicillium* sp. *Aspergillus* sp.	Fungal generation reduced by 0.7 and 1.5 log cycles after using pure rosemary oil	[123]
Clove bud and Oregano	*A. niger* *Penicillium* sp.	Reduced yeast and mold growth for 15 days	[124]
Marjoram and clary sage	*P. chrysogenum* *Rhizopus* spp.	Shelf life 8 days	[127]
Citrus peel	*General fungi*	Shelf life 4 days	[128]
Cinnamon and mustard	*P. roqueforti*	100% reduction of the targeted mold growth	[120]

6.4. Animal-Derived Products

Cheesemaking generates a waste stream with a high biochemical oxygen demand, whey, which is the liquid fragment produced after milk protein coagulation and is also a contaminant to the environment [129]. In recent times, there has been a gush of attraction in investigating and promoting natural antimicrobial compounds produced from food industry by-products that prevent the production of fungi in food [103]. The antimicrobial or antifungal compounds cause target cell membranes to permeabilize, resulting in holes, cell leakage, and cell death [148]. This two-pronged strategy addresses the health problems accompanied by chemical food additives by redressing them with natural preservatives, thereby encouraging better food items and contributing to the prevention and reduction of food waste [130,131]. Whey can be an intriguing technique for bread bio-preservation. A study conducted by [132] exhibited that whey's application as a bio-preservation agent in bread improved shelf life by almost 2 to 15 days compared to bread that contained 0.3% calcium propionate and controlled untreated bread. Bread produced with goat whey hydrolysate (HGW) and treated with toxigenic fungi was included in a shelf life study by [133]. This study determined the effect of calcium propionate on fungal growth and mycotoxin formation in bread. It was proven that bread containing HGW inhibited fungal growth, with minimal inhibitory and fungicidal concentrations of 3.9–62.5 and 15.8–250 g

HGW/L, respectively. In addition, HGW showed a 1-log reduction in fungal production, 85–100% mycotoxin generation, and a 2-day shelf life extension.

6.5. Nanoparticles

Efforts to provide effective bioactive packaging action and to prevent most biopreservatives from degradation under harsh conditions, including high temperatures and high humidity, may improve bakery products by implementing nanotechnology into the food business, specifically by incorporating nanomaterials [75].

Diseta et al. [134] carried out a recent experiment to assess the antifungal efficacy of nanocomplexes based on egg white protein nanoparticles (EWPn) and carvacrol (CAR), bioactive compounds (BC), trans-cinnamaldehyde (CIN), and thymol (THY), as well as their use as edible coatings on preservative-free bread. It was found that EWPn-CAR and EWPn-THY nanocomplex coatings had higher antifungal efficacy, allowing the bread to last an additional 7 days after application. Another study by [75] evaluated starch/carvacrol nanofibers where nanofibers containing 30 or 40% carvacrol showed restriction zones with limited generation and were successful in suppressing both fungi tested in the study. Besides the fact, only bread evaluated with starch/carvacrol nonwovens with 30% carvacrol had lower CFU values and no fungal development after 7 days (0 CFU). As well, incorporating nanomaterials into chitosan-based food wrapping techniques can help to inhibit spoilage and pathogenic microorganism generation, enhance food quality and safety, and lengthen the food shelf life. Based on the findings of study [135], it appears that chitosan-based films, coatings (or treatment) have been applied to prolong the shelf life of fresh produce, meat, bread, and dairy products. It could be a novel food packing system [136]. More recently, [137] studied the potential of expanding the shelf life of white bread by employing paper packages modified with Au/TiO_2, $Ag/N-TiO_2$, and Ag/TiO_2-SiO_2. They discovered that packaging with $Ag/N-TiO_2$, and Ag/TiO_2-SiO_2 paper prolonged the shelf life of bread by 2 days, while utilizing Au/TiO_2 paper had no effect.

However, though nanoparticle-based packaging materials are increasing its wide acceptance, there are still possibilities of migrating nanoparticles from packaging materials to foodstuff. Considering human health effects, it is also essential to consider short- and long-term toxicity studies. Nevertheless, it is not so easy to predict the NPs' mode of action due to their vast range of physicochemical and biological behaviors [145]. Different countries are taking several robust regulatory approaches to cover all these issues. In the US, FDA (Food and Drug Admininstration) looks into the size of the NPs ranging from 1 nm to 100 nm, external dimension(s) (up to 1 μm), properties, etc. [146]. Even in Europe, stakeholders are demanding for greater transparency by either the labeling of products containing NPs or making use of nanotechnology [147].

6.6. Other Novel Technologies

Dispersion from the wrapping material to the food surface is a significant issue when it comes to the choosing and employment of plastic packaging substances for food packaging (2002/72/EC) [138]. The initial purpose of food packaging is to denounce the reactions that deal with the durability of the contents enclosed. Also, effective packaging selection and optimization are critical for food manufacturers.

Nowadays, various packaging materials with varying barrier qualities are available for food packaging, making the problem of selecting the best packaging material for a specific food product more challenging than ever. We will be discussing a few novel technologies of bread preservation below.

The existing packaging materials serve as a barrier to protect the bread from an adverse environment and any spoilage. Among several novel technologies for bread preservation, active packaging can either increase or observe the shelf life by actively interacting with it, which usually necessitates the application of chemical compounds [139]. Active packaging methods reduce bread spoilage by utilizing ethanol emitters, essential oil, and oxygen absorbers with other antimicrobial factors as a coating in the packaging or edible films, or by inserting them into the packaging, for example, as sachets [46,139]. They offer several benefits, along with the capacity to control the inner conditions of the package headspace, the partial or total distribution of other chemical preservatives, as well as the expansion of mold-free shelf life and the conservation of good sensory attributes for more extended periods, allowing for faster stock rotation cycle times and the extension of the distribution channel for bread item distributors [11]. Moreover, [140] studied the influence of active packaging with a cinnamon essential oil label mixed with MAP on the shelf life of gluten-free sliced bread in 2011. They discovered that active packaging extended the shelf life of packed food while retaining the gluten-free bread's sensory qualities.

The rising consumer concern about food preservatives is prompting an expansion in demand for preservative-free goods. Modified Atmosphere Packaging is a substitute to chemical preservatives for governing mold decomposition in bread items, aside from their a_w and pH, which is described as the process of enclosing a food item in a high gas resistance film with the gaseous environment altered or regulated to reduce respiration rate, lower microbiological development, and hinder enzymatic decomposition to extend shelf life [141]. This mechanism involves injecting nitrogen (N_2) and carbon dioxide (CO_2) into the environment where the food packaging is placed. These gases are generally incorporated to appreciate the shelf life of bakery items by preventing fungal growth [138]. To determine which gases are the most successful in maintaining freshness, [142] compared the shelf life of bread prepared and conserved under varying concentrations of gases. They found out that the combination of 50% CO_2 and 50% N_2, with and without calcium propionate, was most dynamic against mold and yeast growth, increasing the shelf life to 117% and 158% at 22–25 °C and 15–20 °C, respectively.

7. Future Aspects and Bio-Preservatives Use in the Food Industry

The fungus problem in bread is alarming, and controlling it is a massive challenge for businesses. In another scenario, the value of functional foods has risen significantly in recent decades, and it is foreseen to continue to grow in importance with the niche market's demanding consumers. So, to resolve this demand for functional and preservative-free safe food, the ongoing experiment has concentrated on bio-preservation techniques to increase the shelf life of bread. While bio-preservation has a long-term positive impact on food, it also has certain limitations. The primary barrier to using essential oils (EO) in foodstuff is that the individual compounds of EOs are insufficient to prevent microbial growth. The antimicrobial activity of EO also depends on the application scheme and the matrix into which it is put.

Furthermore, while spraying EOs to the surface, the active ingredients may get oxidized or volatilized. In addition, the antifungal possibilities and binding abilities of mycotoxins are strain-specific despite the vast abundance of LAB. This should be considered when choosing LAB as starter cultures in foodstuff. Also, the use of antifungal starter strains is confined yet, and additional experimentation is needed to enhance the duration of shelf life of gluten-free bread. In the case of active packaging, adequate labeling information with microbiological ecology must be provided. The price of food products with active packaging must be controlled while keeping in mind consumer acceptances. Finally, adapting and scaling up these technologies for industrial production would be the most effective approach, with the potential to be employed in combination with other technological hurdles, along with safety evaluation and risk analysis initiatives.

8. Conclusions

The pH, preservatives, and a_w, along with the microbial ecology of the food material, the redox state, the permeability attributes of the packing film, the gaseous status surrounding the product, and the storage temperature, are all interconnected factors in food deterioration and food safety. While new and evolving treatment processes can prolong the shelf life of bakery items, they should not endanger the safety of the products. As a result, the on-going study of the impact of these novel treatment methods on the safety and shelf life of high moisture bakery goods is mandatory at the government, industry, and academic levels. This review addressed some of these novel methods, which may be helpful to manage both the food safety and food spoilage problems. Moreover, depending on the information accumulated here so far, these results may emerge to be a breakthrough for producing natural, high-quality bread with slight allergenic components with extended shelf life.

Author Contributions: Conceptualization, M.R. (Mizanur Rahman), A.R. and A.R.S.; designed and performed research, M.R. (Mizanur Rahman); methodology validation, M.R. (Mizanur Rahman), R.I., A.M., S.H., W.Z.; M.R.R., S.A., M.R. (Mukta Roy), A.S., R.P.Z., R.B.A.B., A.R. and A.R.S.; field works, M.R. (Mizanur Rahman); data curation, M.R. (Mizanur Rahman), R.I., S.H. and W.Z.; writing the original draft, M.R. (Mizanur Rahman); reviewing and editing, M.R. (Mizanur Rahman), R.I., A.M., M.R. (Mukta Roy), S.H., W.Z., M.R.R., S.A., M.R. (Mizanur Rahman), A.S., R.P.Z., R.B.A.B., A.R. and A.R.S.; designed research, R.I., S.H., W.Z. and A.R.S; formal analysis, R.I., S.H. and W.Z.; data analysis, R.I., S.H. and W.Z.; visualization, R.I., S.H., W.Z., M.R.R., S.A., A.S., R.P.Z., R.B.A.B., A.R. and A.R.S.; investigation, A.R., M.R.R., S.A., M.R. (Mukta Roy), A.S., R.P.Z. and R.B.A.B.; project administration, A.R.; funding acquisition, A.R. All authors have read and agreed to the published version of the manuscript.

Funding: Supported by FCT-Foundation for Science and Technology, I.P., by the grants UIDB/04567/2020 and UIDP/04567/2020.

Institutional Review Board Statement: Not applicable.

Informed Consent Statement: Written informed consent was obtained from the participants.

Data Availability Statement: Not applicable. The data used to generate the results of this study are included in the article and no additional source is required.

Acknowledgments: We would like to thank the editor of Foods and the anonymous reviewers for their comments and suggestions, which helped improve the paper. We would like to express our deepest gratitude to the respondents who provided valuable information. We acknowledge the logistic and technical support of the Department of Food Engineering and Tea Technology of Shahjalal University of Science and Technology, Sylhet, Bangladesh.

Conflicts of Interest: The authors declare no conflict of interest.

References

1. El Sheikha, A.F. *Bread and Its Fortification: Nutrition and Health Benefits*; CRC Press: Boca Raton, FL, USA, 2015.
2. Garcia, M.V.; Bernardi, A.O.; Copetti, M.V. The fungal problem in bread production: Insights of causes, consequences, and control methods. *Curr. Opin. Food Sci.* **2019**, *29*, 1–6. [CrossRef]
3. Rosell, C.M. The Science of Doughs and Bread Quality. In *Flour and Breads and Their Fortification in Health and Disease Prevention*; Elsevier: Amsterdam, The Netherlands, 2011; pp. 3–14.
4. Norman, N.; Potter, J.H.H. *Food Science*; Springer Science & Business Media: Berlin/Heidelberg, Germany, 2006; Volume XV, p. 608.
5. Khanom, A.; Shammi, T.; Kabir, M.S. Determination of microbiological quality of packed and unpacked bread. *Stamford J. Microbiol.* **2016**, *6*, 24–29. [CrossRef]
6. Cauvain, S.P.; Young, L.S. *Technology of Breadmaking*; Springer: Berlin/Heidelberg, Germany, 2007.
7. Saranraj, P.; Geetha, M. Microbial spoilage of bakery products and its control by preservatives. *Int. J. Pharm. Biol. Arch.* **2012**, *3*, 38–48.
8. Garcia, M.V.; Bernardi, A.O.; Parussolo, G.; Stefanello, A.; Lemos, J.G.; Copetti, M. Spoilage fungi in a bread factory in Brazil: Diversity and incidence through the bread-making process. *Food Res. Int.* **2019**, *126*, 108593. [CrossRef]
9. Ju, J.; Xu, X.; Xie, Y.; Guo, Y.; Cheng, Y.; Qian, H.; Yao, W. Inhibitory effects of cinnamon and clove essential oils on mold growth on baked foods. *Food Chem.* **2018**, *240*, 850–855. [CrossRef]
10. Brashears, M.M.; Durre, W.A. Antagonistic action of Lactobacillus lactis toward Salmonella spp. and Escherichia coli O157:H7 during growth and refrigerated storage. *J. Food Prot.* **1999**, *62*, 1336–1340. [CrossRef]
11. Melini, V.; Melini, F.J.F. Strategies to extend bread and GF bread shelf-life: From Sourdough to antimicrobial active packaging and nanotechnology. *Fermentation* **2018**, *4*, 9. [CrossRef]
12. Pinilla, C.M.B.; Thys, R.C.S.; Brandelli, A. Antifungal properties of phosphatidylcholine-oleic acid liposomes encapsulating garlic against environmental fungal in wheat bread. *Int. J. Food Microbiol.* **2019**, *293*, 72–78. [CrossRef]
13. Melikoglu, M.; Webb, C. Use of waste bread to produce fermentation products. In *Food Industry Wastes: Assessment and Recuperation of Commodities*; Elsevier: Amsterdam, The Netherlands, 2013; pp. 63–76.

14. Pitt, J.I.; Hocking, A.D. *Fungi and Food Spoilage*; Springer: Berlin/Heidelberg, Germany, 2009; Volume 519.
15. Quattrini, M.; Liang, N.; Fortina, M.G.; Xiang, S.; Curtis, J.M.; Gänzle, M. Exploiting synergies of sourdough and antifungal organic acids to delay fungal spoilage of bread. *Int. J. Food Microbiol.* **2019**, *302*, 8–14. [CrossRef]
16. Ju, J.; Xie, Y.; Yu, H.; Guo, Y.; Cheng, Y.; Zhang, R.; Yao, W. Synergistic inhibition effect of citral and eugenol against Aspergillus niger and their application in bread preservation. *Food Chem.* **2020**, *310*, 125974. [CrossRef]
17. Alpers, T.; Kerpes, R.; Frioli, M.; Nobis, A.; Hoi, K.I.; Bach, A.; Jekle, M.; Becker, T. Impact of Storing Condition on Staling and Microbial Spoilage Behavior of Bread and Their Contribution to Prevent Food Waste. *Foods* **2021**, *10*, 76. [CrossRef]
18. Freire, F.C.O.A. A Deterioração Fúngica de Produtos de Panificação no Brasil. Embrapa Agroindústria Tropical-Comunicado Técnico (INFOTECA-E). 2011. Available online: https://www.infoteca.cnptia.embrapa.br/bitstream/doc/907492/1/COT11010.pdf (accessed on 7 December 2021).
19. Rodrigues, I.; Chin, L. A comprehensive survey on the occurrence of mycotoxins in maize dried distillers' grain and solubles sourced worldwide. *World Mycotoxin J.* **2012**, *5*, 83–88. [CrossRef]
20. Rodrigues, I.; Handl, J.; Binder, E. Mycotoxin occurrence in commodities, feeds and feed ingredients sourced in the Middle East and Africa Part B Surveillance. *Food Addit. Contam. Part B* **2011**, *4*, 168–179. [CrossRef]
21. Rodrigues, I.; Naehrer, K. Prevalence of mycotoxins in feedstuffs and feed surveyed worldwide in 2009 and 2010. *Phytopathologia Mediterranea* **2012**, *51*, 175–192. [CrossRef]
22. Rodrigues, I.; Naehrer, K. A three-year survey on the worldwide occurrence of mycotoxins in feedstuffs and feed. *Toxins* **2012**, *4*, 663–675. [CrossRef]
23. Wagacha, J.; Muthomi, J.W. Mycotoxin problem in Africa: Current status, implications to food safety and health and possible management strategies. *Int. J. Food Microbiol.* **2008**, *124*, 1–12. [CrossRef]
24. Goryńska-Goldmann, E.; Gazdecki, M.; Rejman, K.; Kobus-Cisowska, J.; Łaba, S.; Łaba, R. How to Prevent Bread Losses in the Baking and Confectionery Industry?—Measurement, Causes, Management and Prevention. *Agriculture* **2021**, *11*, 19. [CrossRef]
25. Luz, C.; Calpe, J.; Saladino, F.; Luciano, F.B.; Fernandez-Franzón, M.; Mañes, J.; Meca, G. Preservation, Antimicrobial packaging based on ε-polylysine bioactive film for the control of mycotoxigenic fungi in vitro and in bread. *J. Food Processing Preserv.* **2018**, *42*, e13370. [CrossRef]
26. Ju, J.; Chen, X.; Xie, Y.; Yu, H.; Cheng, Y.; Qian, H.; Yao, W. Preservation, Simple microencapsulation of plant essential oil in porous starch granules: Adsorption kinetics and antibacterial activity evaluation. *J. Food Processing Preserv.* **2019**, *43*, e14156. [CrossRef]
27. Dengate, S.; Ruben, A. Controlled trial of cumulative behavioural effects of a common bread preservative. *J. Paediatr. Child Health* **2002**, *38*, 373–376. [CrossRef] [PubMed]
28. Priftis, K.N.; Panagiotakos, D.B.; Antonogeorgos, G.; Papadopoulos, M.; Charisi, M.; Lagona, E.; Anthracopoulos, M.B. Factors associated with asthma symptoms in schoolchildren from Greece: The Physical Activity, Nutrition and Allergies in Children Examined in Athens (PANACEA) study. *J. Asthma* **2007**, *44*, 521–527. [CrossRef] [PubMed]
29. Erickson, M.C.; Doyle, M.P. The challenges of eliminating or substituting antimicrobial preservatives in foods. *Annu. Rev. Food Sci. Technol.* **2017**, *8*, 371–390. [CrossRef] [PubMed]
30. Belz, M.C.; Mairinger, R.; Zannini, E.; Ryan, L.A.; Cashman, K.D.; Arendt, E.K. Biotechnology, the effect of sourdough and calcium propionate on the microbial shelf-life of salt reduced bread. *Appl. Microbiol. Biotechnol.* **2012**, *96*, 493–501. [CrossRef] [PubMed]
31. Da Cruz Cabral, L.; Pinto, V.F.; Patriarca, A. Application of plant derived compounds to control fungal spoilage and mycotoxin production in foods. *Int. J. Food Microbiol.* **2013**, *166*, 1–14. [CrossRef]
32. Ghabraie, M.; Vu, K.D.; Tata, L.; Salmieri, S.; Lacroix, M. Technology, Antimicrobial effect of essential oils in combinations against five bacteria and their effect on sensorial quality of ground meat. *LWT-Food Sci. Technol.* **2016**, *66*, 332–339. [CrossRef]
33. Weegels, P.L. The future of bread in view of its contribution to nutrient intake as a starchy staple food. *Plant Foods Hum. Nutr.* **2019**, *74*, 1–9. [CrossRef]
34. Legan, J.D. Biodegradation, Mould spoilage of bread: The problem and some solutions. *Int. Biodeterior. Biodegrad.* **1993**, *32*, 33–53. [CrossRef]
35. Cauvain, S. Bread: The product. In *Technology of Breadmaking*; Springer: Berlin/Heidelberg, Germany, 2015; pp. 1–22.
36. De Boni, A.; Pasqualone, A.; Roma, R.; Acciani, C. Traditions, health and environment as bread purchase drivers: A choice experiment on high-quality artisanal Italian bread. *J. Clean. Prod.* **2019**, *221*, 249–260. [CrossRef]
37. Gębski, J.; Jezewska-Zychowicz, M.; Szlachciuk, J.; Kosicka-Gębska, M. Preference, Impact of nutritional claims on consumer preferences for bread with varied fiber and salt content. *Food Qual. Prefer.* **2019**, *76*, 91–99. [CrossRef]
38. Edwards, W.P. *The Science of Bakery Products*; Royal Society of Chemistry: London, UK, 2007.
39. Mason, N.M.; Jayne, T.S.; Shiferaw, B. Africa's rising demand for wheat: Trends, drivers, and policy implications. *Dev. Policy Rev.* **2015**, *33*, 581–613. [CrossRef]
40. Cho, I.H.; Peterson, D.G. Biotechnology, Chemistry of bread aroma: A review. *Food Sci. Biotechnol.* **2010**, *19*, 575–582. [CrossRef]
41. Sunny, A.R.; Mithun, M.H.; Prodhan, S.H.; Ashrafuzzaman, M.; Rahman, S.M.A.; Billah, M.M.; Hussain, M.; Ahmed, K.J.; Sazzad, S.A.; Alam, M.T.; et al. Fisheries in the Context of Attaining Sustainable Development Goals (SDGs) in Bangladesh: COVID-19 Impacts and Future Prospects. *Sustainability* **2021**, *13*, 9912. [CrossRef]
42. Statista Bread and Cereal Products. Available online: https://www.statista.com/outlook/cmo/food/bread-cereal-products/asia (accessed on 1 October 2021).

43. Statista Production Volume of Bread, Buns and Croissant across India from Financial Year 2015 to 2020. Available online: https://www.statista.com/statistics/762043/india-bread-buns-and-croissant-production-volume/ (accessed on 17 May 2021).
44. Statista Europe: Bread and Bakery Consumption Volume 2010–2021, by Category. Available online: https://www.statista.com/statistics/806319/europe-bread-and-bakery-production-volume-by-category/ (accessed on 26 July 2021).
45. Statista Latin America: Per Capita Bread Consumption 2017, by Country. Available online: https://www.statista.com/statistics/802811/per-capita-bread-consumption-latam/ (accessed on 1 July 2021).
46. Axel, C.; Zannini, E.; Arendt, E.K. Nutrition, Mold spoilage of bread and its biopreservation: A review of current strategies for bread shelf life extension. *Crit. Rev. Food Sci. Nutr.* **2017**, *57*, 3528–3542. [CrossRef]
47. Fadda, C.; Sanguinetti, A.M.; Del Caro, A.; Collar, C.; Piga, A. Bread staling: Updating the view. *Compr. Rev. Food Sci. Food Saf.* **2014**, *13*, 473–492. [CrossRef]
48. Nicoli, M.C. An introduction to food shelf life: Definitions, basic concepts, and regulatory aspects. In *Shelf Life Assessment of Food*; CRC Press: Boca Raton, FL, USA, 2012; pp. 1–16.
49. Gobbetti, M.; De Angelis, M.; Di Cagno, R.; Calasso, M.; Archetti, G.; Rizzello, C.G. Novel insights on the functional/nutritional features of the sourdough fermentation. *Int. J. Food Microbiol.* **2019**, *302*, 103–113. [CrossRef]
50. Luz, C.; D'Opazo, V.; Mañes, J.; Meca, G. Preservation, Antifungal activity and shelf life extension of loaf bread produced with sourdough fermented by Lactobacillus strains. *J. Food Processing Preserv.* **2019**, *43*, e14126. [CrossRef]
51. Smith, J.P.; Daifas, D.P.; El-Khoury, W.; Koukoutsis, J.; El-Khoury, A. Nutrition, Shelf life and safety concerns of bakery products—a review. *Crit. Rev. Food Sci. Nutr.* **2004**, *44*, 19–55. [CrossRef]
52. Sun, L.; Li, X.; Zhang, Y.; Yang, W.; Ma, G.; Ma, N.; Hu, Q.; Pei, F. A novel lactic acid bacterium for improving the quality and shelf life of whole wheat bread. *Food Control* **2020**, *109*, 106914. [CrossRef]
53. Pateras, I.M. Bread spoilage and staling. In *Technology of Breadmaking*; Springer: Berlin/Heidelberg, Germany, 2007; pp. 275–298.
54. Pareyt, B.; Finnie, S.M.; Putseys, J.A.; Delcour, J.A. Lipids in bread making: Sources, interactions, and impact on bread quality. *J. Cereal Sci.* **2011**, *54*, 266–279. [CrossRef]
55. Taglieri, I.; Macaluso, M.; Bianchi, A.; Sanmartin, C.; Quartacci, M.F.; Zinnai, A.; Venturi, F. Agriculture, Overcoming bread quality decay concerns: Main issues for bread shelf life as a function of biological leavening agents and different extra ingredients used in formulation. *A review. J. Sci. Food Agric.* **2021**, *101*, 1732–1743. [CrossRef]
56. Needham, R.; Williams, J.; Beales, N.; Voysey, P.; Magan, N.J.S.; Chemical, A.B. Early detection and differentiation of spoilage of bakery products. *Sens. Actuators B Chem.* **2005**, *106*, 20–23. [CrossRef]
57. Cauvain, S. Bread spoilage and staling. In *Technology of Breadmaking*; Springer: Berlin/Heidelberg, Germany, 2015; pp. 279–302.
58. Smtth, J.; Simpson, B. Modified atmosphere packaging of bakery and pasta products. In *Principles of Modified-Atmosphere Amd Sous Vide Product Packaging*; Routledge: London, UK, 2018; pp. 207–242.
59. Blakistone, B.A. *Principles and Applications of Modified Atmosphere Packaging of Food*; Blackie Academic & Professional: London, UK, 1998.
60. Lavermicocca, P.; Valerio, F.; De Bellis, P.; Sisto, A.; Leguérinel, I. Sporeforming bacteria associated with bread production: Spoilage and toxigenic potential. In *Food Hygiene and Toxicology in Ready-to-Eat Foods*; Elsevier: Amsterdam, The Netherlands, 2016; pp. 275–293.
61. Valerio, F.; De Bellis, P.; Di Biase, M.; Lonigro, S.; Giussani, B.; Visconti, A.; Lavermicocca, P.; Sisto, A. Diversity of spore-forming bacteria and identification of Bacillus amyloliquefaciens as a species frequently associated with the ropy spoilage of bread. *Int. J. Food Microbiol.* **2012**, *156*, 278–285. [CrossRef] [PubMed]
62. Rosenkvist, H.; Hansen, Å. Contamination profiles and characterisation of Bacillus species in wheat bread and raw materials for bread production. *Int. J. Food Microbiol.* **1995**, *26*, 353–363. [CrossRef]
63. FAO. *FAOSTAT Crops Production*; FAO: Rome, Italy, 2013.
64. Gänzle, M.; Gobbetti, M. *Handbook on Sourdough Biotechnology*; Springer: New York, NY, USA, 2013.
65. Veršilovskis, A.; Bartkevičs, V. Stability of sterigmatocystin during the bread making process and its occurrence in bread from the Latvian market. *Mycotoxin Res.* **2012**, *28*, 123–129. [CrossRef]
66. Giménez, I.; Blesa, J.; Herrera, M.; Ariño, A. Effects of bread making and wheat germ addition on the natural deoxynivalenol content in bread. *Toxins* **2014**, *6*, 394–401. [CrossRef]
67. Brashears, M.M.; Reilly, S.S.; Gilliland, S.E. Antagonistic action of cells of Lactobacillus lactis toward Escherichia coli O157:H7 on refrigerated raw chicken meat. *J. Food Prot.* **1998**, *61*, 166–170. [CrossRef]
68. Adetuniji, M.; Atanda, O.; Ezekiel, C.; Dipeolu, A.; Uzochukwu, S.; Oyedepo, J.; Chilaka, C.A. Technology, Distribution of mycotoxins and risk assessment of maize consumers in five agro-ecological zones of Nigeria. *Eur. Food Res. Technol.* **2014**, *239*, 287–296. [CrossRef]
69. Reddy, K.; Salleh, B. Co-occurrence of moulds and mycotoxins in corn grains used for animal feeds in Malaysia. *J. Animal Vet. Adv.* **2011**, *10*, 668–673. [CrossRef]
70. Ravimannan, N.; Sevvel, P.; Saarutharshan, S. Study on fungi associated with spoilage of bread. *Int. J. Adv. Res. Biol. Sci.* **2016**, *3*, 165–167.
71. Unachukwu, M.; Nwakanma, C. The fungi associated with the spoilage of bread in Enugu state. *Int. J. Curr. Microbiol. Appl. Sci.* **2018**, *4*, 989–995.
72. Legan, J.; Voysey, P.A. Yeast spoilage of bakery products and ingredients. *J. Appl. Bacteriol.* **1991**, *70*, 361–371. [CrossRef] [PubMed]

73. Garcia, M.; Copetti, M.J. Alternative methods for mould spoilage control in bread and bakery products. *Int. Food Res. J.* **2019**, *26*, 737–749.
74. Sofos, J.; Busta, F.F. Antimicrobial activity of sorbate. *J. Food Prot.* **1981**, *44*, 614–622. [CrossRef] [PubMed]
75. Fonseca, L.M.; Souza, E.J.D.; Radünz, M.; Gandra, E.A.; da Rosa Zavareze, E.; Dias, A.R.G. Suitability of starch/carvacrol nanofibers as biopreservatives for minimizing the fungal spoilage of bread. *Carbohydr. Polym.* **2021**, *252*, 117166. [CrossRef]
76. Chavan, P.S.; Tupe, S.G. Antifungal activity and mechanism of action of carvacrol and thymol against vineyard and wine spoilage yeasts. *Food Control.* **2014**, *46*, 115–120. [CrossRef]
77. Vargas, M.C.A.; Simsek, S.J.F. Clean Label in Bread. *Foods* **2021**, *10*, 2054. [CrossRef]
78. King, B.D. Microbial Inhibition in Bakery Products- A Review. *Baker's Dig.* **1981**, *55*, 8–12.
79. Rollán, G.; Gerez, C.; Dallagnol, A.; Torino, M.; Font, G. Update in bread fermentation by lactic acid bacteria. *Appl. Microbiol. Microb. Biotechnol.* **2010**, *2*, 1168–1171.
80. Gerez, C.L.; Torres, M.J.; De Valdez, G.F.; Rollán, G. Control of spoilage fungi by lactic acid bacteria. *Biolog. Control* **2013**, *64*, 231–237. [CrossRef]
81. Pontonio, E.; Nionelli, L.; Curiel, J.A.; Sadeghi, A.; Di Cagno, R.; Gobbetti, M.; Rizzello, C.G. Iranian wheat flours from rural and industrial mills: Exploitation of the chemical and technology features, and selection of autochthonous sourdough starters for making breads. *Food Microbiol.* **2015**, *47*, 99–110. [CrossRef]
82. Cizeikiene, D.; Gaide, I.; Basinskiene, L.J.F. Effect of Lactic Acid Fermentation on Quinoa Characteristics and Quality of Quinoa-Wheat Composite Bread. *Food Microbiol.* **2021**, *10*, 171. [CrossRef]
83. Catzeddu, P. Sourdough breads. In *Flour and Breads and Their Fortification in Health and Disease Prevention*; Elsevier: Amsterdam, The Netherlands, 2019; pp. 177–188.
84. De Angelis, M.; Cassone, A.; Rizzello, C.G.; Gagliardi, F.; Minervini, F.; Calasso, M.; Di Cagno, R.; Francavilla, R.; Gobbetti, M. Mechanism of degradation of immunogenic gluten epitopes from Triticum turgidum L. var. durum by sourdough lactobacilli and fungal proteases. *Appl. Environ. Microbiol.* **2010**, *76*, 508–518. [CrossRef]
85. Gerez, C.L.; Font de Valdez, G.; Rollan, G.C. Functionality of lactic acid bacteria peptidase activities in the hydrolysis of gliadin-like fragments. *Lett. Appl. Microbiol.* **2008**, *47*, 427–432. [CrossRef]
86. Khorshidian, N.; Yousefi Asli, M.; Hosseini, H.; Shadnoush, M.; Mortazavian, A.M. Potential anticarcinogenic effects of lactic acid bacteria and probiotics in detoxification of process-induced food toxicants. *Iran. J Cancer Prev.* **2016**, *9*, e7920. Available online: http://eprints.semums.ac.ir/908/ (accessed on 7 December 2021).
87. Pawlowska, A.M.; Zannini, E.; Coffey, A.; Arendt, E.K. "Green preservatives": Combating fungi in the food and feed industry by applying antifungal lactic acid bacteria. *Adv. Food Nutr. Res.* **2012**, *66*, 217–238.
88. Saranraj, P.; Sivasakthivelan, P. Microorganisms involved in spoilage of bread and its control measures. *Res. Gate Electron. Sci. J.* **2015**. Available online: https://www.researchgate.net/publication/282156531 (accessed on 7 December 2021).
89. Schnürer, J.; Magnusson, J. Technology, Antifungal lactic acid bacteria as biopreservatives. *Trends Food Sci. Technol.* **2005**, *16*, 70–78. [CrossRef]
90. Buck, B.; Azcarate-Peril, M.; Klaenhammer, T.R. Role of autoinducer-2 on the adhesion ability of Lactobacillus acidophilus. *J. Appl. Microbiol.* **2009**, *107*, 269–279. [CrossRef]
91. Corsetti, A.; Perpetuini, G.; Tofalo, R. Biopreservation effects in fermented foods. In *Advances in Fermented Foods and Beverages*; Elsevier: Amsterdam, The Netherlands, 2015; pp. 311–332.
92. Guimarães, A.; Santiago, A.; Teixeira, J.A.; Venâncio, A.; Abrunhosa, L. Anti-aflatoxigenic effect of organic acids produced by Lactobacillus plantarum. *Int. J. Food Microbiol.* **2018**, *264*, 31–38. [CrossRef] [PubMed]
93. Axel, C.; Brosnan, B.; Zannini, E.; Furey, A.; Coffey, A.; Arendt, E.K. Antifungal sourdough lactic acid bacteria as biopreservation tool in quinoa and rice bread. *Int. J. Food Microbiol.* **2016**, *239*, 86–94. [CrossRef]
94. Bartkiene, E.; Bartkevics, V.; Lele, V.; Pugajeva, I.; Zavistanaviciute, P.; Zadeike, D.; Juodeikiene, G. Technology, Application of antifungal lactobacilli in combination with coatings based on apple processing by-products as a bio-preservative in wheat bread production. *J. Food Sci. Technol.* **2019**, *56*, 2989–3000. [CrossRef]
95. Zhang, G.; Tu, J.; Sadiq, F.A.; Zhang, W.; Wang, W. Prevalence, genetic diversity, and technological functions of the Lactobacillus sanfranciscensis in sourdough: A review. *Compr. Rev. Food Sci. Food Saf.* **2019**, *18*, 1209–1226. [CrossRef] [PubMed]
96. Minervini, F.; Di Cagno, R.; Lattanzi, A.; De Angelis, M.; Antonielli, L.; Cardinali, G.; Cappelle, S.; Gobbetti, M. Lactic acid bacterium and yeast microbiotas of 19 sourdoughs used for traditional/typical Italian breads: Interactions between ingredients and microbial species diversity. *Appl. Environ. Microbiol.* **2012**, *78*, 1251–1264. [CrossRef] [PubMed]
97. Ajith, J.; Sunita, M. A study on bread mould spoilage by using lactic acid bacteria and yeast with antifungal properties. *J. Nutr. Ecol. Food Res.* **2017**, *4*, 128–130. [CrossRef]
98. Sadeghi, A.; Ebrahimi, M.; Mortazavi, S.A.; Abedfar, A. Application of the selected antifungal LAB isolate as a protective starter culture in pan whole-wheat sourdough bread. *Food Control.* **2019**, *95*, 298–307. [CrossRef]
99. Axel, C.; Röcker, B.; Brosnan, B.; Zannini, E.; Furey, A.; Coffey, A.; Arendt, E.K. Application of Lactobacillus amylovorus DSM19280 in gluten-free sourdough bread to improve the microbial shelf life. *Food Microbiol.* **2015**, *47*, 36–44. [CrossRef]
100. Cizeikiene, D.; Juodeikiene, G.; Paskevicius, A.; Bartkiene, E. Antimicrobial activity of lactic acid bacteria against pathogenic and spoilage microorganism isolated from food and their control in wheat bread. *Food Control* **2013**, *31*, 539–545. [CrossRef]

101. Rizzello, C.G.; Cassone, A.; Coda, R.; Gobbetti, M. Antifungal activity of sourdough fermented wheat germ used as an ingredient for bread making. *Food Chem.* **2011**, *127*, 952–959. [CrossRef]
102. Zhang, C.; Brandt, M.J.; Schwab, C.; Gänzle, M.G. Propionic acid production by cofermentation of Lactobacillus buchneri and Lactobacillus diolivorans in sourdough. *Food Microbiol.* **2010**, *27*, 390–395. [CrossRef]
103. Ribes, S.; Fuentes, A.; Talens, P.; Barat, J.M. Nutrition, Prevention of fungal spoilage in food products using natural compounds: A review. *Crit. Rev. Food Sci. Nutr.* **2018**, *58*, 2002–2016. [CrossRef]
104. Coda, R.; Cassone, A.; Rizzello, C.G.; Nionelli, L.; Cardinali, G.; Gobbetti, M. Antifungal activity of Wickerhamomyces anomalus and Lactobacillus plantarum during sourdough fermentation: Identification of novel compounds and long-term effect during storage of wheat bread. *Appl. Environ. Microbiol.* **2011**, *77*, 3484–3492. [CrossRef]
105. Mo, E.K.; Sung, C.K. Biotechnology, Production of white pan bread leavened by Pichia anomala SKM-T. *Food Sci. Biotechnol.* **2014**, *23*, 431–437. [CrossRef]
106. Gardner, N.; Champagne, C.; Gélinas, P. Effect of yeast extracts containing propionic acid on bread dough fermentation and bread properties. *J. Food Sci.* **2002**, *67*, 1855–1858. [CrossRef]
107. Gerez, C.L.; Fornaguera, M.J.; Obregozo, M.D.; de Valdez, G.F.; Torino, M.I. Antifungal starter culture for packed bread: Influence of two storage conditions. *Revista Argentina de Microbiologia* **2015**, *47*, 118–124. [CrossRef]
108. Black, B.A.; Zannini, E.; Curtis, J.M.; Gänzle, M.G. Antifungal hydroxy fatty acids produced during sourdough fermentation: Microbial and enzymatic pathways, and antifungal activity in bread. *Appl. Environ. Microbiol.* **2013**, *79*, 1866–1873. [CrossRef]
109. Schieber, A.; Saldaña, M.D. Potato peels: A source of nutritionally and pharmacologically interesting compounds—A review. *Foods* **2009**, *3*, 23–29.
110. Maldonado, A.F.S.; Schieber, A.; Gänzle, M.G. Technology, Plant defence mechanisms and enzymatic transformation products and their potential applications in food preservation: Advantages and limitations. *Trends Food Sci. Technol.* **2015**, *46*, 49–59. [CrossRef]
111. Wei, X.; Zhao, L.; Zhong, J.; Gu, H.; Feng, D.; Johnstone, B.; March, K.; Farlow, M.; Du, Y. Adipose stromal cells-secreted neuroprotective media against neuronal apoptosis. *Neurosci. Lett.* **2009**, *462*, 76–79. [CrossRef]
112. Sahan, Y.A. Effect of Prunus laurocerasus L. (cherry laurel) leaf extracts on growth of bread spoilage fungi. *Bulg. J. Agric. Sci.* **2011**, *17*, 83–92.
113. Rizzello, C.G.; Lavecchia, A.; Gramaglia, V.; Gobbetti, M.J.A. Long-term fungal inhibition by Pisum sativum flour hydrolysate during storage of wheat flour bread. *Appl. Environ. Microbiol.* **2015**, *81*, 4195–4206. [CrossRef]
114. Burt, S. Essential oils: Their antibacterial properties and potential applications in foods—A review. *Int. J. Food Microbiol.* **2004**, *94*, 223–253. [CrossRef]
115. Jayasinghe, J.; Ranasinghe, M.; Tharshini, G. Botanicals as biopreservatives in foodsa mini review. *J. Agric. Eng. Food Technol.* **2016**, *3*, 4–7.
116. Bakkali, F.; Averbeck, S.; Averbeck, D.; Idaomar, M. Biological effects of essential oils–A review. *Food Chem. Toxicol.* **2008**, *46*, 446–475. [CrossRef] [PubMed]
117. Debonne, E.; Van Bockstaele, F.; Samapundo, S.; Eeckhout, M.; Devlieghere, F. The use of essential oils as natural antifungal preservatives in bread products. *J. Essent. Oil Res.* **2018**, *30*, 309–318. [CrossRef]
118. Suhr, K.I.; Nielsen, P.V. Antifungal activity of essential oils evaluated by two different application techniques against rye bread spoilage fungi. *J. Appl. Microbiol.* **2003**, *94*, 665–674. [CrossRef]
119. Horváth, G.; Jenei, J.T.; Vágvölgyi, C.; Böszörményi, A.; Krisch, J. Effects of essential oil combinations on pathogenic yeasts and moulds. *Acta Biol. Hung.* **2016**, *67*, 205–214. [CrossRef]
120. Nielsen, P.V.; Rios, R. Inhibition of fungal growth on bread by volatile components from spices and herbs, and the possible application in active packaging, with special emphasis on mustard essential oil. *Int. J. Food Microbiol.* **2000**, *60*, 219–229. [CrossRef]
121. Guynot, M.; Ramos, A.; Seto, L.; Purroy, P.; Sanchis, V.; Marin, S. Antifungal activity of volatile compounds generated by essential oils against fungi commonly causing deterioration of bakery products. *J. Appl. Microbiol.* **2003**, *94*, 893–899. [CrossRef] [PubMed]
122. Debonne, E.; Van Bockstaele, F.; De Leyn, I.; Devlieghere, F.; Eeckhout, M. Validation of in-vitro antifungal activity of thyme essential oil on Aspergillus niger and Penicillium paneum through application in par-baked wheat and sourdough bread. *LWT* **2018**, *87*, 368–378. [CrossRef]
123. Teodoro, R.A.R.; de Barros Fernandes, R.V.; Botrel, D.A.; Borges, S.V.; de Souza, A.U. Characterization of microencapsulated rosemary essential oil and its antimicrobial effect on fresh dough. *Food Bioprocess Technol.* **2014**, *7*, 2560–2569. [CrossRef]
124. Otoni, C.G.; Pontes, S.F.; Medeiros, E.A.; Soares, N.D.F. Edible films from methylcellulose and nanoemulsions of clove bud (Syzygium aromaticum) and oregano (Origanum vulgare) essential oils as shelf life extenders for sliced bread. *J. Agric. Food Chem.* **2014**, *62*, 5214–5219. [CrossRef]
125. Mani López, E.; Valle Vargas, G.P.; Palou, E.; López Malo, A. Penicillium expansum inhibition on bread by lemongrass essential oil in vapor phase. *J. Food Prot.* **2018**, *81*, 467–471. [CrossRef]
126. Estévez, M.; Ventanas, S.; Cava, R. Protein oxidation in frankfurters with increasing levels of added rosemary essential oil: Effect on color and texture deterioration. *J. Food Sci.* **2005**, *70*, c427–c432. [CrossRef]
127. Krisch, J.; Rentskenhand, T.; Horváth, G.; Vágvölgyi, C. Activity of essential oils in vapor phase against bread spoilage fungi. *Acta Biol. Szeged.* **2013**, *57*, 9–12.
128. Salim-ur-Rehman, S.H.; Nawaz, H.; Ahmad, M.M.; Murtaza, M.A.; Rizvi, A.J. Inhibitory effect of citrus peel essential oils on the microbial growth of bread. *Pak. J. Nutr.* **2007**, *6*, 558–561. [CrossRef]

129. Ryan, M.P.; Walsh, G. The biotechnological potential of whey. Rev. Environ. *Sci. Bio/Technol.* **2016**, *15*, 479–498. [CrossRef]

130. Gamba, R.R.; Ni Colo, C.; Correa, M.; Astoreca, A.L.; Alconada Magliano, T.M.; De Antoni, G.; León Peláez, Á. *Antifungal Activity Against Aspergillus Parasiticus of Supernatants from Whey Permeates Fermented with Kéfir Grains*; Scientific Research: Atlanta, GA, USA, 2015. [CrossRef]

131. Hati, S.; Patel, N.; Sakure, A.; Mandal, S. Therapeutics, Influence of whey protein concentrate on the production of antibacterial peptides derived from fermented milk by lactic acid bacteria. *Int. J. Pept. Res. Ther.* **2018**, *24*, 87–98. [CrossRef]

132. Luz, C.; Quiles, J.M.; Romano, R.; Blaiotta, G.; Rodriguez, L.; Meca, G. Technology, Application of whey of Mozzarella di Bufala Campana fermented by lactic acid bacteria as a bread biopreservative agent. *Int. J. Food Sci. Technol.* **2021**, *56*, 4585–4593. [CrossRef]

133. Luz, C.; Izzo, L.; Ritieni, A.; Mañes, J.; Meca, G. Antifungal and antimycotoxigenic activity of hydrolyzed goat whey on Penicillium spp: An application as biopreservation agent in pita bread. *LWT* **2020**, *118*, 108717. [CrossRef]

134. Deseta, M.L.; Sponton, O.E.; Erben, M.; Osella, C.A.; Frisón, L.N.; Fenoglio, C.; Piagentini, A.M.; Santiago, L.G.; Perez, A.A. Nanocomplexes based on egg white protein nanoparticles and bioactive compounds as antifungal edible coatings to extend bread shelf life. *Food Res. Int.* **2021**, *148*, 110597. [CrossRef]

135. Kumar, S.; Mukherjee, A.; Dutta, J. Chitosan based nanocomposite films and coatings: Emerging antimicrobial food packaging alternatives. *Trends Food Sci. Technol.* **2020**, *97*, 196–209. [CrossRef]

136. Amor, G.; Sabbah, M.; Caputo, L.; Idbella, M.; De Feo, V.; Porta, R.; Fechtali, T.; Mauriello, G. Basil essential oil: Composition, antimicrobial properties, and microencapsulation to produce active chitosan films for food packaging. *Foods* **2021**, *10*, 121. [CrossRef] [PubMed]

137. Peter, A.; Mihaly-Cozmuta, L.; Mihaly-Cozmuta, A.; Nicula, C.; Ziemkowska, W.; Basiak, D.; Danciu, V.; Vulpoi, A.; Baia, L.; Falup, A.; et al. Changes in the microbiological and chemical characteristics of white bread during storage in paper packages modified with Ag/TiO2-SiO2, Ag/N-TiO2 or Au/TiO2. *Food Chem.* **2016**, *197 Pt A*, 790–798. [CrossRef]

138. Galić, K.; Ćurić, D.; Gabrić, D. Shelf life of packaged bakery goods—A review. *Crit. Rev. Food Sci. Nutr.* **2009**, *49*, 405–426. [CrossRef]

139. Pasqualone, A. Bread packaging: Features and functions. In *Flour and Breads and Their Fortification in Health and Disease Prevention*; Elsevier: Amsterdam, The Netherlands, 2019; pp. 211–222.

140. Gutiérrez, L.; Batlle, R.; Andújar, S.; Sánchez, C.; Nerín, C. Evaluation of antimicrobial active packaging to increase shelf life of gluten-free sliced bread. *Packag. Technol. Sci.* **2011**, *24*, 485–494. [CrossRef]

141. Young, L.; Reviere, R.; Cole, A.B. Fresh red meats: A place to apply modified atmospheres. *Food Technol.* **1988**, *42*, 65–69.

142. Rodriguez, M.; Medina, L.; Jordano, R. Effect of modified atmosphere packaging on the shelf life of sliced wheat flour bread. *Food/Nahrung* **2000**, *44*, 247–252. [CrossRef]

143. Radji, M.; Agustama, R.A.; Elya, B.; Tjampakasari, C.R. Antimicrobial activity of green tea extract against isolates of methicillin–resistant Staphylococcus aureus and multi–drug resistant Pseudomonas aeruginosa. *Asian Pac. J. Trop.* **2013**, *3*, 663–667. [CrossRef]

144. Rana, I.S.; Rana, A.S.; Rajak, R.C. Evaluation of antifungal activity in essential oil of the Syzygium aromaticum (L.) by extraction, purification and analysis of its main component eugenol. *Braz. J. Microbiol.* **2011**, *42*, 1269–1277. [CrossRef]

145. Alfei, S.; Marengo, B.; Zuccari, G. Nanotechnology application in food packaging: A plethora of opportunities versus pending risks assessment and public concerns. *Food Res. Int.* **2020**, *137*, 109664. [CrossRef]

146. Amenta, V.; Aschberger, K.; Arena, M.; Bouwmeester, H.; Moniz, F.B.; Brandhoff, P.; Gottardo, S.; Marvin, H.J.; Mech, A.; Pesudo, L.Q. Pharmacology, Regulatory aspects of nanotechnology in the agri/feed/food sector in EU and non-EU countries. *Regul. Toxicol. Pharmacol.* **2015**, *73*, 463–476. [CrossRef] [PubMed]

147. Aschberger, K.; Rauscher, H.; Crutzen, H.; Rasmussen, K.; Christensen, F.M.; Sokull-Klüttgen, B.; Stamm, H. Considerations on information needs for nanomaterials in consumer products. In *Discussion of a Labelling and Reporting Scheme for Nanomaterials in Consumer Products in the EU*; Publications Office of the European Union: Luxembourg, 2014.

148. Ciociola, T.; Giovati, L.; Conti, S.; Magliani, W.; Santinoli, C.; Polonelli, L. Natural and synthetic peptides with antifungal activity. *Future Med. Chem.* **2016**, *8*, 1413–1433. [CrossRef] [PubMed]

149. Rima, H.; Steve, L.; Ismail, F. Antimicrobial and probiotic properties of yeasts: From fundamental to novel applications. *Front. Microbiol.* **2012**, *3*, 421.

150. Muccilli, S.; Restuccia, C.J.M. Bioprotective role of yeasts. *Microorganisms* **2015**, *3*, 588–611. [CrossRef]

Review

Pigmented Potatoes: A Potential Panacea for Food and Nutrition Security and Health?

Callistus Bvenura , Hildegard Witbooi and Learnmore Kambizi *

Department of Horticulture, Faculty of Applied Sciences, Cape Peninsula University of Technology, Bellville 7535, South Africa; Callistusb@gmail.com (C.B.); hildegardwitbooi@gmail.com (H.W.)
* Correspondence: kambizil@cput.ac.za

Abstract: Although there are over 4000 potato cultivars in the world, only a few have been commercialized due to their marketability and shelf-life. Most noncommercialized cultivars are pigmented and found in remote regions of the world. White-fleshed potatoes are well known for their energy-enhancing complex carbohydrates; however, pigmented cultivars are potentially high in health-promoting polyphenolic compounds. Therefore, we reveal the comprehensive compositions of pigmented cultivars and associated potential health benefits, including their potential role in ameliorating hunger, food, and nutrition insecurity, and their prospects. The underutilization of such resources is a direct threat to plant-biodiversity and local traditions and cultures.

Keywords: anthocyanins; carotenoids; food and nutrition security; pigmented potatoes; polyphenols; sustainability

Citation: Bvenura, C.; Witbooi, H.; Kambizi, L. Pigmented Potatoes: A Potential Panacea for Food and Nutrition Security and Health? *Foods* **2022**, *11*, 175. https://doi.org/10.3390/foods11020175

Academic Editors: António Raposo, Renata Puppin Zandonadi and Raquel Braz Assunção Botelho

Received: 8 November 2021
Accepted: 27 December 2021
Published: 10 January 2022

Publisher's Note: MDPI stays neutral with regard to jurisdictional claims in published maps and institutional affiliations.

1. Introduction

The COP26 Glasgow meeting could not have come at a better time, when extreme weather events such as floods, uncontrollable fires resulting from heat waves, heavy downpours, droughts, hurricanes, and heavy snow have become the new norm the world over. In fact, the warmest temperatures, the highest sea level rise as well as warming, and the consequent sea acidification over the past seven years have been the highest on record [1]. Perhaps one of the major concerns about global warming is its negative impact on livelihoods and, consequently, food and nutrition security. The COVID-19 pandemic also came with devastating effects on the state of food security in the world due to strict lockdown regulations. Reports indicate that world hunger increased in 2020 due to the pandemic. More specifically, in the year 2020, undernourishment increased by 1.5%, with 218 and 418 million of these cases residing in Africa and Asia, respectively [2]. More than a third of the world's population did not have access to adequate food in 2020 and 12% were severely food-insecure. Reports further indicate that due to financial and other factors, healthy diets were out of reach for 3 billion people in the world in 2020 [2]. Unhealthy diets cost governments billions of USD in health care annually. Therefore, climate heating, extreme weather events leading to the loss of livelihoods and droughts, and increasing global populations, among other factors, are making it nearly impossible to achieve the UN SDGs to feed the world by 2030.

Malnutrition resulting from unhealthy diets, food insecurity, and other factors have also been on the rise, with statistics indicating that in the year 2020 alone, overweight, wasting, and wasting affected 5.7%, 6.7%, and 22% of children under 5 years, respectively [2]. Most of these cases emanate from Africa and Asia. Malnutrition resulting from diets that do not supply a healthy amount of nutrients has become a major problem in the world today. These diets are normally ladened with highly processed foods that are high in free sugars, saturated fats, and salt and are very low in the recommended daily intakes of fruits and vegetables. Unhealthy diets are the major cause of noncommunicable diseases (NCDs). NCDs, which were associated for a long time with more developed and industrialized countries of the world, have become epidemics the world over, including in those of the poor. Over 70% of global mortalities have been linked to NCDs, and over 77% of these figures have been reported in low- to middle-income countries [3]. Diabetes (1.5 million), respiratory diseases (4.1 million), cancer (9.3 million), and cardiovascular diseases (17.9 million) dominate the global NCD deaths [3]. Although some metabolic risk factors (such as hypertension, obesity, and glycemia) and lifestyle behavior (including smoking, lack of physical activity, and stress) do contribute to NCD, it must be noted that unhealthy unbalanced diets remain the major cause. The WHO and many other organizations are teaming up to encourage the consumption of healthy diets and making use of some indigenous underutilized foods. Most of these foods have been proven to be healthy and have the potential to ameliorate the current unhealthy societies that have been bred over the past few decades. Staple food crops (potatoes, maize/corn, wheat, rice, cassava, sweet potatoes, sorghum, and yams) are an important aspect of the diet because they supply the much-needed carbohydrates for energy. However, these need to be nutritionally balanced and in the right quantities to breed healthy nations. To fully appreciate and increase the pace toward a food- and nutrition-secure world, these crops need to be understood and studied at length. Therefore, in this review, we pay attention to the potato, with special focus on the little-known and underutilized pigmented cultivars.

In general, the potato (*Solanum tuberosum* L.) is one of the most important food crops that feed the world, after rice and wheat, with over a billion people being sustained by the crop and production now exceeding 350 MMT annually harvested on 17 million hectares of land, as shown in Figure 1 [4]. In the past ten years, statistics indicate that production has increased by 11%. However, this is a decline in comparison with a 21% increase that was reported by [5] between 1991 and 2007. Although there has been an increase in potato production in the world, a 7% decline in total harvested area between 2002 and 2019 has been reported, as shown in Figure 2 [6]. Up-to-date statistics also indicate that the world per capita consumption of potatoes was 32.3 kg in 2018, and this was a decline by 3.76% from 2017 and 2.22% decline from 2008 [7]. Belarus (182 kg), Ukraine, Rwanda, Latvia, Kazakhstan, Russia, Poland, Romania, Kyrgyzstan, and Peru are the top ten highest-potato-consuming countries of the world per capita [7].

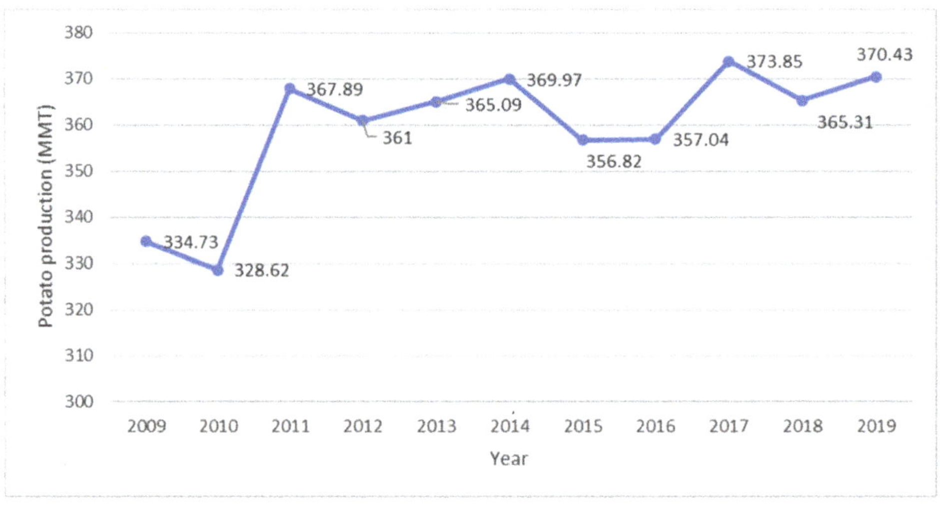

Figure 1. World potato production from 2009 to 2019 [4].

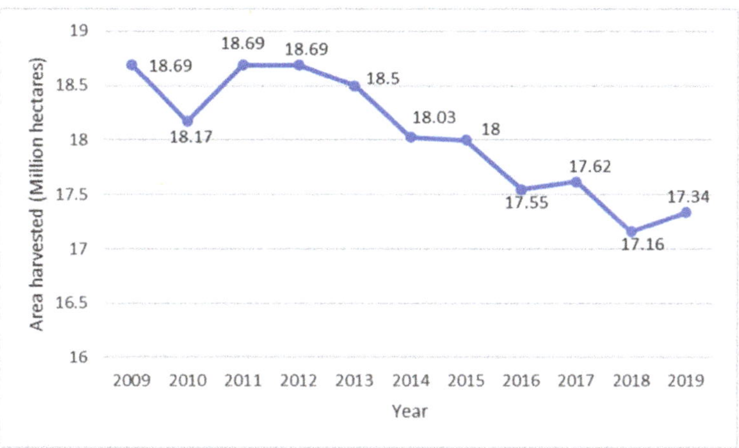

Figure 2. Total area under potato production in the world from 2009 to 2019 [6].

In addition to being the largest vegetative-propagated crop worldwide, the potato tuber has become an important staple in parts of the world where there is a limited but increasing purchasing power, an increasing pressure on scarce land, and an increasing demand for food [8]. This is because the potato is a short season crop and is often replanted as a seed potato, in addition to requiring minimum inputs [4]. First used as food and domesticated from the wild in the Andes region of South America over 8000 years ago, the potato has adapted very well to all regions of the world, except in Antarctica where the soil and conditions are untenable for crop production [5]. The crop was first introduced to Europe in the 1570s after which it spread to the rest of the world in the late 17th century [5]. As shown in Figure 3, China, India, and Russia are the current top three potato producing countries in the world [9].

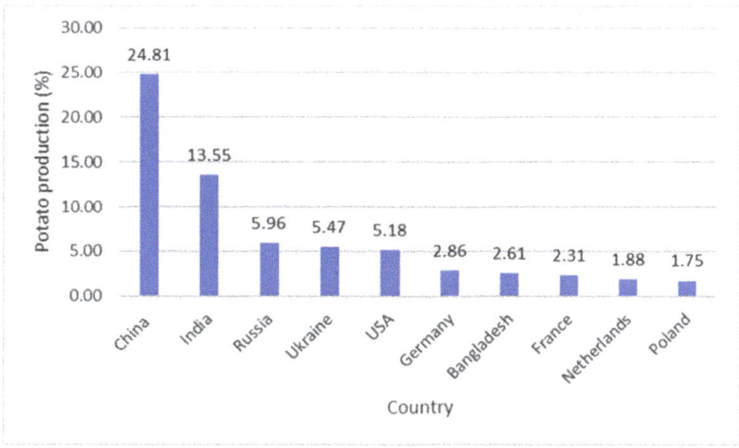

Figure 3. World's leading potato producers in 2019 [9].

Although over 4000 potato cultivars are known today, only a few have been successfully commercialized because of their shelf-life stability and marketability, and none of these are pigmented [10]. For producers and sellers of potatoes, the ability to be stored for long periods of time, a regular-sized tuber, multipurpose use, high production ratio, and customer acceptance are some of the critical characteristics to consider to turn a profit and avoid wasting of the product [11]. Common potato cultivars are white-fleshed, while the pigmented cultivars are rare. According to the same authors, potato farmers in the world are reluctant to cultivate these pigmented potatoes due to their low yielding capacity, as well as their susceptibility to diseases during seed multiplication. Therefore, in this review, we explore pigmented potato cultivars and their possible role in food and nutrition security. We bring to the fore the phytochemical characteristics that have been reported, their biological activities, and how these can be harnessed into breeding healthier nations of the world.

2. Pigmented Potato Cultivars

The literature does not give a specific total number of pigmented cultivars, although a general number exceeding 4000 for all potatoes, including the white-fleshed cultivars, has been reported in the world [12], and, according to [13], nearly 1000 Andean species, many of which are pigmented. An exact number of pigmented species will most likely remain elusive because most of these species are cultivated in remote regions of the world for subsistence purposes and will remain hidden and/or be eventually driven to extinction if nothing is done to preserve them. Variations of red, purple, blue, and yellow skin and/or flesh have been reported in the literature as the dominant colors across the world (Table 1). Pictorial examples of some pigmented cultivars from South Africa are shown in Figure 4. These colors are mainly due to the presence of anthocyanins and other pigments. Although it is assumed that at least 50% of the potato tuber is consumed fresh, this number is conceivably higher for pigmented potatoes [5]. This is because the white-fleshed cultivars can also be developed into products and ingredients, animal feed, and industrial applications such as starch. The white-fleshed cultivars are produced in millions of tons annually and, therefore, multiple uses can be generated. Pigmented cultivars are produced on a small scale and, therefore, food applications dominate their uses. In fact, the internet is awash with gourmet applications of the pigmented cultivars in high-end restaurants. Food lovers are perhaps attracted by their unique and beautiful colors and, least of all, by their nutritional and health benefits. Ref. [14] worked on extracts of red and purple cultivars as potential food colorants. Such studies would bring to the fore the potential of these cultivars and help to improve on their size, shelf life, and their marketability. Several genetic studies have been undertaken on these species, including those of [15,16]. More research needs to be channeled toward the development of better cultivars.

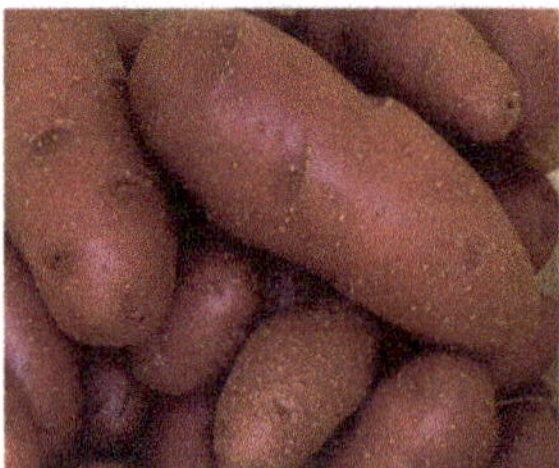

Figure 4. Some pigmented cultivars from South Africa.

Table 1. Distribution of pigmented potato cultivars in various regions of the world.

	Cultivar and Colour	Country	Ref.
1	Shepody, Desirée, 457-CON-1157, Corazón de buey, 302-UA-1634A, 304-UA-1135, (Boyo de chancho, Michuñeroja, Meca de gato, Cacho azu, Corazón azul, 239-UA-1388, Chona negra, and Bruja	Chiloe Island and Valdivia, Chile	[17]
2	NA	Louvain-La-Neuve, Belgium	[13]
3	NA	Louvain-La-Neuve, Belgium	[18]
4	704,429-Guincho Negra, 700,347-SS-2613, 702,535-Sipancachi, 701,997-Sullu, and 703,905-Huata Colorada	Louvain-La-Neuve, Belgium	[19]
5	Waicha—Reddish Moradita—Purple	Argentina	[20]
6	Spunta, CN1—Pink	Sfax, Tunisia	[21]
7	No data	Global	[22]
8	Purple (PA97B29-2, PA97B29-4, PA97B29-5, and PA97B29-6) Red (PA97B29-3, PA97B35-1, PA97B35-2, PA97B36-3, PA97B37-2, PA97B37-3, PA97B37-7, PA97B39-2, and NDOP5847-1)	Washington, DC, USA	[23]
9	Light Yellow (Adora, Divina, Fabula, Ilona, Morning gold, Provento, Satino, Yukon Gold, and POR00PG4-2 Dark yellow (91E22, PA99P11-2, PA99P1-2, PA99P2-1, and POR00PG4-1) Red and yellow (PO00PG9-1, PO00PG9-2, PO00PG9-3, PO00PG9-5, and PO00PG9-6)	Washington, DC, USA	[24]
10	38 Cultivars	Washington, DC, USA	[25]
11	Purple (PA97B29-2, PA97B29-4, PA97B29-5, and PA97B29-6) Red (PA97B29-3, PA97B35-1, PA97B35-2, PA97B36-3, PA97B37-2, PA97B37-3, PA97B37-7, PA97B39-2, and NDOP5847-1)	Washington, DC, USA	[26]
12	NA	Colorado, CO, USA	[27]
13	Hongyoung, Jayoung, and Atlantic	Dae-GwalLyeong, Korea	[15]
14	Hermanns Blaue, Highland Burgundy Red, Shetland Black, and Vitelotte	Barum, Germany 0.6–46 mg	[28]
15	NA	Sevilla, Spain	[29]
16	Congo—Purple	Bergen, Norway	[30]
17	Yellow (Innovator, Bintje, Challenger, Yukon, AR2009-10) Purple (AR2, ADB) White flesh, red skin (Norland) Blue (Adirondack Blue) Red (Adirondack Red, ADR),	New Brunswick, Canada	[31]
18	Purple (705,534, 703,640, 706,726, 704,733, 703,862, and 704,133) Red (702,556, 706,630, 700,234, 705,841, 703,752, 703,695, 705,820, 704,537, 705,946, 705,500, 702,464, and 703,782) Yellow (706,884 and 704,481)	Lima, Peru	[32]
19	Purple (Blue Congo, Blaue Veltlin, Blaue Schweden, Synkeä Sakari).	Finland	[33]
20	Purple (Blaue Elise, Blaue St. Galler, Blue Congo, Valfi, Violette, and Vitelotte) Red (Herbie 26, Highland B. Red, Rosalinde, and Rote Emma)	Prague, Czech Republic	[34]
21	Light purple—KM Medium-dark purple—H92	Obihiro, Hokkaido, Japan	[35]

Table 1. *Cont.*

	Cultivar and Colour	Country	Ref.
22	Yellow (Agria, Russet Burbank, Lady Balfour, and Mayan Gold)	Czech Republic	[36]
	Purple (Violette, Vitelotte, Violetta, Valfi, Blue Congo, Blaue St. Galler, Olivia, and Blaue Anneliese)		
	Red (Rosemarie, Rote Emmalie, Highland Burgundy Red, and Herbie 26)		
	Yellow with red spots (Mayan Queen)		
23	Salad Blue, Shetland Black, Blue Congo, Blaue St. Galler, Highland Burgundy Red, Violette, and Valfi, Vitelotte	Přerov nad Labem, Suchdol, Valečov and Stachy, Czech Republic	[37]
24	Hermanns Blaue, Vitelotte, Shetland Black, and Valfi	Braunschweig, Germany	[38]
25	Blue violet (Blaue Schweden, British Columbia Blue, Violettfleischige, 1.81.203–92N, Blaue Mauritius, Bleu, Blaue Utwill, Peru Purple, Mesabi Purple Smith's Purple, UAC NEG 61, UAC CON 917, Weinberger Blaue, Mesabi Purple, Bells Purple, Magdeburger Blaue, Shetland Black, Caribe, Purple and White, Mrs. Moerles Purple Baker, Long Blue, Edzell Blue, Odenwa¨lder Blaue, Schwarze Ungarin, Arran Victory, UAC 1258, Purple Fiesta, Viola, and Blaue Zimmerli). Red (Sangre, Kefermarkter Zuchtstamm, and Red Cardinal).	Lenzen, Germany 0.01–1.57 g/kg	[39]
26	Review	Review	[40]
27	Purple—Hongyoung, Red—Jayoung	Korea	[41]
28	WP (Ranger Russet), YP (PORO3PG6–3), and PP (PORO4PG82–1)	Washington, USA	[42]
29	White (Russet Burbank), yellow (PORO3PG6-3), and purple-flesh (PORO4PG82-1)	Toppenish, Washington, USA	[43]
30	Red—Hongyoung	Korea	[44]
	Purple—Jayoung, clones Jje08-11,		
	DJ12X-5, and Jje08-43		
31	Purple (Salad Blue, Vitelotte, Valfi, Blue Congo)	Prerov nad Labem, Czech Republic	[45]
	Red (Rosalinde, Herbie 26, Highland Burgundy Red)		
32	Purple (Blaue Elise, Blaue St. Galler, Blue Congo, Valfi, and Vitelotte)	Přerov nad Labem, Czech Republic	[46]
	Red (Highland Burgundy Red, Herbie 26, Rosalinde, and Rote Emma)		
33	Yellow (Agria, Russet Burbank, Valy, Salome, Bohemia, Axa, Jelly, Ditta, Bionta, Ker˘kovsky' rohlíc˘ek, Dali, and Mayan Gold)	Valečov, Czech Republic	[47]
	Red (Rosara, Rosemarie, Königspurpur, Highland Burgundy Red, Herbie 26, and Red Emmalie)		
	Purple (Valfi, Violeta, Blaue Anneliese, and Vitelotte)		
34	CO97226-2R/R, CO99364-3R/R, CO97215-2P/P, CO97216-3P/P, CO97227-2P/P, CO97222-1R/R, Purple Majesty, Mountain Rose, and All Blue), and yellow (Yukon Gold)	Colorado, USA	[48]
35	Blue Congo, Highland Burgundy Red, Salad Blue, Shetland Black, Valf, Vitelotte, Violette, Blaue St. Galler, Blaue Hindel Bank, Blaue Ludiano, Blaue Mauritius, Blaue Schweden, British Columbia Blue, Farbe Kartoffel, Hafija, and Salad Red	Czech Republic	[49]
36	HB Red, Rote Emma, Blaue St Galler, Valfi, Violette and Agria	Czech Republic	[50]
37	Vitelotte Noire and Highland Burgundy Red	Milan, Italy	[51]
38	Yellow (Agrie Dzeltenie, Prelma, Lenora, Brasla, Anuschka, Gundega, S04009-37)	Priekuli, Latvia	[52]
	Light yellow (S99108-8)		
	Purple (Fenton, Purple Fiesta, British Columbia		
	Blue, Purple Peru, and Blue Congo)		
39	Red (Red Rodeo) Yellow (Yukon Gold), and Purple Majesty (CO94165-3P/P)	Washington, USA	[53]

Table 1. *Cont.*

	Cultivar and Colour	Country	Ref.
40	Red—Herbie 26	Poland	[54]
	Purple—Valfi, Blue Congo and Salad Blue		
41	Purple (Violettfleischige x Blue Marker-B, (Violettfleischige)-A1, (Blue Marker)-B, Purple, Violettfleischige x Blue Marker-D, (Violettfleischige)-A2, Vitelotte, and Blaue Ajanhuiri)	Groß Lüsewitz, Germany	[55]
	Yellow (Bangladesh, Desiree, Early Rose, and Shetland Blau I,)		
	Red (Ko¨nigspurpur, Rote Emmalie, and Rosemarie)		
42	Red (CO99256-2R, CO98012-5R, Colorado Rose, VC0967-2R/Y, CO97222-1R/R, and CO97226-2R/R)	Colorado, USA	[56]
	Purple (CO01399-10P/Y, AC99329-7PW/Y, and Purple Majesty)		
43	Purple (All Blue and CO94165-3P/P)	Texas, USA	[14]
	Red (NDC4069-4 and CO94183-1R/R)		
44	All Blue, NDC4069-4, Russian Blue, Purple Peruvian, COl11F2-1, COl12F1-1, COl12F1-2, CO141F2-1, CO142F2-1, RC2003-2	Colorado and Texas, USA	[57]
45	Purple (All Blue, Purple Peruvian, and RC2003-2)	Texas and Colorado, USA	[58]
	Red (NDC4069-4, CO94183-1R/R, and CO94183-1R/)		
46	NA	Oregon, USA	[59]
47	NA	Oregon, USA	[60]
48	NA	Oregon, USA	[61]
49	Blue—Valfi, Blaue Elise, Bore Volley, and Blue Congo	Přerov nad Labem, Czech Republic	[62]
50	Red—Rosemarie, Herbie 26, and Rote Emma	Wrocław, Poland	[63]
51	Red (Rosemary, Red Emmalie, and Red Cardinal)	Velestino, Greece	[64]
	Purple (Purple, Violetta, and Kefermarkter Blaue)		
52	Purple (Moradita)	Argentina	[65]
	Yellow (Waicha)		
	Red (Santa Marıa)		
53	Rio Grande, Mountain Rose, R Burbank, R Nugget, Yukon Gold, and Purple Majesty	Colorado, USA	[66]
54	Yellow (Satina and Tajfun, and Jelly)	Central Poland	[67]
55	Purple Majesty, Yukon Gold, Mountain Rose	Colorado, USA	[68]
56	Zheshu 13, 33, 75, 81, 132, 259, 6025	An'ji, Zhejiang Province, China	[69]
57	Red (Red Emmalie, Tornado, and Laura)	Širvintos district, Lithuania	[70]
	Dark purple (Violetta)		
	Dark-blue purple (Salad Blue)		
58	50 cultivars	Argentina	[71]
59	Salad Blue, Pink Fir Apple, and Highland Burgundy Red	Cape Town, South Africa	[11]
60	Salad Blue, Pink Fir Apple, and Highland Burgundy Red	Cape Town, South Africa	[72]
61	Agata, Cherie, Kennebec, Monalisa, Red Pontiac and Spirit	Barcelona, Spain	[10]
62	Purple Cloud No. 1, Red Cloud No. 1, Yunnan Potato 303, Yunnan Potato 603, S03-2677, S03-2685, S03-2796, S05-603, S06-277, and S06-1693	Chengdu, China 60–294 mg	[73]

NA: Not Available.

3. Antioxidant Activity in Pigmented Potatoes

Over the last decade, there has been a growing interest and preference of consumers for food containing natural antioxidants for health and nutritional-related reasons. In any case, food regulatory institutions demand that food products should contain labels that specify the ingredients but should not include any health claims. However, whether the consumers understand what antioxidants are remains an open question. In human physiology, normal cellar metabolism produces reactive oxygen species (ROS), which play a positive role in physiological processes but only in low concentrations [74]. In high concentrations, these ROS become toxic and cause damage to proteins, DNA, and lipids [75]. Therefore, antioxidants stop the damage caused by these ROS. Antioxidants are therefore a good indicator of the health benefits that can be derived from the foods. A simple schematic presentation of antioxidants is given in Figure 5, showing an overview of these compounds.

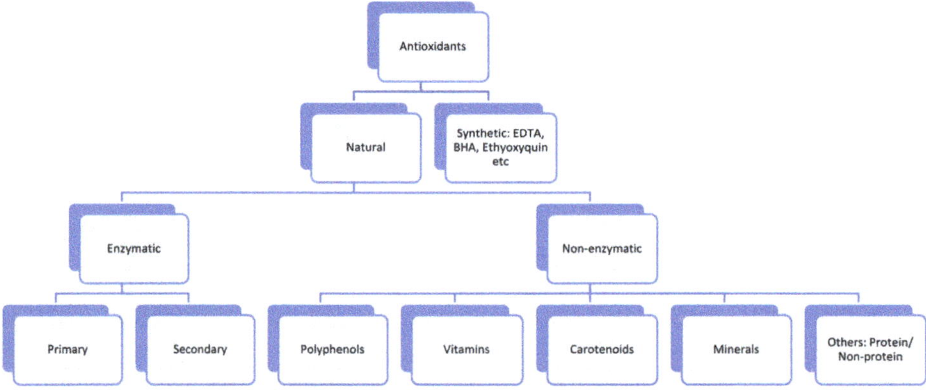

Figure 5. Schematic representation of antioxidants.

The rich antioxidant potential of pigmented potatoes in comparison with white-fleshed cultivars has been scarcely reported in the literature. In fact, the study of [42] showed a reduction in chronic disease susceptibility, inflammation, and cellular oxidative damage in adult males. These authors' study showed the antioxidant potential of pigmented potatoes. Furthermore, the in vivo study of [35] revealed the high antioxidant activity of purple potato flakes through RNA and linoleic acid oxidation inhibition. These authors further showed that hepatic Cu/Zn-SOD, GSH-Px mRNA, and Mn-SOD expression is further improved by the purple potato flakes' antioxidant activity. More studies have shown that pigmented potatoes possess high antioxidant activity and possess the potential to reduce oxidative stress [17–19,25,35,48,49,53,66,73]. Therefore, the consumption of pigmented potato cultivars could potentially offer better health benefits in comparison with the traditional white-fleshed cultivars. Health benefits associated with antioxidants such as anticancer, antiaging, and anti-inflammatory activities could potentially be derived from the consumption of these species.

4. Carotenoids

Several studies have reported on carotenoids in pigmented potato cultivars. This is because carotenoids are the basic source of orange, yellow, and red pigments in plants and are also widely distributed in nature, including in algae, bacteria, and photo as well as non-photosynthetic organism tissues [76]. The basic structure of a carotenoid is shown in Figure 6. Although α and β carotenes are known for provitamin A activity, they cannot be synthesized by the body and therefore needs to be sourced from food. Carotenoids are very common in vegetables and fruits such as peaches, spinach, apricots, carrots, butternut, and sweet potatoes. Although over 700 carotenoids are known to exist in nature, about 24 are common in the food that we consume and only four have been exhaustively studied, viz., α-carotene, β-carotene, lycopene, lutein, and zeaxanthin [77].

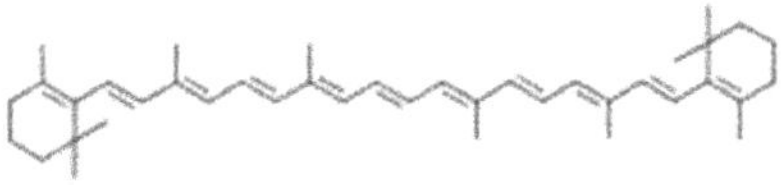

Figure 6. General structure of carotenoids—β-carotene [78].

Based on their chemical structures, the role of dietary carotenoids in quenching harmful reactive oxygen species (ROS) and intercepting toxic free radicals is well reported [79]. Carotenoids are therefore important natural antioxidants that form a strong defense against harmful biological processes. The defensive role of carotenoids against muscular degeneration, cardiovascular diseases, and a wide range of cancers including uterine, lung, colorectal, prostate, and breast have been reported [80]. Reports also reveal the protective effects of lycopene, β-carotene, and lutein against the formation of erythema, which is caused by exposure to UV light [81]. Therefore, it is not surprising that carotenoids play a significant role in the cosmetics industry, which is worth billions of USD in the world today [82]. Some of the carotenoids that the pharmaceutical industry has been able to synthesize in the laboratory for the cosmetic industry include astaxanthin, β-carotene, canthaxanthin, lycopene, and zeaxanthin [76].

Nevertheless, the current literature search revealed several studies that have documented carotenoids in pigmented potato cultivars. The study of [26] found that yellow-pigmented cultivars possessed between 280% and 2000% more carotenoids than the white-fleshed cultivars depending on pigment intensity. The intensely yellow cultivars possessed more carotenoids, while lutein and xanthophyll were the dominant compounds. Some authors profiled 60 pigmented potatoes and found that the major carotenoids to dominate the cultivars were neoxanthin, lutein, and violaxanthin in 7, 16, and 37 cultivars, respectively, and the total compositions ranged between 50.0 and 1552.0 μg/100 g [29]. Minor carotenoids such as zeaxanthin, antheraxanthin, β-carotene, and β-cryptoxanthin were also reported by the previous authors. In another study, Ref. [52] found that total polyphenols and carotenoids increased in purple potatoes in response to conventional cultivation as opposed to organic cultivation. These authors reported a positive correlation between the color of the potatoes and polyphenol and carotenoid contents, although the carotenoids ranged between 0.012 and 0.085 mg/100 g. The two-year study of [36] in the Czech Republic found that genotype, soil type, and year of growth in pigmented cultivars affected the carotenoid contents and ranged between 0.779 and 13.3 mg/kg. Lutein dominated the carotenoids in their study, while β-carotene, zeaxanthin, neoxanthin, and violaxanthin were also found but in marginal proportions in all the cultivars. Additionally, the yellow cultivar Agria possessed higher levels than the blue and red types. To corroborate this study, [67] found that location, genotype, period of growth, and their interactions in Poland significantly affected carotenoid contents in some pigmented cultivars grown in four different locations for three consecutive years. The contents were higher than in the Czech Republic and ranged between 5.57 and 20.20 mg/kg, and lutein dominated (2.92 to 6.66 mg/kg) more than zeaxanthin (1.44 to 3.05 mg/kg). The Belgian study of [18] revealed that zeaxanthin and lutein contents in 23 pigmented cultivars ranged between 0 and 17.7 μg/g and between 1.12 and 17.69 μg/g, respectively, while β-carotene ranged between 0.42 and 2.19 μg/g in 16 cultivars. β-carotene is rarely reported in potato tubers, perhaps a strong indication of the elevated carotenoid contents in these species. The study of [70] revealed that the carotenoids lutein and β-carotene were more elevated in potatoes cultivated under a biodynamic system, in comparison to those that were organically produced. These studies indicate that carotenoids in pigmented potatoes are cultivar-, pigment-, intensity-, and environment-dependent. The yellow-pigmented cultivars are generally higher in carotenoids in comparison to other cultivars, and more so, the white-fleshed type and lutein appear to dominate these compounds. In human physiology, lutein is found concentrated in the macula of the retina in the eyes. Although the precise lutein role is not yet fully known and understood, its association with vision cannot be farfetched. Therefore, the consumption of these carotenoid-rich potatoes could potentially act as a buffer against the threat of lifestyle and diet-induced diseases and conditions and help in the management and treatment of vision-related issues.

5. Anthocyanins in Pigmented Cultivars

Most phytochemical studies that have been conducted on pigmented cultivars are on anthocyanins. This is because of the color of these species' skin and/or flesh. Anthocyanins are water-soluble phenolic compounds that fall under the flavonoid subgroup [83]. Anthocyanins occur in plants as glycosides attached to a sugar group, as shown in Figure 7. Located in the vacuole, these compounds give plants their distinctive color and are widely distributed in vegetables and fruits, such as strawberries, blueberries, blackberries, currents, mulberries, blackcurrant, and red/blue grapes [84]. The purple to blue and red pigments given off by the action of anthocyanins function to attract animals to consume the fruits and aid in seed dispersal, while in flowers, pollinators are also attracted [83]. In addition, the bright colors help to absorb radioactive ultraviolet and blue-green light and act as a plant sunscreen. Peonidin, petunidin, cyanidin, malvidin, delphinidin, and pelargonidin are six widely distributed anthocyanins from a total of 17 that are known to be found in nature [84]. However, [85] had earlier argued that there are nearly a thousand anthocyanins in Kingdom Plantae. Nevertheless, when the pH is acidic, anthocyanins are thought to exhibit a positive charge in their structure [84]. Furthermore, at pH < 2, anthocyanins are said to exist as flavylium, a basic and stable compound. Consequently, anthocyanin's bioavailability, metabolism, absorption, and biological responses are thought to be influenced by these unique chemical structures [84]. According to the previous authors, neutral and alkaline pH have a degrading effect on anthocyanins, with a bioavailability as low as 0.1%. Understanding the behavior of this antioxidant is critical in understanding its role in health and disease management, especially NCDs, which are often negatively affected by poor and unhealthy diets. Several studies have documented the anthocyanin antidiabetic, antiproliferative, and antioxidant, as well as improved insulin resistance in nonalcoholic fatty liver disease (NAFLD), risk factor reduction in cardiovascular diseases and an improvement in eyesight among other benefits [28,85,86].

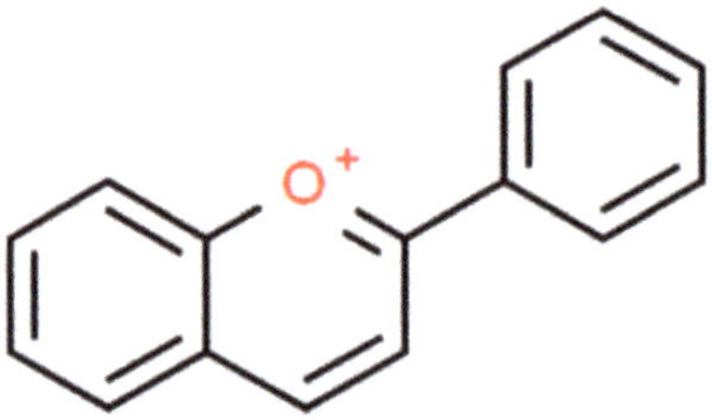

Figure 7. Basic anthocyanin structure—flavylium [86].

As shown in Table 2, the current literature search was able to reveal petunidin, cyanidin, delphinidin, petunidin, pelargonidin, peonidin, and malvidin that have been profiled and quantified in these species. The study of [73] showed that anthocyanins were more elevated in the skin of pigmented potatoes in comparison to the flesh. Jansen and Flamme [39] also found the skins to contain more anthocyanins than the flesh, while the whole tuber compositions lied between the skin and the flesh. The previous results have also been corroborated by [71]. Therefore, the current literature search is a good indicator of the potential of pigmented potato cultivars as health-bearing underutilized foods. Anthocyanins are, in fact, the highest-consumed flavonoids with an estimated intake of 200 mg/d, therefore making them valuable components of the diet [85]. The use of the skin for both food and industrial applications or color dyes should therefore not be ignored, as anthocyanins are more elevated in these organs.

Table 2. Anthocyanins in pigmented potato cultivars.

	Anthocyanin	Ref.
Blue	Petunidin, Cyanidin, Delphinidin, Petunidin, Pelargonidin, Peonidin, Malvidin	[28,32,38,73]
Red	Pelargonidin, Cyanidin, Delphinidin, Petunidin, Peonidin, Malvidin	[28,32,38,73]
Purple	Peonidin, Petunidin, Delphinidin, Petunidin, Pelargonidin, Peonidin, Malvidin	[28,32,38,73]
Yellow	Malvidin, petunidin, Cyanidin, Delphinidin, Petunidin, Pelargonidin, Peonidin	[28,32,38,73]
Pink	Cyanidin, Delphinidin, Petunidin, Pelargonidin, Peonidin, Malvidin	[32,38,73]

6. Pigmented Potato Biological Activity

The antioxidant activity and anthocyanins contained in pigmented potatoes are a good indicator of some potential biological activity. Few studies have been conducted to determine the biological activities of these species; however, a few that have been conducted open insights into endless possibilities on the applications of these species in diverse areas of life. Anticancer studies of pigmented potato cultivars are scarce but look promising. Although the in vitro study of [72] showed that the ethanolic extracts were minimally active against the hepatocellular carcinoma cell lines, these need to be tested against other cell lines to confirm these findings. The study of [68] showed a reduction in cancer incidence and multiplicity, but in vivo, when they used the extracts of pigmented potatoes. The study of [20] also revealed that the extracts of some pigmented potatoes were lethal against the hepatocarcinoma (Hep3B) cell lines. Silveyra et al. [65] also showed that the skin and flesh extracts of some cultivars possessed some antibacterial properties against *E. coli*. These authors reported that the skin possessed more activities than the flesh extracts. These studies are promising and should further be addressed toward other yet-to-be-researched cell lines. Pigmented potatoes could potentially hold the key to solving the seemingly never-ending cancer plague that the world is facing today. If polyphenols are profiled, some of the compounds could be useful as anticancer compounds. Unfortunately, exploratory studies to exhaustively profile polyphenols in these species are lacking. This knowledge gap needs to be filled.

7. The Impact of Environmental, Genotypic, and Soil Types on Pigmented Potato Characteristics

The effects of environmental factors on the growth of plants have been widely studied and reported. A few studies of this nature have been conducted on pigmented cultivars, and the results offer some insights into commercialization prospects of these species. Witbooi et al. [11] subjected pigmented potatoes to various root zone temperatures and found that 24 °C significantly increased plant height and tuber weight in all cultivars, while 28 °C increased polyphenols. Gutiérrez-Quequezana et al. [33] reported that 13 and 18 °C did not affect either anthocyanins or polyphenolics; however, the increase in anthocyanins at maturity in purple cultivars and high polyphenolics in the blue cultivars were highly cultivar-dependent. In another study, Ref. [39] showed that nitrogen fertilizer did not affect pigmentation in potatoes. Meanwhile, Ref. [26] reported that the growing environment, including altitude and soil factors, significantly increase anthocyanin production but have no effect on carotenoids. A Korean study [41] showed that anthocyanins responded well to high-altitude areas in comparison to those cultivated in a low-lying area over a two-year period and in 14 different locations. These authors also reported a negative correlation between soil acidity and anthocyanin contents. However, in Poland, the effect of environmental factors, genotypes, the year, and interaction of these factors had a significant effect on carotenoid contents [67]. Andre et al. [19] showed that the effect of drought stress on potato phytochemicals was highly cultivar-dependent and varied between cultivars.

However, antioxidant contents were weakly affected by drought stress in yellow cultivars, while polyphenols and anthocyanins were drastically reduced in purple and red cultivar flesh. The study of [70] showed that organically and biodynamically produced cultivars had higher anthocyanin and polyphenolic compounds, although biodynamic cultivars had significantly higher carotenoids in comparison with the control. In a similar study, [52] found that the conventional cultivation of pigmented cultivars increased total polyphenols and carotenoids in comparison to those that were organically produced. In Texas and Colorado USA, anthocyanins and polyphenols decreased in the tubers as they matured, but the yield and compounds increased [58]. Harvesting time is thus useful if the maximum compound yield is the objective. Some breeding studies have also been conducted on high-anthocyanin-yielding clones to understand their breeding behavior. For example, [16] characterized loci and genetically mapped the traits that influence anthocyanin pigmentation in potatoes. This study revealed that 21 pigmented out of 53 white cultivars shared a common bHLH allele, and this indicates the contribution of this allele, but this is also not adequate for completely pigmented cultivars. Studies on growth and physiological response, and breeding lack coordination and are therefore difficult to conclude. Yield data and the breeding of better yielding cultivars to aid in pigmented cultivar revitalization are severely lacking. This again shows the lack of data and the knowledge gap that needs to be filled by researchers around the world.

8. The Effect of Processing on Phytochemical Contents in Pigmented Potatoes

Potatoes including the pigmented cultivars need to be cooked (boiled, steamed, or fried) or processed into products such as crisps before consumption. As such, phytochemical compositions that are present before processing may change postprocessing and these may further change during absorption in the small intestines [87]. The antioxidant vitamin C, for example, is known to be heat-labile and, therefore, drastically deteriorates during cooking [87]. Therefore, vitamin C can perhaps be extracted more from fruits as they do not usually need to be cooked. In general, heat application is known to destabilize anthocyanins [88]. For example, frying reduced anthocyanins by 38–70% [45]. An earlier study of [46] had shown that anthocyanins were almost totally degraded by frying, while polyphenols remained stable, and antioxidants were significantly reduced. In contrast to these previous studies, [26] showed that anthocyanins increased in pigmented cultivars during cooking, although carotenoids decreased. In the previous authors' study, boiling and microwaving significantly increased anthocyanins, while baking and frying did not significantly lead to an increase in the compounds. Boiling, on the other hand, increased antioxidant activity, while the other methods decreased these activities. However, the study of [51] showed that there was no change in phenolic acids in pigmented potatoes during cooking using the microwave or by boiling, while anthocyanins decreased by 16–29%. In another study, [53] showed that, and refractive window drying, drum drying, and freeze drying led to 23, 41, and 45% reductions in anthocyanin contents in pigmented cultivars, respectively. However, phenolic acids and antioxidant activity were not affected by drying in the previous study. The study of [47] reported that boiling and baking, respectively, led to 92 and 88% reductions in carotenoids. However, lutein and β-carotene were the most stable carotenoids in their study. The study of [54] also showed that snacks produced using pigmented potatoes were 20–30% richer in antioxidants and anthocyanins than the controls were, while polyphenols increased by up to 40%. Rytel et al. [63] reported that pre-drying and peeling pigmented potatoes led to significant losses in anthocyanins (75%), while polyphenols were significantly reduced (69%) by drying, blanching, and pre-drying during industrial processing. However, Furrer et al. [31] had earlier reported that the industrial processing of pigmented cultivars retained anthocyanins by 79–129%, while phenolics were retained by 49–85%. Although these studies are variable, carotenoids appear to be negatively affected by thermal processing, while polyphenols appear stable, and anthocyanins generally vary. As promising as these findings look, studies on the absorption and bioavailability of the compounds derived from pigmented potatoes need to be conducted to understand their real effect on human physiology.

9. Conclusions and Prospects

Pigmented potato cultivars are rare, and few studies have been conducted to document their characteristics. This review was able to reveal about 62 studies that have been conducted on pigmented cultivars around the world. Therefore, the dearth in the literature on physiological growth and yield, breeding, exhaustive secondary metabolite profiles, nutritional contents, absorption, and bioavailability studies and many other important studies is glaring. These studies would play an important role in revitalization efforts and the eventual commercialization of these important species. More studies need to be conducted to fill the missing knowledge gaps. The polyphenolic and antioxidant (anthocyanins and carotenoids) studies that have been conducted so far reveal the superior nature of these tuber crops in comparison with white-fleshed cultivars that are common throughout the world, and that have gained popularity in other applications that are too many to mention. Breeding studies need to concentrate on increasing the yield without compromising the phytochemical constituents. Yield has been shown to be one of the major factors limiting the commercialization of these species regardless of their promising health attributes. The antiproliferative studies that have been conducted so far look promising, but more could be done to reveal their full potential. The skins clearly contain higher phytochemical and biological constituents than the flesh parts and, therefore, consumers should be encouraged to explore the skins more. In addition, industrial applications of the skins, for example, in colorants, is clearly a viable option. What is clear from the work that has been conducted so far and the results is that pigmented potatoes are heavily underutilized, and they need to be revitalized and consequently commercialized. Their nutraceutical applications are glaring, and their functions potentially yield better results than the traditional, white-fleshed types. Some of the fears of the continued neglect of such species is extinction, and, if they go extinct, so do their potential food and nutrition security, as well as health/pharmaceutical benefits. Pigmented potatoes should not remain in the shadows, in high-end restaurants as gourmet foods, but should now trickle down to the wider populace for consumption and with the possibility to eventually replace the white-fleshed cultivars. Pigmented cultivars are healthier, and their consumption will potentially help to breed healthier nations through decreased lifestyle diseases and conditions. The proliferation of lifestyle diseases is costing governments billions in USD annually the world over and this problem needs to be nipped in the bud before it gets out of hand.

Author Contributions: C.B., H.W. and L.K. all equally contributed to the writing of the manuscript. LK sourced the funding. All authors have read and agreed to the published version of the manuscript.

Funding: This research was funded by the Cape Peninsula University of Technology and the National Research Foundation of South Africa.

Institutional Review Board Statement: Not applicable.

Informed Consent Statement: Not applicable.

Data Availability Statement: Not applicable.

Acknowledgments: The authors of this article are grateful to the Cape Peninsula University of Technology and the National Research Foundation of South Africa for funding this work.

Conflicts of Interest: The authors declare no conflict of interest.

References

1. Gehrels, R.; Garrett, E. Chapter 11—Rising sea levels as an indicator of global change. In *Climate Change*, 3rd ed.; Observed Impacts on Planet Earth; Elsevier: Amsterdam, The Netherlands, 2021; pp. 205–217.
2. FAO/IFAD/UNICEF/WFP/WHO. *The State of Food Security and Nutrition in the World 2021. Transforming Food Systems for Food Security, Improved Nutrition and Affordable Healthy Diets for All*; FAO: Rome, Italy, 2021. [CrossRef]

3. WHO. Noncommunicable Diseases: Key Facts. 2021. Available online: https://www.who.int/news-room/fact-sheets/detail/noncommunicable-diseases (accessed on 1 November 2021).
4. FAO. Potato Production Worldwide from 2002 to 2019 (in Million Metric Tons). Statista Inc. 2021. Available online: https://www-statista-com.libproxy.cput.ac.za/statistics/382174/global-potato-production (accessed on 9 June 2021).
5. Birch, P.R.J.; Bryan, G.; Fenton, B.; Gilroy, E.M.; Hein, I.; Jones, J.T.; Prashar, A.; Taylor, M.A.; Torrance, L.; Toth, I.K. Crops that feed the world 8: Potato: Are the trends of increased global production sustainable? *Food Secur.* **2012**, *4*, 477–508. [CrossRef]
6. FAO. Potato Area Harvested Worldwide from 2002 to 2019 (in Million Hectares): Statista Inc. 2021. Available online: https://www-statista-com.libproxy.cput.ac.za/statistics/625120/global-potato-area-harvested (accessed on 9 June 2021).
7. FAOSTAT. Potato Consumption Per Capita. 2021. Available online: https://www.helgilibrary.com/indicators/potato-consumption-per-capita (accessed on 1 November 2021).
8. Devaux, A.; Kromann, P.; Ortiz, O. Potatoes for sustainable global food security. *Potato Res.* **2014**, *57*, 185–199. [CrossRef]
9. FAO. Potato Production Worldwide in 2019, by Leading Country (in Million Metric Tons). Statista Inc. 2021. Available online: https://www-statista-com.libproxy.cput.ac.za/statistics/382192/global-potato-production-by-country (accessed on 9 June 2021).
10. Yang, Y.; Achaerandio, I.; Pujolà, M. Classification of potato cultivars to establish their processing aptitude. *J. Sci. Food Agric.* **2016**, *96*, 413–421. [CrossRef] [PubMed]
11. Witbooi, H.; Bvenura, C.; Oguntibeju, O.O.; Kambizi, L. The role of root zone temperature on physiological and phytochemical compositions of some pigmented potato (*Solanum tuberosum* L.) cultivars. *Cogent Food Agric.* **2021**, *7*, 1905300. [CrossRef]
12. Camire, M.E.; Kubow, S.; Donnelly, D.J. Potatoes and human health. *Crit. Rev. Food Sci. Nutr.* **2009**, *49*, 823–840. [CrossRef] [PubMed]
13. Andre, C.M.; Ghislain, M.; Bertin, P.; Oufir, M.; Herrera, M.; Hoffmann, L.; Hausman, J.F.; Larondelle, Y.; Evers, D. Andean potato cultivars (*Solanum tuberosum* L.) as a source of antioxidant and mineral micronutrients. *J. Agric. Food Chem.* **2007**, *55*, 366–378. [CrossRef]
14. Reyes, L.F.; Cisneros-Zevallos, L. Degradation kinetics and colour of anthocyanins in aqueous extracts of purple- and red-flesh potatoes (*Solanum tuberosum* L.). *Food Chem.* **2007**, *100*, 885–894. [CrossRef]
15. Cho, K.; Cho, K.; Sohn, H.; Ha, I.J.; Hong, S. Network analysis of the metabolome and transcriptome reveals novel regulation of potato pigmentation. *J. Exp. Bot.* **2016**, *67*, 1519–1533. [CrossRef]
16. Zhang, Y.; Jung, C.S.; De Jong, W.S. Genetic analysis of pigmented tuber flesh in potato. *Theor. Appl. Genet.* **2009**, *119*, 143–150. [CrossRef]
17. Ah-Hen, K.S.; Fuenzalida, C.; Hess, S.; Contreras, A.; Vega-Gálvez, A.; Lemus-Mondaca, R. Antioxidant capacity and total phenolic compounds of twelve selected potato landrace clones grown in southern Chile. *Chilean J. Agric. Res.* **2012**, *72*, 3–9. [CrossRef]
18. Andre, C.M.; Oufir, M.; Guignard, C.; Hoffmann, L.; Hausman, J.F.; Evers, D.; Larondelle, Y. Antioxidant profiling of native Andean potato tubers (*Solanum tuberosum* L.) reveals cultivars with high levels of beta-carotene, alpha-tocopherol, chlorogenic acid, and petanin. *J. Agric. Food Chem.* **2007**, *55*, 10839–10849. [CrossRef]
19. Andre, C.M.; Schafleitner, R.; Guignard, C.; Oufir, M.; Aliaga, C.A.; Nomberto, G.; Hoffmann, L.; Hausman, J.F.; Evers, D.; Larondelle, Y. Modification of the health-promoting value of potato tubers field grown under drought stress: Emphasis on dietary antioxidant and glycoalkaloid contents in five native andean cultivars (*Solanum tuberosum* L.). *J. Agric. Food Chem.* **2009**, *57*, 599–609. [CrossRef]
20. Balbina, A.A.; Julia, M.M.; Luciana, B. Chemical characterization of polyphenol extracts from Andean and industrial *Solanum tuberosum* tubers and their cytotoxic activity on human hepatocarcinoma cells. *J. Food Sci. Technol.* **2017**, *2*, 205–217.
21. Ben Jeddou, K.; Kammoun, M.; Hellström, J.K.; Gutiérrez-Quequezana, L.; Rokka, V.; Gargouri-Bouzid, R.; Ellouze-Chaâbouni, S.; Nouri-Ellouz, O. Profiling beneficial phytochemicals in a potato somatic hybrid for tuber peels processing: Phenolic acids and anthocyanins composition. *Food Sci. Nutr.* **2021**, *9*, 1388–1398. [CrossRef]
22. Brown, C.R. Antioxidants in potato. *Am. J. Potato Res.* **2005**, *62*, 163–172. [CrossRef]
23. Brown, C.R.; Wrolstad, R.E.; Durst, R.W.; Yang, C.P.; Clevidence, B.A. Breeding studies in potatoes containing high concentrations of anthocyanins. *Am. J. Potato Res.* **2003**, *80*, 241–250. [CrossRef]
24. Brown, C.R.; Culley, D.E.; Yang, C.P.; Durst, R.W.; Wrolstad, R.E. Variation of anthocyanin and carotenoid contents and associated antioxidant values in potato breeding lines. *J. Am. Soc. Hortic. Sci.* **2005**, *130*, 174–180. [CrossRef]
25. Brown, C.R.; Culley, D.E.; Bonierbale, M.W.; Amorós, W. Anthocyanin, carotenoid content, and antioxidant values in native South American potato cultivars. *Hortscience* **2007**, *42*, 1733–1736. [CrossRef]
26. Brown, C.R.; Durst, R.W.; Wrolstad, R.E.; Jong, W.S. Variability of phytonutrient content of potato in relation to growing location and cooking method. *Potato Res.* **2008**, *51*, 259–270. [CrossRef]
27. Chaparro, J.M.; Holm, D.G.; Broeckling, C.D.; Prenni, J.E.; Heuberger, A.L. Metabolomics and ionomics of potato tuber reveals an influence of cultivar and market class on human nutrients and bioactive compounds. *Front. Nutr.* **2018**, *5*, 36. [CrossRef]
28. Yamane, T. Beneficial Effects of Anthocyanin from Natural Products on Lifestyle-Related Diseases Through Inhibition of Protease Activities. *Stud. Nat. Prod. Chem.* **2018**, *58*, 245–264.

29. Fernandez-Orozco, R.; Gallardo-Guerrero, L.; Hornero-Méndez, D. Carotenoid profiling in tubers of different potato (*Solanum* sp.) cultivars: Accumulation of carotenoids mediated by xanthophyll esterification. *Food Chem.* **2013**, *141*, 2864–2872. [CrossRef] [PubMed]

30. Fossen, T.; Andersen, Ø.M. Anthocyanins from tubers and shoots of the purple potato, *Solanum tuberosum*. *J. Hortic. Sci. Biotechnol.* **2000**, *75*, 360–363. [CrossRef]

31. Furrer, A.; Cladis, D.P.; Kurilich, A.C.; Manoharan, R.; Ferruzzi, M.G. Changes in phenolic content of commercial potato varieties through industrial processing and fresh preparation. *Food Chem.* **2017**, *218*, 47–55. [CrossRef]

32. Giusti, M.M.; Polit, M.F.; Ayvaz, H.; Tay, D.B.; Manrique, I. Characterization and quantitation of anthocyanins and other phenolics in native Andean potatoes. *J. Agric. Food Chem.* **2014**, *62*, 4408–4416. [CrossRef] [PubMed]

33. Gutiérrez-Quequezana, L.; Vuorinen, A.L.; Kallio, H.P.; Yang, B. Impact of cultivar, growth temperature and developmental stage on phenolic compounds and ascorbic acid in purple and yellow potato tubers. *Food Chem.* **2020**, *326*, 126966. [CrossRef]

34. Hamouz, K.; Lachman, J.; Pazderu, K.; Tomášek, J.; Hejtmánková, K.; Pivec, V. Differences in anthocyanin content and antioxidant activity of potato tubers with different flesh colour. *Plant Soil Environ.* **2018**, *57*, 478–485. [CrossRef]

35. Han, K.; Sekikawa, M.; Shimada, K.; Hashimoto, M.; Hashimoto, N.; Noda, T.; Tanaka, H.; Fukushima, M. Anthocyanin-rich purple potato flake extract has antioxidant capacity and improves antioxidant potential in rats. *Br. J. Nutr.* **2006**, *96*, 1125–1134. [CrossRef]

36. Hejtmánková, K.; Kotikova, Z.; Hamouz, K.; Pivec, V.; Vacek, J.; Lachman, J. Influence of flesh colour, year and growing area on carotenoid and anthocyanin content in potato tubers. *J. Food Compos. Anal.* **2013**, *32*, 20–27. [CrossRef]

37. Hejtmánková, K.; Pivec, V.; Trnková, E.; Hamouz, K.; Lachman, J. Quality of coloured varieties of potatoes. *Czech J. Food Sci.* **2018**, *27*, S310–S313. [CrossRef]

38. Hillebrand, S.; Heike, B.; Kitzinski, N.; Köhler, N.; Winterhalter, P. Isolation and characterization of anthocyanins from blue-fleshed potatoes (*Solanum tuberosum* L.). *Food* **2009**, *3*, 96–101.

39. Jansen, G.; Flamme, W. Coloured potatoes (*Solanum tuberosum* L.)—Anthocyanin content and tuber quality. *Genet. Resour. Crop Evol.* **2006**, *53*, 1321. [CrossRef]

40. Jaromír, L.; Karel, H.; Matyáš, O. Chapter 9—Colored potatoes. In *Advances in Potato Chemistry and Technology*, 2nd ed.; Elsevier: Amsterdam, The Netherlands, 2016; pp. 249–281.

41. Jeong, J.; Kim, S.; Hong, S.; Nam, J.; Sohn, H.; Kim, Y.; Mekapogu, M. Growing environment influence the anthocyanin content in purple- and red-fleshed potatoes during tuber development. *Korean J. Crop Sci.* **2015**, *60*, 231–238. [CrossRef]

42. Kaspar, K.L.; Park, J.S.; Brown, C.R.; Mathison, B.D.; Navarre, D.A.; Chew, B.P. Pigmented potato consumption alters oxidative stress and inflammatory damage in men. *J. Nutr.* **2011**, *141*, 108–111. [CrossRef] [PubMed]

43. Kaspar, K.; Park, J.; Brown, C.R.; Weller, K.; Ross, C.; Mathison, B.; Chew, B. Sensory evaluation of pigmented flesh potatoes (*Solanum tuberosum* L.). *Food Nutr. Sci.* **2013**, *4*, 77–81.

44. Kim, H.; Kim, S.R.; Lee, Y.M.; Jang, H.; Kim, J.B. Analysis of variation in anthocyanin composition in Korean coloured potato cultivars by LC-DAD-ESI-MS and PLS-DA. *Potato Res.* **2017**, *61*, 1–17. [CrossRef]

45. Kita, A.; Bakowska-Barczak, A.M.; Hamouz, K.; Kułakowska, K.; Lisińska, G. The effect of frying on anthocyanin stability and antioxidant activity of crisps from red- and purple-fleshed potatoes (*Solanum tuberosum* L.). *J. Food Compos. Anal.* **2013**, *32*, 169–175. [CrossRef]

46. Kita, A.; Bakowska-Barczak, A.M.; Lisińska, G.; Hamouz, K.; Kułakowska, K. Antioxidant activity and quality of red and purple flesh potato chips. *LWT-Food Sci. Technol.* **2015**, *62*, 525–531. [CrossRef]

47. Kotíková, Z.; Šulc, M.; Lachman, J.; Pivec, V.; Orsák, M.; Hamouz, K. Carotenoid profile and retention in yellow-, purple- and red-fleshed potatoes after thermal processing. *Food Chem.* **2016**, *197*, 992–1001. [CrossRef]

48. Külen, O.; Stushnoff, C.; Holm, D.G. Effect of cold storage on total phenolics content, antioxidant activity and vitamin C level of selected potato clones. *J. Sci. Food Agric.* **2013**, *93*, 2437–2444. [CrossRef]

49. Lachman, J.; Hamouz, K.; Šulc, M.; Orsák, M.; Pivec, V.; Hejtmánková, A.; Dvořák, P.; Čepl, J.J. Cultivar differences of total anthocyanins and anthocyanidins in red and purple-fleshed potatoes and their relation to antioxidant activity. *Food Chem.* **2009**, *114*, 836–843. [CrossRef]

50. Lachman, J.; Hamouz, K.; Musilová, J.; Hejtmánková, K.; Kotíková, Z.; Pazderů, K.; Domkářová, J.; Pivec, V.; Cimr, J. Effect of peeling and three cooking methods on the content of selected phytochemicals in potato tubers with various colour of flesh. *Food Chem.* **2013**, *138*, 1189–1197. [CrossRef] [PubMed]

51. Mulinacci, N.; Ieri, F.; Giaccherini, C.; Innocenti, M.; Andrenelli, L.; Canova, G.; Saracchi, M.; Casiraghi, M.C. Effect of cooking on the anthocyanins, phenolic acids, glycoalkaloids, and resistant starch content in two pigmented cultivars of *Solanum tuberosum* L. *J. Agric. Food Chem.* **2008**, *56*, 11830–11837. [CrossRef]

52. Murniece, I.; Tomsone, L.; Skrabule, I.; Vaivode, A. Carotenoids and total phenolic content In potatoes with different flesh colour. *Foodbalt* **2014**, 206–211.

53. Nayak, B.; Berrios, J.D.; Powers, J.R.; Tang, J.; Ji, Y. Colored potatoes (*Solanum tuberosum* L.) dried for antioxidant-rich value-added foods. *J. Food Process. Preserv.* **2011**, *35*, 571–580. [CrossRef]

54. Nemś, A.; Pęksa, A.; Kucharska, A.Z.; Sokół-Łętowska, A.; Kita, A.; Drożdż, W.; Hamouz, K. Anthocyanin and antioxidant activity of snacks with coloured potato. *Food Chem.* **2015**, *172*, 175–182. [CrossRef]

55. Oertel, A.; Matros, A.; Hartmann, A.; Arapitsas, P.; Dehmer, K.J.; Martens, S.; Mock, H.P. Metabolite profiling of red and blue potatoes revealed cultivar and tissue specific patterns for anthocyanins and other polyphenols. *Planta* **2017**, *246*, 281–297. [CrossRef]
56. Perla, V.; Holm, D.G.; Jayanty, S.S. Effects of cooking methods on polyphenols, pigments and antioxidant activity in potato tubers. *LWT-Food Sci. Technol.* **2012**, *45*, 161–171. [CrossRef]
57. Reyes, L.F.; Miller, J.C.; Cisneros-Zevallos, L. Antioxidant capacity, anthocyanins and total phenolics in purple-and red-fleshed potato (*Solanum tuberosum* L.) genotypes. *Am. J. Potato Res.* **2005**, *82*, 271–277. [CrossRef]
58. Reyes, L.F.; Miller, J.C.; Cisneros-Zevallos, L. Environmental conditions influence the content and yield of anthocyanins and total phenolics in purple- and red-flesh potatoes during tuber development. *Am. J. Potato Res.* **2008**, *81*, 187–193. [CrossRef]
59. Rodriguez-Saona, L.; Giusti, M.; Wrolstad, R. Anthocyanin pigment composition of red-fleshed potatoes. *J. Food Sci.* **1998**, *63*, 458–465. [CrossRef]
60. Rodriguez-Saona, L.E.; Giusti, M.M.; Wrolstad, R.E. Color and pigment stability of red radish and red-fleshed potato anthocyanins in juice model systems. *Food Sci.* **2006**, *64*, 451–456. [CrossRef]
61. Rodriguez-Saona, L.E.; Wrolstad, R.E.; Pereira, C.B. Glycoalkaloid content and anthocyanin stability to alkaline treatment of red-fleshed potato extracts. *J. Food Sci.* **2006**, *64*, 445–450. [CrossRef]
62. Rytel, E.; Tajner-Czopek, A.; Kita, A.; Aniołowska, M.; Kucharska, A.Z.; Sokół-Łętowska, A.; Hamouz, K. Content of polyphenols in coloured and yellow fleshed potatoes during dices processing. *Food Chem.* **2014**, *161*, 224–229. [CrossRef] [PubMed]
63. Rytel, E.; Tajner-Czopek, A.; Kita, A.; Soko, A.B.; Towska Kucharska, A.Z.; Hamouz, K. Effect of the production process on the content of anthocyanins in dried red-fleshed potato cubes. *Ital. J. Food Sci.* **2019**, *31*, 150–160.
64. Sampaio, S.L.; Lonchamp, J.; Dias, M.I.; Liddle, C.; Petropoulos, S.A.; Glamočlija, J.; Alexopoulos, A.A.; Santos-Buelga, C.; Ferreira, I.C.; Barros, L. Anthocyanin-rich extracts from purple and red potatoes as natural colourants: Bioactive properties, application in a soft drink formulation and sensory analysis. *Food Chem.* **2020**, *342*, 128526. [CrossRef] [PubMed]
65. Silveyra, M.X.; Lanteri, M.L.; Damiano, R.B.; Andreu, A.B. Bactericidal and cytotoxic activities of polyphenol extracts from *Solanum tuberosum* spp. *tuberosum* and spp. *andigena* cultivars on *Escherichia coli* and human Neuroblastoma SH-SY5Y Cells In Vitro. *J. Nutr. Metab.* **2018**, *2018*, 8073679.
66. Stushnoff, C.; Holm, D.; Thompson, M.D.; Jiang, W.; Thompson, H.J.; Joyce, N.I.; Wilson, P. Antioxidant properties of cultivars and selections from the Colorado potato breeding program. *Am. J. Potato Res.* **2008**, *85*, 267–276. [CrossRef]
67. Tatarowska, B.; Milczarek, D.; Wszelaczyńska, E.; Pobereżny, J.; Keutgen, N.; Keutgen, A.J.; Flis, B. Carotenoids variability of potato tubers in relation to genotype, growing location and year. *Am. J. Potato Res.* **2019**, *96*, 493–504. [CrossRef]
68. Thompson, M.D.; Thompson, H.J.; McGinley, J.N.; Neil, E.S.; Rush, D.K.; Holm, D.G.; Stushnoff, C. Functional food characteristics of potato cultivars (*Solanum tuberosum* L.): Phytochemical composition and inhibition of 1-methyl-1-nitrosourea induced breast cancer in rats. *J. Food Compos. Anal.* **2009**, *22*, 571–576. [CrossRef]
69. Tong, C.; Ru, W.; Wu, L.; Wu, W.; Bao, J. Fine structure and relationships with functional properties of pigmented sweet potato starches. *Food Chem.* **2020**, *311*, 126011. [CrossRef] [PubMed]
70. Vaitkevičienė, N.; Kulaitienė, J.; Jarienė, E.; Levickienė, D.; Danillčenko, H.; Średnicka-Tober, D.; Rembiałkowska, E.; Hallmann, E. Characterization of bioactive compounds in colored potato (*Solanum tuberosum* L.) cultivars grown with conventional, organic, and biodynamic methods. *Sustainability* **2020**, *12*, 2701. [CrossRef]
71. Valiñas, M.A.; Lanteri, M.L.; Ten Have, A.; Andreu, A.B. Chlorogenic acid, anthocyanin and flavan-3-ol biosynthesis in flesh and skin of Andean potato tubers (*Solanum tuberosum* subsp. andigena). *Food Chem.* **2017**, *229*, 837–846. [CrossRef]
72. Witbooi, H.; Bvenura, C.; Reid, A.-M.; Lall, N.; Oguntibeju, O.O.; Kambizi, L. The potential of elevated root zone temperature on the concentration of chlorogenic, caffeic, ferulic acids and biological activity of some pigmented *Solanum tuberosum* L. cultivar extracts. *Appl. Sci.* **2021**, *11*, 6971. [CrossRef]
73. Yin, L.; Chen, T.; Li, Y.; Fu, S.; Li, L.; Xu, M.; Niu, Y. A comparative study on total anthocyanin content, composition of anthocyanidin, total phenolic content and antioxidant activity of pigmented potato peel and flesh. *Food Sci. Technol. Res.* **2016**, *22*, 219–226. [CrossRef]
74. Mamta Misra, K.; Dhillon, G.S.; Brar, S.K.; Verma, M. Antioxidants. In *Biotransformation of Waste Biomass into High Value Biochemicals*; Brar, S., Dhillon, G., Soccol, C., Eds.; Springer: New York, NY, USA, 2014.
75. Birben, E.; Sahiner, U.M.; Sackesen, C.; Erzurum, S.; Kalayci, O. Oxidative stress and antioxidant defense. *World Allergy Organ. J.* **2012**, *5*, 9–19. [CrossRef]
76. Prakash, D.; Gupta, C. Carotenoids: Chemistry and Health Benefits. In *Phytochemicals of Nutraceutical Importance*; Prakash, D., Sharma, G., Eds.; CAB International: Wallingford, UK, 2014; pp. 181–195.
77. Marhuenda-Muñoz, M.; Hurtado-Barroso, S.; Tresserra-Rimbau, A.; Lamuela-Raventós, R.S. A review of factors that affect carotenoid concentrations in human plasma: Differences between Mediterranean and Northern diets. *Eur. J. Clin. Nutr.* **2019**, *72*, 18–25. [CrossRef]
78. ChemSpider. β-Carotene. Available online: https://www.chemspider.com/Chemical-Structure.4444129.html (accessed on 5 November 2021).
79. Faraloni, C.; Torzillo, G. Synthesis of antioxidant carotenoids in microalgae in response to physiological stress. In *Carotenoids*; Cvetkovic, D.J., Nikolic, G.S., Eds.; IntechOpen: London, UK, 2017. [CrossRef]

80. Kiokias, S.; Proestos, C.; Oreopoulou, V. Effect of natural food antioxidants against LDL and DNA oxidative changes. *Antioxidants* **2018**, *7*, 133. [CrossRef]

81. Stahl, W.; Sies, H. Carotenoids and flavonoids contribute to nutritional protection against skin damage from sunlight. *Mol. Biotechnol.* **2007**, *37*, 26–30. [CrossRef]

82. Kambizi, L.; Bvenura, C. Medicinal plants for strong immune system and traditional skin therapy in South Africa. In *Medicinal Plants and the Human Immune System*; Husen, A., Ed.; CRC/Taylor & Francis: Boca Raton, FL, USA, 2021; pp. 123–172.

83. Bvenura, C.; Ngemakwe Nitcheu, P.H.; Chen, L.; Sivakumar, D. Nutritional and health benefits of temperate fruits. In *Postharvest Biology and Technology of Temperate Fruits*; Mir, S.A., Shah, M.A., Mir, M.M., Eds.; Springer: Cham, Switzerland, 2018; pp. 51–75.

84. Reyes, B.A.; Dufourt, E.C.; Ross, J.A.; Warner, M.J.; Tanquilut, N.C.; Leung, A.B. Selected Phyto and Marine Bioactive Compounds: Alternatives for the Treatment of Type 2 Diabetes. *Stud. Nat. Prod. Chem.* **2018**, *55*, 111–143.

85. Valavanidis, A.; Vlachogianni, T. Chapter 8: Plant Polyphenols. Recent Advances in Epidemiological Research and Other Studies on Cancer Prevention. *Stud. Nat. Prod. Chem. Bioact. Nat. Prod.* **2013**, *39*, 269–295.

86. ChemSpider. Flavylium. Available online: https://www.chemspider.com/Chemical-Structure.128674.html?rid=18e11ee5-0ecf-4264-bc49-ed7446fbe36e (accessed on 4 November 2021).

87. Bvenura, C.; Sivakumar, D. The role of wild fruits and vegetables in delivering a healthy and balanced diet. *Food Res. Int.* **2017**, *99*, 15–30. [CrossRef] [PubMed]

88. Patras, A.; Brunton, N.P.; O'Donnell, C.; Tiwari, B.K. Effect of thermal processing on anthocyanin stability in foods; mechanisms and kinetics of degradation. *Trends Food Sci. Technol.* **2010**, *21*, 3–11. [CrossRef]

Review

Experiences and Lessons from Agri-Food System Transformation for Sustainable Food Security: A Review of China's Practices

Yujia Lu [1], Yongxun Zhang [1], Yu Hong [1], Lulu He [2] and Yangfen Chen [1,*]

[1] Institute of Agricultural Economics and Development, Chinese Academy of Agricultural Sciences, Beijing 100081, China; 82101201652@caas.cn (Y.L.); zhangyongxun@caas.cn (Y.Z.); hongyu@caas.cn (Y.H.)
[2] College of Humanities and Development, China Agricultural University, No. 2 Yuanmingyuan West Road, Haidian District, Beijing 100094, China; helulu@cau.edu.cn
* Correspondence: chenyangfen@caas.cn; Tel.: +86-0108-2106-162

Abstract: Food system transformation has been a widely discussed topic in international society over time. For the last few decades, China has made remarkable achievements in food production and has contributed greatly to the reduction in global hunger and poverty. Examining experiences and lessons from China's food security practices over the years is helpful to promote a national food system transformation for China, as well as other developing countries. This study systematically reviews the literature on Chinese food security studies, with the aim of assessing China's food security achievements and examining the remaining and emerging issues in the pursuit of food system transformation. The results show that China has continuously promoted food system transformation in land consolidation, agri-food production technologies, management and organization modes, food reserves, trade governance, and food consumption. These transformations ensure not only food availability, timeliness, and nutrition, but also in terms of the ecological, social, and economic sustainability, feasibility, and justice of food security. However, China is also confronting new challenges in food security, for example, malnutrition, environmental unsustainability, and reductions in diversified agri-food. In the future, China is expected to be committed to promoting healthy diets, sustainable agricultural production, climate change mitigation, and the reduction of food waste and loss to enhance its agri-food system's resilience.

Keywords: agri-food system; food security; transformation; China; sustainability

Citation: Lu, Y.; Zhang, Y.; Hong, Y.; He, L.; Chen, Y. Experiences and Lessons from Agri-Food System Transformation for Sustainable Food Security: A Review of China's Practices. *Foods* **2022**, *11*, 137. https://doi.org/10.3390/foods11020137

Academic Editors: António Raposo, Raquel Braz Assunção Botelho and Renata Puppin Zandonadi

Received: 7 December 2021
Accepted: 2 January 2022
Published: 6 January 2022

Publisher's Note: MDPI stays neutral with regard to jurisdictional claims in published maps and institutional affiliations.

1. Introduction

Achieving the goal of global food security, together with environmental sustainability and socio-economic justice, is one of the greatest challenges for human beings in the 21st century [1]. There are several factors impeding progress toward achieving global food security and nutrition targets. For instance, starting from 2019, it is estimated that the outbreak of the COVID-19 pandemic has led to an increase in undernourishment from 8.4 to 9.9% worldwide, leaving nearly 2.37 billion people without access to adequate food supplies, while 2.2 billion adults are overweight [2–4]. Human activities directly affect more than 70% of global ice-free land at present, and up to one-third of the terrestrial net primary productivity is being used for food, feed, fiber, timber, and energy [5]. Food systems that rely heavily on chemical fertilizers, pesticides, and antibiotics [6] have driven the widespread degradation of land, water and ecosystems, high greenhouse gas (GHG) emissions, biodiversity losses [7], and the spread of pathogens [8], which in turn lead to food insecurity in many countries [9]. It is projected that climate change will cause about 250,000 deaths each year, resulting from malnutrition, malaria, diarrhea, and heat stress, between 2030 and 2050 [10].

Food security itself is a multi-dimensional concept, with goals ranging from ensuring survival and health to gradually considering food preferences, involving food accessibility, availability, utilization, and stability [11], and it operates within social, economic, environmental, and political contexts [12]. It is inconclusive how analysts should rank the various indicators that influence food security [13], which can only be achieved when the specific objective is clearly stated [14]. It will be more valuable for policy guidance to identify clearly with food security's meaning, limitations, and interaction approaches with non-food factors, from a systematic perspective [15]. Globally, the in-depth structural transformation of agri-food systems is expected to be achieved in the future in a more inclusive way [1], i.e., delivering healthy food for human beings, and that it should be resilient to shocks from climate change, economic risk, and market failures [16].

Current trends in poverty, malnutrition, climate change and economic turmoil have driven global food insecurity [17], suggesting widespread failures in food systems [1]. The COVID-19 pandemic continues to expose the weaknesses of current global food systems and heighten concerns over food security risks, especially in low- and middle-income countries [18]. It has been observed that the evolution of our food systems should shift from exclusively boosting production to nourishing people in a more inclusive and sustainable approach, to ensure that future generations will be better provided for [19]. As food system transformation is expected to reflect the approach of pursuing social, environmental, nutritional and health outcomes and potential influences, ensuring global food security to achieve the Sustainable Development Goals will depend on the potential of food systems to increase the efficiency and resilience of the food supply chain.

The ability to deliver affordable and nutritional food and the resilience to adapt to a new normal of regional food systems varied considerably as a result of numerous interacting factors, such as agricultural resources, socio-economic context, and policy orientation [20]. As mentioned earlier, China has made tremendous progress in reducing hunger and food insecurity in recent decades. The total number of undernourished people in China fell from 289 million in 1990–1992, to 125 million in 2016–2017 [21]. Chinese food systems may have global implications, as the country is the largest producer, importer, and exporter of many food commodities, and is the biggest GHG emitter from agri-food systems [22]. More importantly, China has increasingly become a key global player and is an increasingly influential actor in global agri-food network governance, through participating in world cooperation, initiatives, and activities [23]. China and India are considered the countries that are best equipped to instigate food system changes, because they have the necessary "technological capacities, state-regulatory systems and socio-economic need" [24]. Therefore, it would be valuable for other developing countries to analyze the experience of China's food system transformation, although there is no one-size-fits-all solution for all countries. Policymakers will need to assess their own context-specific challenges to achieve the required transformations [25].

Considering a food system captures all the elements and activities that relate to the food supply chain and the outputs of these activities, including the socio-economic and environmental outcomes [26] (Figure 1). This paper will systematically review China's experiences from the perspective of the food supply chain, including food production and management, food reserves, food trade and circulation, as well as food demand and consumption, and then summarize the challenges it faces, the achievements and outcomes of food security in China. It is expected to provide details of real experiences and lessons for global poverty reduction and malnutrition elimination in developing countries.

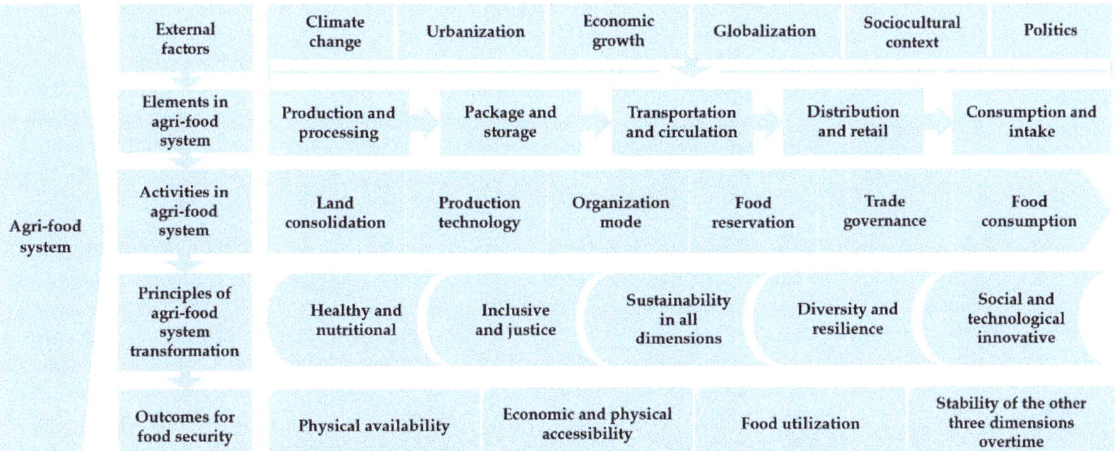

Figure 1. Research framework.

2. Materials and Methods

We conducted extensive literature reviews constituting four steps: identification, screening, eligibility, and inclusion [27], in order to identify peer-reviewed journal articles and gray literature documents that reported the transformation of the food system in China. At the identification step, in the WOS Core Collection, we set the themes as "food security", "food system", and "China". In total, 1323 documents were found, of which 469 belonged to the environmental sciences category. Then, by resetting the themes as "food security", "food system transformation" and "China", 43 documents were found. In the CNKI-CSSCI collection, we set the themes as "food security" and "food system", resulting in 36 documents. For the sake of the comprehensiveness of the included body of literature [28], grey literature was retrieved by searching Google Scholar and the websites of five organizations, including the UN Food and Agriculture Organization (FAO), the International Panel of Experts on Sustainable Food Systems (IPES-Food), the Intergovernmental Panel on Climate Change (IPCC), the World Health Organization (WHO), and the Global Nutrition Report. The objective of this paper was not specifically to assess the effect of China's food policy, but instead to analyze the experience of China's food system transformation from the perspective of the supply chain, including both policy innovation and informal trends, and to obtain enlightenment on global food security governance, including experience and lessons. During the screening and eligibility steps, we reviewed the titles and abstracts of identified articles and examined citations from documents for additional references, after which 96 articles were retained. At the inclusion step, the latter articles were carefully reviewed and all the results presented in this study are based on the analysis of these 96 articles.

3. Results

3.1. China's Food Security Challenges

3.1.1. Increasing Pressure on Resources

Firstly, the deterioration in the quantity and quality of natural resources in China has been attributed to the main constraints on food production. According to the third national land survey, the total area of cultivated land in China was about 0.13 billion ha in 2019, which is 7.53 million ha less than that in 2013. The per capita of arable land decreased from 0.106 ha in 1996 to 0.085 ha in 2018, approximating 46.3% of the world's average. Regarding the current state of land productivity, it was reported that high-quality farmland and reserve land have little room to improve in terms of productivity, while there have been increasing restrictions on low- and medium-level field transformation. These changes have increased concerns about the future development of agricultural production [29]. Secondly, population growth, rising household incomes, and changing dietary patterns will contribute to an increase in the demand for food. With the two-child policy scenario in China, it is estimated that food demand will reach its peak at 39.26 million tons by 2030 [30]. The food consumption pattern in China has gradually transformed from a vegetable-based dietary structure to an animal-based one [31,32]. With the expansion of urbanization, Chinese meat consumption is anticipated to arrive at its peak of 123.44 million tons in 2030, leaving great scope for growth in demand for dairy and aquatic products [33].

3.1.2. Smallholder Vulnerability

Small-scale farmers still play a critical role in reducing rural poverty and ensuring national food and nutrition security [34]. One problem is that with China's aging population, many formerly rural laborers have moved to urban areas, and most rural areas face prominent non-agricultural livelihood transformations, leaving questions such as: "Who will cultivate the land in China?", exerting a negative impact on the domestic food supply [35,36]. In 2018, 59.3% of the rural employed population were engaged in primary industry. Since 2015, the wage incomes of farmers have exceeded the operating income, accounting for 41% of the total income in 2018 [37]. Another issue is that resources are re-allocated from low-profit industries (agriculture) to high-profit industries [38]. The average cost of cultivating crops (16633.5 yuan/ha) in 2019 was 2.2 times above that in 2001; the average net profit was a negative value from 2016 of minus 457.5 yuan/ha in 2019 [39]. In addition, global socio-economical change, and low levels of adaptive capacity caused by inequality and poverty, will exacerbate the challenges that smallholders already face [40].

3.1.3. Increasing Ecological Impact

Ecological problems resulting from the misallocation of resources and food loss in the supply chain are increasingly prominent in China. On the one hand, the rapid changes in grain production patterns in China have increased the vulnerability of the food system. The gravity center of China's grain production has gradually shifted to the north away from the south, and the role of the northern part of China has changed from a major grain-consuming area to a major grain-producing zone [41]. However, water resources in northern regions account for approximately 19% of the total water resources in China, while cultivated land area accounts for 60% [42]; the food provision in Northeast China is approaching its maximum potential capacity. Therefore, it is not wise to rely solely on food provision increases in North China [43], which may increase risks such as water shortages and ecosystem deterioration, and eventually reduce the comprehensive production capacity of grain as well as increase the cost of access to food [33,43]. On the other hand, approximately one-third of the food produced globally for humans is lost or wasted along the food supply chain [44]. Between 2014 and 2018, approximately 27% of China's total food production was lost or wasted annually, making up about one-quarter of the world's total food losses [45].

3.1.4. Dynamic Global Agricultural and Food Policies

With the growth of globalization, the depth of the impact of the global economic and trade environment on China's food security has increased significantly. China's agricultural imports reached USD 170.8 billion in 2020, a 14-fold increase over that in 2001. The import of grain, cotton, oil, sugar, and milk was equivalent to the output of more than 66.7 million ha of cultivated sown area of agricultural land, accounting for 40% of the total planting area of domestic crops [46]. Of these, soybean imports exceeded 100 million tons in 2020, equivalent to 5.1 times that of domestic production. Global food supply chains expose vulnerability as a result of the high concentration of natural resources, unchangeable marine transportation, and unpredictable international situations. In addition, China's participation in global food governance is challenged by international society, with opinions such as "neocolonialism" and "land grabbing" being voiced, and domestic policies are strictly bound by international regulations [47]. These challenges work together to limit the feasible space for China to use the international market to ensure food security.

3.2. China's Food Security Practices

3.2.1. Land Consolidation

Rural development has long been the priority of national planning in Chinese policy. Land consolidation is recognized as a positive factor with economic, social, ecological, and psychological effects around the world [24], one that involves land redistribution and infrastructure construction, and that could be an ideal tool to improve the efficiency of land use and the revitalization of rural areas [48,49]. After establishing a system for separating the ownership rights, contract rights, and management rights for contracted rural land, the Chinese government has implemented an overall plan for land use throughout the country. Since 2004, with the implementation of land remediation measures such as "Permanent basic farmland protection", "Balancing the occupation and replenishment of arable land", "Building High-Standard Basic Farmland", and a compensation system for the people responsible for making it work, China has gradually formed a complete system of farmland protection policy, which has basically achieved no reduction in the cultivated land area [50]. Abiding by the principles of basic food self-sufficiency, based on domestic grain production, China practices the strictest farmland protection system and employs a strategy of sustainable farmland use to increase farmland productivity [51]. Recently, China announced the implementation of "Preventing the use of arable land for non-farming purposes" in the 2020 Annual Central Economic Work Conference to protect the soil fertility of cultivated land by means of crop rotation, according to regional climate, crop types and soil characteristics.

Land policy in China is now focused on the trinity of quantity control, quality management, and eco-environmental protection [52]. At the same time, a series of research projects on land consolidation was carried out following policy orientation, including "non-grain" production behavior [53], agricultural layout optimization [54], land-function zoning [55], and spatial feasibility [56]. Most of these studies were aimed at exploring theoretical support and feasibility evaluations for policymaking from diverse perspectives and multi-scale approaches, and recently began to focus on the relationship between land ecology and food security [57,58]; however, most of the studies were limited to describing the subject as well lack specific trade-offs between policy objectives. Meanwhile, it is argued that many practices did not do well in terms of environmental protection, although China has turned to comprehensive land consolidation in terms of policies and ideology, for which food security seems to be the main goal of land policy in China, rather than promoting ecological protection [59,60].

3.2.2. Production Technologies

At first, traditional agriculture focused on improving resource utilization efficiency and intensification to increase domestic food supply, to satisfy the diverse agri-food demand, resulting in high environmental costs and large-scale resource pressures. It is less likely that "high input–high output" farming will be the future of China's agriculture. After experiencing severe climatic warming and resources limitations, China has begun to explore a sustainable agricultural transformation, based on the principles of "high crop productivity and high resource-use efficiency" (that is, the so-called "double-high" technology system) since 2005 [61], setting national mitigation targets to reach peak carbon dioxide emissions before 2030 and as a means to achieve carbon-neutrality by 2060 [62]. The innovative trajectories of "double-high" agriculture have effectively improved the soil quality, nutrient use efficiency, and adaptation to climate change. As food security is and will be part of the primary agenda, research and policies have shown a gradual emphasis on environmental protection and ecological diversity.

Firstly, much effort has been made in past decades to increase the soil carbon content, including crop straw incorporation, conservation farming, and organic fertilizer application.

Soil organic carbon content has increased by 1.5–2 $g\cdot kg^{-1}$ in most cropping areas in China except the northeast, which means that Chinese success in grain production has also contributed greatly to carbon emission mitigation by improving soil fertility during the past two or three decades [62]. Secondly, a series of technological and policy innovation in nutrient management have been developed since 2004 to increase nutrient-use efficiency in China's grain production [63], including the Zero Increase Action Plan, a project comprising 10,000 mu ($\approx$667 ha) high-yield demonstration areas, farmland protection and quality improvement project, etc. The increase in grain production in China indicates a shift from a quantity-driven, highly resource-dependent, extensive development model to a sustainable intensification model [64]. Thirdly, the Chinese government proposed a new agricultural pattern, namely, Climate-Smart Agriculture (CSA). This was promoted by the FAO to develop agricultural strategies for sustainable food security under the threat of climatic warming, involving policy and mechanism creation, technology integration and demonstration, and institutional capacity-building. China has carried out a large-scale demonstration of the application of climate-smart agriculture in its main grain-producing areas. In the project areas, grain production has increased by more than 5%, with the soil organic carbon content of farmland increasing by 10% [65].

3.2.3. Organization Mode

Existing literature has discussed repeatedly the disconnection and limitations of smallholders, focusing on the causes, necessity, and production vulnerability of a small-scale peasant economy, exploring the path needed for smallholders to integrate into a sustainable agri-food system from political, economic, and ecological perspectives. Several social innovations concerning the smallholder gap between urban markets and rural regions have been developed, to rebuild a more inclusive and just food system [66]. In recent decades in China, new operators of agriculture, including family farms, cooperatives, and "dragon-head" enterprises are emerging. They have the potential to construct large-scale and intensive production bases, disseminate technology to farmers to a large extent through a radiation-driven system, and obtain more market information and policy support [67], although many scholars consider that smallholders are concealed by the capitalist markets that differentiate, destabilize, and transform the meaning of household production [38].

It is clear that e-commerce, as a new mode of alternative food network (AFN), has become a new and effective way to help smallholders to get access to markets. The meteoric rise of "Taobao villages", representing highly concentrated rural e-commerce centers, has been enabled in part by land consolidation projects [68]. With the implementation of a national rural development project since 2013, a series of government-support measures, like the provision of relevant village-level training courses, have been tackling the problem of connecting smallholders to the market, so as to reduce poverty and rural inequality [69], to improve vulnerable groups' green credentials and safe production [70], and to make a revolutionary change to the supply chain through eliminating profit-sapping intermediaries.

Since smallholders have a willingness to protect themselves from the lack of food safety from an increasingly fragmented food system, a "bottom-up force of social self-protection" has gradually developed [71]. Some rural producers have adopted a "one family–two system" approach to farming. Agricultural outputs, produced by conventional methods relying on intensive agrochemicals, are usually sold to the market while those cultivated using organic fertilizers (e.g., manure) are left for self-consumption [72]. This represents an emerging folk practice in China's food system transformation, to ensure food security and food safety. These changes need policy support for organizing farmers to connect with larger consumer groups in the future.

Smallholders play multiple roles in China's food system. Firstly, smallholders have a resilient social foundation as an essential provider of domestic food output. Secondly, the low income of rural households, caused by inefficient agricultural benefits, restricts their access to adequate, nutritious, and healthy food. Considering the specific characteristics of smallholders in the agri-food system, it is important to explore an efficient mechanism to integrate them, to improve production efficiency and household income. In recent years, this topic has been gradually linked with emerging technology, such as the Internet and e-commerce. Thirdly, there has also been a spontaneous reform force from bottom to top, creating a positive economic and ecological impact. In the future, the digital gap, connection mechanisms, and risk management may be an important field of research, and the role of government policy in the connections among various subjects will also become an explorable theme, to find ways for smallholders to bridge the gap.

3.2.4. Food Reserves

The Chinese government has built food reserve systems consisting of a central reserve and a local reserve with different functions. The central grain reserves are designed to satisfy the basic needs of consumers, to respond to disasters and wars; this is the "ballast stone" of national food security. The national reserves can generally meet demotic food consumption for one year [73]. Local grain reserves are expected to counter emergencies in the regional market, stabilize grain prices, and guarantee supply. Meanwhile, China has an established processed grain reserve system for emergency supply for 10 to 15 days in large and medium-sized cities and areas that are prone to price fluctuations; this covers an emergency reserve, processing, and distribution system [74]. Since the implementation of minimum purchase prices and temporary purchase and storage policies in 2004, a modern grain reserve system integrating central, local, and social resources has been established, which can not only resist the fluctuation of grain prices but also calmly deal with natural disasters and the global food crisis.

The food reserve and circulation systems in China have shown flexibility and effectiveness when responding to major events. Since the COVID-19 pandemic, China has established a coordination mechanism to ensure food supply that involved nine provinces and 500 enterprises for prioritizing the shipment of supplies in times of emergency. The mechanism involved coordination between the central government and local governments, and joint actions by the government and private enterprises [75]. It boosted the supply of grain and cooking oil, released central government reserves of frozen pork, and raised the vegetable supply capacity in some provinces [76]. Local governments launched emergency reserves to further expand the scale of grain reserves by coordinating the transfer of other cities' grain resources, supporting the resumption of grain enterprises, encouraging enterprises to expand production capacity, and increasing stocks. The government also set up a "green channel" for food transportation, to resolve the difficulties caused by "lockdown" affecting grain transportation. Other measures, e.g., food monitoring systems, public opinion guidance, and temporary subsidies were implemented to guarantee food consumption and ensure food safety in terms of epidemic prevention [77].

The food reserve policy reflects China's top-down control and management of food security. It is a key means by which to solve the problem of the imbalance between provincial production and demand and the internal food shortage. Many researchers have deconstructed the behavior, layout, and structure of food reserve systems for the reflection and optimization of policymaking, mostly based on qualitative analysis [78,79]. The security of the food supply chain has suffered from multiple challenges, such as natural disasters, health events and policy intervention, highlighting the importance of a reasonable reserve scale and the smooth circulation of the food system. At the same time, cooperation in the whole supply chain will become more complex, requiring multidisciplinary methods to explore risk-response strategies, and the quantitative analysis of food reserve scales and ecological impact needs to be incorporated into the analysis framework of the food system.

3.2.5. Trade Governance

Changes in the diet of residents may lead to a large gap between supply and consumption in feed grain [80], while the global food trade provides an opportunity to alleviate the pressure [81]. In 2013, the Chinese government established a national food security strategy, characterized by self-sufficiency, based on domestic grain production, guaranteed food production capacity, moderate imports, and technological support. Since 2015, China has signed agricultural cooperation agreements with 86 countries along the "Belt and Road", and has invested in more than 820 agricultural projects. The total investment stock was more than USD 17 billion, and the total agricultural trade within the co-construction countries was around USD 95.79 billion in 2020 [82].

Among the countries along the "Belt and Road", Russia, Ukraine, Kazakhstan, Romania, Bulgaria, and other emerging grain exporters are not under the comprehensive control of international grain traders. China has strengthened its in-depth cooperation with these countries in all aspects of the industrial chain, including breeding, processing, warehousing, and logistics, etc. The Chinese government has also participated in the construction of its food industry systems and supply chains through investment, trade, and technology transfer [83]. China has included trade in services within the category of "comprehensive agricultural trade", providing developing countries with full-cycle solutions for industrial chains through foreign aid channels, and realizing dislocation competition with multinational giant companies [84].

Many studies have examined the economic and ecological impacts of the moderate import strategy of China. They suggest moderately reducing the pursuit of a high self-sufficiency rate in terms of food, because this reduces the environmental cost of domestic production and gives consumers more diverse and low-cost food. To protect the safety of the global supply chain while expanding food imports, many researchers have focused on location choices, the layout optimization of global investment, and the "Belt and Road" cooperation promotion strategy. Through practices that include taking part in the initiatives of international organizations, promoting international food regulations, and providing international food aid, China's role in world food security governance has changed from that of a passive participant to an active builder, which has played a positive role in promoting the transformation of world food security governance in a direction more appropriate for developing countries.

3.2.6. Food Consumption

Food waste and loss have become a matter of public concern because they have led to nutrient losses and the inefficient use of resources, including farmland, energy, water, and fuel for food production [85]. Xue et al. found that 17% of food loss and waste was generated at the consumption stage, of which 77% occurs in out-of-home activities; these figures are much higher than those in industrialized countries. One possible reason is the increase in dining out in China. They also pointed out that halving food waste at the consumption stage would reduce almost the same amount of GHG emissions as halving food loss and waste from the production to retail stage [45]. In responding to food waste and loss, China has taken several measures. For example, the Chinese Anti-Food Waste Law has been issued and was implemented in April 2021. More recently, an action plan was issued in November 2021, proposing that food should be saved throughout the entire food chain and establishing a standard and a monitoring system for reducing food loss and waste. However, compared with the food production section, Chinese policies and practices have paid less attention to the trend of increasing nutrition and healthy demand in terms of food consumption, which is a major source of food waste [86]. There are a number of studies on food waste in China from the macro-level that have assessed its economic and environmental impacts, while few studies have been conducted at the household level. These micro-level studies suggested that educating consumers to foster a green diet and improving dietary knowledge would help to reduce chronic disease and food waste [87,88]. There are several challenges in reducing food waste, apparent from international experiences. These are food waste measurement, quantifying the benefits and costs, examining external connections in the food value chain, and learning how to balance the trade-offs between efficiency, food security, and environmental objectives [89]. Considering that the study of food waste may be a good way to combine production and consumption for systematic research, it is necessary to integrate multidisciplinary methods from fields such as geography, resources and the environment, economy, and society, etc. Lastly, including factors such as institutional design, economic behavior, technical constraints, and social cognition would be helpful in future research to examine the underlying reasons behind food waste.

As China's food production and consumption systems have rapidly transformed, sustainable food consumption has become a key issue of global concern [90]. The problem of unreasonable dietary structure has gradually been exposed, while the upgrading of food demand from residents has led to the concept of "hidden hunger", from the perspective of nutrition [73]. Food availability does not guarantee its access [91], which also depends on various factors, such as food allocation, intra-household behavior, and individual food preferences [92]. Thus, anthropometry and household observation aimed at nutritional monitoring can be regarded as an essential supplement of food security measurement. To effectively promote nutritional improvement, China has launched a series of nutrition improvement projects, including nutrition intervention for children and infants, popularizing nutrition knowledge, and supporting the nutrition industry [93]. The heights and weights of children in the pilot areas of the nutrition improvement projects in 2019 have increased significantly, according to the Chinese Center for Disease Control and Prevention [94]. Overall, China still pays more attention to food quantity security, but there is a lack of pursuit of inclusiveness and sustainability on the consumption side compared to developed countries, although in recent years, some scholars have called for more attention to be paid to nutritional health and dietary structure at the household level, to establish a nutrition-oriented food production system and a healthy dietary structure [90]. Therefore, many issues, such as food supply and demand aimed at an overall nutritional goal, the trade-off between food quantity security and nutrition orientation, and long-term nutrition monitoring at both household and individual levels, need to be further studied.

3.3. China's Food Security Achievements and Food System Transformation

China's food security has made great progress in terms of four dimensions: availability, accessibility, affordability, and stability. Firstly, China has achieved self-sufficiency in grain supply in a relatively short period. In 2020, its total grain output was 2.2 times more than that in 1978 (Figure 2). Per capita grain output was around 474 kg, well above the world average of 400 kg [95]. Secondly, the per capita consumption of grains (rice, wheat, corn, barley, oats, etc.) in 2020 was 128.1 kg, approximately 7.8% less than that in 2013, while the per capita consumption of poultry, eggs, aquatic products, and milk and dairy products in 2020 had increased by 76.4, 56.1, 33.7 and 11.1%, respectively (Figure 3). The average daily energy intake of an adult was 2,172 kcal, and the intakes of protein, fat and carbohydrate were 65 g, 80 g, and 301 g, respectively, according to the National Health Commission's standard. Thirdly, remarkable achievements have been realized in terms of targeted poverty alleviation in China. The per capita disposable income of the rural poor increased from 6,079 yuan in 2013 to 12,588 yuan in 2020, with an annual increase rate of 11.6% [96], making food in the market more affordable. Lastly, extensive (mal)nutritional improvement programs implemented in poor areas have contributed tremendously to the improvement of local residents' nutrition and health conditions [95].

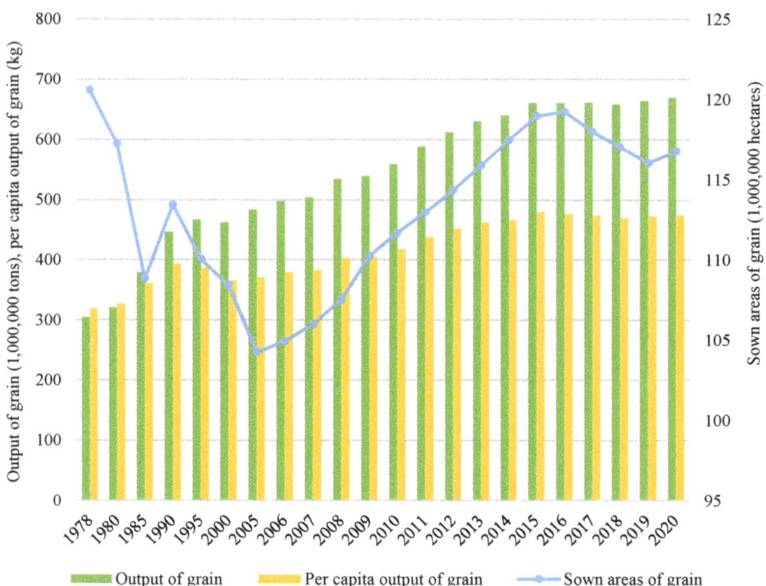

Figure 2. China's output of grain in 1978–2020. Data source: China Statistical Yearbook.

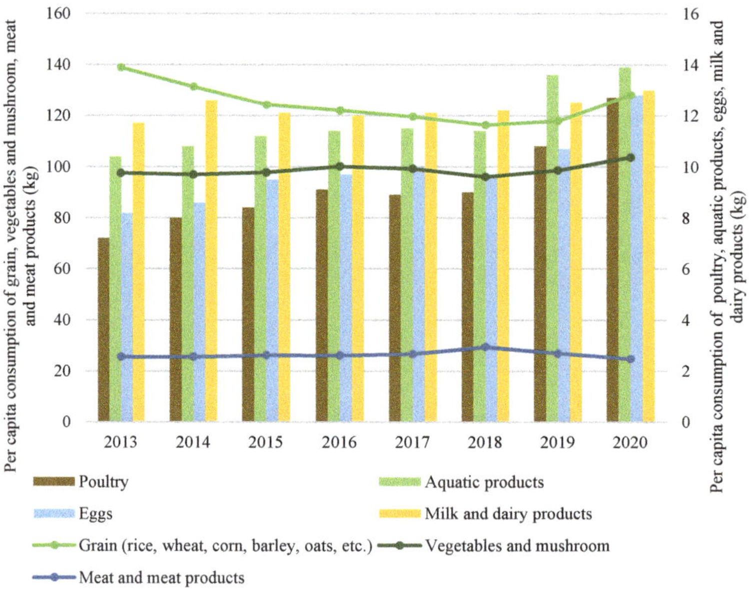

Figure 3. China's per capita consumption of major foods in 2013–2020. Data source: China Statistical Yearbook.

At present, the transformation of China's food system is still in the exploratory stages, with certain characteristic principles:

(1) Health and nutrition. The supply-side reform promotes healthier and nutritious food, according to the changing demand of residents, and supplements the shortage of diverse agri-food through international trade and investment;

(2) Inclusivity and fairness. Poverty alleviation policies improve the ability of the poor to obtain food and expand the employment opportunities of vulnerable small farmers through skills training, knowledge studies, and the cultivation of new subjects, so as to ensure the fairness and justice of food security on both the food supply side and the demand side;

(3) Sustainability in all dimensions. China has reduced carbon emissions and optimized the allocation of global resources through ecological innovation, the reduction of food loss and waste, and international resources usage to promote agricultural green development;

(4) Diversity and resilience. China's food systems have been reoriented around the principles of diversity, multi-functionality, and resilience, to sustain yields and agro-ecosystems in the longer term, and support diverse and nutritious diets;

(5) Social and technological innovation. China's eco-technological innovation has provided more healthy and nutritious food, and the development of digital agriculture provides opportunities for small farmers to innovate their business models.

An agri-food system transformation around these principles can gradually ensure sustainability in environmental, health, social, cultural, and economic dimensions, and will eventually reinforce the ability to support national food security.

4. Conclusions and Recommendations

This paper summarizes China's food security practices from the perspective of the supply chain, including land consolidation, production technology, organization modes, food reserves, trade governance and food consumption. The experiences and lessons to be drawn from these policies may contribute to the transformation of the global agri-food system in the following ways:

(1) Top-down and bottom-up forces are collectively shaping the transformation of China's agri-food system as a whole. The standards in the utilization and allocation of agricultural resources, established through top-down power, have set the baseline for domestic food supply. The bottom-up reform force, involving the spontaneous evolution of business models, benefit mechanisms and alternative markets, has increased the possibilities of realizing food safety and environmental sustainability;

(2) China has significantly strengthened its food security and ecological sustainability through expanding food imports. An external food supply eliminates some negative ecological impacts on agricultural production but also increases risks from foreign markets and global economic crises. Therefore, it is crucial for developing countries to improve the efficiency of foreign agricultural investment and participate in global policymaking;

(3) The vulnerability of the global supply chain exposed by the COVID-19 pandemic enlightens us as to the necessity of improving food reserves and rapidly responding to a sudden shock. The food supply in China during the COVID-19 pandemic offers a good demonstration that the resilience of the agri-food system matters.

At present, China's food system still faces many challenges, including resource and environmental pressure from, e.g., groundwater overexploitation; pesticide and fertilizer overuse; socio-economic challenges, for instance, smallholders having less motivation to grow grain and the lack of innovativeness in the seed industry; trade challenges, such as pressure from international public opinion and tighter import and export regulations. These challenges have directly and/or indirectly led to malnutrition and an unsustainable and inefficient food system. To maximize the potential of agri-food system transformation, more efforts are suggested regarding further studies: (1) to seek integrated solutions through holistic and systematic thinking; (2) to conduct multi-disciplinary cooperation from economics, ecology, politics, geography and other disciplines; (3) to design multi-scale research, e.g., a global-national-local-individual approach is needed to identify the potential challenges and outcomes; (4) to take the mutual support of policy innovation and civil reform into consideration, since the decision-making of actors and power relationships in the system should not be neglected.

Author Contributions: Conceptualization, Y.C. and Y.Z.; methodology, Y.L. and L.H.; software, Y.L.; validation, Y.L., Y.H., L.H., Y.C. and Y.Z.; formal analysis, Y.Z. and Y.C.; investigation, Y.L., Y.H. and L.H.; resources, Y.C. and Y.Z.; data curation, Y.C.; writing—original draft preparation, Y.L.; writing—review and editing, Y.C. and Y.Z.; visualization, Y.L.; supervision, Y.C. and Y.Z.; project administration, Y.C.; funding acquisition, Y.C. and Y.Z. All authors have read and agreed to the published version of the manuscript.

Funding: This research was funded by the National Natural Science Foundation of China, grant number 41871109; the Foundational Research Fund for Central Non-profit Scientific Institution, grant number 161005202116; and The Agricultural Science and Technology Innovation Program, grant number ASTIP-IAED-2022-06.

Institutional Review Board Statement: Not applicable.

Informed Consent Statement: Not applicable.

Data Availability Statement: The public datasets analyzed in this study can be found here: http://www.stats.gov.cn/tjsj/ndsj/2021/indexch.htm (accessed on 1 January 2022).

Acknowledgments: Authors would like to thank the editors and anonymous referees for their kind work and constructive comments.

Conflicts of Interest: The authors declare no conflict of interest.

References

1. Pereira, L.M.; Drimie, S.; Maciejewski, K.; Tonissen, P.B.; Biggs, R. Food system transformation integrating a political–economy and social-ecological approach to regime shifts. *Int. J. Environ. Res. Public Health* **2020**, *17*, 1313. [CrossRef] [PubMed]
2. International Panel of Experts on Sustainable Food Systems. A Long Food Movement: Transforming Food Systems by 2045. Available online: https://www.ipes-food.org/_img/upload/files/LongFoodMovementEN.pdf (accessed on 29 December 2021).
3. Food and Agriculture Organization of the United Nations; International Fund for Agricultural Development; The United Nations International Children's Emergency Fund; The World Food Programme; World Health Organization. *The State of Food Security and Nutrition in the World 2021: Transforming Food Systems for Food Security, Improved Nutrition and Affordable Healthy Diets for All*; Food and Agriculture Organization of the United Nations: Rome, Italy, 2021; p. 8.
4. 2021 Global Nutrition Report: The State of Global Nutrition. Available online: https://globalnutritionreport.org/reports/2021-global-nutrition-report/ (accessed on 23 November 2021).
5. Intergovernmental Panel on Climate Change. Special Report on Climate Change and Land. Available online: https://www.ipcc.ch/srccl/chapter/summary-for-policymakers/ (accessed on 29 December 2021).
6. International Panel of Experts on Sustainable Food Systems. From Uniformity to Diversity: A Paradigm Shift from Industrial Agriculture to Diversified Agroecological Systems. Available online: https://www.ipes-food.org/_img/upload/files/UniformityToDiversity_FULL.pdf (accessed on 2 June 2016).
7. Environmental Impacts of Food Production. Available online: https://ourworldindata.org/environmental-impacts-of-food (accessed on 29 December 2021).
8. International Panel of Experts on Sustainable Food Systems. COVID-19 and the Crisis in Food Systems: Symptoms, Causes, and Potential Solutions. Available online: https://www.ipes-food.org/_img/upload/files/COVID-19_CommuniqueEN(2).pdf (accessed on 29 December 2021).
9. International Panel of Experts on Sustainable Food Systems. Breaking away from Industrial Food and Farming Systems: Seven Case Studies of Agroecological Transition. Available online: https://www.ipes-food.org/_img/upload/files/CS2_web.pdf (accessed on 15 October 2018).
10. World Health Organization. Climate Change and Health. Available online: https://www.who.int/news-room/fact-sheets/detail/climate-change-and-health (accessed on 30 October 2021).
11. Tyfield, D. Food systems transition and disruptive low carbon innovation: Implications for a food security research agenda. *J. Exp. Bot.* **2011**, *62*, 3701–3706. [CrossRef] [PubMed]
12. Rutten, L.F.; Yaroch, A.L.; Story, M. Food systems and food security: A conceptual model for identifying food system deficiencies. *J. Hunger Environ. Nutr.* **2011**, *6*, 239–246. [CrossRef]
13. Santeramo, F.G. On the composite indicators for food security: Decisions matter! *Food Rev. Int.* **2015**, *31*, 63–73. [CrossRef]
14. Haddad, L.; Kennedy, E.; Sullivan, J. Choice of indicators for food security and nutrition monitoring. *Food Policy* **1994**, *19*, 329–343. [CrossRef]
15. Pinstrup-Andersen, P. Food security: Definition and measurement. *Food Secur.* **2009**, *1*, 5–7. [CrossRef]
16. Food and Agriculture Organization of the United Nations. Sustainable Food Systems: Concept and Framework. Available online: www.fao.org/3/ca2079en/CA2079EN.pdf (accessed on 29 December 2021).
17. Liu, Y.; Zhou, Y. Reflections on China's food security and land use policy under rapid urbanization. *Land Use Policy* **2021**, *109*, 105699. [CrossRef]

18. Carducci, B.; Keats, E.C.; Ruel, M.; Haddad, L.; Osendarp, S.J.M.; Bhutta, Z.A. Food systems, diets and nutrition in the wake of COVID-19. *Nat. Food* **2021**, *2*, 68–70. [CrossRef]
19. Caron, P.; Ferrero, Y.; De Loma-Osorio, G.; Nabarro, D.; Hainzelin, E.; Guillou, M.; Andersen, I.; Arnold, T.; Astralaga, M.; Beukeboom, M.; et al. Food systems for sustainable development: Proposals for a profound four-part transformation. *Agron. Sustain. Dev.* **2018**, *38*, 41. [CrossRef]
20. Fan, S.; Teng, P.; Chew, P.; Smith, G.; Copeland, L. Food system resilience and COVID-19—Lessons from the Asian experience. *Glob. Food Sec.-Agric. Policy* **2021**, *28*, 100501. [CrossRef]
21. Zhong, T.; Si, Z.; Scott, S.; Crush, J.; Yang, K.; Huang, X. Comprehensive food system planning for urban food security in Nanjing, China. *Land* **2021**, *10*, 1090. [CrossRef]
22. Fan, S. Economics in food systems. *Nat. Food* **2021**, *2*, 218–219. [CrossRef]
23. Fan, S.; Brzeska, J. Feeding more people on an increasingly fragile planet: China's food and nutrition security in a national and global context. *J. Integr. Agric.* **2014**, *13*, 1193–1205. [CrossRef]
24. Martindale, L. From land consolidation and food safety to Taobao Villages and alternative food networks: Four components of China's dynamic agri-rural innovation system. *J. Rural Stud.* **2021**, *82*, 404–416. [CrossRef]
25. Food and Agriculture Organization of the United Nations; International Fund for Agricultural Development; The United Nations International Children's Emergency Fund; The World Food Programme; World Health Organization. *The State of Food Security and Nutrition in the World 2020: Transforming Food Systems for Affordable Healthy Diets*; Food and Agriculture Organization of the United Nations: Rome, Italy, 2020; pp. 38–42.
26. 2020 Global Nutrition Report: Action on Equity to End Malnutrition. Available online: https://globalnutritionreport.org/reports/2020-global-nutrition-report/ (accessed on 29 December 2021).
27. Nosratabadi, S.; Khazami, N.; Abdallah, M.B.; Lackner, Z.S.; Band, S.S.; Mosavi, A.; Mako, C. Social capital contributions to food security: A comprehensive literature review. *Foods* **2020**, *9*, 1650. [CrossRef]
28. Candel, J.J.L. Food security governance: A systematic literature review. *Food Secur.* **2014**, *6*, 585–601. [CrossRef]
29. Chen, Y.; Wang, J.; Zhang, F.; Liu, Y.; Cheng, S.; Zhu, J.; Si, W.; Fan, S.; Gu, S.; Hu, B.; et al. New patterns of globalization and food security. *J. Nat. Resour.* **2021**, *36*, 1362–1380. [CrossRef]
30. Liu, Q.; Liu, X.; Wang, S. Estimating China's food grains demand from 2020 to 2050 based on reasonable dietary pattern. *Syst. Eng.-Theory Pract.* **2018**, *38*, 616–622.
31. He, P.; Baiocchi, G.; Hubacek, K.; Feng, K.; Yu, Y. The environmental impacts of rapidly changing diets and their nutritional quality in China. *Nat. Sustain.* **2018**, *1*, 122–127. [CrossRef]
32. Zhu, Y. International trade and food security: Conceptual discussion, WTO and the case of China. *China Agric. Econ. Rev.* **2016**, *8*, 399–411. [CrossRef]
33. China's Total Grain Demand Will Reach a Peak in 2030. Available online: https://jjsb.cet.com.cn/show_480319.html (accessed on 1 December 2016).
34. Dramatic Changes Needed in Global Food Systems to Address Nutrition Disparity, Poverty: IFAD. Available online: https://www.downtoearth.org.in/news/agriculture/dramatic-changes-needed-in-global-food-systems-to-address-nutrition-disparity-poverty-ifad-79120 (accessed on 21 September 2021).
35. Du, H.; Chen, J.; Liu, B.; Gong, J. The logic innovation of agricultural production trusteeship to promote smallholder's production modernization. *Chin. J. Agric. Resour. Reg. Plan.* 2021, in press.
36. Peng, C.; Liu, H. Agriculture and rural modernization during the 14th Five-Year Plan Period: The situation, problems and countermeasure. *Reform* **2020**, *2*, 20–29.
37. National Bureau of Statistics. *China Statistical Yearbook*; China Statistics Press: Beijing, China, 2020; p. 176.
38. Yan, H.; Chen, Y. Agrarian Capitalization without capitalism? Capitalist dynamics from above and below in China. *J. Agrar. Change* **2015**, *15*, 366–391.
39. National Development and Reform Commission. *Compilation of Cost-Benefit Data of National Agricultural Products*; China Statistics Press: Beijing, China, 2020; p. 3.
40. Burnham, M.; Ma, Z. Climate change adaptation: Factors influencing Chinese smallholder farmers' perceived self-efficacy and adaptation intent. *Reg. Environ. Change* **2017**, *17*, 171–186. [CrossRef]
41. Liu, Y.; Li, J.; Yang, Y. Strategic adjustment of land use policy under the economic transformation. *Land Use Policy* **2018**, *74*, 5–14. [CrossRef]
42. Chen, L.; Chang, J.; Wang, Y.; Guo, A.; Liu, Y.; Wang, Q.; Zhu, Y.; Zhang, Y.; Xie, Z. Disclosing the future food security risk of China based on crop production and water scarcity under diverse socioeconomic and climate scenarios. *Sci. Total Environ.* **2021**, *790*, 148110. [CrossRef]
43. Wang, Q.; Liu, X.; Yue, T.; Wang, C.; Wilson, J.P. Using models and spatial analysis to analyze spatio-temporal variations of food provision and food potential across China's agro-ecosystems. *Ecol. Model.* **2015**, *306*, 152–159. [CrossRef]
44. Food and Agriculture Organization of the United Nations. *Global Food Losses and Food Waste: Extent, Causes and Prevention*; Food and Agriculture Organization of the United Nations: Rome, Italy, 2020; p. 4.
45. Xue, L.; Liu, X.; Lu, S.; Cheng, G.; Hu, Y.; Liu, J.; Dou, Z.; Cheng, S.; Liu, G. China's food loss and waste embodies increasing environmental impacts. *Nat. Food* **2021**, *2*, 519–528. [CrossRef]
46. Ni, H. China's agricultural supply-side structural reform from an opening-up perspective. *Issues Agric. Econ.* **2019**, *2*, 9–15.

47. Zhu, J.; Li, T.; Zang, X. Emerging challenges and coping strategies in China's food security under the High-level Opening Up. *Issues Agric. Econ.* **2021**, *1*, 27–40.
48. Jiang, Y.; Long, H.; Tang, Y. Land consolidation and rural vitalization: A perspective of land use multifunctionality. *Prog. Geogr.* **2021**, *40*, 487–497. [CrossRef]
49. Jiang, Y.; Tang, Y.; Long, H.; Deng, W. Land consolidation: A comparative research between Europe and China. *Land Use Policy* 2022, in press. [CrossRef]
50. Yu, H.; Zeng, S.; Wang, Q.; Dai, J.; Bian, Z. Forecast on China's cultivated land protection baseline in the new era by multi-scenario simulations. *Resour. Sci.* **2021**, *43*, 1222–1233.
51. Huang, S. Discussion on China agricultural security: From both sides of supply and demand. *Issues Agric. Econ.* **2021**, *8*, 4–11.
52. Yan, J.; Xia, F.; Bao, H.X.H. Strategic planning framework for land consolidation in China: A top-level design based on SWOT analysis. *Habitat Int.* **2015**, *48*, 46–54. [CrossRef]
53. Chen, F.; Liu, J.; Chang, Y.; Zhang, Q.; Yu, H.; Zhang, S. Spatial pattern differentiation of non-grain cultivated land and its driving factors in China. *China Land Sci.* **2021**, *35*, 33–43.
54. Zhou, K.; Li, J.; Wang, Q. Evaluation on agricultural production space and layout optimization based on resources and environmental carrying capacity: A case study of Fujian Province. *Sci. Geol. Sin.* **2021**, *41*, 280–289.
55. Zhao, Q.; Jiang, G.; Xiong, C.; Yan, G.; Su, S. Study on land consolidation function zoning and its consolidation direction. *Chin. J. Agric. Resour. Reg. Plan.* **2021**, *42*, 52–60.
56. Duan, J.; Ren, C.; Wang, S.; Zhang, X.; Reis, S.; Xu, J.; Gu, B. Consolidation of agricultural land can contribute to agricultural sustainability in China. *Nat. Food* 2021, in press. [CrossRef]
57. Xue, X.; Ma, L. Analysis on the coupling and coordination of land ecological and food security in main grain producing areas. *Chin. J. Agric. Resour. Reg. Plan.* 2021, in press.
58. Liu, Y.; Fu, B.; Wang, S.; Li, Y.; Zhao, W.; Li, C. Review and prospect of the water-food-ecosystem nexus in dryland's Coupled Human-Earth System. *Geogr. Res.* **2021**, *40*, 541–555.
59. Scott, S.; Si, Z.; Schumilas, T.; Chen, A. Contradictions in state-and civil society-driven developments in China's ecological agriculture sector. *Food Policy* **2014**, *45*, 158–166. [CrossRef]
60. Tang, H.; Yun, W.; Liu, W.; Sang, L. Structural changes changes in the development of China's farmland consolidation in 1998-2017: Changing ideas and future framework. *Land Use Policy* **2019**, *89*, 104212. [CrossRef]
61. Shen, J.; Cui, Z.; Miao, Y.; Mi, G. Transforming agriculture in China: From solely high yield to both high yield and high resource use efficiency. *Glob. Food Sec.-Agric. Policy* **2013**, *2*, 1–8. [CrossRef]
62. Deng, A.; Chen, C.; Feng, J.; Chen, J.; Zhang, W. Cropping system innovation for coping with climatic warming in China. *Crop J.* **2017**, *5*, 136–150. [CrossRef]
63. Zhang, F.; Cui, Z.; Zhang, W. Managing nutrient for both food security and environmental sustainability in China: An experiment for the world. *Front. Agric. Sci. Eng.* **2014**, *1*, 53–61.
64. Jiao, X.; He, G.; Cui, Z.; Shen, J.; Zhang, F. Agri-environment policy for grain production in China: Toward sustainable intensification. *China Agric. Econ. Rev.* **2018**, *10*, 78–92. [CrossRef]
65. Reinforcing Ability to Sequestrate Carbon, Reduce Emission, Stabilize Grain Production and Increase Farmers' Income. Available online: https://szb.farmer.com.cn/2020/20200918/20200918_007/20200918_007_1.htm (accessed on 18 September 2020).
66. Huang, G.; Tsai, F. Social innovation for food security and tourism poverty alleviation some examples from China. *Front. Psychol.* **2021**, *12*, 712709. [CrossRef]
67. Schneider, M. Dragon head enterprises and the state of agribusiness in China. *J. Agrar. Change* **2017**, *17*, 3–21. [CrossRef]
68. Ma, W.; Jiang, G.; Zhang, R.; Li, Y.; Jiang, X. Achieving rural spatial restructuring in China: A suitable framework to understand how structural transitions in rural residential land differ across peri-urban interface? *Land Use Policy* **2018**, *75*, 583–593. [CrossRef]
69. Li, A.H.F. E-Commerce and Taobao Villages. A promise for China's rural development? *China Perspect.* **2017**, *3*, 57–62. [CrossRef]
70. Zhao, Q.; Pan, Y.; Xia, X. Internet can do help in the reduction of pesticide use by farmers: Evidence from rural China. *Environ. Sci. Pollut. Res.* **2021**, *28*, 2063–2073. [CrossRef] [PubMed]
71. Zhang, L.; Qi, G. Bottom-up self-protection responses to China's food safety crisis. *Can. J. Dev. Stud.* **2019**, *40*, 113–130. [CrossRef]
72. Si, Z.; Li, Y.; Fang, P.; Zhou, L. "One family, two systems": Food safety crisis as a catalyst for agrarian changes in rural China. *J. Rural Stud.* **2019**, *69*, 87–96. [CrossRef]
73. Zhou, L.; Luo, J.Z.; Fang, P. Who will feed China in the 21st Century? Pandemic crisis, glocalization and organized responsibility. *Int. Econ. Rev.* **2021**, *5*, 53–80.
74. Domestic Food Security Provides A Solid Foundation for the Realization of Moderate Prosperity throughout the Country. Available online: http://paper.ce.cn/jjrb/html/2020-08/17/content_425909.htm (accessed on 17 August 2020).
75. Zhan, Y.; Chen, K.Z. Building resilient food system amidst COVID-19: Responses and lessons from China. *Agric. Syst.* **2021**, *190*, 103102. [CrossRef]
76. The State Council Information Office of the People's Republic of China. Fighting Covid-19: China in Action. Available online: http://www.scio.gov.cn/ztk/dtzt/42313/43142/index.htm (accessed on 7 June 2020).
77. Li, X.; Han, X.; Qi, H. The situation of China's grain supply and price stability: Status quo, challenges and government's measures—Based on the perspective of COVID-19 pandemic. *Prices Mon.* **2021**, *1*, 9–15.
78. Han, J.; Zou, Y. Spatial differences and scale determination of regional grain reserves. *J. Nat. Resour.* **2019**, *34*, 464–472. [CrossRef]

79. Wei, X.; Shi, Q. Household grain storage and food security: An analysis based on data from Shanxi, Zhejiang and Guizhou. *Chin. Rural Econ.* **2020**, *9*, 86–104.

80. Xin, L. Dietary structure upgrade of China's residents, international trade and food security. *J. Nat. Resour.* **2021**, *36*, 1469–1480. [CrossRef]

81. He, R.; Zhu, D.; Chen, X.; Cao, Y.; Chen, Y.; Wang, X. How the trade barrier changes environmental costs of agricultural production: An implication derived from China's demand for soybean caused by the US-China trade war. *J. Clean Prod.* **2019**, *227*, 578–588. [CrossRef]

82. Promote Agricultural Cooperation and Share Development Achievements. Available online: http://paper.people.com.cn/rmrb/html/2021-11/25/nw.D110000renmrb_20211125_2-06.htm (accessed on 25 November 2021).

83. Zhang, J. Beyond the 'Hidden agricultural revolution' and 'China's overseas land investment': Main trends in China's agriculture and food sector. *J. Contemp. China* **2019**, *28*, 746–762. [CrossRef]

84. Sun, Z.; Zhang, D. China's increased utilization of grain markets and resources in countries along the "One-Belt and One-Road" under the new situation. *Res. Agric. Mod.* **2021**, *42*, 827–840.

85. Lang, L.; Wang, Y.; Chen, X.; Zhang, Z.; Yang, N.; Han, W. Awareness of food waste recycling in restaurants: Evidence from China. *Resour. Conserv. Recycl.* **2020**, *161*, 104949. [CrossRef]

86. Cheng, S.; Bai, J.; Jin, Z.; Wang, D.; Liu, G.; Gao, S.; Bao, J.; Li, X.; Li, R.; Jiang, N.; et al. Reducing food loss and food waste: Some personal reflections. *J. Nat. Resour.* **2017**, *32*, 529–538.

87. Ma, L.; Qin, W.; Garnett, T.; Zhang, F. Review on drivers, trends and emerging issues of the food wastage in China. *Front. Agric. Sci. Eng.* **2015**, *2*, 159–167. [CrossRef]

88. Min, S.; Wang, X.; Yu, X. Does dietary knowledge affect household food waste in the developing economy of China? *Food Policy* **2021**, *98*, 101896. [CrossRef]

89. Cattaneo, A.; Sanchez, M.V.; Torero, M.; Vos, R. Reducing food loss and waste: Five challenges for policy and research. *Food Policy* **2021**, *98*, 101974. [CrossRef] [PubMed]

90. Zhang, C.; Qiang, W.; Niu, S.; Wang, R.; Zhang, H.; Cheng, S.; Li, F. Options of Chinese dietary pattern based on multi-objective optimization. *Resour. Sci.* **2021**, *43*, 1140–1152. [CrossRef]

91. Pinstrup-Andersen, P.; Pandya-Lorch, R. Food Security: A Global Perspective. In *Food Security, Diversification and Resource Management: Refocusing the Role of Agriculture?* Peters, G.H., von Braun, J., Eds.; Routledge: New York, NY, USA, 2018; pp. 51–76. ISBN 0429854390.

92. Jones, A.D.; Ngure, F.M.; Pelto, G.; Young, S.L. What are we assessing when we measure food security? A compendium and review of current metrics. *Adv. Nutr.* **2013**, *4*, 481–505. [CrossRef]

93. Qing, P.; Liao, F.; Min, S.; Pei, M.; You, L. Nutrition poverty alleviation: A new mode to help health poverty alleviation and promote targeted poverty alleviation—Literature review based on domestic and foreign research. *Issues Agric. Econ.* **2020**, *5*, 4–16.

94. Zhou, L.; Wang, J.; Jiang, B. The impact of rural compulsory education nutrition improvement program on students' health. *China Rural Surv.* **2021**, *2*, 97–114.

95. The State Council Information Office of the People's Republic of China. Food Security in China. Available online: http://www.gov.cn/zhengce/2019-10/14/content_5439410.htm (accessed on 14 October 2019).

96. The State Council Information Office of the People's Republic of China. Poverty Alleviation: China's Experience and Contribution. Available online: https://english.www.gov.cn/archive/whitepaper/202104/06/content_WS606bc77ec6d0719374afc1b9.html (accessed on 6 April 2019).

Article

Effectiveness of Triple Benefit Health Education Intervention on Knowledge, Attitude and Food Security towards Malnutrition among Adolescent Girls in Borno State, Nigeria

Ruth Charles Shapu [1,2], Suriani Ismail [1,*], Poh Ying Lim [1], Norliza Ahmad [1], Hussaini Garba [3] and Ibrahim Abubakar Njodi [3]

[1] Department of Community Health, Faculty of Medicine and Health Science, University Putra Malaysia, Serdang 43400, Selangor, Malaysia; ruthyshapu52@gmail.com (R.C.S.); pohying_my@upm.edu.my (P.Y.L.); lizaahmad@upm.edu.my (N.A.)
[2] College of Nursing and Midwifery, Damboa Road, Maiduguri 600252, Borno State, Nigeria
[3] Department of Physical and Health Education, University of Maiduguri, Maiduguri 600230, Borno State, Nigeria; garbahussaini1759@gmail.com (H.G.); ibrahimnjodi@gmail.com (I.A.N.)
* Correspondence: si_suriani@upm.edu.my; Tel.: +60-19-2249828

Abstract: Knowledge and attitude are essential components of food security as malnutrition remains a critical public health concern among adolescents. The study evaluates the effectiveness of a Triple Benefit Health Education Intervention on knowledge, attitude and food security towards malnutrition among adolescent girls. This was a cluster randomized controlled trial among 417 randomly selected adolescent girls aged 10 to 19 years old in Maiduguri, Borno state, Nigeria from October 2019 to March 2020. About 208 respondents were assigned to experimental while 209 to control group, respectively, using an opaque sealed envelope. A structured questionnaire using KoBo Collect Toolbox was used for the collection of data at baseline, three and six-months post intervention while the data collected were analyzed using generalized estimating equation (GEE). The outcome of the baseline shows no statistically significant difference in sociodemographic characteristics, knowledge, attitude and food security between experimental and control groups. The study reveals a statistically significant difference between experimental and control groups for knowledge ($p < 0.001; p < 0.001$), attitude ($p < 0.001; p < 0.001$) and food security ($p = 0.026; p = 0.001$) at three and six-months post intervention, respectively. The triple benefit health education intervention package employed in this study can serve as an intervention tool to combat malnutrition among adolescent girls in Nigeria at large.

Keywords: triple benefit health education intervention; adolescent girls; knowledge; attitude; food security; KoBo Collect Toolbox; generalized estimating equation

Citation: Shapu, R.C.; Ismail, S.; Lim, P.Y.; Ahmad, N.; Garba, H.; Njodi, I.A. Effectiveness of Triple Benefit Health Education Intervention on Knowledge, Attitude and Food Security towards Malnutrition among Adolescent Girls in Borno State, Nigeria. *Foods* **2022**, *11*, 130. https://doi.org/10.3390/foods11010130

Academic Editors: Theodoros Varzakas, António Raposo, Renata Puppin Zandonadi and Raquel Braz Assunção Botelho

Received: 6 September 2021
Accepted: 20 October 2021
Published: 5 January 2022

1. Introduction

Adolescents are the nation's future, so there is a need for them to have an adequate knowledge of, and attitude towards, the importance of balanced diet and the importance of food security in terms of its availability, accessibility, affordability and utilization, as it forms the basis for a healthy life, mental development, intellectual abilities, physical growth and strength to cope with daily activities [1,2]. The effect of malnutrition and food security among adolescents, and especially during their childhood years, is apparent. Food insecurity at the adolescent stage is the possible cause of numerous chronic diseases during adult life. Preventive effort to minimize both short and long term consequences of malnutrition should start at the early childhood stage, all through adolescence and even beyond, to the benefit of health and later life [3–5]. The adverse effect of malnutrition with an increasing trend among adolescents globally reflect poor knowledge, attitude and insufficient food intake [6]. The various forms of malnutrition are persistent and co-exist within various families, communities, regions and countries of the world. Nearly one third of the world population are facing one or more forms of malnutrition, and with this rising trend malnutrition may rise to one half by 2030, rendering the objective of ending malnutrition by 2030 unachievable. The negative health effect of malnutrition, the risk of premature death and the prevalence of non-communicable diseases is increasingly becoming unbearable in middle and low income countries [7–9].

Globally, it has been reported that many children and adolescents (young people) are not able to develop and grow to reach their full potential and become productive in life due to an inadequate diet to meet their daily nutritional needs. In order for children and adolescents to have good nutritional intake, it is important for them, especially adolescent girls, to have a nutritious, affordable, safe, and sustainable diet to meet their everyday needs. The triple burden of malnutrition (undernutrition, over nutrition and micronutrient deficiency) has become a threat to the growth, development and survival of children and adolescents, further endangering the productivity of nations, as a result of the poor quality and quantity of diet consumed [10]. Knowledge and attitude are significant components in the transformation of food security especially among children and adolescents who will have to be accountable for the implementation of most of the nutrition related policies as they advance towards their maturity. Healthy and sufficient nutrition is indispensable for children and adolescents at all stages of their growth and development, from playing to learning and being happy [11]. Findings from a study on high-return investments in school health regarding increased participation and learning suggests that the first 1000 days is important in a child's life, but not sufficient for the survival of the child. Investment in the 8000 days of childhood through adolescence requires intense intervention in three possible stages. Phase one (5 to 9 years) is when mortality rate is higher due to malnutrition and infection, thereby restraining growth and development, referred to as middle childhood growth and consolidation stage. Phase two (10–14 years) is the growth spurt stage when adolescents require an adequate and balanced diet for their healthy growth and development. Phase three (15 to 19 years), also called the growth and consolidation stage, supports the maturation of the brain, emotional control and socialization. The suggested 8000 days will enable a catch up from growth failure at the early stage [12–14]. Poor diet as a result of food insecurity, poor knowledge and attitude are the major underlying causes of malnutrition, with high economic costs, placing the burden of intergenerational malnutrition across generations to come in the future [7].

Health education intervention for adolescents in a school setting provides them with the required skills, knowledge and attitudes for healthy selection and utilization of food. Unhealthy selection of food, inadequate information on the importance of food and food groups, malnutrition and its consequences and lack of a kitchen garden among adolescents can increase the burden of malnutrition, which can be mitigated through health education intervention. Previous studies have revealed that health education intervention lays a foundation for the improvement of nutritional knowledge and attitudes relevant to healthy living [1,15–17]. Addressing the root causes of malnutrition requires a common understanding of the problems and the possible solutions therein, health education intervention across all age groups, both in school and out of school, and the political will in terms of governance, implementation and alignment of policies [8]. The rising trend of poverty and low income has remained a complex, chronic and pervasive problem; an estimated 40.1% of the Nigerian population live below the poverty line where children from the poorest economic quartile were reported to be four times more likely to be malnourished compared with children from the richest households. Malnutrition is a serious consequence of food insecurity [18,19]. The level and dimension of hunger and food insecurity have become a public health concern in Nigeria. Agricultural production in Nigeria is largely dependent on rainfall at a subsistent level on a small scale. Government investment in agricultural production has not contributed to the reduction of malnutrition significantly to meet the national development goal, as the inadequate storage system, crop seasonality, and inadequate transport system has significantly influenced food distribution in Nigeria [18,20,21].

Persistent humanitarian crises especially in the north-eastern part of Nigeria have greatly exposed the people to untold hardship.

Since there are high burdens of malnutrition among children and adolescents that may heighten health related problems, there is need for nutrition sensitive intervention, especially in low and middle-income countries, which must be addressed as a priority. The sustainable development goals have pointed out ways of reducing the determinant of malnutrition, including no poverty, no hunger, quality education, and gender equality, but for the adolescent age group social determinants and nutrition sensitive intervention must be addressed to improve their health and wellbeing. Healthy adolescents who are protected from morbidity and early pregnancy are less likely to develop all forms of malnutrition during adolescence and adulthood, and more likely to reduce the occurrence of non-communicable diseases, have optimal maternal and birth outcomes, and enjoy increased productivity and work capacity [6].

There are no existing data on the overall current prevalence of malnutrition, knowledge, attitudes and food security related issues among adolescents in Nigeria. The trend of malnutrition among women aged 15–49 years, with adolescents included, has been stable for the past 10 years, reporting 12% from 2008 to 2018 [22]. Furthermore, the prevalence of acute malnutrition among older adolescents was 19% and about four times higher compared to 4% among adult women aged 20–49 years in Nigeria. About 6.1% out of 14% of women between 15–49 years pregnant particularly in the northeast and northwest Nigeria were adolescent girls aged 15–19 years old. More so, anemia is also a trending concern among 58% of women aged 15–49 years leading to increased burden of maternal mortality, poor birth outcome and reduced productivity [22,23]. Childbearing among adolescent girls is significantly associated with a high risk of pregnancy complication outcome [24]. The findings from the 2018 national survey call for urgency in the development of an intervention to improve nutrition related knowledge and attitude among adolescent girls for better health, birth outcome and nutrition throughout their life cycle. Taming malnutrition among adolescent girls is key to improving the nutritional status of the family and the entire population.

There are several policies and programs put in place by the Nigerian government to address the problem of malnutrition among children including the national policy on food and nutrition in Nigeria [25,26], the national strategic plan of action (health sector response), the food security bill, the National Plan of Action on Food and Nutrition in Nigeria and the micronutrients control program, among others, to address the issue of malnutrition and food insecurity at all levels in Nigeria [27]. Nevertheless, the implementation of these programs and policies continues to be a challenge, with a persistently high level of malnutrition among children [27–29]. There are no existing nutrition programs and policies for adolescents as the national food and nutrition policy in Nigeria made little or no reference to nutrition-related issues regarding this important specific population (adolescents). The strategic plan of action and most of the intervention target children that are below the age of five years, pregnant and lactating women, failing to notice the plight of adolescents. There is a need to revisit the national strategic plan of action in Nigeria to include interventions targeting adolescents for the good of the future, in all its seriousness. Nonetheless, amid the trending burden of malnutrition, there is no existing comprehensive health education program on malnutrition targeting adolescents in Nigeria; though there are policies to reduce the burden of malnutrition, the implementation is still a challenge. The triple benefit health education intervention was introduced to look beyond the 1000 days of the child's life (from conception to the second birthday), before the preconception period to productive adult life in the future, and the health and wellbeing of their offspring.

Usually, researchers incorporate a theory-based approach for their research to be successful. The information, motivation, behavioral skills model (IMB) has been used in promoting nutrition-related practices and the development of preventive intervention [30]. Although the IMB model works at the individual level, the construct of the theory is useful in presenting health behavioral changes through making individuals well informed by using their attitude and perception as motivation to make them comply and act, thus possessing the essential behavioral skills for effective achievement. The information, motivation, behavioral skills model was used in developing the Triple Benefit Health Education Intervention to improve knowledge, attitude, and food security among adolescent girls. The study is named Triple Benefit because it will improve the health of adolescent girls now, improve their productivity and well-being when she fully-grown in the future and minimize health threats and improve nutritional status and wellbeing of future offspring. Knowledge, attitude and food security towards malnutrition and nutrition-related studies among adolescents have not yet been studied in the north-eastern part of Nigeria comprising Borno, Yobe, Adamawa, Bauchi, Gombe and Taraba states from the 36 states in Nigeria. The study tends to look at what is the effectiveness of the Triple Benefit Health Education Intervention on knowledge, attitude, and food security regarding malnutrition among adolescent girls in Maiduguri Metropolitan Council, Borno State, Nigeria.

2. Materials and Methods

2.1. Study Design and Population

A school-based single blinded cluster randomized controlled trial designed was used to evaluate the effectiveness of the triple benefit health education intervention among adolescent girls in four randomly selected secondary schools in Maiduguri, Borno state, Nigeria. A two population proportion formula was used to calculate the population size [31], based on previous study [32], considering effect size = 1.3, and non-response rate of 20%, giving a total population size of 424.

2.2. Inclusion and Exclusion Conditions

Inclusion criteria were government secondary schools, schools with full secondary school, schools with boys and girls or girls school only, schools within Maiduguri only and adolescent girls between 10 and 19 years old. Schools not owned by the government, primary schools and schools with women and not girls were excluded.

2.3. Duration of Study and Recruitment of Participants

This study was conducted between October 2019 and March 2020. A two-stage sampling method was adopted to randomly select four schools using a software number generator, and a total of 424 were randomly selected from six arms of the schools (JSS1 to SS3) to partake in the study. About 212 participants were recruited into the experimental group and 212 into the control group; each group was allocated using the simple random method, the allocation was performed as a cluster into the experimental and control groups using opaque sealed envelopes. However, 417 were eligible for participation in the study. Seven participants were excluded from the study, of which three were more than 19 years old, while parents of the four students did not give their consent. About 417 respondents brought back the signed consent form. Participants were interviewed using KoBo Toolbox at baseline with 208 in the intervention group and 209 in the control group, respectively.

2.4. Intervention Stratebt

The triple benefit health education intervention module was guided by the information, motivation, behavioral skill (IMB) theory to educate adolescent girls on knowledge, attitudes and food security. The experimental group in this study received the triple benefit health education intervention twice in a month for three months. The module was developed with the clear intention of reducing the burden of malnutrition and ultimately improving the health, well-being and the health of the future offspring of adolescent girls. The Triple Benefit Health Education Intervention module was subdivided into six modules and was conducted for three months with the experimental group by the facilitator. Each session lasted for 1 h:30 min, and the topics covered in the sessions are presented in Table 1. The control group received education twice a month for three months on malaria prevention.

Table 1. Illustration of Triple Benefit Intervention Module using Information Motivation Behavioral Skill Model.

Module	Theory Construct	Content	Strategy	Estimated Time
Module 1	Information and behavioral skills	Definitions, forms and causes of malnutrition	Lecture, brainstorming and discussion	1 h 30 min
Module 2	Information and behavioral skills	Sign, consequences and prevention of malnutrition	Lectures, brainstorming and role play	1 h 30 min
Module 3	Information and behavioral skills	Food Groups (macro and micronutrient).	matching food	1 h 30 min
Module 4	Motivation	Prevention of malnutrition, participant's experiences and those of other adolescent girls.	Brainstorming, discussion	1 h 30 min
Module 5	Information and behavioral skills	Intergenerational Cycle of Malnutrition, Food Groups by Food and Agriculture Organisation of the United Nations (FAO),	Lectures, discussion and kitchen/backyard garden.	1 h 30 min
Module 6	Motivation	Precautionary measures, the norms of their community, and best ways they can spread what they have learnt to their families and peers for sustainability.	Brainstorming, discussion	1 h 30 min

2.5. Study Instruments and Variables

A questionnaire was used for the collection of data, using KoBo Toolbox, on sociodemographic characteristics, knowledge, attitude and food security. The questionnaire was developed by the researcher in English.

2.5.1. Demographic Features

This segment comprises 15 questions on sociodemographic characteristics including age, ethnicity, class in school, religion, household size, place of residence, household income, head of household, age of father, father's education, father's occupation, age of mother, mother's education, mother's occupation, and family type. This section has both nominal and continuous scale variables.

2.5.2. Knowledge, Attitude and Food Security

The knowledge questionnaire was adapted from the Food and Agricultural Organization of the United Nations [33] and consists of 28 questions with yes, no, or don't know as options; yes responses were scored one while no or don't know were scored zero. Total score for knowledge was 28; scores less than 50% were considered as poor knowledge, while scores greater or equal to 50% were considered good knowledge [34].

In this study the attitude questionnaire was adapted from the Food and Agricultural Organization of the United Nations [33]. Attitude questions consist of 17 statements on a five point scale from strongly disagree to strongly agree (1 to 5). The total attitude score was 85; scores less than the mean score were considered poor attitude while scores equal or greater than the mean were considered as showing good attitude [35].

Food security questions were adapted from those for older children [36] and consist of nine statements with options as never, sometimes and a lot, with scores of 1, 2 and 3, respectively; never was recoded as '0', sometimes and a lot were recorded as '1'. Food security has a total score of nine; 0 to 1 were considered food secure, 2 to 5 low food secured and 6 to 9 very low food security, respectively [37].

2.6. Data Collection and Analysis

Data were collected at baseline, three and six-month post intervention using KoBo Toolbox and were analyzed using SPSS version 25. Variables that are continuous were presented as mean and standard deviation before being categorized, while categorical variables were presented in the form of frequency and percentage. The differences between experimental and control group for each demographic feature were determined using Chi-square at baseline, whereas for differences in overall scores between experimental and control group for knowledge, attitude and food security at baseline, independent *t*-test was used. A generalized estimating equation was used to evaluate the changes between experimental and control group at three and six-months post intervention. Statistical significance was determined using p-value < 0.05 for all the comparisons in this study.

2.7. Ethics

Approval for ethics was obtained from Universiti Putra Malaysia Ethical Committee UPM/TNCPI/RMC/JKEUPM/1.4.18.2 (JKEUPM). Permission was obtained from the Ministry of Education. Written consent from parent/guardian and respondents was obtained before the intervention. The study was also registered with Pan African Clinical Trials Registry (PACTR201905528313816).

3. Results

This study evaluates the effectiveness of a triple benefit health education intervention in improving knowledge, attitude and food security among adolescent girls. Out of a total of 424 participants, 417 gave their consent to partake in the study. A total of 208 were assigned to the experimental group, while 209 to the control group. There was a response rate of 100% at baseline, 98.1% at three-months and 96.6% at six-months post-intervention in both groups, respectively. Intention to treat analysis was employed and all randomized participants were included in the analysis by the imputation of the mean of the responses at three and six-months post-intervention [38]. The proportion of missing data due to attrition was 4% in the experimental group and 2% in the control group.

3.1. Baseline Results

3.1.1. Sociodemographic Characteristics at Baseline

Table 2 reveals the baseline demographic features of respondents in the experimental and control groups. There was no statistically significant difference in socio-demographic characteristics between experimental and control groups at baseline.

Table 2. Baseline Comparison of Sociodemographic Characteristics Between Intervention and Control Groups.

Variables	Experimental n (%)/Mean $\pm$ SD	Control n (%)/Mean $\pm$ SD	Total n (%)	X^2/t	p-Value
Age of adolescent girls (Years)	15.0 ± 2.0	15.0 ± 2.0		-3.390 [e]	0.697
Early adolescents (10–13)	55 (26.4)	52 (24.9)	107 (25.7)	0.171 [d]	0.918
Middle adolescents (14–16)	101 (48.6)	102 (48.8)	203 (48.7)		
Late adolescents (17–19)	52 (25.0)	55 (26.3)	107 (25.7)		
Class [a]				0.704 [d]	0.983
JSS1	36 (17.3)	31 (14.8)	67 (16.1)		
JSS2	39 (18.8)	41 (19.6)	80 (19.2)		
JSS3	32 (15.4)	30 (14.4)	62 (14.9)		
SS1	29 (13.9)	30 (14.4)	59 (14.1)		
SS2	45 (21.6)	47 (22.5)	92 (22.1)		
SS3	27 (13.0)	30 (14.4)	57 (13.7)		
Ethnicity				10.494 [d]	0.232
Bura	35 (16.8)	16 (7.7)	51 (12.2)		
Kanuri	61 (29.3)	76 (36.4)	137 (32.9)		
Hausa	13 (6.3)	13 (6.2)	26 (6.2)		
Marghi	18 (8.7)	21 (10.0)	39 (9.4)		
Shuwa	13 (6.3)	18 (8.6)	31 (7.4)		
Fulani	17 (8.2)	20 (9.6)	37 (8.9)		

Table 2. *Cont.*

Variables	Experimental *n* (%)/Mean ± SD	Control *n* (%)/Mean ± SD	Total *n* (%)	X^2/t	*p*-Value
Chibok	14 (6.7)	11 (5.3)	25 (6.0)		
Gwoza	20 (9.6)	18 (8.6)	38 (9.1)		
Other ethnic groups [b]	17 (8.2)	16 (7.7)	33 (7.9)		
Religion				0.035 [d]	0.851
Christianity	57 (27.4)	59 (28.2)	116 (27.8)		
Islam	151 (72.6)	150 (71.8)	301 (72.2)		
Place of residence				1.991 [d]	0.158
Rural	30 (14.4)	41 (19.6)	71 (17.0)		
Urban	178 (85.6)	168 (80.4)	346 (83.0)		
Household size				0.021 [d]	0.989
≤5 members	15 (7.2)	15 (7.2)	30 (7.2)		
6–8 members	77 (37.0)	76 (36.4)	153 (36.7)		
≥9 members	116 (55.8)	118 (56.5)	234 (56.1)		
Monthly income				1.578 [d]	0.664
Less than ₦18,000	53 (25.5)	44 (21.1)	97 (23.3)		
₦18,000–₦30,000	65 (31.3)	69 (33.0)	134 (32.1)		
₦31,000–₦50,000	38 (18.3)	36 (17.2)	74 (17.7)		
₦51,000 and above	52 (25.0)	60 (28.7)	112 (26.9)		
Head of household				2.466 [d]	0.291
Father	185 (88.9)	176 (84.2)	361 (86.6)		
Mother	16 (7.7)	20 (9.6)	36 (8.6)		
Relations	7 (3.4)	13 (6.2)	20 (4.8)		
Age group of the father (Years)	54.1 ± 9.6	54.6 ± 9.9		−0.534 [e]	0.593
35 to 44	28 (13.5)	25 (12.0)	53 (12.7)	0.211 [d]	0.646
≥45	180 (86.5)	184 (88.0)	364 (87.3)		
Education of father				7.482 [d]	0.112
No education	25 (12.0)	25 (12.0)	50 (12.0)		
Informal education	17 (8.2)	33 (15.8)	50 (12.0)		
Primary education	6 (2.9)	8 (3.8)	14 (3.4)		
Secondary education	78 (37.5)	61 (29.2)	139 (33.3)		
Tertiary education	82 (39.4)	82 (39.2)	164 (39.3)		
Occupation of fathers				6.324 [d]	0.097
Civil service	75 (36.1)	90 (43.1)	165 (39.6)		
Trading/business	95 (45.7)	98 (46.9)	193 (46.3)		
There Farming	21 (10.1)	12 (5.7)	33 (7.9)		
Other occupation [c]	17 (8.2)	9 (4.3)	26 (6.2)		
Age group of the mother (Years)	41.6 ± 8.1	40.9 ± 8.3		0.850 [e]	0.396
≤34	29 (13.9)	39 (18.7)	68 (16.3)	1.990 [d]	0.370
35 to 44	102 (49.0)	92 (44.0)	194 (46.5)		
≥45	77 (37.0)	78 (37.3)	155 (37.2)		
Education of mothers				7.743 [d]	0.101
No education	39 (18.8)	41 (19.6)	80 (19.2)		
Informal education	26 (12.5)	44 (21.1)	70 (16.8)		
Primary education	13 (6.3)	9 (4.3)	22 (5.3)		
Secondary education	81 (38.9)	80 (38.3)	161 (38.6)		
Tertiary education	49 (23.6)	35 (16.7)	84 (20.1)		
Occupation of mothers				0.378 [d]	0.945
Civil service	56 (26.9)	55 (26.3)	111 (26.6)		
Trading/business	68 (32.7)	64 (30.6)	132 (31.7)		
Farming	15 (7.2)	15 (7.2)	30 (7.2)		
House wives	69 (33.2)	75 (35.9)	144 (34.9)		
Family type				0.194 [d]	0.660
Monogamy	123 (59.1)	128 (61.2)	251 (60.2)		
Polygamy	85 (40.9)	81 (38.8)	166 (39.8)		
single parenting					

[a] Junior secondary school (JSS), Senior secondary school (SS) [b] Karekare, Kilba, Minchika, Manga, Tambai, Yoruba, Mandara, Basaye, Angas, Terawa, Kanakuru, Nupe. [c] Malami (Voluntary Quranic teacher), [d] Chi-square, [e] t-value, SD = Standard deviation.

3.1.2. Knowledge, Attitude and Food Security at Baseline

Table 3 indicates that the mean ± SD for knowledge score between experimental and control groups at baseline was 8.87 ± 4.3. The scores range from 0 to 19. There was no statistically significant difference between experimental and control group at baseline for knowledge (*p* = 0.589). The mean ± SD for attitude score between experimental and control group at baseline was 53.36 ± 5.0. The scores range from 41 to 70. There was no statistically significant difference between experimental and control group for attitude at baseline for attitude ($p = 0.744$). The mean ± SD for food security score between experimental and control group at baseline was 5.49 ± 3.5. The scores range from 0 to 9. There was no statistically significant difference between experimental and control groups for food security at baseline ($p = 904$).

Table 3. Mean Scores for Knowledge, Attitude and Food Security at Baseline.

Variable	Experimental Mean ± SD (*n* = 208)	Control Mean ± SD (*n* = 209)	Overall Sample Mean ± SD (*n* = 417)	Minimum-Maximum	t	*p*-Value
Knowledge	8.75 ± 4.2	8.98 ± 4.3	8.87 ± 4.3	0–19	0.540	0.589
Attitude	53.28 ± 5.1	53.45 ± 4.9	53.36 ± 5.0	41–70	0.327	0.744
Food security	5.51 ± 3.4	5.47 ± 3.6	5.49 ± 3.5	0–9	−0.120	0.904

t = independent *t*-test.

3.2. Effectiveness of Triple Benefit Health Education Intervention on Knowledge, Attitude and Food Security among Respondents

The participants in the experimental and control groups were compared on knowledge, attitude and food security towards malnutrition at three and six-month post intervention using generalized estimating equation (GEE). The following are the results of the changes in knowledge, attitude and food security at 3 and 6-month post-intervention.

3.2.1. Comparison of Knowledge between Experimental and Control Groups and Time Points (Baseline, 3 and 6-Months Post-Intervention) Respectively

Table 4 shows that participants in the experimental group have higher odds of having good knowledge compared to control group (AOR = 6.380, 95% CI: 4.665–8.725, $p < 0.001$). Participants at 3 and 6-month post-intervention have higher odds of having good knowledge compared to participants at baseline respectively (AOR = 9.595, 95% CI: 6.371–14.449, $p < 0.001$; AOR = 14.993, 95% CI: 9.919–22.662, $p < 0.001$).

Table 4. Comparison of Knowledge of Malnutrition between Groups and Time Points using GEE.

Variables	B	SE	Crude Odd Ratio Exp (B)	Wald Chi-Square	95% CI Lower Bound	95% CI Upper Bound	*p*-Value
Groups							
Control	Ref						
Experimental	1.853	0.160	6.380	134.597	4.665	8.725	<0.001 *
Time points							
Baseline	Ref						
3-months Post-intervention	2.261	0.209	9.595	117.194	6.371	14.449	<0.001 *
6-months Post-intervention	2.708	0.211	14.993	165.012	9.919	22.662	<0.001 *

* Significant < 0.05; SE = Standard error; CI = Confidence Interval; Ref = Reference category; B = Unstandardized beta.

GEE was used to assess the effectiveness of the Triple Benefit Health Education Intervention on knowledge from baseline to three and six-months post-intervention adjusted with covariates. Factors with $p < 0.25$ at a univariate level were tested as the covariates in the final model. Table 5 shows the knowledge of participants between experimental and control groups from baseline to three and six-month post-intervention adjusted with covariates. There was no significant difference between experimental and control group for knowledge (AOR = 1.063, 95% CI: 0.594–1.901, $p = 0.837$). Participants at three and six-months post-intervention have higher odds of having good knowledge compared to participants at baseline respectively (AOR = 4.164, 95% CI: 2.321–7.471, $p < 0.001$; AOR = 5.805, 95% CI: 3.204–10.515, $p < 0.001$).

Table 5. Effectiveness of Triple Benefit Health Education Intervention on Knowledge towards Malnutrition between Groups and Time Points (baseline, 3 and 6-month post-intervention) adjusted with Covariates using GEE.

Variables	B	SE	Adjusted Odds Ratio Exp (B)	Wald Chi-Square	95% CI Lower Bound	95% CI Upper Bound	*p*-Value
Intercept	−3.078	0.413					
Groups							
Control	Ref						
Experimental	0.061	0.297	1.063	0.042	0.594	1.901	0.837
Time points							
Baseline	Ref						
3-months Post-intervention	1.426	0.298	4.164	22.882	2.321	7.471	<0.001 *
6-months Post-intervention	1.759	0.303	5.805	33.656	3.204	10.515	<0.001 *
Interaction							
Control * baseline	Ref						
Experimental * 3-months Post-intervention	1.483	0.413	4.407	12.929	1.964	9.893	<0.001 *
Experimental * 6-months Post-intervention	2.238	0.446	9.379	25.212	3.915	22.471	<0.001 *
Class [a]							
JSS1	Ref						
JSS2	0.458	0.269	1.580	2.905	0.934	2.675	0.088
JSS3	−0.023	0.295	0.977	0.006	0.548	1.743	0.937
SS1	0.586	0.296	1.797	3.928	1.007	3.208	0.047 *
SS2	0.631	0.257	1.879	6.019	1.135	3.109	0.014 *
SS3	0.454	0.285	1.575	2.537	0.901	2.755	0.111
Ethnicity							
Bura	Ref						
Kanuri	−0.021	0.252	0.979	0.007	0.597	1.605	0.932
Hausa	−0.672	0.393	0.510	2.927	0.236	1.103	0.087
Marghi	−0.080	0.343	0.923	0.054	0.471	1.809	0.816
Shuwa	−0.761	0.396	0.467	3.665	0.214	1.018	0.056
Fulani	0.351	0.367	1.420	0.913	0.691	2.918	0.339
Chibok	−0.776	0.357	0.460	4.717	0.228	0.927	0.030 *
Gwoza	−0.078	0.093	0.925	0.063	0.505	1.697	0.802
Other ethnic groups	−0.295	0.377	0.745	0.611	0.356	1.559	0.434
Food security [a]							
Very low food secured	Ref						
Low food secured	0.246	0.197	1.279	0.632	0.870	1.881	0.210
Food secured	0.656	0.197	1.926	1.042	1.309	2.835	0.001 *
Information [a]							
Poor information	Ref						
Good information	0.812	0.175	2.252	21.644	1.600	3.170	<0.001 *
Motivation [a]							
Poor motivation	Ref						
Good motivation	0.680	0.184	1.973	13.688	1.377	2.829	<0.001 *
Behavioral skills [a]							
Poor behavioral skill	Ref						
Good behavioral skill	0.383	0.173	1.466	4.902	1.045	2.057	0.027 *

* Significant < 0.05; SE = Standard error; CI = Confidence Interval; Ref = Reference category; B = Unstandardized beta; QIC = 1150.370; QICC = 1148.529, [a] Covariates.

There was a significant interaction at three and six-month post-intervention; participants at three and six-month post-intervention have higher odds of having good knowledge compared to control group at baseline, respectively (AOR = 4.407, 95% CI: 1.964–9.893, *p* < 0.001; AOR = 9.397, 95% CI: 3.915–22.471, *p* < 0.001).

Participants in SS1 and SS2 have higher odds of having good knowledge compared to those JSS 1 (AOR = 1.797, 95% CI: 1.007–3.208, *p* = 0.047; AOR = 1.879, 95% CI: 1.135–3.109, *p* = 0.014). Participants from Chibok ethnic group have higher odds of having good knowledge compared to those from Bura ethnic group (AOR = 0.460, 95% CI: 0.228–0.927, *p* = 0.030). Participants in food secured level have higher odds of having good knowledge compared to those in very low food secured level (AOR = 1.926, 95% CI: 1.309–2.835, *p* = 0.001). Participants with good information have higher odds of having good knowledge compared to those with poor information (AOR = 2.252, 95% CI: 1.600–3.170, *p* < 0.001).

Participants with good motivation have higher odds of having good knowledge compared to those with poor motivation (AOR = 1.973, 95% CI: 1.377–2.829, $p < 0.001$). Participants with good behavioral skills have higher odds of having good knowledge compared to those with poor behavioral skills (AOR = 1.466, 95% CI: 1.045–2.057, $p = 0.027$).

3.2.2. Comparison of Attitude between Experimental and Control Groups and Time Points (Baseline, Three and Six-Months Post-Intervention) Respectively

Table 6 shows that participants in the experimental group have higher odds of having good attitude compared to control group (AOR = 2.002, 95% CI: 1.619–2.476, $p < 0.001$). Participants at three and six-months post-intervention have higher odds of having good attitude compared to those at baseline respectively (AOR = 1.949, 95% CI: 1.451–2.616, $p < 0.001$; AOR = 2.276, 95% CI: 1.692–3.060, $p < 0.001$).

Table 6. Comparison of Attitude towards Malnutrition between Groups and Time Points (baseline to three and six-months post-intervention) respectively using GEE.

Variables	B	SE	Crude Odd Ratio Exp (B)	Wald Chi-Square	95% CI Lower Bound	95% CI Upper Bound	*p*-Value
Groups							
Control	Ref						
Experimental	0.694	0.109	2.002	40.970	1.619	2.476	<0.001 *
Time points							
Baseline	Ref						
3-months Post-intervention	0.667	0.150	1.949	19.702	1.451	2.616	<0.001 *
6-months Post-intervention	0.822	0.151	2.276	29.597	1.692	3.060	<0.001 *

* Significant < 0.05; SE = Standard error; CI = Confidence Interval; Ref = Reference category; B = Unstandardized beta.

GEE was used to assess the effectiveness of the Triple Benefit Health Education Intervention on the attitude of participants from baseline to three and six-months post-intervention adjusted with covariates. Factors with $p < 0.25$ at a univariate level were tested as the covariates in the final model. Table 7 shows the attitude of participants between experimental and control groups from baseline to three and six-months post-intervention, adjusted with covariates. There was no significant difference between experimental and control groups for attitude (AOR = 1.000, 95% CI: 0.674–1.482, $p = 0.998$). Participants at three-months post-intervention have higher odds of having good attitude towards malnutrition compared to those at baseline (AOR = 1.624, 95% CI: 1.807–2.428, $p = 0.018$).

There was a significant interaction at three and six-months post-intervention; participants at three and six-months post-intervention have higher odds of having good attitude compared to control group at baseline (AOR = 1.367, 95% CI: 1.747–2.501, $p = 0.017$; AOR = 3.076, 95% CI: 1.636–5.785, $p < 0.001$), respectively.

Islamic participants have higher odds of having good attitude compared to those from the Christian religion (AOR = 1.505, 95% CI: 1.189–1.903, $p = 0.001$). Participants with normal BMI have higher odds of having good attitude compared to underweight (AOR = 1.480, 95% CI: 1.107–1.978, $p = 0.008$). Participants in low food secured level have higher odds of having good attitude compared to very low food secured level (AOR = 1.414, 95% CI: 1.031–1.939, $p = 0.032$). Participants with good information on malnutrition have higher odds of having good attitude compared to those with poor information on malnutrition (AOR = 1.645, 95% CI: 1.271–2.128, $p < 0.001$).

3.2.3. Comparison of Food Security between Experimental and Control Groups and Time Points (Baseline, Three and Six-Months Post-Intervention) Respectively

Table 8 shows that participants in the experimental group have higher odds of being in food secured level compared to control group (AOR = 4.688, 95% CI: 3.654–6.015, $p < 0.001$). Participants at three and six-months post-intervention have higher odds of being in food secured level compared to those at baseline respectively (AOR = 1.356, 95% CI: 1.037–1.771, $p = 0.026$; AOR = 1.589, 95% CI: 1.223–1.064, $p = 0.001$).

Table 7. Effectiveness of Triple Benefit Health Education Intervention on Attitude towards Malnutrition between Groups and Time Points (baseline, three and six post-intervention) adjusted with Covariates using GEE.

Variables	B	SE	Crude Odd Ratio Exp (B)	Wald Chi-Square	95% CI Lower Bound	95% CI Upper Bound	*p*-Value
Intercept	−1.237	0.210					
Groups							
Control	Ref						
Experimental	0.000	0.201	1.000	0.000	0.674	1.482	0.998
Time points							
Baseline	Ref						
3-months Post-intervention	0.248	0.205	1.624	5.596	1.087	2.428	0.018 *
6-months Post-intervention	0.485	0.216	1.282	1.324	0.840	1.957	0.250
Interaction							
Control * baseline	Ref						
Experimental * 3-months Post-intervention	1.124	0.308	1.367	5.027	1.747	2.501	0.017 *
Experimental * 6-months Post-intervention	0.312	0.322	3.076	12.160	1.636	5.785	<0.001 *
Religion [a]							
Christianity	Ref						
Islam	0.409	0.120	1.505	1.027	1.189	1.903	0.001 *
Nutritional status (BMI percentile) [a]							
Underweight	Ref						
Normal	0.349	0.148	1.480	7.012	1.107	1.978	0.008 *
Overweight	0.214	0.494	1.238	0.187	0.470	3.263	0.666
Obese	0.392	0.148	1.418	0.213	0.321	6.256	0.645
Food security [a]							
Very low food secured	Ref						
Low food secured	0.346	0.161	1.414	4.622	1.031	1.939	0.032 *
Food secured	0.208	0.144	1.231	2.089	0.929	1.632	0.148
Information [a]							
Poor information	Ref						
Good information	0.498	0.131	1.645	14.344	1.271	2.128	<0.001 *

* Significant < 0.05; SE = Standard error; CI = Confidence Interval; Ref = Reference category; B = Unstandardized beta; BMI = Body mass index, QIC = 1611.735; QICC = 1612.245, [a] Covariates.

Table 8. Comparison of Food Security towards Malnutrition between Groups and Time Points (baseline to three and six-months) respectively using GEE.

Variables	B	SE	Crude Odd Ratio Exp (B)	Wald Chi-Square	95% CI Lower Bound	95% CI Upper Bound	*p*-Value
Groups							
Control	Ref						
Experimental	1.545	0.127	4.688	147.643	3.654	6.015	<0.001 *
Time points							
Baseline	Ref						
3-months Post-intervention	0.303	0.137	1.356	4.968	1.037	1.771	0.026 *
6-months Post-intervention	0.463	0.134	1.589	12.019	1.223	2.064	0.001 *

* Significant < 0.05; SE = Standard error; CI = Confidence Interval; Ref = Reference category; B = Unstandardized beta.

GEE was used to assess the effectiveness of the Triple Benefit Health Education Intervention on the food security level of respondents from baseline to three and six-months post-intervention adjusted with covariates. Factors with $p < 0.25$ at a univariate level were tested as the covariates in the final model. Table 9 shows the food security level of participants between experimental and control group from baseline to three and six-months post-intervention, adjusted with covariates. There was no significant difference between experimental and control groups for food security level (AOR = 0.887, 95% CI: 0.606–1.229, $p = 0.539$). There were no significant differences at three and six-months post-intervention for food security level compared to those at baseline (AOR = 0.914, 95% CI: 0.616–1.358, $p = 0.658$; AOR = 0.869, 95% CI: 0.581–1.299, $p = 0.492$), respectively.

Table 9. Effectiveness of Triple Benefit Health Education Intervention on Food Security towards Malnutrition between Groups (intervention and control) and Time Points (baseline, 3 and 6-months post-intervention) adjusted with Covariates using GEE.

Variables	B	SE	Crude Odd Ratio Exp (B)	Wald Chi-Square	95% CI Lower Bound	95% CI Upper Bound	p-Value
Groups							
Control	Ref						
Experimental	−0.120	0.195	0.887	0.378	0.606	1.299	0.539
Time points							
Baseline	Ref						
3-months Post-intervention	−0.089	0.202	0.914	0.472	0.616	1.358	0.658
6-months Post-intervention	−0.141	0.205	0.869	0.196	0.581	1.299	0.492
Interaction							
Control * baseline	Ref						
Experimental * 3-months Post-intervention	0.749	0.291	2.116	6.651	1.197	3.740	0.010
Experimental * 6-months Post-intervention	0.937	0.283	2.552	10.995	1.467	4.440	0.001
Place of residence [a]							
Rural	Ref						
Urban	0.427	0.159	1.532	7.220	1.122	2.091	0.007 *
Monthly income [a]							
Less than ₦18,000	Ref						
₦18,000–₦30,000	0.027	0.158	1.311	2.950	0.963	1.785	0.086
₦31,000–₦50,000	0.431	0.172	1.539	6.309	1.099	2.153	0.012 *
₦51,000 and above	0.805	0.154	2.236	27.171	1.652	3.025	<0.001 *
Education of mothers [a]							
No education	Ref						
Informal education	0.023	0.183	1.023	0.016	0.715	1.646	0.900
Primary education	0.553	0.227	1.739	5.946	1.115	2.712	0.015 *
Secondary education	0.150	0.152	1.162	0.983	0.863	1.564	0.321
Tertiary education	0.374	0.169	1.454	4.915	1.044	2.025	0.027 *
Source of information [a]							
Mass/social media	Ref						
Family/friends	−0.287	0.245	1.361	1.583	0.842	2.200	0.208
Health worker/clinic	−0.495	0.271	0.507	6.294	0.298	0.862	0.012 *
School teacher	−0.679	0.242	0.610	4.181	0.379	0.980	0.041 *
Health education program	0.308	0.258	0.750	1.237	0.452	1.245	0.266
Protein [a]	0.003	0.001	1.003	4.605	1.000	1.005	0.032 *

* Significant < 0.05; SE = Standard error; CI = Confidence Interval; Ref = Reference category; B = unstandardized beta; [a] Covariates.

There was a significant interaction at three and six-months post-intervention; participants at three and six-months post-intervention have higher odds of being in food secured level compared to control group at baseline (AOR = 2.116, 95% CI: 1.197–3.740, p = 0.010; AOR = 2.552, 95% CI: 1.467–4.440, p = 0.001), respectively.

Participants in urban areas have higher odds of being in food secured level compared to those in rural areas (AOR = 1.532, 95% CI: 1.122–2.091, p = 0.007). Participants whose monthly income was between ₦31,000–₦50,000 and ₦51,000 and above have higher odds of being in food secured level compared to those with monthly income less than ₦18,000 (AOR = 1.539, 95% CI: 1.099–2.153, p = 0.012; AOR = 2.236, 95% CI: 1.652–3.025, p < 0.001). Participants whose mothers' education were primary and tertiary education have higher odds of being in food secured level compared to those whose mothers had no education (AOR = 1.739, 95% CI: 1.115–2.712, p = 0.015; AOR = 1.454, 95% CI: 1.044–2.025, p = 0.027). Participants whose sources of information were from health worker/clinic and schoolteachers have lower odds of being in food secured level compared to those whose sources of information were from social media (AOR = 0.507, 95% CI: 0.298–0.862, p = 0.012, AOR = 0.610, 95% CI: 0.379–0.980, p = 0.041). For every increase in protein, participants have higher odds of being in food secured level compared to very low food secured level (AOR = 1.003, 95% CI: 1.000–1.005, p = 0.032).

4. Discussion

Acquiring adequate nutritional knowledge and attitude will help children, adolescents and entire families to make better choices regarding foods that are convenient, available, affordable and desirable for healthy living. There is need to invest in children and adolescents globally to achieve the sustainable development goal by 2030 [10].

The findings in this study showed that there was no significant difference in the age of respondents between the experimental and control group at baseline, similar to studies in U.S.A and China [39,40]. Class of participants between experimental and control groups were not significant, concurring with a study in California [41]. The ethnicity of participants showed no significant difference between experimental and control group, in line with a study in Canada [42]. Religion, place of residence, household size, head of household, and education of father and mother were not significant between experimental and control, agreeing with a study in Ethiopia [43,44]. This study further revealed that there was no significant difference in age of fathers and mother, or occupation of father and mother between experimental and control, similar to a study in Iran [45]. The study overall reveals no significant differences in sociodemographic characteristics of respondents between intervention and control group at baseline, indicating that the two groups are comparable in their characteristics due to a good randomization process, Table 2.

This study hypothesized that there were no significant differences in knowledge of malnutrition between the experimental and control group at baseline; there was similar agreement with other studies in Canada, China, Los Angeles and California [32,41,42,46]. The study reveals that there was no significant difference in the attitude of participants towards malnutrition between experimental and control group at baseline. There are similarities with experimental studies conducted in Canada, South Africa, the southwestern state of Malaysia and China [3,32,42,47]. There was no significant difference in all the food security statement between the experimental and control group at baseline. Food security can be said to exist if all people at all times have cost-effective and physical access to safe, adequate and nourishing food to meet their nutritional needs for an energetic and healthy life. Globally, more than 820 million people are still hungry, and about 2 billion people experience moderate or severe food insecurity. The severe impact of food insecurity on malnutrition has been identified as a problem to the overall health status of people, and food security can be a contributing factor to malnutrition [48,49].

The findings from this study reveal a statistically significant difference between the experimental and control groups for knowledge ($p < 0.001$). There was a statistically significant improvement after the Triple Benefit Health Education Intervention on knowledge of participants towards malnutrition at three ($p < 0.001$) and six-months ($p < 0.001$) post-intervention compared to the baseline result. This significant improvement might be attributed to the information obtained from the Triple Benefit Health Education Intervention module. Therefore, the outcome of this study supports the effectiveness of Triple Benefit Health Education Intervention in improving the knowledge of adolescent girls. The findings of the current study regarding the effectiveness of the Triple Benefit Health Education Intervention was significantly supported by educational interventions in India ($p < 0.05$), Palestine ($p < 0.001$), Bangladesh ($p < 0.001$), Iran ($p < 0.05$), Baltimore ($p < 0.001$) and Urbana city ($p < 0.05$) [39,50–55], but contrary to a study in the U.S.A that revealed no significant difference after the intervention program ($p = 0.45$), which may be due to the absence of a theory of behavioral change, the methodological nature of the intervention program and the narrow focus of the intervention program [41]. The study by Shaaban et al. (2014) [56] further supports the view that education intervention on malnutrition is effective in increasing knowledge towards nutrition-related issues, as knowledge on nutrition-related issues is an integral achievement of healthy attitudes and practice and consequently leads to improvement of diet quality for better well-being. By using this educational intervention, knowledge significantly increased ($p < 0.001$). Similarly, the educational intervention by Shesha et al. (2018) [57] consisted of lectures, presentations, interactive discussions and the distribution of information booklets which was effective in improving the knowledge of respondents towards nutrition-related issues. The result showed significant improvement in overall knowledge ($p < 0.05$) after the educational intervention, as it has proven to bridge the gap in knowledge among adolescent girls. Knowledge is the corner stone of attitude, and limited access to knowledge has been assumed to be among the causes of malnutrition [58]. Adolescents with good nutritional knowledge have higher odds of following healthy eating habits and lifestyle, as nutrition education intervention has been promising in improving these for this age group. The use of flip chart, pictures, brainstorming, discussion, role play and group work among adolescents is less expensive, easy to use, cost effective and it holds the attention of participants since it is interactive and makes use of pictures that explain key messages which can be easily used to cater for the local needs of the people [59,60]. Improvement in nutrition related knowledge is essential in balancing the intake of food containing carbohydrate, protein, fats, mineral and vitamins. Nutritionally related dietary habits and knowledge are very significant for a healthy lifestyle, although women, children and adolescents are vulnerable because of their exceptional physiological and socioeconomic characteristics as a result of poor nutrition [61]. Adolescence is the period of growth and development in humans that occurs after childhood and before adulthood from 10 to 19 years old. Specific and unique changes occur during this phase; furthermore growth occurs in the skeleton, muscles and in every part of the body. both in systems and organs, except the brain and the head in adolescence [62]. The nutrition education intervention is a basic program to target adolescents and young children and is considered as a critical factor required to promote positive changes in individual diet, health and healthy wellbeing. Providing adolescents with nutrition related information in a multisectoral context will significantly improve adolescent knowledge towards nutrition, better dietary practice and improved health and nutritional indicators among children and adolescents [63]. Other factors such as class (SS2 $p = 0.010$) were significantly associated with good knowledge, which suggests that the triple benefit health education intervention was more effective among respondents in the higher class SS1 and SS2 compared to those in lower class JSS1, indicating that the higher the social class, the better the understanding [64]. Ethnicity ($p < 0.025$) was significantly associated with good knowledge, and socio-demographic characteristics and cultural norms and differences prevalent in the various communities could be the contributing factors to good knowledge. This study also reveals that food security (food secured; $p = 0.001$) was significantly associated with good knowledge. Food secured adolescents are more likely to have good dietary intake, contributing to healthy and good nutritional status, supporting these girls during their first pregnancy and, in turn, a positive impact on the future generation. Food insecurity or hunger reduces learning capacity, school attendance, and earning capacity, thereby increasing the risk of hunger and poverty in the future [65]. Information (good information; $p < 0.001$), motivation (good motivation; $p < 0.001$), and behavioral skills (good behavioral skills; $p = 0.027$) further reveal the significant association between knowledge of malnutrition and the IMB model, indicating that the incorporation of the IMB model in the intervention study would improve knowledge [66].

The results reveal a statistically significant difference in attitude between the experimental and control groups ($p < 0.001$). There was also a statistically significant difference at three ($p < 0.001$) and six-months ($p < 0.001$) post-intervention after the Triple Benefit Health Education Intervention. The findings showed that the content of the Triple Benefit Health Education Intervention module and the methodological concept had positively influenced the attitude of adolescent girls. Similarly, education intervention studies in Canada ($p < 0.001$), Palestine ($p < 0.001$), Shahr-e-kord city, Iran ($p < 0.001$), Bangladesh ($p < 0.05$), and China ($p < 0.023$) reported significant changes in post-intervention and follow up [32,42,50,51,57,67], but contrary to the interaction effect in Laram et al., 2017 [42]. In contrast, studies in Urbana ($p > 0.05$) and Bangalore ($p > 0.05$) showed no significant result in post-intervention. This may be due to a small sample size of less than 100 with less than 50 respondents in each group, inadequate training on some specific aspect of the study, presentation skills, and, more importantly, a large number of dropouts in the control group during post-intervention could be a contributing factor to its non-significance [55,57]. More so, the education intervention study showed a significant increase in attitude from baseline to post-intervention in Nigeria ($p < 0.001$) [68]. Further improvement was reported in a study in India ($p = 0.007$) from baseline to post-intervention and follow up. It appears that education interventions that concentrates on developing health-related skills and imparting health-related attitudes are more likely to benefit adolescent girls in health-enhancing practices [69]. School based nutrition intervention has proven to improve attitude among adolescents in the prevention of malnutrition. Nutrition education intervention has a long-lasting approach to building a good nutritional status among adolescents. Although the participants are from different socioeconomic contexts, the triple benefit health education intervention was effective in improving their attitude by providing valuable information as a practical solution in enhancing their attitude towards positive living [70]. Attitude can be said to be a determinant in individual choice of food and healthy living as good attitude towards malnutrition in adolescence immensely contribute to healthy lifestyle [71]. Religion (Islam; $p = 0.001$) was significantly associated with attitude. Religion has its unique way of disseminating information to its adherents and this information is transmitted through generations to meet the basic needs of its groups and for survival, shaping the attitude of individuals in the form of norms and values [72]. In terms of nutritional status (normal; $p = 0.008$), respondents with normal nutritional status significantly improved from baseline, and there was also a significant decrease in underweight from baseline as a result of the triple benefit health education intervention, giving the adolescent child hope for future survival. Food security (low food secured; $p = 0.032$) was significantly associated with attitude towards malnutrition. Food security can have impact on attitude and nutritional status thereby influencing the growth and development of adolescents as it becomes woven in an intergenerational cycle of malnutrition [73]. As for information (good information; $p < 0.001$), good information is likely to improve attitude, as acquired information on malnutrition (definition, causes, consequences and future implications) tends to change beliefs and personal values associated with various health behaviors capable of changing attitudes towards malnutrition and knowing fully the future consequences.

Findings from this study reveal a statistically significant difference for food security between the experimental and control group ($p < 0.001$). There was a statistically significant difference at three ($p = 0.026$) and six-month ($p = 0.001$) post-intervention for food security after the Triple Benefit Health Education Intervention compared to the control group at baseline. This indicated that Triple Benefit Health Education Intervention improved the food security situation of participants. The effects of malnutrition in women are borne throughout their lifecycle and through generations. Nutritional inadequacy through food insecurity during the period of adolescence can affect present and future health and well-being, as it is intrinsically linked to the health and well-being of offspring [73,74]. The interaction between food security knowledge and attitude creates a cycle whereby poor diet due to poor knowledge and attitude compromises the immune system exposing the individual to infection, poor health and wellbeing, decreased absorption of nutrients and worsened nutritional status. Most episodes of malnutrition and morbidity are linked to infection leading to wasting, stunting and underweight. The origin of wavering growth usually begins as early as during pregnancy, so there is a need for adequate nutrition and health status in the prevention of malnutrition through food security to reduce the short and long-term consequences of malnutrition. This includes the intergenerational cycle of malnutrition extended from malnourished girls becoming malnourished mothers at risk of giving birth to low birth weight infants, placing them at disadvantaged development throughout life [75]. The more knowledge and the better attitude the individuals have, the more they will prefer to acquire more information on nutritious food and be willing to buy and consume it. Knowledge of and attitude towards malnutrition influence perceptions on food security in the aspect of the quality and the quantity food and its utilization [76]. Since adolescence is a dynamic stage, diet at this stage is driven by immediate needs, and choice of food is influenced by energy needs and what will satisfy their hunger, more so because adolescents have more control, choice and responsibility regarding the household diet [77]. Eradicating food insecurity depends entirely on agricultural productivity, in line with the United Nations Millennium Development Goals to eradicate poverty and hunger, but this can be feasible in Borno state, Nigeria if the rural dwellers are back in their respective communities rather than camping in the urban areas as refugees with limited land for agricultural production, due to the prolonged insurgency threatening the peace and unity of the country [78–80]. Furthermore, other factors associated with food security include place of residence (urban; $p = 0.007$). This study reveals that urban residence has higher odds of being food secure compared to rural dwellers, perhaps due to non-availability of sufficient farm produce and the financial resources caused by displacement as a result of the lingering humanitarian crises that have limited farm production and food availability in the rural areas and the outskirts of the urban areas. As regards education [78–81] and monthly income (₦31,000–₦50,000; $p = 0.012$, ₦51,000 and above; $p < 0.001$), this study reveals that households with better income have higher odds of being food secured compared to lower income households. Income is an important determinant of food insecurity. Higher household income reduces the severity of food insecurity, as lack of financial security can lead to food insecurity. High household income means earning more, making more food choices and better management of the household [82,83]. As regards education of mother (primary education; $p = 0.015$, tertiary education; $p = 0.027$), mothers who are educated might have less tendency to become food insecure, as since they are educated there is tendency for them to be working or in menial business, earning income that will help in cushioning the family needs in term of food choice, availability and utilization. Mothers with higher level of education are more likely to earn income, adopt technology and be engaged in livelihood activities that can contribute to better food production, thereby reducing the risk of household food insecurity [84].

One of the strengths of this study is that it was a randomized control trial with single blinding and a validated questionnaire. The study design and its tools allowed the assessment of knowledge, attitude and food security. To the best of our knowledge, there has been no previous intervention study among adolescent girls aged 10 to 19 years on nutrition-related knowledge, attitude and food security in the northeastern part of Nigeria and no previous research available for these age groups using a randomized controlled trial. This was the first intervention module on malnutrition developed based on the IMB model. Subsequently, findings from this study can be used as fundamental data for future studies in Borno State, the northeastern part of Nigeria and Nigeria at large.

Another limitation of this study was that the intervention was conducted in only four governments secondary schools in Maiduguri Metropolitan Council (MMC). The study was conducted in only one out of the 27 districts in the state. Another limitation of the intervention includes non-inclusion of schools with married women, schools with boys only, private schools, primary schools with early adolescents, school dropouts, and adolescents not attending any school at all. Other limitations of our intervention include limited time, lack of funding, and inability to strategize for a parent teacher's association (PTA) meeting before the onset of the intervention to enable full participation of all eligible participants.

5. Conclusions

Globally, humanitarian and developmental agencies have acknowledged the importance of recognizing that adolescents belong to a separate age group, from children to young adults. Adolescents, especially girls, poorly transit into adulthood at the risk of being pushed into the vicious cycle of the intergenerational cycle of malnutrition, deficiency and poverty. In this study the Triple Benefit Health Education Intervention was found to be effective in improving knowledge, attitudes and food security towards malnutrition among adolescent girls in Maiduguri Metropolitan Council, Borno State, Nigeria. Participants in the intervention group had higher odds of having good knowledge and attitude and also had higher odds of being in the food secured level at three and six-months post intervention compared to those at baseline due to the impact of the Triple Benefit Health

Education Intervention. This health education intervention towards malnutrition among adolescents will go a long way in giving them a better future life. There is need for more awareness among adolescent girls for both school based and community based intervention in the future to ensure all adolescents across the community benefit from health education intervention for the betterment of their future family.

Author Contributions: Conceptualization, R.C.S. and P.Y.L.; methodology, R.C.S. and P.Y.L.; software, R.C.S. and P.Y.L.; validation, R.C.S., S.I., P.Y.L., N.A., H.G. and I.A.N.; formal analysis, R.C.S., P.Y.L. and S.I.; investigation, R.C.S.; resources, R.C.S.; data curation, R.C.S. and P.Y.L.; writing—original draft preparation, R.C.S.; writing—review and editing, R.C.S., P.Y.L., S.I., N.A., H.G. and I.A.N.; visualization, R.C.S., P.Y.L., S.I., N.A., H.G. and I.A.N.; supervision, S.I., P.Y.L., N.A., H.G. and I.A.N. project administration, S.I. All authors have read and agreed to the published version of the manuscript.

Funding: This research received no external funding.

Institutional Review Board Statement: The study was conducted according to the guidelines of the Declaration of Helsinki, and approved by Universiti Putra Malaysia Ethical Committee UPM/TNCPI/RMC/JKEUPM/1.4.18.2 (JKEUPM). The study was also registered with Pan African Clinical Trials Registry (PACTR201905528313816).

Informed Consent Statement: Informed consent was obtained from all subjects involved in the study.

Data Availability Statement: Not applicable.

Acknowledgments: The authors wish to appreciate Hussaini Garba Dibal, (Department of Physical and Health Education, University of Maiduguri, Borno State, Nigeria) and Elizabeth Tanko (Department of Nutrition, College of Nursing and Midwifery, Maiduguri, Borno State, Nigeria) who assisted in validating the study instruments. The authors want to acknowledge Majinga Samuel Amos (Forestry Research Institute, Nigeria) for his technical input. The authors also want to appreciate the management of the four Secondary Schools (Government Girls College Maiduguri, Government Girls Secondary School Yerwa, Zajeri Day Secondary School and Bulabulin Day Secondary School), their enumerators and the respondents who participated in the study.

Conflicts of Interest: The authors declare no conflict of interest.

References

1. Ashraf, Z.; Hussain, M.; Majeed, I.; Afzal, M.; Parveen, K.S.A.G. Effectiveness of Health Education on Knowledge and Practice Regarding the Importance of Well-Balanced Nutrition Among School Students. *J. Health Med. Nurs.* **2019**, *69*, 1–7.
2. Sharma, M.; Watode, B.; Srivastava, A. Nutritional Status of Primary School Children through Anthropometric Assessment in Rural Areas of Moradabad. *Ann. Int. Med. Dent. Res.* **2017**, *3*, 2–6. [CrossRef]
3. Shariff, Z.M.; Bukhari, S.S.; Othman, N.; Hashim, N.; Ismail, M.; Kasim, S.M.; Paim, L.; Samah, B.A.; Azhar, Z.; Hussein, M. Nutrition Education Intervention Improves Nutrition Knowledge, Attitude and Practices of Primary School Children: A Pilot Study. *Int. Electron. J. Health Educ.* **2008**, *11*, 119–132.
4. Pushpa, K.S.; Rani, D.J. Nutritional profile of children (0–5 years) in the service villages of Gandhigram rural institute. *Curr. Res. Nutr. Food Sci.* **2015**, *3*, 81–88. [CrossRef]
5. Yeganeh, S.; Motamed, N.; Najafpourboushehri, S.; Ravanipour, M. Assessment of the knowledge and attitude of infants' mothers from Bushehr (Iran) on food security using anthropometric indicators in 2016: A cross-sectional study. *BMC Public Health* **2018**, *18*, 1–9. [CrossRef]
6. World Health Organization. *Guidelines: Implementing Effective Actions for Improving Adolescents Nutrition*; Nutrition for Health and Development World Health Organization: Geneva, Switzerland, 2018.
7. HLPE. *Nutrition and Food Systems. A Report by the High Level Panel of Experts on Food Security and Nutrition of the Committee on World Food Security*; FAO: Rome, Italy, 2017.
8. FAO; WHO; UNICEF. *Sustainable Food Systems for Healthy Diets in Europe and Central Asia a Joint FAO/WHO Regional Symposium and Initiative in Collaboration of UNICEF and WFP*; FAO: Rome, Italy, 2017.
9. FAO. *Influencing Food Environments for Healthy Diets*; FAO: Rome, Italy, 2016; ISBN 9789251095188.
10. United Nations Children's Fund (UNICEF). *The State of the World's Children 2019: Children, Food and Nutrition. Growing Well in a Changing World*; UNICEF: New York, NY, USA, 2019.
11. Caldeira, S.; Storcksdieckgennant, B.; Bakogianni, I.; Gauci, C.; Calleja, A.; Furtado, A. *Public Procurement of Food for Health: Technical Report on School Setting*; JRC: Ispra, Italy, 2017; ISBN 9789995710880.

12. Hyska, J.; Burazeri, G.; Menza, V.; Dupouy, E. Assessing nutritional status and nutrition-related knowledge, attitudes and practices of Albanian schoolchildren to support school food and nutrition policies and programmes. *Food Policy* **2020**, *96*, 101888. [CrossRef]

13. Bundy, D.; de Silva, N.; Horton, S.; Jamison, D.; Patton, G. *Re-Imagining School Feeding: A High-Return Investment in Human Capital and Local Economies*; World Food Programme: Rome, Italy, 2018; Volume 8.

14. Bundy, D.; de Silva, N.; Horton, S.; Jamison, D.; Patton, G. *Optimizing Education Outcomes: High-Return Investments in School Health for Increased Participation and Learning*; World Bank Group: Washington, DC, USA, 2018; Volume 8.

15. Coppoolse, H.L.; Seidell, J.C.; Dijkstra, S.C. Impact of nutrition education on nutritional knowledge and intentions towards nutritional counselling in Dutch medical students: An intervention study. *BMJ Open* **2020**, *10*, e034377. [CrossRef] [PubMed]

16. Kıvrak, A.O.; Altın, M. Nutrition Knowledge and Attitude Change of Students Studying in State and Private Secondary Schools. *J. Educ. Train. Stud.* **2018**, *6*, 63. [CrossRef]

17. Rajikan, R.; Esmail, S. Nutrition Education to Improve Nutrition Knowledge, Attitude and Practice among Yemeni School Children. *Trends Appl. Sci. Res.* **2020**, *15*, 207–213. [CrossRef]

18. National Bureau of Statistics. *2019 Poverty and Inequality in Nigeria: Executive Summary*; National Bureau of Statistics: Minna, Nigeria, 2020.

19. Owoo, N.S. Demographic considerations and food security in Nigeria. *J. Soc. Econ. Dev.* **2020**, *23*, 128–167. [CrossRef]

20. Nwozor, A.; Olanrewaju, J.S.; Ake, M.B. National insecurity and the challenges of food security in Nigeria. *Acad. J. Interdiscip. Stud.* **2019**, *8*, 9–20. [CrossRef]

21. The Federal Republic of Nigeria. *The Federal Republic of Nigeria Agricultural Sector Food Security and Nutrition Strategy 2016–2025*; The Federal Republic of Nigeria: Abuja, Nigeria, 2016.

22. National Population Commission (NPC); ICF. *Nigeria Demographic Health Survey 2018*; The Federal Republic of Nigeria: Abuja, Nigeria, 2019.

23. National Beareu of Statistics (NBS). *National Nutrition and Health Survey*; UNICEF: New York, NY, USA, 2018.

24. Parul, C.; Smith, R.E. Adolescent Undernutrition: Global Burden, Physiology, and Nutritional Risks. *Ann. Nutr. Metab.* **2018**, *72*, 316–328.

25. Ministry of Budget and National Planning. *National Policy on Food and Nutrition in Nigeria*; Ministry of Budget and National Planning: Abuja, Nigeria, 2016.

26. International Food Policy Research Institute (IFPRI). *Nutrition Policy in Nigeria*; IFPRI: Washington, DC, USA, 2019.

27. Federal Ministry of Health. *National Plan of Action on Food and Nutrition in Nigeria*; Federal Ministry of Health: Abuja, Nigeria, 2005; Volume 1.

28. Adinma, J.I.B.; Adinma, A.E.; Umeononihu, O.S. Nutrition Policy and Practice Landscape on Adolescent, Pre-Pregnancy and Maternal Nutrition in Nigeria. *JOJ Nurs. Health Care* **2017**, *1*, 28–33. [CrossRef]

29. Save the Children. *Scaling Up Nutrition Interventions for Children Left Behind in Nigeria*; Save the Children International: London, UK, 2016.

30. Peyman, N.; Abdollahi, M. Using of Information-Motivation-Behavioral Skills Model on Nutritional Behaviors in Controlling Anemia among Girl Students. *J. Res. Health* **2016**, *12*, 1–9.

31. Lemeshow, S.; Hosmer, D.W., Jr.; Klar, J.; Lwanga, S.K. *Adequacy of Sample Size in Health Studies*; WHO: Geneva, Switzerland, 1990; ISBN 0471925179.

32. Wang, D.; Stewart, D.; Chang, C.; Shi, Y. Effect of a school-based nutrition education program on adolescents' nutrition-related knowledge, attitudes and behaviour in rural areas of China. *Environ. Health Prev. Med.* **2015**, *20*, 271–278. [CrossRef] [PubMed]

33. Macías, Y.F.; Glasauer, P. *Guidelines for Assessing Nutrition-Related Knowledge, Attitudes and Practices Manual*; WHO: Geneva, Switzerland, 2014; ISBN 9789251080979.

34. Govender, D.; Naidoo, S.; Taylor, M. Knowledge, Attitudes and Peer Influences related to Pregnancy, Sexual and Reproductive Health among Adolescents using Maternal Health Services in Ugu, KwaZulu-Natal, South Africa. *BMC Public Health* **2019**, *19*, 928. [CrossRef] [PubMed]

35. Chanyalew, W.; Kassahun, A.G.M. Knowledge, Attitude, Practices and their Associated Factors towards Diabetes Mellitus among non Diabetes Community members of Bale Zone Administrative Towns, South East. *PLoS ONE* **2017**, *12*, e0170040.

36. Carol, L.C.; Nord, M.; Kristi, L.L.; Yadrick, K. Self-Administered Food Security Survey Module for Children Ages 12 Years and Older. *J. Nutr.* **2006**, *134*, 2566–2572.

37. Shtasel-Gottlieb, Z.; Palakshappa, D.; Yang, F.; Goodman, E. The Relationship Between Developmental Assets and Food Security in Adolescents from a Low-Income Community. *J. Adolesc. Health* **2016**, *56*, 215–222. [CrossRef] [PubMed]

38. Dziura, J.D.; Post, L.A.; Zhao, Q.; Fu, Z.; Peduzzi, P. Strategies for Dealing with Missing Data in Clinical Trials: From Design to Analysis. *Yale J. Biol. Med.* **2013**, *86*, 343–358.

39. Shin, A.; Surkan, P.J.; Coutinho, A.J.; Suratkar, S.R.; Campbell, R.K.; Rowan, M.; Sharma, S.; Dennisuk, L.A.; Karlsen, M.; Gass, A.; et al. Impact of Baltimore Healthy Eating Zones: An Environmental Intervention to Improve Diet Among African American Youth. *Health Educ. Behav.* **2015**, *42*, 97S–105S. [CrossRef]

40. Zhou, W.; Xu, X.; Li, G.; Sharma, M.; Qie, Y.; Zhao, Y. Effectiveness of a school-based nutrition and food safety education program among primary and junior high school students in Chongqing, China. *Glob. Health Promot.* **2016**, *23*, 37–49. [CrossRef] [PubMed]

41. Robert, G.; LaChausse, A. Clustered Randomized Controlled Trial to Determine Impacts of the Harvest of the Month Program. *Health Educ. Res.* **2017**, *32*, 375–383.
42. Laramee, C.; Drapeau, V.; Valois, P.; Goulet, C.; Jacob, R.; Provencher, V.; Lamarche, B. Evaluation of a Theory-Based Intervention Aimed at Reducing Intention to Use Restrictive Dietary Behaviors among Adolescent Female Athletes. *J. Nutr. Educ. Behav.* **2017**, *49*, 497–505. [CrossRef] [PubMed]
43. Karimi-Shahanjarini, A.; Rashidian, A.; Omidvar, N.; Majdzadeh, R. Assessing and Comparing the Short-Term Effects of TPB Only and TPB Plus Implementation Intentions Interventions on Snacking Behavior in Iranian Adolescent Girls: A Cluster Randomized Trial. *Am. J. Health Promot.* **2013**, *27*, 152–162. [CrossRef]
44. Tamiru, D.; Argaw, A.; Gerbaba, M.; Ayana, G.; Nigussie, A.; Belachew, T. Effect of integrated school-based nutrition education on optimal dietary practices and nutritional status of school adolescents in Southwest of Ethiopia: A quasi-experimental study. *Int. J. Adolesc. Med. Health* **2016**, *29*, 1–12. [CrossRef]
45. Mokhtari, F.; Kazemi, A.; Ehsanpour, S. Effect of educational intervention program for parents on adolescents' nutritional behaviors in Isfahan in 2016. *J. Educ. Health Promot.* **2017**, *6*, 103. [PubMed]
46. Bogart, L.M.; Cowgill, B.O.; Elliott, M.N.; Klein, D.J.; Hawes-Dawson, J.; Uyeda, K.; Elijah, J.; Binkle, D.G.; Schuster, M.A. A Randomized Controlled Trial of Students for Nutrition and EXercise: A Community-Based Participatory Research Study. *J. Adolesc. Health* **2014**, *55*, 415–422. [CrossRef]
47. Kupolati, M.D.; Macintyre, U.E.; Gericke, G.J.; Becker, P. A Contextual Nutrition Education Program Improves Nutrition Knowledge and Attitudes of South African Teachers and Learners. *Front. Public Health* **2019**, *7*, 258. [CrossRef]
48. Food and Agriculture Organization of the United Nations. *An Introduction to the Basic Concepts of Food Security*; FAO: Rome, Italy, 2008.
49. FAO; IFAD; UNICEF. *The State of Food Security and Nutrition in the World 2019. Safeguarding against Economic Slowdowns and Downturns*; FAO: Rome, Italy, 2019.
50. Razzak, M.A.; Al Hasan, S.M.; Rahman, S.S.; Asaduzzaman, M. Role of nutrition education in improving the nutritional status of adolescent girls in North West areas of Bangladesh. *Int. J. Sci. Eng. Res.* **2016**, *7*, 1340–1346.
51. Jalambo, M.O.; Sharif, R.; Naser, I.A.; Karim, N.A. Improvement in Knowledge, Attitude and Practice of Iron Deficiency Anaemia among Iron-Deficient Female Adolescents after Nutritional Educational Intervention. *Glob. J. Health Sci.* **2017**, *9*, 15. [CrossRef]
52. Saraf, D.S.; Gupta, S.K.; Pandav, C.S. Effectiveness of a School Based Intervention for Prevention of Non-communicable Diseases in Middle School Children of Rural North India: A Randomized Controlled Trial. *Indian J. Pediatr.* **2014**, *82*, 354–362. [CrossRef] [PubMed]
53. Bandyopadhyay, L.; Maiti, M.; Dasgupta, A.; Paul, B. Intervention for Improvement of Knowledge on Anemia Prevention: A School-based Study in a Rural area of West Bengal. *Int. J. Health Allied Sci.* **2017**, *6*, 69–74.
54. Kaveh, M.H.; Darabi, F.; Khalajabadi-Farahani, F.; Yaseri, M.; Kaveh, M.H.; Mohammadi, M.J.; Reza, H.; Behrooz, A.; Shojaeizadeh, A.R.D. The Impact of a TPB-Based Educational Intervention on Nutritional Behaviors in Iranian Adolescent Girls: A Randomized Controlled Trial. *Fresenius Environ. Bull.* **2018**, *27*, 4349–4356.
55. Oakley, A.R.; Nelson, S.A.; Nickols-Richardson, S.M. Peer-Led Culinary Skills Intervention for Adolescents: Pilot Study of the Impact on Knowledge, Attitude, and Self-efficacy. *J. Nutr. Educ. Behav.* **2017**, *49*, 852–857. [CrossRef]
56. Shaaban, S.Y.; Nassar, M.F.; Shatla, R.H.; Deifallah, S.M. Nutritional Knowledge, Attitude and Practice of Predominantly Female Preschool Teachers: Effect of Educational Intervention. *Br. J. Med. Med. Res.* **2014**, *4*, 1739–1749.
57. Shesha, T.; Chaluvaraj, I.; Satyanarayana, P.T. Change in Knowledge, Attitude and Practice Regarding Anaemia among High School Girls in Rural Bangalore: An Health Educational Interventional Study. *Natl. J. Community Med.* **2018**, *9*, 358–362.
58. Agustina, R.; Wirawan, F.; Sadariskar, A.A.; Ked, S.; Setianingsing, A.A.; Nadiya, K.; Prafiantini, E.; Asri, E.K.; Purwanti, T.S.; Kusyuniati, S.; et al. Associations of Knowledge, Attitude, and Practices toward Anemia with Anemia Prevalence and Height-for-Age Z-Score among Indonesian Adolescent Girls. *Food Nutr. Bull.* **2021**, *42*, S92–S108. [CrossRef]
59. Kameshwary, R.; Thakur, A.; Mangal, A.; Vaghela, J.F.; Banerjee, S.; Gupta, V. A study to assess the effectiveness of a nutrition education session using flipchart among school-going adolescent girls. *J. Educ. Health Promot.* **2020**, *9*, 1–7.
60. Bharti, R.; Marwaha, A.; Badshah, T.; Sengupta, R. Effectiveness of a Nutritional Education Intervention Focussed on Iron among School Children in National Capital Region and Mumbai. *J. Clin. Diagn. Res.* **2021**, *15*, 31–36.
61. World Health Organization. *Adolescent Nutrition: A Review of the Situation in Selected South-East Asian Countries*; WHO: Geneva, Switzerland, 2006; ISBN 0025-7125.
62. Bhuiyan, F.R.; Barua, J.L.; Kalam, K.A. Knowledge, Attitude and Practices Regarding Nutrition among Adolescent Girls in Dhaka City: A Cross-sectional Study. *Nutr. Food Sci. Int. J.* **2021**, *10*, 1–9.
63. Hewett, P.C.; Willig, A.L.; Digitale, J.; Soler-Hampejsek, E.; Behrman, J.R.; Austrian, K. Assessment of an adolescent-girl-focused nutritional educational intervention within a girls' empowerment programme: A cluster randomised evaluation in Zambia. *Public Health Nutr.* **2020**, *24*, 651–664. [CrossRef]
64. Viggiano, A.; Viggiano, E.; di Costanzo, A.; Viggiano, A.; Scianni, G.; Santangelo, C.; Battista, R.; Monda, M.; Viggiano, A.; de Luca, B.; et al. Kaledo, a board game for nutrition education of children and adolescents at school: Cluster randomized controlled trial of healthy lifestyle promotion. *Eur. J. Pediatr.* **2015**, *174*, 217–228. [CrossRef] [PubMed]
65. FAO. *The State of Food Insecurity in the World: Eradicating World Hunger—Key to Achieving the Millennium Development Goals*; FAO: Rome, Italy, 2005.

66. Shapu, R.C.; Ismail, S.; Ahmad, N.; Ying, L.P.; Abubakar, I. Knowledge, Attitude, and Practice of Adolescent Girls Towards Reducing Malnutrition in Maiduguri Metropolitan Council, Borno State, Nigeria: Cross-Sectional Study. *Nutrients* **2020**, *12*, 1681. [CrossRef] [PubMed]
67. Vardanjani, A.E.; Reisi, M.; Javadzade, H.; Pour, Z.G. The Effect of Nutrition Education on Knowledge, Attitude, and Performance about Junk Food Consumption among Students of Female Primary Schools. *J. Educ. Health Promot.* **2015**, *4*, 53. [PubMed]
68. Ogunsile, S.E.; Ogundele, B.O. Effect of game-enhanced nutrition education on knowledge, attitude and practice of healthy eating among adolescents in Ibadan, Nigeria. *Int. J. Health Promot. Educ.* **2016**, *54*, 207–216. [CrossRef]
69. Anand, T.; Ingle, G.K.; Meena, G.S.; Kishore, J.; Yadav, S. Effect of Life Skills Training on Dietary Behavior of School Adolescents in Delhi: A Nonrandomized Interventional Study. *Asia-Pac. J. Public Health* **2015**, *27*, 1616–1626. [CrossRef] [PubMed]
70. Abu-Baker, N.N.; Eyadat, A.M.; Khamaiseh, A.M. The impact of nutrition education on knowledge, attitude, and practice regarding iron deficiency anemia among female adolescent students in Jordan. *Heliyon* **2021**, *7*, e06348. [CrossRef] [PubMed]
71. Aduroja, A.E.; Remo, I. Dietary Knowledge and Attitudes of In-School Adolescents in Private Secondary Schools in Ifako-Ijaye Local Government, Lagos, Nigeria. *Int. J. Innov. Sci. Res. Technol.* **2021**, *6*, 1144–1149.
72. Alonso, E.B. *The Impact of Culture, Religion and Traditional Knowledge on Food and Nutrition Security in Developing Countries*; FoodSecure: Brussels, Belgium, 2015.
73. Das, J.K.; Lassi, Z.S.; Hoodbhoy, Z.; Salam, R.A. Nutrition for the Next Generation: Older Children and Adolescents. *Ann. Nutr. Metab.* **2018**, *72*, 56–64. [CrossRef]
74. United Nations Children's Fund (UNICEF). *Food Systems for Children and Adolescents: Working Together to Secure Nutritious Diets*; UNICEF: New York, NY, USA, 2018.
75. Ghattas, H. *Food Security and Nutrition in the Context of the Global Nutrition Transition*; Technical Paper; WHO: Geneva, Switzerland, 2014.
76. Weerasekara, P.C.; Withanachchi, C.R.; Ginigaddara, G.A.S.; Ploeger, A. Food and Nutrition-Related Knowledge, Attitudes, and Practices among Reproductive-age Women in Marginalized Areas in Sri Lanka. *Int. J. Environ. Res. Public Health* **2020**, *17*, 3985. [CrossRef] [PubMed]
77. Food and Agriculture Organization of the United Nations (FAO). *Strengthening Sector Policies for Better Food Security and Nutrition Results*; WHO: Geneva, Switzerland, 2019.
78. Otaha, J.I. Food Insecurity in Nigeria: Way Forward. *An. Int. Multidiscip. J. Ethiop.* **2013**, *7*, 26–35. [CrossRef]
79. Shapu, R.C.; Ismail, S.; Ahmad, N.; Lim, P.Y.; Njodi, I.A. Food Security and Hygiene Practice among Adolescent Girls in Maiduguri Metropolitan Council, Borno State, Nigeria. *Foods* **2020**, *9*, 1–21.
80. Global Network Against Food Crisis. *2020 Global Report on Food Crises*; WHO: Geneva, Switzerland, 2020.
81. WFP. *Global Report on Food Crisis 2018*; WFP: Rome, Italy, 2018.
82. Roncarolo, F.; Potvin, L. Associations between the local food environment and the severity of food insecurity among new families using community food security interventions in Montreal. *Can. J. Public Health* **2017**, *108*, e49–e55.
83. Ukegbu, P.; Nwofia, B.; Ndudiri, U.; Uwakwe, N. Food Insecurity and Associated Factors Among University Students. *Food Nutr. Bull.* **2019**, *40*, 271–281. [CrossRef]
84. Masa, R.; Chowa, G.; Nyirenda, V. Prevalence and Predictors of Food Insecurity among People Living with HIV Enrolled in Antiretroviral Therapy and Livelihood Programs in Two Rural Zambian Hospitals. *Ecol. Food Nutr.* **2017**, *56*, 1–21. [CrossRef] [PubMed]

Article

Overstocked Agricultural Produce and Emergency Supply System in the COVID-19 Pandemic: Responses from China

Mingzhe Pu , Xi Chen and Yu Zhong *

Institute of Agricultural Economics and Development, Chinese Academy of Agricultural Sciences, Beijing 100082, China; pumingzhe@caas.cn (M.P.); chenxi_alice@163.com (X.C.)
* Correspondence: zhongyu@caas.cn

Abstract: The spread of COVID-19 has affected not only public health but also agriculture, raising global concerns regarding the food system. As an immediate impact of COVID-19, farmers around the globe have had difficulties with sales, resulting in large amounts of overstocked agricultural products and food loss. This further threatens the livelihood of rural, poor farmers and impacts sustainable production. To provide a better understanding of the overstocking situation after the outbreak of the pandemic, this study depicts the distribution characteristics of overstocked agricultural products in China. After analyzing a nationwide data set collected from 3482 individuals/organizations by the Chinese Agri-products Marketing Association after the outbreak of the pandemic, we found that some of the initial prevention and control measures disrupted sales channels, and in turn, caused the farmers to suffer losses. The impact was more severe in perishable products and their production areas, as well as in poverty-stricken regions. Then, we identified China's quick and effective actions to match the supply and demand. These findings suggest that emergency responses should coordinate the relationship between emergency actions and the necessary logistics of agricultural production. To prepare for the possibility of such shock in the future, the government should take actions to clear logistics obstacles for necessary transportation, keep enhancing the fundamental infrastructure and effective mechanism of the food supply chain, and actively include innovative techniques to build a more resilient food system.

Keywords: COVID-19; overstocked agricultural products; food loss; food system; online data; e-commerce; China

Citation: Pu, M.; Chen, X.; Zhong, Y. Overstocked Agricultural Produce and Emergency Supply System in the COVID-19 Pandemic: Responses from China. *Foods* **2021**, *10*, 3027. https://doi.org/10.3390/foods10123027

Academic Editors: António Raposo, Renata Puppin Zandonadi and Raquel Braz Assunção Botelho

Received: 10 November 2021
Accepted: 22 November 2021
Published: 6 December 2021

Publisher's Note: MDPI stays neutral with regard to jurisdictional claims in published maps and institutional affiliations.

1. Introduction

The emergence of coronavirus disease 2019 (COVID-19) has imposed unprecedented challenges to the world economy. In the agricultural sector, the lockdown measures and movement restrictions disrupted food supply chains, raising concerns regarding food security [1–5]. Difficulties in selling agricultural products and highly overstocked situations had been reported around the globe [6–9]. In developing countries, such as China, India, and Ethiopia, farmers had to leave their produce in the field to rot [10–16]. Even in developed countries such as the United States and Canada [17–20], farmers had to dump milk, crush eggs and plow under crops [21,22]. The adverse effects of overstocked agricultural products have been analyzed in the current research. Firstly, overstocked agricultural products aggravate the contradiction between food supply and demand [23]. Restrictions on movement are reducing farmers' ability to sell products. Agricultural produce is accumulating at farms, resulting in huge food losses [24–26]. On the demand side, indoor demand for food is strong under the current quarantine measures [27]. Some store shelves have been kept empty due to panic buying [28]. Secondly, overstocked agricultural products decrease the agricultural income of rural households [29]. In developing countries, a large subpopulation lives on agriculture and informal food markets [15,30,31]. Difficulties in sales directly threaten the livelihoods of rural, poor, and informal sector workers [32]. As estimated by the UN, an additional 83 million people, for a total of 132 million people, will experience hunger this year because of COVID-19 [33]. Thirdly, overstocked agricultural products undermine agricultural production capacity and sustainability [34]. Constrained by limited cash flow and financial liquidity [7], farmers lack incentives to invest in the next season and consequently reduce production [35]. This will increase market volatility and lead to more severe food insecurity shortly. Based on the above concerns, overstocked agricultural products in the pandemic have already aroused many concerns from the public and the government [36].

However, few academic research efforts tracked and analyzed the overstocked problems of agricultural products at the onset of the COVID-19 pandemic. The overstocking situation during the COVID-19 pandemic is quite different from that in the normal market. Previous research has suggested that excess supply caused by asymmetric market information or less organized production is the main reason for overstocking. During the COVID-19 pandemic, the problem has lain with demand and distribution channels [11,37]. On the other hand, the overstocking problem has revealed some shortcomings in the emergency food supply system. The current emergency system remains fragile in terms of logistics and distribution channels [38] and fails to coordinate the supply of agricultural products with overall crisis-response measures [39]. In China, many details concerning the overstocking problem during the spread of the COVID-19 have not been clarified yet. For instance, what are the distribution characteristics of overstocked agricultural products across the country? Are the impacts widespread, even in counties with different levels of pandemic risk, poverty, and e-commerce development? What can we learn from the government responses and the system recovery process?

This paper seeks to fill the gap in knowledge by shedding light on the above questions. To do so, we used online data from the Chinese Agri-products Marketing Association (CHAMA), including 3482 pieces of information regarding overstocking across the whole country from 20 February to 31 March 2020. These unique data allow us to provide a nationwide picture of the distribution characteristics of overstocked agricultural products. By combining information regarding overstocking with county-level data, we further examine the effects of pandemic risk, poverty, and e-commerce on difficulties in sales, providing in-depth detail on the impacts of the COVID-19 pandemic on the agricultural sector and emergency response system. This study contributes to the literature in three ways. First, to the best of our knowledge, it is the first work to use nationwide online data to investigate the impact of COVID-19 on overstocked problems. Most works are based on case studies in a certain city. Second, this article quantifies the effects of COVID-19 on the sales of agricultural products. It enriches current research on overstocked agricultural products by analyzing the problem in an emergency, which has seldom been discussed before. Third, our analysis provides descriptive and empirical evidence for academic research and policy practices to strengthen the emergency response system, and to find a way to consolidate the food supply chain and build a more sustainable food system in China. For other developing countries that are experiencing serious outbreaks, our research will aid in keeping food systems well-functioning and in protecting domestic agricultural production.

The rest of this paper is organized as follows. Section 2 introduces the research methodology. Section 3 reports the descriptive statistics and empirical results. Section 4 introduces the government responses to the overstocked problem and market responses. Section 5 discusses our findings. The final section concludes and derives several policy implications.

2. Research Methodology

This study employs both qualitative and quantitative analysis. For a holistic examination of the overstocking problem of the agricultural products in China after the COVID-19 lockdown, it was first important to understand the distribution characteristics of overstocked produce. Most available overstocking information were certain cases in a region or city without a nationwide representative. On 17 February 2020, the Ministry of Agriculture and Rural Affairs (MARA) called for industry associations and e-commerce enterprises to collect information on overstocking and promote sales. These data platforms allowed us to collect nationwide data and analyze the overstocking problem from a global perspective. Therefore, we collected our data on overstocking information from a large industrial association called the CHAMA. The data were scraped daily and provided information on food categories, overstocked amounts, prices, and locations.

Based on these details, we analyzed the distribution characteristics from perspectives of time-evolution trend, regional distribution, and category distribution. To reveal the potential economic loss behind the overstocked problem, we used the prices reported by the producers to estimate overstocked values.

To find the determinants of the overstocking magnitude, we combine the overstocked data with the data of exposure risk to the pandemic, poverty situation, and e-commerce development. Then we used multiple linear regression to analyze the impacts of pandemic risk, e-commerce development, and poverty on the overstocked quantities.

Government responses were important to help the market back to its normal levels. Governments' interventions were reviewed based on the descriptive analysis, followed by the market rebound data to indicate the recovery performances. Following data analysis, mitigation strategies were identified by combining our observations in China and the literature.

2.1. Data

We chose the CHAMA as the data source for our research. The CHAMA, 18 years of experience in digitized supply chain management and supply-demand information matching, ranked first on the list of online assistant platforms recommended by the MARA. All its 945 association members are part of large-scale wholesale markets of agricultural produce, with national or regional influence. To collect overstocking information across the country, the CHAMA started a special online platform named "matching agricultural production and demand to fight against COVID-19 and assist farmers" on 20 February 2020. In parallel with the web platform, the CHAMA also used WeChat, email, and 24-h hotlines to facilitate information collection. To assist the "point-to-point" connection of producers and buyers, all collected information was open to the public and posted in real time on this platform.

A producer who was experiencing difficulty in sales needs only to fill out an online enrollment form, including the following information: the name of the organization, the category of the organization (government/association/other organization, agrifirm/cooperative, individual), contact person, phone number, location (accurate to the county level), product categories (grain and edible oil, vegetable, fruit, livestock and poultry, aquatic product and so on), product name, overstocked quantity, and expected price (optional). This form helped to provide standardized data about different foods, different organizations, and different regions. All data needed to be reviewed by CHAMA to ensure credibility and quality.

Web crawling was carried out from 20 February to 31 March 2020. The start date was when the CHAMA started running the online platform, and the end date was the amount of overstocked information decreased to zero.

To examine what factors might be correlated with the overstocking of agricultural products, we drew on information from both sample observations and the county from which they originated. We were interested in three key variables: risk of exposure to the pandemic, poverty levels, and e-commerce development.

The risk of exposure to the pandemic was measured by the cumulative number of confirmed COVID-19 cases at the city (district) level during the pandemic. These numbers were collected by Dingxiangyuan and Dr. Dingxiang and were posted on an online platform called "real-time dynamics of COVID-19 outbreak". The cumulative confirmed cases for each city (district) were retrieved on 2 April 2020, two days after the deadline for collecting overstocking information. The time lag did not make a large difference, because the pandemic had been largely controlled by the end of March.

Poverty status was divided into three categories, poverty-stricken county, poverty-relieved county, and ordinary county, according to the official documents of the State Council Leading Group Office of Poverty Alleviation and Development (SCLGOPAD). In 2014, SCLGOPAD announced there were 832 poverty-stricken counties across the country. By the end of 2019, the total amount had decreased to 52 counties in seven provinces. These 52 counties were defined as poverty-stricken counties. The counties which succeeded in alleviating poverty during 2014–2019 were defined as poverty-relieved counties. The remaining counties were considered ordinary counties.

The development of e-commerce was measured by the number of Internet users at the end of the year in a county. These data came from the statistical yearbooks published by the governments of the provinces or the cities, or the national economic statistics bulletins of the counties.

We matched the pandemic risk, poverty alleviation, and e-commerce development to the overstocking information according to the county information. For some pieces of information for which the address could not be accurate to the county level, these three variables were set as vacancies.

Other data regarding government and market responses were from the government's released documents and reports, the National Bureau of Statistics, a big-data information platform called National Data Service Platform of Agricultural and Rural Response to COVID-19.

2.2. Descriptive Statistics

The overstocking data were initially analyzed using descriptive statistics. We tracked changes in the amount of overstocking information over time. The proportion of each product was analyzed, indicating the most vulnerable product during the pandemic. Various subsamples, decomposed according to the categories and regions, are investigated to provide in-depth detail.

Then, the potential economic losses were estimated based on the online information. The overstocked quantities for each product could be calculated by conversion into compatible units. By timing the prices, the economic value of overstocked agricultural produces could be estimated accordingly.

The descriptive statistics method was also used to analyze government responses and market rebound. Several important government intervention measures were reviewed. Then the market transaction amount change compared to 2019 was calculated day by day. The evolution of these daily changes provided dynamic information on market responses.

2.3. Empirical Investigation

An econometric model was used to investigate the impact of risk of exposure to the pandemic, poverty levels, and e-commerce development on the overstocking amount. A multiple linear regression model was set as follows:

$$overstocking_i = \alpha_1 \cdot pandemic\_risk_i + \alpha_2 \cdot poverty\_relieved_i$$
$$+\alpha_3 \cdot poverty\_stricken_i + \alpha_4 \cdot ecommerce_i + \varepsilon$$

The dependent variable $overstocking_i$ represents the overstocking number of agricultural products that a producer i had during the pandemic. Considering the dependent variable is a continuous variable, the OLS method is applicable in this paper.

$pandemic\_risk_i$ represents the exposure risk to the pandemic. It is set as a categorical variable ordered from 1–7 according to the cumulative number of confirmed COVID-19 cases at the city (district) level during the pandemic. The grouping method followed the official reports. The value of $pandemic\_risk_i$ was set to 1 if no confirmed cases were reported; 2 if 1–9 confirmed cases were reported; 3 if 10–99 confirmed cases were reported;

4 if 100–499 confirmed cases were reported; 5 if 500–999 confirmed cases were reported; 6 if 1000–10,000 confirmed cases reported; 7 if over 10,000 confirmed cases reported.

Poverty status was represented by two dummy variables $poverty\_relieved_i$ and $poverty\_stricken_i$. The treatment group is the ordinary counties whose values of $poverty\_relieved_i$ and $poverty\_stricken_i$ were set to 0. $poverty\_relieved_i$ was set to 1 if a county succeeded in alleviating poverty during 2014–2019. $poverty\_stricken_i$ was set to 1 if a county was still in the listed 52 poverty-stricken counties released by SCLGOPAD in 2019. $ecommerce_i$ represented the number of Internet users at the end of the year in a county.

3. Immediate Impacts of COVID-19: Evidence from CHAMA Overstocking Information

By 31 March 2020, the CHAMA online platform had collected 3938 pieces of information regarding overstocked agricultural products. After clearing the duplicate information, there were 3482 valid observations with approximately 85 observations per day. All observations came from three categories of organizations: individual, agricultural companies/cooperatives, government/Association. 52.83% of the information was reported by agricultural companies and cooperatives, and 34.17% from individuals. The information from the government or association accounted for 13%. Relatively speaking, the large-scale producers are easier to be shocked by the pandemic [40].

3.1. Distribution of Overstocked Agricultural Products

The time-evolution trend of overstocking information was the result of the government's efforts to balance COVID-19 restraints and food supply chain protection. There were three peaks in the posted amount of overstocked agricultural products during the data collection period (see Figure 1). The first peak was on 20 February, when 2882 pieces of information were uploaded to the online platform. This accounted for 82.77% of the total information. The second peak was at the end of February, including 61 pieces on 27 February and 67 pieces on 28 February. The third was on 12 March, when 108 pieces of information were released. After the middle of March 2020, the amount of overstocking information decreased rapidly. The first peak was much higher than the others because the overstocking information was accumulating before the online platform was established. The national-wide lockdown measures dated back to 29 January 2020. During the 21 days before the platform was established, a large number of chicken coops, pig pens, fishponds, and farmlands were soon overstocked, and farmers could not find outflow channels. The CHAMA data showed that there was an average of 137 messages per day during this severe period. After the platform opened on 20 February, all information was reported to the platform. After 21 February, the overstocking information followed a trend of fluctuating downward (see the small figure in the upper-right corner of Figure 1). There has been no overstocked information released since April.

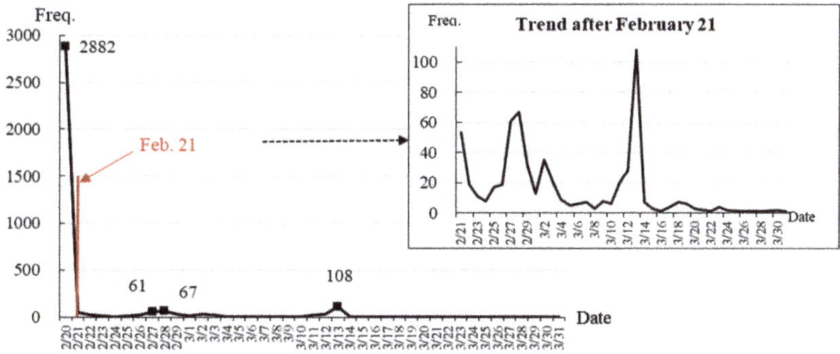

Figure 1. Time-evolution trend of overstocked agricultural products posted by the CHAMA from 20 February to 31 March 2020.

The pandemic struck the food supply chains across the nation. Almost all provinces (autonomous regions and municipalities) reported overstocking information of agricultural produce, especially those which produce perishable products [41]. Among 34 provinces (autonomous regions and municipalities), 32 provinces (autonomous regions and municipalities) reported overstocking cases (see Table 1). The top five provinces suffering the most from the pandemic were Guangxi, Anhui, Yunnan, Shandong, and Shanxi. Guangxi experienced the most severe overstocking situation. The amount of overstocking information in Guangxi was 2198, accounting for 63.12% of the total sample. Guangxi suffered the worst because the pandemic overlapped with its harvesting season of late-ripening citrus products. Some strict lockdown and quarantine measures cut off the traditional offline sales channels directly [32], resulting in an enormous number of overstocked produce. Anhui ranked second, with 486 pieces of information, accounting for 13.96% of the data. The proportions in Yunnan, Shandong, and Shanxi were 4.8%, 4.34%, and 4.22%, respectively. It is noted that all these provinces are the fresh "food baskets" in China, producing fruit, vegetable, livestock, and other perishable products for the rest of the country in winter. From the perspective of economic development, all these provinces were less developed. The poor farmers in these areas relied on the above cash crops to increase incomes, and the local governments to boost the rural economy. Analyzing the overstocking situation and consequent release actions is of importance not only to build a resilient food system but also for protecting the poor rural population and rural economy.

Table 1. Regional distribution of overstocked agricultural products posted by the CHAMA from 20 February to 31 March 2020.

Province	Freq.	Percent	Province	Freq.	Percent
Guangxi	2198	63.12	Heilongjiang	13	0.37
Anhui	486	13.96	Shaanxi	11	0.32
Yunnan	167	4.8	Gansu	9	0.26
Shandong	151	4.34	Sichuan	8	0.23
Shanxi	147	4.22	Qinghai	8	0.23
Hainan	65	1.87	Zhejiang	7	0.2
Jilin	33	0.95	Liaoning	7	0.2
Hebei	28	0.8	Chongqing	4	0.11
Hubei	23	0.66	Jiangsu	3	0.09
Inner Mongolia	20	0.57	Jiangxi	3	0.09
Henan	19	0.55	Fujian	3	0.09
Tianjin	17	0.49	Guangdong	2	0.06
Xinjiang	17	0.49	Shanghai	1	0.03
Hunan	17	0.49	Beijing	1	0.03
Guizhou	13	0.37	Ningxia	1	0.03
Total	3482	100			

Highly perishable and time-sensitive food was highly overstocked after the pandemic. Fruit, vegetables, livestock and poultry were the top three varieties of all overstocked agricultural products. Regardless of geographic distribution, 53% of the unmarketable information was related to fruit, 18% to livestock and poultry, and 16% to vegetables. The grain and edible oil and aquatic products accounted for 7% and 3%, respectively (see Figure 2A). In the Guangxi sample (see Figure 2C), the proportion of fruit, livestock and poultry, and vegetables was 73%, 14%, and 8%, respectively. This distribution characteristic caused the proportion of fruits in the total sample to exhibit a positive bias. After excluding the Guangxi sample, the largest proportion became vegetables, accounting for 30%; the livestock and poultry still ranked second, accounting for 24%; followed by fruit, accounting for 20%. The proportion of grain and edible oil rose to 14%, and aquatic products rose to 7% (see Figure 2B). On the whole, the pandemic had a great impact on perishable agricultural products, such as vegetables, fruits, and livestock and poultry [41,42].

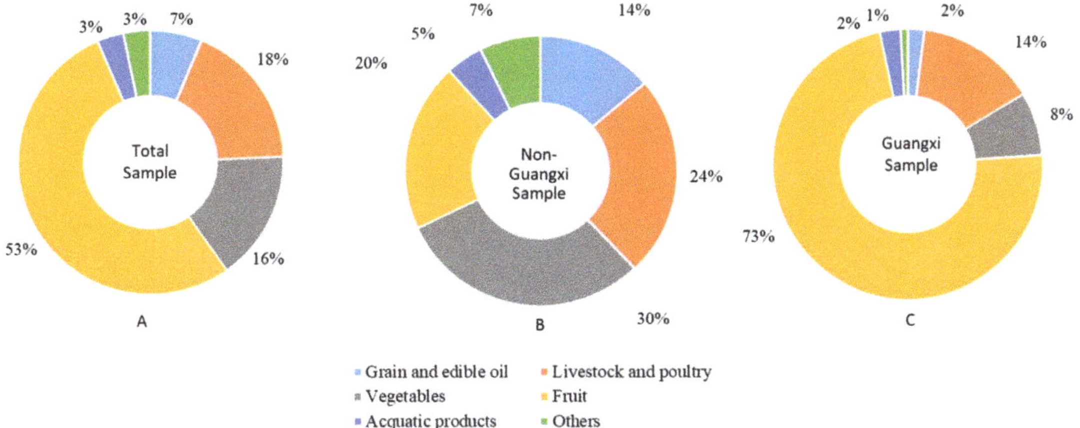

Figure 2. Variety distribution of overstocked agricultural products posted by the CHAMA from 20 February to 31 March 2020.

3.2. Estimation of Potential Economic Losses

The overstocked amount and economic values are directly associated with producers' income and planting motivation for the next season. Reducing current sales not only threatened the livelihood of agricultural producers but also reduced future agricultural production. Both would bring more challenges to food security from the micro and macro levels. Although these potential risks had been identified [38], more specific analyses and calculations regarding the magnitude of agricultural produce were absent. Taking the CHAMA as a case, we could estimate the total amount (in weight) and economic value of overstocked agricultural products reported on the CHAMA online platform. Some producers provided overstocking information in the form of units of quantity. We converted all quantity units to weight units for comparison. To do so, we tried to reveal the economic burdens brought by overstocking agricultural produce during the pandemic, providing a more intuitive understanding of the potential risks behind the surface.

The total overstocked weight recorded by CHAMA was approximately 2.77 million tons (see Table 2). The largest share was fruit, reaching 1.81 million tons, followed by livestock and poultry at approximately 636.2 thousand tons. Vegetables ranked third with approximately 206.9 thousand tons. The weight of grain and edible oil and aquatic products was 73.3 thousand tons and 46.4 thousand tons, respectively. Regarding livestock and poultry, there were 21.65 million chickens, 11.08 million ducks, and 246.6 thousand pigs. From the perspective of geographic distribution, most of the overstocked grain and edible oil was located in the north-eastern region. Most other agricultural products were located in Southern China. However, after excluding the Guangxi sample, most of the overstocked fruit was located in the northern and central regions. Most unsaleable vegetables were located both in the southern and northwestern regions. Most unmarketable livestock and poultry were located in the eastern region, along with aquatic products.

Table 2. Distribution of overstocked agricultural products in different regions by CHAMA (unit: ton).

Region	Grain and Edible Oil	Fruit	Vegetables	Livestock and Poultry	Aquatic Products
North	12,824	155,537	2309	300	0
Northeast	30,365	201	460	1	20
East	15,876	5996	21,383	54,204	29,340
Central	4251	132,806	9213	140	560
South	3971	1,499,299	105,363	540,733	16,466
—excluding Guangxi	200	63,566	36,719	8	0
Southwest	2749	2885	21,660	17,757	0
Northwest	3220	10,326	46,491	23,042	0
Total	73,256	1,806,882	206,879	636,177	46,386

Notes: The North China includes Beijing, Tianjin, Hebei, Shanxi, Inner Mongolia. The Northeast region includes Jilin, Heilongjiang, and Liaoning. The East region includes Shanghai, Jiangsu, Zhejiang, Anhui, Fujian, Jiangxi and Shandong. The Central region includes Henan, Hubei, and Hunan. The South region includes Hainan, Guangdong, and Guangxi. The Southwest region includes Chongqing, Sichuan, Guizhou, Yunnan, and Tibet. The Northwest region includes Shaanxi, Gansu, Qinghai, Ningxia, and Xinjiang. There is no overstocked information reported in other provinces/regions.

A conservative estimation showed that the potential economic value of the overstocked agricultural products (see Table 3) was approximately CNY 8.1 billion (USD 1.14 billion). On average, each organization risked losing 230 thousand CNY during the lockdowns. Livestock and poultry accounted for the largest share, reaching CNY 4.92 billion (USD 693.6 million). The second-largest share was of fruit, at approximately CNY 1.67 billion (USD 235.4 million). The economic value of overstocked grain and edible oil, vegetables, and aquatic products was approximately CNY 5 million (USD 70.5 million). The Guangxi sample still dominated the distribution of economic value. Approximately 83% of the value of livestock and poultry, and 67% of the value of fruit came from Guangxi. For other products, non-Guangxi regions contributed to over 77% for each. The above estimation showed that the breeding industry was facing the greatest financial risk during the lockdowns. Their products were not only highly perishable but also high value. If the entirety of the overstocking situation was realized, the economic losses would be too large to ignore. These losses would shock farmers' livelihoods severely, as well as the rural economy.

Table 3. Estimated economic value of overstocked agricultural products (unit: million CNY).

Products		Grain and Edible Oil	Fruit	Vegetables	Livestock and Poultry	Aquatic Products
Total sample		458.95	1673.09	604.10	4924.75	462.78
Guangxi sample	value	8.93	1124.90	136.15	4108.37	78.28
	percent	2%	67%	23%	83%	17%
Non-Guangxi sample	value	450.02	548.19	467.95	816.38	384.50
	percent	98%	33%	77%	17%	83%

Such a large number of overstocked products was not just the loss of producers and agricultural sector, but also consumers and the whole society. For the consumers, they experienced a shortage of foods due to disrupted supply chains. Panic buying and stockout drove prices high on the demand side. More consumers were in food insecurity. For the whole society, unreasonable lockdown measures resulted in a mismatch between supply and demand, causing efficiency loss to social welfare. This mismatch also led to massive food loss during the pandemic [43,44].

3.3. Factors Affecting the Overstocked Amount

Table 4 reports the regression results. The Variance Inflation Factor (VIF) test suggests that the explanatory variables are not strongly multicollinear. In addition, the F-statistics test shows the significant joint explanatory power of the used independent variables. The statistics validate the specifications of our models.

Table 4. Impacts of various factors on the overstocked amount of different agricultural products.

Variables	Total	Grain and Edible Oil	Livestock and Poultry	Vegetables	Fruit
Epidemic risk	−0.269**	−1.044 ***	−2.182 ***	0.186	0.361 ***
	(−2.53)	(−3.10)	(−5.16)	(0.75)	(2.77)
Poverty-relieved county	2.225 ***	−0.246	0.772	−0.456	2.443 ***
	(20.92)	(−0.51)	(1.45)	(−1.21)	(24.18)
Poverty-stricken county	0.649 ***	0.390	−2.048 **	—	0.740 ***
	(3.33)	(0.48)	(−2.42)	—	(4.45)
Internet user amount	−0.027	1.300 ***	0.608 *	0.392 **	−0.273 ***
	(−0.30)	(3.25)	(1.89)	(2.26)	(−2.74)
Western region	−0.340	0.527	0.476	0.389	−1.003*
	(−1.26)	(0.41)	(0.51)	(1.32)	(−1.65)
Central region	−1.586 ***	1.068 **	−1.777 ***	−1.040 ***	0.372
	(−11.49)	(2.37)	(−5.93)	(−4.62)	(1.48)
Constant	4.341 ***	2.523 ***	5.684 ***	2.461 ***	3.737 ***
	(17.07)	(3.44)	(5.49)	(3.95)	(12.57)
Observations	2564	107	436	356	1587
F statistics	119.4	3.983	14.61	10.23	125.3
R-squared	0.233	0.198	0.207	0.114	0.326

Notes: Robust t-statistics in parentheses, *** $p < 0.01$, ** $p < 0.05$, * $p < 0.1$. The dependent variables and the number of Internet users are logarithm form.

More overstocked agricultural products were located in cities (districts) where the pandemic was mild. This effect varied across different subgroups of products. The coefficients of grain and edible oil and livestock and poultry are negative, while one of the fruits is positive and statistically significant at the 5% level. In our data, about 73% of the overstocking samples were reported in cities (districts) where cumulative confirmed cases were between 10 and 99; 21.19 percent were reported in cities (districts) where cumulative confirmed cases were between 1 and 9. In the subgroup of fruit, the significantly positive coefficient indicates that higher exposure risk to the pandemic experienced more overstocked pressure. This confirms the previous research that the pandemic disrupted food supply chains by imposing various kinds of lockdown measures.

For other subgroups, the negative effects of exposure risk to the pandemic do not mean that difficulty in sales and overstocked agricultural products did not exist in high-risk regions. On the contrary, it was the highest pandemic risk that restricted gathering information on overstocked agricultural products, because medical treatment and rescuing life were given priority. In mild regions, the prevention and control measures were relatively loosened to resume production and boost the rural economy. Thus, the difficulties in sales in these regions could be observed more easily.

The organizations, which produce fruit in poverty-relived/poverty-stricken cities, suffer more from overstocked pressure. In the subgroup of fruit, the coefficients of poverty-stricken counties and poverty-stricken counties are significantly positive at the level of 1%. If the overstocked problem is unsolved, more poor farmers have to bear the economic losses. In the poor areas, most farmers used to plant grain for self-sufficiency due to poor conditions of transportation and less-developed markets. With infrastructure improved by the government, some poor areas tried turning to plant high-value crops such as fruit to earn more money. Many poverty-stricken households manage to shake off poverty by producing fruit. However, the logistics system is still vulnerable in these areas due to long-distance and complex geographical conditions. A systematic shock makes this weakness more salient. This result raises concerns over the performance of the governments' poverty alleviation actions.

The coefficient in the total sample was biased to positive because of the large share of fruit. The coefficient of the poverty-stricken county regarding livestock and poultry is significantly negative. A possible explanation is that the average scale of the breeding industry in poverty-stricken counties is smaller. Most farmers breed livestock and poultry mainly for self-sufficiency. The coefficients of grain and edible oil in both types of counties are not significant, indicating the stability of the grain market during the pandemic [45]. The results concerning vegetables should not be taken into account, because there was no observation of poverty-stricken counties in the vegetable subsample.

The impact of e-commerce development varies across different subgroups of products. The coefficient in the fruit subsample is significantly negative, while others are significantly positive. The possible reason is that the dependence of these products on the Internet is different. Fruit demonstrates the feature of regionally specialized production in China and depends on the Internet to realize cross-region sales. The higher the level of e-commerce development, the more conducive a product is to achieving cross-regional production and demand information matching. Other products rely less on cross-regional e-commerce, because the supply of grain products is guaranteed by the "Cereal Bag" Provincial Governor Responsibility Mechanism, and the supply of livestock, poultry, and vegetables is guaranteed by the "food basket" Mayor Responsibility Mechanism. Grain consumption security is always the top priority in the food system. In addition, areas with more Internet users have better economic development and a higher population density, resulting in stricter pandemic prevention and control measures. This, in turn, impacts the supply of grain and edible oil, livestock and poultry, and vegetables.

4. Government Responses and Market Rebound

With an unprecedented scale of protection and control measures to isolate and control the COVID-19 pandemic, some restrictions had a severe impact on the sales of agricultural products [46–48]. Not long after the national lockdown, China witnessed difficulties in sales and the overstocking problem of agricultural products.

Fortunately, the Chinese government has realized these problems and in February 2020 issued six urgent notices to clear logistics restrictions and to directly match production with demand [49] (see Figure 3). On February 4, the Ministry of Agriculture and Rural Affairs announced an urgent notice to maintain livestock and poultry industry operations and ensure the supply of meat, eggs, and milk. Two days later, the Ministry of Commerce (MC) announced a notice to sustain the delivery and distribution of life necessities, which included agricultural products. On February 11, the MARA announced a notice to ensure sales of agricultural products in poor-stricken regions. In February, there were three notices released by the MARA, MC, and the Ministry of Finance (MF) at the same time. The first announced actions to strengthen linkages between demand and supply of agricultural products. The second announced actions to enhance collaboration and coordination between the agricultural and commercial sectors to improve the food supply chain. The third took actions to ensure stable production and adequate supply of agricultural products.

In addition, the MARA organized industry associations and e-commerce enterprises to collect overstocked information and promote sales on 17 February 2020. As of 24 April, all major e-commerce platforms have sold 882,000 tons of fresh agricultural products and stimulated 19.8 million online transactions. The CHAMA has organized major agricultural product markets across the country to actively connect with 12 key production and marketing regions, including Hainan, Guangxi, Shandong, Hubei, and Xinjiang. A total of 2.245 million tons of agricultural products were sold by the CHAMA, reaching CNY 12.17 billion (USD 1.71 billion). Among them were 808,300 tons (CNY 4.35 billion) in Hainan, 459,500 tons in Guangxi (CNY 2.55 billion), and 281,500 tons in Shandong (CNY 1.39 billion). In addition, the MARA has launched programs to support cold chain logistics and facilitate the online selling of agricultural products.

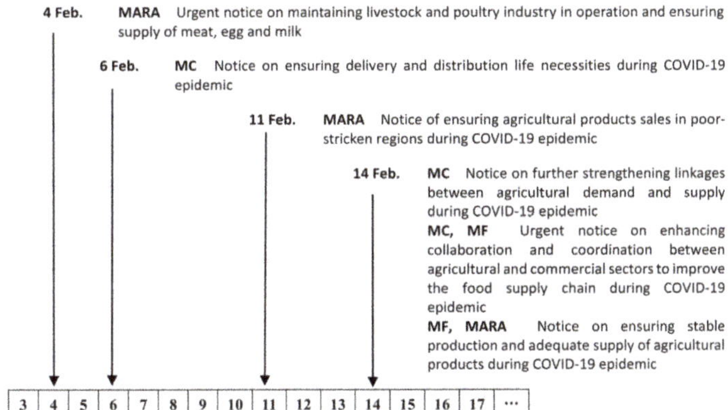

Figure 3. Timeline of Chinese government responses to alleviate overstocked difficulties. Notes: MARA, Ministry of Agriculture and Rural Affairs; MC, Ministry of Commerce; MF, Ministry of Finance.

The agricultural market rebounded soon after the government intervention. After implementation of traffic control and social distancing, the transaction number of agricultural products reduced sharply (see Figure 4). From 22 to 24 January, when grain trading was at its peak in normal years, it plunged by more than 70% compared to 2019, translating into a decline of about 30 million tons. As a result of both the epidemic and the African Swine Fever, the transaction number of livestock and poultry was 20–45% lower than the same period last year. Among them, the transaction amount on 25 January was only 13 million tons, 68.3% lower than that on 22 January, and 71 percent lower than the same period in 2019. The transaction amount recovered slightly after 22 January, but still decreased by 50–70% compared with the same period in 2019. On 9 February, the transaction amount hit the bottom again, at only 11.4 million tons. Since then, the transaction amount increased slightly and steadily, basically remaining around 20 million tons, and recovered to around 28 million tons by the end of February and mid-March. The transaction number of vegetables and fruits fell sharply first, then showed a fluctuation recovery situation. Before 10 February, the transaction number of vegetables and fruits showed a reverse trend from 2019. The trading volume of vegetables dropped from 988 million tons on 21 January to 275 million tons on 27 January, a decrease of 72.2%. Fruit trade fell 80.4% from 182 million tons to 35.8 million tons on 28th January.

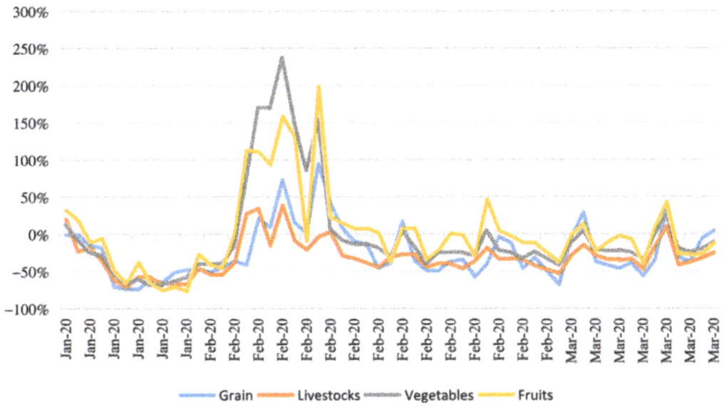

Figure 4. Reduction in marketing quantity of food during the COVID-19 epidemic in China compared to 2019. Note: Numbers are calculated based on a big-data information platform called National Data Service Platform of Agricultural and Rural Response to COVID-19 set by MARA.

5. Discussion

The upstream section of the food supply system suffered a lot from the COVID-19 pandemic due to inelasticity. The overstocked problems in the short term were determined by the biological nature of agricultural production [50,51]. The inevitable time lags in agricultural production weaken its ability to respond to sudden change. After the outbreak of the COVID-19, overstocked problems were observed around the globe. According to the shock transmission theory, the consequences of a shock are borne by the least elastic [44]. In our analysis, producers reported severe overstocked problems during the pandemic. The potential economic and social losses require prompt policy intervention [38].

Government intervention requires timely and accurate identification of risk points in the food supply system. A resilient food supply chain helps to enhance economic and social resilience and accelerates the recovery process. The policymakers should realize that a well-functioned food supply system is essential when coping with a sudden shock [50]. In the case of China, the government took quick lockdown actions to slow the spread of COVID-19. Fortunately, the government observed the shocks to the food supply system and released timely countermeasures. The transportation and distribution part has a vulnerability during the COVID-19 pandemic. The logistics lockdown was the key risk point. It blocked the outflow channels of agricultural products and hindered necessary production inputs and may finally destroy production cycles [51] and undermine production capacity [35]. Ensuring the minimum transportation conditions for agricultural products must be a policy priority during sudden shocks.

Innovative technologies will play a more important role in building a resilient food system [26,49]. In the case of COVID-19, the innovative technology was e-commerce. The lockdowns cut off the traditional offline channels from the producers to consumers, bringing about challenges to smooth operation of the food system. E-commerce allowed information exchange and contract transactions with minimum human-to-human interaction. A lot of retrospective research found that the individuals/organizations that were more familiar with digital tools were able to respond to and resist shocks effectively. Our empirical results also support this argument. Increasing automation and digitalization, online delivery services are changing the food supply system.

Fundamental infrastructure and effective mechanisms of the food system always matter [44]. Except for some disruption in the short term, the food supply system in China performed remarkably well during the pandemic. The policy interventions were timely in practice and the market recovered soon after the shock [35,52,53]. The remarkable performances would not have been achieved without the solid food supply basis enhanced by the Chinese government in the past decades. The "Cereal Bag" Provincial Governor Responsibility Mechanism and the "Food Basket" Mayor Responsibility Mechanism ensure the smooth operation of the food system in normal times. The national emergency food supply system consists of processing enterprises, supply outlets, distribution centers, and emergency storage and transportation enterprises to handle the emergency in abnormal times. In addition to China, the food system in Canada and the United States also worked well during the pandemic due to their resilient food system [45,50,54]. Research in India also showed that a better warehousing infrastructure can build resilience against such supply disruptions [38]. Thus, the policymakers should lay a solid foundation for the food supply system.

6. Conclusions

After the outbreak of COVID-19, the Chinese government has taken timely and effective prevention and control measures to contain the spread of the virus. However, some restrictions have led to difficulties in the sale of agricultural products across the country. By using 3482 pieces of overstocking information collected by the CHAMA, this study is the first to depict the distribution characteristics of overstocked agricultural products from a nationwide perspective. We seek to provide a deeper understanding of the linkage between the overstocking situation and socioeconomic development. The results suggest that the pandemic has a greater impact on perishable products. The provinces which produce fruit, vegetables, and livestock and poultry suffered the most from the shock, such as Guangxi, Anhui, Yunnan, and Shandong. Most overstocked agricultural products were located in regions where the pandemic risk was mild in February and March. The poverty-relieved and poverty-stricken counties which produce fruit faced more severe difficulties in sales. The mitigation effect of e-commerce varied across different products.

The findings of this study reveal the shortcomings in the emergency response system. It has failed to coordinate immediate actions to safeguard the necessary logistics for agricultural production. As a result, the upstream section of the food supply system suffers a lot from the COVID-19 pandemic due to its inelasticity to the market change.

The results generate new policy implications to help protect farmers and build up a more effective supply system. First, quick actions should be taken that clear the logistics bottlenecks and ensure the smooth operation of the food supply system. All unauthorized roadblocks should be prohibited. "Green channels" for agricultural products are encouraged to open across the country. Second, innovative methods should be used to enhance the resilience of the food supply system. Big data technologies can be used to share and match market information. E-commerce, as well as social media and livestreaming events, can help to promote and revive sales of agricultural products. The automation technique could provide contactless and efficient distribution services. Third, fundamental infrastructures of the food system should be consolidated. The basic elements of the food system, such as warehousing facilities, processing facilities, and logistics networks, should be strengthened at all times. Establish contingency plans for the food system for a different emergency, to make sure of timely and orderly responses.

Our study extended the work that uses online data in conjunction with other data sets for policymaking [38]. However, this study is still highly limited by the data availability. First, the large share of fruit in the CHAMA data left limited data to investigate other foods. Second, the overstocking information in the CHAMA was simple. To investigate deeper, we had to match the county-level data to the individual-level overstocking data, which lead to information loss. The simple information limited more in-depth investigation. Third, the CHAMA data is only a part of overstocked information during the pandemic. In addition to CHAMA, China Good Agri-Products Development & Service Association (CGAPA), Pinduoduo, Yimutian, Alibaba, and other organizations also established data platforms and collected information during the pandemic. Actually, the pandemic accelerated the development of e-commerce and relative online service. With more data, even the worldwide data available, a better understanding of the overstocking problem can be provided. Real-time policymaking to enhance the resilience of the food supply system can be realized.

Author Contributions: Conceptualization, M.P. and Y.Z.; methodology, M.P. and Y.Z.; software, M.P. and X.C.; validation, M.P., Y.Z. and X.C.; formal analysis, M.P. and X.C.; investigation, M.P., Y.Z. and X.C.; resources, M.P. and Y.Z.; data curation, M.P. and X.C.; writing—original draft preparation, M.P. and X.C.; writing—review and editing, M.P. and X.C.; visualization, M.P.; supervision, Y.Z.; project administration, Y.Z.; funding acquisition, M.P. and Y.Z.. All authors have read and agreed to the published version of the manuscript.

Funding: This research was funded by the National Natural Science Foundation of China (grant number 71903187), the National Social Science Foundation of China (grant number 21ZDA056, 21BJY131), the Agricultural Science and Technology Collaborative Innovation Key Program (grant number CAAS-ZDRW202012) and the Agricultural Science and Technology Innovation Program (grant number ASTIP-IAED-2021-01) from Chinese Academy of Agricultural Sciences.

Data Availability Statement: Publicly available datasets were analyzed in this study. This data can be found here: [http://zn.agri.cn/index/dataInfo.htm?t=1] (accessed on 1 April 2020) [http://snsj.agri.cn/cockpit-index] (accessed on 15 April 2020).

Acknowledgments: The authors would like to thank the anonymous reviewers and the editor for their support, insightful critiques, and constructive comments.

Conflicts of Interest: The authors declare no conflict of interest.

References

1. Torero, M.; Prepare Food Systems for a Long-Haul Fight against COVID-19. IFPRI Blog. 2 July 2020. Available online: https://www.ifpri.org/blog/prepare-food-systems-long-haul-fight-against-covid-19 (accessed on 7 July 2021).
2. WFP. Global Report on Food Crises. Available online: https://docs.wfp.org/api/documents/WFP-0000114546/download/?_ga=2.110830339.728196019.1594181106-199106846.1593758578 (accessed on 6 July 2021).
3. FAO. Impact of the COVID-19 Pandemic on Food Security and Food Systems. *FAO Council. CL 164/10, 6–10 July 2020*. Available online: http://www.fao.org/3/nd059en/ND059EN.pdf (accessed on 7 July 2020).
4. Stephens, E.C.; Martin, G.; Wijk, M.; Timsina, J.; Snow, V. Editorial: Impacts of COVID-19 on agricultural and food systems worldwide and on progress to the sustainable development goals. *Agric. Syst.* **2020**, *183*, 102873. [CrossRef]
5. Gralak, S.; Spajic, L.; Blom, I.M.; El Omrani, O.; Bredhauer, J.; Uakkas, S.; Mattijsen, J.; Ali, A.O.; Iturregui, R.S.; Ezzine, T.; et al. COVID-19 and the future of food systems at the UNFCCC. *Lancet Planet. Health* **2020**, *4*, e309–e311. [CrossRef]
6. Carducci, B.; Keats, E.C.; Ruel, M.; Haddad, L.; Osendarp, S.J.M.; Bhutta, Z.A. Food systems, diets and nutrition in the wake of COVID-19. *Nat. Food* **2021**, *2*, 68–70. [CrossRef]
7. Dixon, J.M.; Weerahewa, J.; Hellin, J.; Rola-Rubzen, M.F.; Huang, J.; Kumar, S.; Das, A.; Qureshi, M.E.; Krupnik, T.J.; Shideed, K.; et al. Response and resilience of Asian agrifood systems to COVID-19: An assessment across twenty-five countries and four regional farming and food systems. *Agric. Syst.* **2021**, *193*, 103168. [CrossRef]
8. Davila, F.; Bourke, R.; McWilliam, A.; Crimp, S.; Robins, L.; van Wensveen, M.; Alders, R.G.; Butler, J.R. COVID-19 and food systems in Pacific Island Countries, Papua New Guinea, and Timor-Leste: Opportunities for actions towards the sustainable development goals. *Agric. Syst.* **2021**, *191*, 103137. [CrossRef]
9. Galanakis, C.M. The Food Systems in the Era of the Coronavirus (COVID-19) Pandemic Crisis. *Foods* **2020**, *9*, 523. [CrossRef] [PubMed]
10. Ye, X.Q.; Cheng, Y.; Zhou, H.L.; Yin, H.D. Impacts of COVID-19 on rural and agricultural development in 2020 and its countermeasures. *Agric. Econ.* **2020**, *3*, 4–10. (In Chinese)
11. Zhang, H.Y.; Hu, L.X.; Hu, Z.T.; Wang, Y.H.; The impact of COVID-19 on Agriculture and Rural Areas and the Response Measures. The WeChat official account of Farmer's Daily. Available online: https://mp.weixin.qq.com/s/7-sQZR2Y7t-mh-BU4QrhJw (accessed on 13 February 2020).
12. Jiang, H.P.; Yang, D.Q.; Guo, C.R. Impact of novel coronavirus pneumonia on agricultural development in China and its countermeasures. *Reform* **2020**, *3*, 5–13. (In Chinese)
13. Priyadarshini, P.; Abhilash, P.C. Agri-food systems in India: Concerns and policy recommendations for building resilience in post COVID-19 pandemic times. *Glob. Food Secur.* **2021**, *29*, 100537. [CrossRef]
14. Dev, S.M. Addressing COVID-19 Impacts on Agriculture, Food Security, and Livelihoods in India. 8 April 2020. IFPRI Research Post. Available online: https://www.ifpri.org/blog/addressing-covid-19-impacts-agriculture-food-security-and-livelihoods-india (accessed on 23 April 2021).
15. Pothan, E.; Taguchi, M.; Santini, G. Local Food Systems and COVID-19: A Glimpse on India's Responses. A Report Submitted to FAO. Available online: http://www.fao.org/in-action/food-for-cities-programme/news/detail/en/c/1272232/ (accessed on 22 April 2020).
16. Tamru, S.; Hirvonen, K.; Minten, B. Impacts of the COVID-19 Crisis on Vegetable Value Chains in Ethiopia, IFPRI Research Post, 13 April 2020. Available online: https://www.ifpri.org/blog/impacts-covid-19-crisis-vegetable-value-chains-ethiopia (accessed on 23 July 2021).
17. Bruno, E.M.; Richard, J.S.; Daniel, A.S. The coronavirus and the food supply chain. *ARE Update* **2020**, *23*, 1–4.
18. Weersink, A.; von Massow, M.; Bannon, N.; Ifft, J.; Maples, J.; McEwan, K.; McKendree, M.G.; Nicholson, C.; Novakovic, A.; Rangarajan, A.; et al. COVID-19 and the agri-food system in the United States and Canada. *Agric. Syst.* **2021**, *188*, 103039. [CrossRef]
19. Barrett, C.B.; Fanzo, J.; Herrero, M.; Mason-D'Croz, D.; Mathys, A.; Thornton, P.; Wood, S.; Benton, T.G.; Fan, S.; Lawson-Lartego, L.; et al. COVID-19 pandemic lessons for agri-food systems innovation. *Environ. Res. Lett.* **2021**, *16*, 101001. [CrossRef]
20. Weersink, A.; Von Massow, M.; McDougall, B. Economic thoughts on the potential implications of COVID-19 on the Canadian dairy and poultry sectors. *Can. J. Agric. Econ.* **2020**, *68*, 195–200. [CrossRef]
21. Bellany, D.Y.; Corkery, M. Dumped Milk, Smashed Eggs, Plowed Vegetables: Food Waste of the Pandemic. *New York Times*, 11 April 2020. Available online: https://www.nytimes.com/2020/04/11/business/coronavirus-destroying-food.html(accessed on 7 July 2021).
22. Hawkes, C.; Squires, C.G. A double-duty food systems stimulus package to build back better nutrition from COVID-19. *Nature Food* **2021**, *2*, 212–214. [CrossRef]
23. Susanna, K.; Sophoa, M. Equity as both a means and an end: Lessons for resilient food systems from COVID-19. *World Dev.* **2020**, *136*, 105104.
24. FAO. *COVID-19 and the Risk to Food Supply Chains: How to Respond?* FAO: Rome, Italy, 2020; Available online: https://www.fao.org/fsnforum/es/node/4998 (accessed on 7 July 2021).
25. FAO. *Sustainable Crop Production and COVID-19*; FAO: Rome, Italy, 2020; Available online: https://www.fao.org/3/ca8807en/CA8807EN.pdf (accessed on 7 July 2021).

26. FAO. *Local Food Systems and COVID-19: A Look into China's Responses*; FAO: Rome, Italy, 8 April 2020; Available online: http://www.fao.org/in-action/food-for-cities-programme/news/detail/en/c/1270350/ (accessed on 2 May 2020).

27. Dou, Z.; Stefanovski, D.; Galligan, D.; Lindem, M.; Rozin, P.; Chen, T.; Chao, A.M. Household Food Dynamics and Food System Resilience Amid the COVID-19 Pandemic: A Cross-National Comparison of China and the United States. *Front. Sustain. Food Syst.* **2021**, *4*, 577153. [CrossRef]

28. Trollman, H.; Jagtap, S.; Garcia-Garcia, G.; Harastani, R.; Colwill, J.; Trollman, F. COVID-19 demand-induced scarcity effects on nutrition and environment: Investigating mitigation strategies for eggs and wheat flour in the United Kingdom. *Sustain. Prod. Consum.* **2021**, *27*, 1255–1272. [CrossRef] [PubMed]

29. Blazy, J.; Causeret, F.; Guyader, S. Immediate impacts of COVID-19 crisis on agricultural and food systems in the Caribbean. *Agric. Syst.* **2021**, *190*, 103106. [CrossRef]

30. Resnick, D. COVID-19 Lockdowns Threaten Africa's Vital Informal Urban Food Trade. *IFPRI Research Post*, 31 March 2020. Available online: https://www.ifpri.org/publication/covid-19-lockdowns-threaten-africas-vital-informal-urban-food-trade(accessed on 5 July 2021).

31. Ebata, A.; Nisbett, N.; Gillespie, S. Food Systems after Covid-19, Building a Better World: The Crisis and Opportunity of Covid-19. *IDS Bull.* **2021**, *52*, 73–94. [CrossRef]

32. Adjognon, G.S.; Bloem, J.R.; Sanoh, A. The coronavirus pandemic and food security: Evidence from Mali. *Food Policy* **2021**, *101*, 102050. [CrossRef]

33. UN. Policy Brief: The Impact of COVID-19 on Food Security and Nutrition. June 2020. Available online: https://www.un.org/sites/un2.un.org/files/sg_policy_brief_on_covid_impact_on_food_security.pdf (accessed on 7 July 2020).

34. Nchanji, E.B.; Lutomia, C.K. COVID-19 Challenges to Sustainable Food Production and Consumption: Future Lessons for Food Systems in Eastern and Southern Africa from a gender lens. *Sustain. Prod. Consum.* **2021**, *27*, 2208–2220. [CrossRef]

35. Pu, M.; Zhong, Y. Rising concerns over agricultural production as COVID-19 spreads: Lessons from China. *Glob. Food Secur.* **2020**, *26*, 100409. [CrossRef] [PubMed]

36. Fan, S.; Teng, P.; Chew, P.; Smith, G.; Copeland, L. Food system resilience and COVID-19—Lessons from the Asian experience. *Glob. Food Secur.* **2021**, *28*, 100501. [CrossRef]

37. Zhu, M.J. Reflections and Inspirations on Overstocked Agricultural Products under COVID-19 Outbreak. *The People's Daily Online*, 25 February 2020. Available online: http://yuqing.people.com.cn/GB/n1/2020/0225/c429781-31603808.html(accessed on 4 June 2021).

38. Mahajan, K.; Tomar, S. COVID-9 and Supply Chain Disruption: Evidence from Food Markets in India. *Am. J. Agric. Econ.* **2021**, *103*, 35–52. [CrossRef]

39. Cheng, G.Q. Strengthen Risk Management and Improve the Supply and Security System for Important Agricultural Products. *The Guangming Daily*, 3 March 2020. Available online: http://news.gmw.cn/2020-03/03/content_33612587.htm(accessed on 9 June 2021).

40. Nordhagen, S.; Igbeka, U.; Rowlands, H.; Shine, R.S.; Heneghan, E.; Tench, J. COVID-19 and small enterprises in the food supply chain: Early impacts and implications for longer-term food system resilience in low- and middle-income countries. *World Dev.* **2021**, *141*, 105405. [CrossRef]

41. Hayes, D.J.; Schulz, L.L.; Hart, C.E.l; Jacobs, K. A descriptive analysis of the COVID-19 and small enterprises in the food supply. *Agribusiness* **2021**, *37*, 122–141.

42. Chenarides, L.; Richards, T.J.; Rickard, B. COVID-19 impact on fruit and vegetable markets: One year later. *Can. J. Agric. Econ.* **2021**, *69*, 203–214. [CrossRef]

43. Aldaco, R.; Hoehn, D.; Laso, J.; Margallo, M.; Ruiz-Salmón, J.; Cristobal, J.; Kahhat, R.; Villanueva-Rey, P.; Bala, A.; Batlle-Bayer, L.; et al. Food waste management during the COVID-19 outbreak: A holistic climate, economic and nutritional approach. *Sci. Total. Environ.* **2020**, *742*, 140524. [CrossRef] [PubMed]

44. Di Marcantonio, F.; Twum, E.K.; Russo, C. Covid-19 Pandemic and Food Waste: An Empirical Analysis. *Agronomy* **2021**, *11*, 1063. [CrossRef]

45. Brewin, D.G. The impact of COVID-19 on the grains and oilseeds sector: 12 months later. *Can. J. Agric. Econ.* **2021**, *69*, 197–202. [CrossRef]

46. Reardon, T.; Bellemare, M.F.; Zilberman, D. How COVID-19 May Disrupt Food Supply Chains in Developing Countries. 2 April 2020. Available online: https://www.ifpri.org/blog/how-covid-19-may-disrupt-food-supply-chains-developing-countries (accessed on 6 August 2020).

47. Reardon, T.; Zilberman, D. Climate smart food supply chains in developing countries in an era of rapid dual change in agrifood systems and the climate. In *Climate Smart Agriculture—Building Resilience to Climate Change*; Lipper, L., McCarthy, N., Zilberman, D., Asfaw, S., Branca, G., Eds.; Springer: New York, NY, USA, 2017.

48. Reardon, T.; Echeverriab, R.; Berdeguéc, J.; Mintend, B.; Tasiea, S.L.; Tschirleya, D.; Zilberman, D. Rapid transformation of food systems in developing regions: Highlighting the role of agricultural research & innovations. *Agric. Syst.* **2019**, *172*, 47–59.

49. Zhan, Y.; Chen, K.Z. Building resilient food system amidst covid-19: Responses and lessons from China. *Agric. Syst.* **2021**, *190*, 103102. [CrossRef]

50. Hobbs, J.E. Food supply chain resilience and the COVID-19 pandemic: What have we learned? *Can. J. Agric. Econ.* **2021**, *69*, 189–196. [CrossRef]

51. Beckman, J.; Countryman, A.M. The Importance of Agriculture in the Economy: Impacts from COVID-19. *Am. J. Agric. Econ.* **2021**, *103*, 1595–1611. [CrossRef]
52. Wang, Y.; Wang, J.; Wang, X. COVID-19, supply chain disruption and China's hog market: A dynamic analysis. *China Agric. Econ. Rev.* **2020**, *12*, 427–443. [CrossRef]
53. Zhang, Y.; Diao, X.S.; Chen, K.Z.; Robinson, S.; Fan, S. Impact of COVID-19 on China's macroeconomy and agri-food system—An economy-wide multiplier model analysis. *China Agric. Econ. Rev.* **2020**, *12*, 387–407. [CrossRef]
54. Deaton, B.J. Food security and Canada's agricultural system challenged by COVID-19: One year later. *Can. J. Agric. Econ.* **2021**, *69*, 161–166. [CrossRef]

 *foods*

Article

Impact of Trade Openness on Food Security: Evidence from Panel Data for Central Asian Countries

Zhilu Sun * and Defeng Zhang

Institute of Agricultural Economics and Development, Chinese Academy of Agricultural Sciences, Beijing 100081, China; 17863905481@163.com
* Correspondence: sunzhilu@caas.cn

Abstract: The problem of food insecurity has become increasingly critical across the world since 2015, which threatens the lives and livelihoods of people around the world and has historically been a challenge confined primarily to developing countries, to which the countries of Central Asia, as typical transition countries, cannot be immune either. Under this context, many countries including Central Asian countries have recognized the importance of trade openness to ensure adequate levels of food security and are increasingly reliant on international trade for food security. Using the 2001–2018 panel data of Central Asian countries, based on food security's four pillars (including availability, access, stability, and utilization), this study empirically estimates the impact of trade openness and other factors on food security and traces a U-shaped (or inverted U-shaped) relationship between trade openness and food security by adopting a panel data fixed effect model as the baseline model, and then conducts the robustness test by using the least-squares (LS) procedure for the pooled data and a dynamic panel data (DPD) analysis with the generalized method of moments (GMM) approach, simultaneously. The results show that: (1) a U-shaped relationship between trade openness and the four pillars of food security was found, which means that beyond a certain threshold of trade openness, food security status tends to improve in Central Asian countries; (2) gross domestic product (GDP) per capita, GDP growth, and agricultural productivity have contributed to the improvement of food security. Employment in agriculture, arable land, freshwater withdrawals in agriculture, population growth, natural disasters, and inflation rate have negative impacts on food security; and (3) this study confirms that trade policy reforms can finally be conducive to improving food security in Central Asian countries. However, considering the effects of other factors, potential negative effects of trade openness, and vulnerability of global food trade network, ensuring reasonable levels of food self-sufficiency is still very important for Central Asian countries to achieve food security. Our research findings can provide scientific support for sustainable food system strategies in Central Asian countries.

Keywords: food security; trade openness; countries of Central Asia; sustainable food system

Citation: Sun, Z.; Zhang, D. Impact of Trade Openness on Food Security: Evidence from Panel Data for Central Asian Countries. *Foods* **2021**, *10*, 3012. https://doi.org/10.3390/foods10123012

Academic Editors: António Raposo, Renata Puppin Zandonadi and Raquel Braz Assunção Botelho

Received: 9 November 2021
Accepted: 3 December 2021
Published: 5 December 2021

1. Introduction

The problem of food insecurity has become increasingly critical across the world since 2015, which threatens the lives and livelihoods of people around the world, particularly the most vulnerable classes [1,2]. According to the latest estimates, at the global level, it is projected that in 2020, between 720 million and 811 million people suffered from hunger, and nearly 2.37 billion people did not obtain enough food, considering the median value, as many as 161 million and 320 million people more than in 2019, respectively [2]. At the same time, all regions of the world will be affected, taking Central Asia for example, in 2020, 2.6 million people were suffering from hunger, being 18.18% higher than in 2019 and one of the regions with a significant growth rate [2]. The COVID-19 pandemic—that broke out at the end of 2019—has further aggravated current trends and policy reactions have triggered a massive economic recession and major disruptions in national and global food supply chains [3–6], and the World Food Programme (WFP) estimated that 272 million people have already been or will soon become acutely food insecure in countries where the COVID-19 pandemic is present [7]. The World Food Summit in 1996 set a goal for halving undernourished populations all around the world by 2015, being one of the eight Millennium Development Goals (MDGs) [8]. In order to offer a new global approach for building sustainable food and nutrition system from 2015 to 2030, the 17 Sustainable Development Goals (SDGs) built on the MDGs, were adopted at the Sustainable Development Summit in 2015, among which the SDG2 is aimed at ending hunger, improving food security status, and promoting sustainable agriculture and food [9]. Therefore, in recent years, the world has not been generally progressing towards the SDG2, making it more difficult to realize the Zero Hunger goal by 2030, so does the Central Asia region.

According to the Food and Agriculture Organization (FAO) of the United Nations (UN), the most widely used and authoritative definition of food security today is described as a condition in which all people, at all times, have physical, social, and economic access to sufficient, safe, and nutritious food that meets their dietary needs and food preferences for an active and healthy life [10–12]. Therefore, based on this perspective, food security means regularly having enough food to eat, not just for today or tomorrow, but also for the next month and next year, and is a global issue affecting every human being's daily life, both in developing and developed countries [13,14]. The FAO suggests food security is generally the composite of four pillars, including availability, access, stability, and utilization [15], based on which food security could be better understood when presented through a series of indicators corresponding to each pillar; furthermore, it is possible to conduct a scientific assessment of the food security status from the local level to the global level.

The phenomenon of food insecurity has historically been an issue confronted primarily by developing countries [16], where millions of people are suffering from serious undernourishment due to food insecurity [14], and the transition countries cannot be spared either. After independence from the Soviet Union, five Central Asian countries—Kazakhstan, Kyrgyzstan, Tajikistan, Turkmenistan, and Uzbekistan, underwent a series of transitions from centrally planned systems to market-orientated economies [17]. However, after independence, due to a number of challenges, such as continuing economic instability and household incomes decline, poor design and slow implementation of economic and political reforms for reorienting of agricultural production from collective farms to private farms, inefficient agricultural infrastructure, and low levels of commercial agricultural inputs [17–19], food insecurity remained at a high level in Central Asian countries, then great concerns over food security emerged, therefore, achieving food security is at the forefront of these countries' economic strategies and policies [20,21]. Despite minor positive adjustments at the break of this century, the agricultural sector in most Central Asian countries was hit hard yet again by the 2007–2008 food price crisis, which further aggravated the food insecurity problem in the region [22]. In recent years, Central Asian countries have experienced several transitions in institutions and governance, each with consequences for agriculture in the region, and have made considerable progress towards increasing domestic food production [23,24], but food security in the region is still threatened by a series of geopolitical, socio-economic and environmental challenges, such as lack of support from institutions to advance farming, lack of efficient and reliable transport systems and adequate infrastructure, lack of research and extension of support to farmers, low capacity for designing evidence-based policies, small-scale farming of land use, and land degradation aggravated by climate change [25–27], contributing to limited productivity in agriculture which is unable to fully satisfy domestic food demand, especially in Tajikistan and Kyrgyzstan both of which are highly dependent on the import of basic food commodities [28,29].

Using 2001–2018 panel data for Central Asian countries, from the four pillars of food security, including availability, access, stability, and utilization, this research empirically estimates the effect of trade openness and other influences on food security by performing the following work. First, we empirically estimate the effect of trade openness and other influences on food security and trace a U-shaped (or inverted U-shaped) relationship between trade openness and food security by adopting a panel data fixed effect model as the baseline model. Second, we conduct a robustness test by using the least-squares (LS) procedure for pooled data and a dynamic panel data (DPD) analysis with a generalized method of moments (GMM) approach, simultaneously. Then, this study makes policy implications for improving the food security status in Central Asian countries based on research findings.

2. Literature Review

Food security is traditionally analyzed in terms of food self-sufficiency or food self-reliance [30]. Self-sufficiency requires domestic food consumption mainly depends on domestic production rather than imports as the main supply source, while self-reliance requires more diverse supply sources, including imports [30,31]. In recent years, international food trade has become an increasingly integral part of the global food supply system [32], and more countries have recognized the significance of trade openness to maintain adequate food security status [14] and are increasingly dependent on trade to maintain food security due to climate change, population growth and many other factors. It is estimated that one in every six people around the world presently depends almost entirely on international trade for the food they eat, a proportion that could rise to 50% by 2050 [33].

In light of the fact that trade openness plays an important role in each country's food security, the effect of trade openness on food security deserves deep consideration. However, long-standing debates over this issue remain unresolved [12], and it is found in the relevant literature that the impact is mainly twofold. On one hand, based on the trade policy reform from the World Trade Organization (WTO) multilateral negotiations, and regional and bilateral trade negotiations, deterministic interests from trade attained through the removal or reduction in tariff and nontariff barriers to trade and relaxation of market access is conducive to making up for the shortages of the domestic food supply by increasing the accessibility to the international market and reducing the price of imported goods, and improving the availability and diversity of foods [2,30,34–36]. The WTO and World Bank are continuously promoting the concept that international market access plays an important role in its clients' development, especially developing countries, and is increasingly attracting attention for the removal of trade restrictions mainly imposed by developed countries to developing countries [37]. With adequate domestic social safety-net policies for vulnerable classes, trade openness has been consistent with food security by lowering food price which benefits everyone including those suffering from most serious food insecurity, and a more diversified diet has made poor classes less vulnerable to food unavailability or price fluctuation in specific foods [38]. Although tariffs have increased the spread of global price fluctuation and exposed consumers to momentary price spikes, the duration of which is likely to be short, unlike the time with low prices may be endured, but also allowed them to benefit from extended periods of low prices [2]. Therefore, trade openness would increase host countries' economic welfare and decrease its economic volatility [39], especially the protection of domestic agricultural market would harm, rather than ensure their food security, which has been observed in many countries [34,38].

On the other hand, stochastic gains and losses caused by supply-side impacts such as fluctuating agricultural productivity, violent conflict, and war, natural disasters including drought and embargoes can be magnified or reduced depending on trade openness level, and can endanger food security [40]. Trade openness has lowered the level of food self-sufficiency, and has made the food supply more dependent on imports [34], which is supposed to make the food supply less secure, especially considering that agriculture is incompatible with free trade because of its innate role in managing ecological/natural resources at both national and global levels and the uneven playing field that was created from how agriculture has been treated (protected/taxed) differently across countries [41,42]. Liberalizing trade, especially in agriculture, is not conducive to unleashing agricultural production potential, considering the market competitiveness of smallholder farmers in developing economies is significantly weaker than large farmers in middle-income economies or subsidized farmers in developed economies [41]. Additionally, reducing foreign dependency on food could significantly abate the conveyance of international market shocks to indigenous markets [31]. It has also been found in some relevant literature that there was a U-shaped relationship between trade openness and food security in less developed countries (LDCs), and food security has decreased in the initial stages of trade openness expansion but increased beyond a given threshold [43].

The relationship between trade openness and food security is likely to provide the empirical evidence necessary to address and respond to one of the key issues during the WTO's international negotiations, and regional or bilateral trade negotiations [12,38], although the prospects for positive and beneficial solutions are still uncertain [44]. According to the relevant literature, various factors can affect trade openness's impact on food security, especially considering that food security is a multipillarsal issue, and make it difficult to determine whether trade openness's impact on food security is positive or negative for specific countries only by qualitative and descriptive analysis performed by the majority of previous studies [12,36,45]. As typical transition economies, Central Asia represents an important strategic junction and its economic development has important connotations for Central and Eastern Europe, China, and the Middle East and West Asia, and improving food security is vital for sustainable economic development in Central Asia [17]. However, quantitatively empirical studies that assess trade openness's potential impact on food security in Central Asian countries are still rare. An additional problem is that most previous studies used poverty indicators rather than direct food security indicators for representing food security [14,36].

Therefore, this study aims to quantitatively explore trade openness's impact on food security in Central Asian countries. Using four different direct food security indicators that correspond to the four pillars of food security, including availability, access, stability, and utilization, respectively, a panel data fixed effect model has been established to empirically analyze trade openness's impact on food security in Central Asian countries. We explore the following questions: (1) Is trade openness a threat or an opportunity for food security in Central Asian countries? Is there a U-shaped (or inverted U-shaped) relationship between trade openness and the four pillars of food security? (2) Do other economic and non-economic factors chosen as control variables have positive or negative effects on food security in Central Asian countries?

3. Materials and Methods

3.1. Methods

3.1.1. Baseline Model

As indicated previously, this study aims to empirically analyze trade openness's impact on food security in Central Asian countries by adopting a panel data fixed effect model as the baseline model. Considering the trade openness–food security relationship and the effect of other important potential determinants, and based on the model developed by Kang [43], Dithmer and Abdulai [36] and Fusco et al. [14], we built the panel data fixed effect model as:

$$FS_{i,t} = \alpha_0 + \alpha_1 TO_{i,t} + \alpha_2 TO_{i,t}^2 + \alpha_3 CV_{i,t} + T_{i,t} + C_{i,t} + \varepsilon_{i,t} \tag{1}$$

where i and t denote Central Asian countries and time periods, respectively.

As the dependent variable, FS represents food security and is denoted by following four indicators that respectively correspond to the four pillars of food security, including availability, access, stability, and utilization, considering the data availability across Central Asian years and countries of all food security indicators in the Food and Agriculture Organization Corporate Statistical Database (FAOSTAT; http://www.fao.org/faostat/en/#data/FS (accessed on 16 March 2021)): (1) dietary energy supply, used as a proxy for availability, which is computed by comparing a probability distribution of habitual daily dietary energy consumption with the minimum dietary energy; (2) rail lines density, used as a proxy for access, corresponds to the ratio between the length of railway route available for train service, irrespective of the number of parallel tracks within the country; (3) food supply variability, used as a proxy for stability, represents the per capita variability of the "food supply in kcal/caput/ day" as disseminated in the FAOSTAT; (4) population using safely managed drinking water services, used as a proxy for utilization, corresponds to the ratio between population using safely managed drinking water services and total population. These four pillars are necessary for food security to be understood as a continuum that progresses from uncertainty and anxiety regarding access to sufficient and appropriate food at the household level to the extreme condition of hunger among children because they do not have enough to eat [46].

TO_{it} is the trade openness measure, which is computed as the share of trade value (the sum of exports and imports of goods and services) over the gross domestic product (GDP) from the World Bank's World Development Indicators (WDI; https://data.worldbank.org/indicator?tab=all (accessed on 16 May 2021)). This trade openness measure is the measure usually employed in impact studies of trade openness and is arguably better than de jure measures (e.g., tariffs) to the extent that the latter are difficult to summarize in a single indicator [36] and has great advantages in that it can reflect the entirety of a country's commercial transactions, including exports and imports [47]. $TO_{i,t}^2$ is the quadratic form of TO_{it}, which is also treated as an independent variable for testing whether there is a U-shaped (or inverted U-shaped) relationship. If α_2 is statistically significant, then it is demonstrated as a U-shaped (or inverted U-shaped) relationship. CV is a set of control variables that might be major determinants in Central Asian countries' food security and are explained in more detail below. $T_{i,t}$ and $C_{i,t}$ are the period fixed effects (such as geographic characteristics or unobserved cultural and institutional factors) and cross-section fixed effects (such as changes in world prices and shocks that are identical to all countries), respectively; α_0 is the constant term; α_1, α_2 and α_3 are the estimated values of independent variables' coefficients; and $\varepsilon_{i,t}$ is the random error term.

3.1.2. Robustness Analysis Models

For verifying the validity of panel data fixed effect model's estimation results, we used an LS procedure for pooled data and a DPD analysis with the GMM approach, simultaneously.

(1)   LS procedure for pooled data

The estimation equation form of the LS procedure for pooled data is:

$$FS_{i,t} = \beta_0 + \beta_1 TO_{i,t} + \beta_2 TO_{i,t}^2 + \beta_3 CV_{i,t} + \varepsilon_{i,t} \tag{2}$$

where β_0 is the constant term; and β_1, β_2 and β_3 are the estimated values of independent variables' coefficients.

(2)   DPD analysis with GMM approach

Considering the potential effect of past food security levels and testing for the dynamic effects and the endogeneity problem, based on Nickell [48], Holtz-Eakin et al. [49], and Arellano and Bond [50], the estimation equation form of DPD analysis with the GMM approach is:

$$FS_{i,t} = \phi_1 FS_{i,t-1} + \phi_2 TO_{i,t} + \phi_3 TO_{i,t}^2 + \phi_4 CV_{i,t} + T_{i,t} + C_{i,t} + \varepsilon_{i,t} \tag{3}$$

where $FS_{i,t-1}$ is the lagged food security and treated as predetermined to the current period for modeling current food security levels as a function of past food security levels (also capturing the effects of past status), and current determinants [36]. The lagged dependent variable would control for longer-term effects of all our included control variables and would also pick up effects of other potentially omitted variables as it will be highly correlated with the current values of the food security variables [14,36]. ϕ_1, ϕ_2, ϕ_3 and ϕ_4 are the estimated values of independent variables' coefficients.

3.2. Data

Based on the FAO's conceptual analysis framework for trade policy reforms and food security [11,38], experiences of previous literature, and data availability, we used panel data composed of five Central Asian countries over the period 2001–2018. Most of the data used in this study come from the FAO's FAOSTAT and World Bank's WDI. According to FAO [11,38], Dithmer and Abdulai [36], and Fusco et al. [14], we considered the following five categories of food security determinants: the first category is trade openness measure; the second category accounts for country context and characteristics; the third category describes the development of economy and population; the fourth category captures non-economic events; the fifth category denotes domestic macroeconomic policies except trade policy. The trade openness measure, as the first category, is the core independent variable in this study and has already been explained in detail earlier. Subsequently, we focus on the last four groups of food security determinants, which represent the control variables included in Equations (1)–(3).

Referring to the second category, we take into account the following three aspects to represent the country's context and characteristics: macroeconomic scale, importance of agriculture, and availability of resource endowment for agricultural production in the country. Gross domestic product (GDP) per capita in constant 2010 USD is used as the variable to measure the macroeconomic scale and data on which comes from the WDI. A positive impact of GDP per capita on food security is projected, as strong income growth is crucial to enhance the acquisition for buying food [36]. The importance of agriculture is measured by the employment in agriculture, which refers to the share of persons of working age who were engaged in the agricultural industry of total employment, and data on which comes from the WDI. The first important element of national resource endowment for agricultural production is arable land per person and data on which comes from the WDI. In the context of agricultural land use, the function to produce biomass is the most relevant to sustain food production and respond to cultural practices conducive to sustainable agricultural land management [51]. The second important element of national resource endowment for agricultural production is freshwater withdrawals in agriculture, which refers to annual freshwater withdrawals in agriculture divided by total freshwater withdrawal, and data on which comes from the WDI. The amount and timing of water available for irrigation have a direct effect on crop yields, which is vital for ensuring food production, and water security is the basis for food security [52,53].

Referring to the third category, three variables are used to reflect the agricultural and overall development of the economy and population: agricultural productivity, GDP growth, and population growth. Cereal yield is used as a proxy for agricultural productivity, and data on which comes from the WDI. Despite increasing internationalization, many developing countries' food security still mainly depends on domestic agricultural production [54], then agricultural productivity growth is one of the important approaches to enhance domestic food availability and decrease their prices [55,56]. GDP growth is used to capture cyclical fluctuations in total output, and data on which comes from the WDI. Increases in GDP are associated with increasing income per capita and changes in demands, often reflected in stronger demand for food and resulting in increased consumption [57]. Population growth is used to describe population pressures on food security, and data on which comes from the WDI. Population pressures as denoted by the rapid population growth rate always result in increasing food demands for the total population and reduce food availability per capita [36].

Referring to the fourth category, one variable is chosen to represent non-economic events: natural disasters. The natural disasters data are from the Emergency Events Database (EM-DAT; https://public.emdat.be/ (accessed on 28 May 2021)) of Belgium launched by the Centre for Research on the Epidemiology of Disasters (CRED), which is calculated by the ratio between the total number of injured, affected, and homeless populations as a direct result of natural disasters (including geophysical disasters, meteorological disasters, hydrological disasters, climatological disasters, biological disasters, and extra-terrestrial disasters) and total population. When natural disasters affect the agriculture sector, they can have far-reaching negative consequences beyond physical damage, they: lower production and productivity; decrease exports of agricultural commodities, and increase food imports, causing disequilibrium in the balance of trade and in the balance of payments in affected countries; and arrest agriculture sector growth and the sustainable development of the sector [58].

Referring to the fifth category, one variable is chosen to capture domestic macroeconomic policies except for trade policy: inflation rate (GDP deflator). The inflation rate (GDP deflator) calculated by the annual growth rate of GDP implicit deflator, denotes the price's changing rate at a national level, and data on which comes from the WDI. The inflation rate is an important indicator of macroeconomic stability [59,60]. Considering national food security, macroeconomic policies that affect certain vital macroeconomic factors can have both a direct and indirect impact on a country's rate of food production and food inflation [61], and high efficiency in development and economic growth will be conducive to reducing poverty and hunger and finally eliminating food insecurity in every country [62].

The variables and their data sources, and their descriptive statistics used in this study are presented in Tables 1 and 2, respectively.

Table 1. Variables and their data sources.

Variables		Unit	Symbol	Data Sources
Dependent variable	Dietary energy supply	kcal/capita/day	*DES*	FAOSTAT
	Rail lines density	per 10^2 square km of land area	*RLD*	FAOSTAT
	Food supply variability	kcal/capita/day	*PCF*	FAOSTAT
	Population using safely managed drinking water services	%	*PUS*	FAOSTAT
Core independent variable	Trade openness	%	*TO*	WDI
	GDP per capita	Constant 2010 USD	*GDPC*	WDI
	Employment in agriculture	%	*EA*	WDI
	Arable land	Hectares per person	*AL*	WDI
Control variables	Freshwater withdrawals in agriculture	%	*FWA*	WDI
	Agricultural productivity	kg per hectare	*AP*	WDI
	GDP growth	Annual %	*GDPG*	WDI
	Population growth	Annual %	*PG*	WDI
	Natural disasters	%	*ND*	EM-DAT
	Inflation rate	Annual %	*IR*	WDI

Note: *DES*—dietary energy supply; *RLD*—rail lines density; *PCF*—food supply variability; *PUS*—population using safely managed drinking water services; *TO*—trade openness; *TO²*—quadratic form of trade openness; *GDPC*—gross domestic product per capita; *EA*—employment in agriculture; *AL*—arable land; *FWA*—freshwater withdrawals in agriculture; *AP*—agricultural productivity; *GDPG*—domestic product growth; *PG*—population growth; *ND*—natural disasters; *IR*—inflation rate. FAOSTAT—Food and Agriculture Organization Corporate Statistical Database; WDI—World Development Indicators; GDP—gross domestic product; EM-DAT—Emergency Events Database.

Table 2. Variables' descriptive statistics.

Variables	Mean	Standard Deviation	Minimum	Maximum
Dietary energy supply	2684.71	372.77	1945.00	3240.00
Rail lines density	0.56	0.31	0.20	1.60
Food supply variability	47.11	34.53	9.00	188.00
Population using safely managed drinking water services	63.83	14.95	37.00	94.80
Trade openness	84.33	30.82	29.75	146.66
GDP per capita	3325.87	3286.36	447.96	11,165.57
Employment in agriculture	34.63	10.82	15.77	58.69
Arable land	0.53	0.64	0.08	1.98
Freshwater withdrawals in agriculture	86.30	12.36	58.38	96.52
Agricultural productivity	2529.38	1166.92	804.10	4851.70
GDP growth	6.72	3.25	−0.47	14.70
Population growth	1.55	0.54	−0.17	2.82
Natural disasters	0.97	5.68	0.00	38.84
Inflation rate	13.95	11.33	−5.15	59.74

Note: GDP—gross domestic product.

3.3. Descriptive Analysis on the Food Security Status in Central Asian Countries

Taking the "availability" pillar of food security as an example, which is denoted by the dietary energy supply in this study, we first conducted a descriptive analysis of the evolutionary characteristics of food security status in Central Asian countries and also conducted a comparison with two major transition countries (China and the Russian Federation) and the world average. Then, we further briefly discussed the ongoing COVID-19 pandemic's impact on food security status in Central Asian countries. Figure 1 shows the evolutionary trends of the dietary energy supply in five Central Asian countries, China, the Russian Federation, and the world average from 2001–2018.

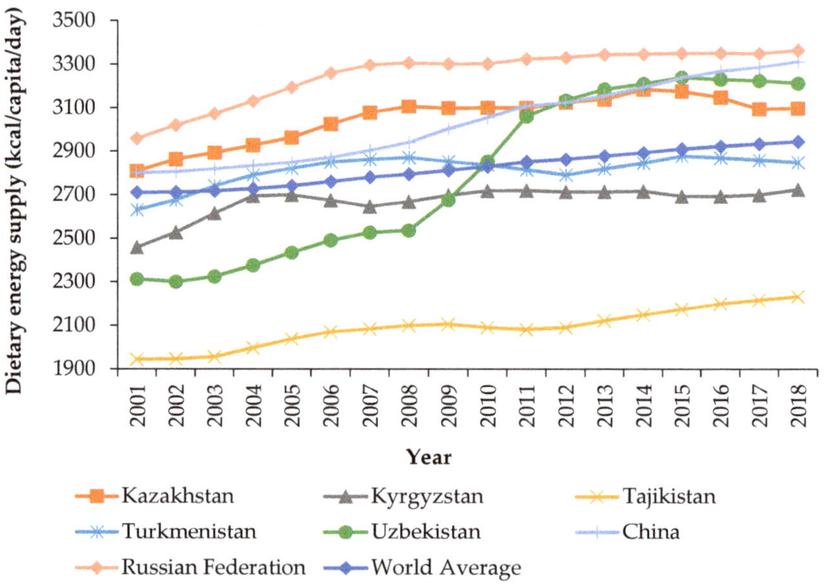

Figure 1. The "availability" pillar of food security in Central Asian countries, China, the Russian Federation, and the world average from 2001–2018. The x-axis represents the year. The y-axis refers to food security's "availability" pillar, which is denoted by the dietary energy supply.

After independence, Central Asian countries underwent a series of transitions from centrally planned economies to a market-orientated system [18]. During the initial years of transition, all the Central Asian countries experienced rising poverty, food insecurity, and malnutrition [18,63]. Therefore, Central Asian countries have attached great importance to improving their food security status by adopting food self-sufficiency policies, and for implementing such an approach; legislation including laws and strategies to support agricultural producers by various subsidies for input factors in production, such as fertilizers, seeds, plant protection pesticides, fuel and electricity, extension technicians, and agricultural machinery services have been developed [64]. Then, the dietary energy supply in most Central Asian countries totally showed positive trends in the 21st century. Land reform and farm restructuring also contributed to the improvement of food security in Central Asian countries. Kazakhstan and Kyrgyzstan were relatively faster in privatization reform of land. Taking Kazakhstan as an example, the share of non-state enterprises reached 95% of arable land and 91% of animal husbandry by 1999, and the private business had dominated the agro-industry [23,65]. In Uzbekistan and Turkmenistan, governments adopted slower paths for the transition to a market-orientated economy, and preserved state ownership and controlled most agricultural land, and land reform mainly focused on fragmentation of large collective and state farms but without transferring land ownership to farmers [23,66]. In Tajikistan, the civil war immediately after independence in 1992–1997 had strong damaging effects on its agricultural production, and privatization reform of land was initiated later than in other Central Asian countries [67].

Since 2007, the dietary energy supply in Kazakhstan has been the highest among Central Asian countries from 2001–2011, and Uzbekistan overtook Kazakhstan to become the highest from 2012–2018, while Tajikistan has always been the lowest. By 2015, as a region, Central Asia achieved MDGs 1C goal (halving the proportion of people suffering hunger from 1990–2015), but the progress on reducing hunger incidence differed from country to country. Kazakhstan achieved the MDG 1C goal as early as 2006; Kyrgyzstan, Turkmenistan, and Uzbekistan achieved the goal in 2014–2016; Tajikistan has consistently had the highest rate of undernourishment in the region (the percentage of the population with a caloric intake below the minimum dietary energy requirement). The prevalence of undernourishment in Tajikistan was estimated at 28.1 percent in 1990–1992, climbing to 33.2 percent in 2014–2016 [64], showing serious food insecurity issues. The year 2016 and beyond have marked the transition from the MDGs to the SDGs, which recognized the unfinished nature of efforts to eradicate poverty and hunger, but also expanded the coverage of programs to include both developing and developed countries and widen the range of the goals to include sustainability as a fundamental leitmotif. In line with Goal 2 of the SDGs, calling for the eradication of hunger, achieving food security, and the elimination of all forms of malnutrition, national authorities in Central Asia will build on progress made in the region in sharply improving food and nutrition security since 1990 by identifying their national priorities to achieve this goal [68]. However, since 2016, the dietary energy supply in Kazakhstan, Uzbekistan, and Turkmenistan has tended to decline—mainly due to insufficient quantities of vegetables, fruits, and fish. This trend has pointed to the urgent demand for more efforts towards raising the availability of several food items through nutrition-sensitive policies in both production and trade areas [69]. In 2018, the dietary energy supply in all Central Asian countries was still significantly lower than that in China and the Russian Federation, especially Tajikistan is only 67.5% and 66.4% of China and Russia Federation, respectively; when compared to the world average, only the dietary energy supply in Kazakhstan and Uzbekistan was higher than the former. Therefore, the food security status in Central Asian countries needs to be further improved, not to mention the nutrition issues in these countries, and larger efforts from within and outside the region need to be carried out comprehensively to eliminate food insecurity as soon as possible.

The COVID-19 pandemic—still ongoing all around the world as of 2021—is also present in Central Asia, which is further worsening the food insecurity status in Central

Asian countries, in particular, it is estimated that the number of food-insecure people in Central Asia is expected to grow significantly, which will become worse as the COVID-19 continues to spread. A survey conducted by the FAO Regional Office for Europe and Central Asia reveals that COVID-19 has led to disruptions in transportation, storage, output deliveries, input supplies, and operational financing throughout the food supply chains and has continued to pose operational challenges that affect crop farmers, livestock farmers, and traders/processors in most countries in the region [70]. As COVID-19 has disrupted markets, food prices have risen, the incomes of many rural and urban families have fallen, especially low-income classes, and the diversity of food has decreased [4]. Moreover, COVID-19 has negatively impacted Central Asian countries' food security, and the innutrition status is also likely to deteriorate, which will significantly heighten the difficulty of realizing the Zero Hunger goal on schedule by 2030 in Central Asia.

4. Results and Discussion

4.1. Estimation Results of Panel Data Fixed Effect Model

4.1.1. Data Test

Correlation Analysis

Considering the possible multicollinearity issues in the models we have built, which could negatively affect the reliability of estimation results, a correlation analysis was computed between independent variables. According to the correlation analysis results presented in Table 3, we found that the vast majority of absolute values of the Pearson correlation coefficient between the independent variables were less than 0.5, then based on the determination criteria for the degrees of correlation and the values of Pearson correlation coefficient [71], we can confirm that there was no strong multicollinearity in the equations, therefore, we can continue to conduct the empirical analysis based on these equations.

Table 3. Correlation analysis results between independent variables.

Variables	TO	TO^2	GDPC	EA	AL	FWA	AP	GDPG	PG	ND	IR
TO	1.000	0.985	−0.264	0.361	−0.082	0.104	−0.347	0.107	−0.095	0.216	0.063
TO^2	0.985	1.000	−0.306	0.369	−0.143	0.151	−0.284	0.105	−0.058	0.219	0.078
GDPC	−0.264	−0.306	1.000	−0.614	0.848	−0.398	−0.677	−0.012	−0.236	−0.129	−0.205
EA	0.361	0.369	−0.614	1.000	−0.396	0.309	0.109	0.154	0.189	0.154	0.108
AL	−0.082	−0.143	0.848	−0.396	1.000	−0.152	−0.644	−0.003	−0.469	−0.094	−0.084
FWA	0.104	0.151	−0.398	0.309	−0.152	1.000	0.532	0.071	0.293	0.070	0.044
AP	−0.347	−0.284	−0.677	0.109	−0.644	0.532	1.000	−0.127	0.290	0.001	0.194
GDPG	0.107	0.105	−0.012	0.154	−0.003	0.071	−0.127	1.000	−0.056	−0.049	0.154
PG	−0.095	−0.058	−0.236	0.189	−0.469	0.293	0.290	−0.056	1.000	0.036	−0.106
ND	0.216	0.219	−0.129	0.154	−0.094	0.070	0.001	−0.049	0.036	1.000	0.034
IR	0.063	0.078	−0.205	0.108	−0.084	0.044	0.194	0.154	−0.106	0.034	1.000

Panel Cointegration Test

In the macroeconomic panel data, due to identical impacts or spillover effects, cross-sectional dependence would arise, and for which if not accounted for, biased estimated results may be obtained [72]. Therefore, since the early 2000s, aiming to analyze the existence or otherwise of long-run stationary relationships among integrated economic variables, several kinds of panel cointegration test methods have been developed [73], such as the Pedroni (Engle-Granger Based) cointegration test [74,75], the Kao (Engle–Granger Based) cointegration test [76], and combined individual tests [77]. In order to examine the relationship among the dependent variable (the selected four proxy variables corresponding to food security's four pillars), core independent variable, and control variables in this study, the Kao cointegration test was employed here. Under the null hypothesis of no cointegration, the Kao cointegration test specifies cross-section-specific intercepts and homogeneous coefficients on the first-stage regressors [76]. When we conducted the Kao cointegration test, the individual intercept was selected for deterministic trend specification, lag length was automatically selected by using Schwarz information criterion (SIC), Bartlett kernel was selected for spectral estimation, and bandwidth was automatically selected by using Newey–West. According to the Kao cointegration test results presented in Table 4, we found all the values of the augmented Dickey–Fuller (ADF) t-Statistic were statistically significant at a 1% significance level, and all the values of probability corresponding to ADF t-Statistic were less than 1%. Therefore, the null hypothesis of no cointegration was strongly rejected, and long-run stationary relationships existed among the dependent variable, core independent variable, and control variables in this study.

Table 4. Results of Kao cointegration test.

Variables	ADF		Residual Variance	HAC Variance
	t-Statistic	*p*-Value		
DES, TO, TO^2, GDPC, EA, AL, FWA, AP, GDPG, PG, ND, IR	−3.791 ***	0.001	14.932	65.142
RLD, TO, TO^2, GDPC, EA, AL, FWA, AP, GDPG, PG, ND, IR	−3.034 ***	0.001	0.006	0.007
PCF, TO, TO^2, GDPC, EA, AL, FWA, AP, GDPG, PG, ND, IR	−3.801 ***	0.0001	52.450	45.883
PUS, TO, TO^2, GDPC, EA, AL, FWA, AP, GDPG, PG, ND, IR	−3.218 ***	0.001	0.556	1.001

Note: ADF—augmented Dickey-Fuller; *p*-value—value of probability; HAC—heteroscedasticity-autocorrelation consistent. ***—1% significance level.

4.1.2. Estimation Results

The specification for the panel data fixed effect model has period and cross-section fixed effects, based on which three specifications could be estimated: one only with period fixed effects, one only with cross-section fixed effects, and one only with a common intercept. The redundant fixed effects test should be used to determine the optimal specification for the panel data fixed effect model, and specifically using the likelihood function (chi-square test). Cross-section chi-square is used to evaluate cross-section fixed effects' significance, period chi-square is used to evaluate the period fixed effects' significance, and cross-section/period chi-square is used to evaluate both of the effects' joint significance, respectively. According to the redundant fixed effects test results presented in Table 5, all the three statistic values corresponding to the above three tests in columns (1)–(4) were statistically significant. Therefore, both null hypotheses suggest that cross-section fixed effects are redundant and no period fixed effects were strongly rejected, and the optimal specification of panel data fixed effect model in this study should include period and cross-section fixed effects, simultaneously. Additionally, considering that period and cross-section specific effects' presence may also be handled by using random effect methods, this study further used the Hausman test to compare the results from the estimated random effect specification and a corresponding fixed effect specification, with the null hypothesis that the random effect is correlated with the right-hand side variables in the panel equation setting. Random effect estimation requires the number of coefficients for between estimator for an estimate of random effect innovation variance is less than the number of cross-sections [78], which is not applicable in this study, therefore, when we used random effect methods, cross-section random effects are excluded, and only period random effects are included. According to the Hausman test results presented in Table 5, all the statistic values were statistically significant, which means that the null hypothesis was strongly rejected, and fixed effect specification is more appropriate than random effect specification in this study.

Table 5 presents estimation results of the panel data fixed effect model which takes the selected four proxy variables corresponding to food security's four pillars as dependent variables. The results in columns (1)–(4) show that *TO* has negative coefficients and *TO^2* has positive coefficients, and both are statistically significant. Therefore, the estimation results of the panel data fixed effect model show a U-shaped relationship between trade openness and food security's four pillars in Central Asian countries, which is similar to the results of

Kang [43], but different from the findings of Fusco et al. [14] and Dithmer and Abdulai [36]. The early stages of trade openness negatively impact food security, which implies that increased trade openness contributes to the redistribution of world production based on a comparative advantage due to trade and globalization [43,79]. Considering that trade openness means a change in relative prices of traded and non-traded commodities [38], the more a country becomes reliant on traded food, the more it will be inducting the "global food price" for related goods, and global inflation will more negatively influence its low-income groups, which spend most of their family income on food, resulting in increased food security risk [80], especially in transition economies such as Central Asian countries, where a perfect market-oriented economic system has not yet been established since achieving independence from the Soviet Union [24,81]. Additionally, benefits from a nation's larger openness to food security might be neutralized by greater unsteadiness or ecological costs [82,83], not to mention the negative effect of the vulnerability of the global food trade network [84,85]. Beyond a certain threshold of trade openness, food security status tends to improve. This indicates that relative price changes will finally result in changes in resources allocation on various sectors followed by changes in production' subsectoral levels, and in turn, changes in income levels (which are expected to increase in aggregate as resources are used more efficiently) have the potential both to decrease poverty levels and improve food security by improving poor's food acquisition [38]. Imported food supply—generally with relatively higher quality and cheaper prices—has intertemporal substitution effects or income effects, which means an increase in real income, and induces stronger domestic demand [86,87]. Therefore, participation in the world markets through international trade can finally improve food security in Central Asian countries.

Table 5. Estimation results of panel data fixed effect model.

Variables	(1): *DES*	(2): *RLD*	(3): *PCF*	(4): *PUS*
Constant	15.501	0.495	38.880	17.259 ***
	(1.160)	(0.481)	(0.952)	(6.738)
TO	−14.769 ***	−0.008 ***	−1.818 **	−0.167 ***
	(−4.599)	(−2.707)	(−2.116)	(−2.963)
TO^2	0.080 ***	0.00001 **	0.007 *	0.001 ***
	(5.038)	(2.203)	(1.965)	(3.191)
GDPC	0.013 **	0.0001 ***	0.005 *	0.002 *
	(2.426)	(4.829)	(1.989)	(1.904)
EA	−1.841 ***	−0.017 ***	−1.545 ***	−0.658 ***
	(−3.357)	(−3.161)	(−5.297)	(−4.239)
AL	−8.064 *	−3.969	−4.498 **	−4.337 *
	(−1.902)	(−0.756)	(−2.027)	(−1.834)
FWA	−1.873 ***	−0.002	−5.160	−1.024 ***
	(−3.227)	(−0.191)	(−1.310)	(−3.674)
AP	0.061 *	0.0001	0.008 **	0.002 ***
	(1.874)	(0.806)	(2.644)	(2.848)
GDPG	3.915	0.006 *	0.336	0.085 *
	(0.755)	(1.848)	(0.269)	(1.917)
PG	−8.595 *	−0.031	−9.509	−0.612
	(−1.912)	(−0.773)	(−0.525)	(−0.488)
ND	−0.487 *	−0.0002	−0.570 *	−0.018
	(−1.906)	(−0.129)	(−1.910)	(−0.362)
IR	−1.208 **	−0.003 **	−0.444 *	−0.012
	(−2.034)	(−2.385)	(−1.882)	(−0.327)
Cross-section fixed effects	Yes	Yes	Yes	Yes
Period fixed effects	Yes	Yes	Yes	Yes
R-squared	0.944	0.936	0.864	0.986
F-statistic	29.762 ***	26.093 ***	21.303 ***	125.039 ***
Redundant Fixed Effect Tests				
Cross-section chi-square	62.506 ***	137.896 ***	20.376 ***	74.972 ***
Period chi-square	47.892 ***	11.825 ***	23.503 ***	33.327 ***
Cross-section/period chi-square	76.625 ***	151.049 ***	42.727 ***	80.968 ***
Hausman Test for Random Effect				
Period chi-square	69.174 ***	74.263 ***	90.854 ***	86.721 ***

Note: Numbers in parentheses are t-statistics values. ***—1% significance level; **—5% significance level; *—10% significance level.

The empirical findings in Table 5 further confirm that other economic factors and non-economic factors chosen as control variables in this study can also be important determinants of food security. Both GDP per capita and GDP growth have positive coefficients—most of which are statistically significant—indicating that sustained economic development is a major factor in improving food security in Central Asian countries. Employment in agriculture has negative coefficients and all of which are statistically significant, meaning that there is likely to be a labor surplus in Central Asian countries' agriculture. Both arable land and freshwater withdrawals in agriculture have negative coefficients, most of which are statistically significant, indicating that the utilization efficiency of arable land and water is low in Central Asian countries' agriculture, which is not conducive to improving food security in these countries. It has been calculated that the average consumption coefficients of arable land denoted by the ratio of arable land area and total cereal production in Central Asia are much higher than the world average value of up to 7.74 m^2/kg, which is 3.6 times that of China, and the average agricultural water consumption denoted by the ratio of water used in agriculture and total cereal production in Central Asia is about 9.43 m^3/kg, or nearly 9.3 times the average value elsewhere in Asia [88]. Agricultural productivity has positive coefficients, most of which are statistically significant, indicating that an increase in agricultural productivity is conducive to food security's improvement. Increased agricultural productivity often means an increase in food production and supply capacity, and thus especially improving availability, access, and stability from a domestic food supply perspective. Population growth has negative coefficients, which shows that population growth is not conducive to the improvement of food security. A larger population always means greater food demand, which in turn increases the pressure to ensure food security. Natural disasters have negative coefficients, which suggest that natural disasters are not conducive to the improvement of food security. Natural disasters usually cause considerable damage to the production and transport of food [89,90], and then availability, access, stability, and utilization of food will all be adversely affected. The inflation rate has negative coefficients, most of which are statistically significant, showing that the inflation rate is not conducive to the improvement of food security. Most countries in Central Asia have experienced persistently unstable and high inflation since independence [91], which has worsened their food security status, considering that inflation often means a decline in real income levels and purchasing power, especially for low-income classes.

4.2. Robustness Test

For testing the robustness of estimation results of the panel data fixed effect model, an LS procedure for pooled data and a DPD analysis with the GMM approach are used in this study, simultaneously.

4.2.1. Estimation Results of LS Procedure for Pooled Data

The number of observations analyzed in this study is not very large, and there may be complicated panel error structure problems in the observations' data, such as synchronous correlation, heteroscedasticity, and series correlation [92]. For dealing with these problems effectively, Beck and Katz [93] developed the weighted panel corrected standard errors (PCSE) method. According to weight covariance processing methods' differences, PCSE can be further divided into cross-section weights, cross-section seemingly unrelated regression (SUR), period weights, and period SUR [92]. After comparison, based on cross-section SUR, we found both of independent variables' t-statistics and models' R-squared statistics were generally larger than those based on the other three approaches, indicating a total fitting effect of cross-section SUR is optimal. Consequently, in this study, we used the weighted PCSE cross-section SUR approach to analyze the LS procedure for pooled data. According to DW-statistics' values, AR terms with appropriate lagged periods are also supplemented for removing probable self-correlation problems. It is found that the effect direction and statistical significance of most independent variables' estimated coefficients in Table 6 were totally the same as the panel data fixed effect model's estimation results in Table 5, which implies that the LS procedure for pooled data supports the consistency and validity of the baseline model estimators.

Table 6. Estimation results of LS procedure for pooled data.

Variables	(5): *DES*	(6): *RLD*	(7): *PCF*	(8): *PUS*
TO	−17.203 *** (−3.417)	−0.008 *** (−3.930)	−1.332 ** (−2.147)	−0.146 *** (−3.575)
TO^2	0.042 *** (3.743)	0.00003 ** (2.773)	0.002 * (1.866)	0.002 *** (4.625)
GDPC	0.026 * (1.816)	0.0001 *** (3.188)	0.003 * (1.894)	0.001 *** (5.215)
EA	−1.215 *** (−5.465)	−0.003 ** (−2.343)	−1.078 ** (−2.198)	−0.631 *** (−4.552)
AL	−9.963 ** (−2.145)	−3.146 (−0.703)	−2.851 (−1.106)	−6.992 ** (−2.160)
FWA	−2.853 *** (−3.961)	0.009 (0.810)	−1.654 ** (−2.162)	−0.830 * (−1.856)
AP	0.104 *** (5.341)	0.0001 (0.775)	0.013 *** (2.901)	0.003 *** (2.672)
GDPG	1.522 (0.524)	0.008 ** (2.308)	0.413 (0.797)	0.051 (0.779)
PG	−7.472 *** (−3.489)	−0.097 (−0.857)	−3.200 *** (−4.025)	0.149 (0.801)
ND	−0.094 * (−1.881)	−0.0002 (−0.195)	−0.214 (−0.855)	−0.020 (−0.635)
IR	−0.020 (−0.820)	−0.005 ** (−2.103)	−0.189 (−0.987)	−0.037 ** (−2.118)
AR(1)	1.690 (1.288)	0.783 *** (2.945)	1.035 (0.934)	1.014 (1.368)
AR(2)		0.223 (1.408)		
R-squared (weighted)	0.963	0.981	0.845	0.973
DW-statistic (weighted)	1.920	1.897	1.939	1.964

Note: Numbers in parentheses are values of cross-section SUR correction t-statistics; AR—autoregressive; DW—Durbin–Watson; ***—1% significance level; **—5% significance level; *—10% significance level.

4.2.2. Estimation Results of DPD Analysis with GMM Approach

Table 7 presents the estimation results of DPD analysis with the GMM approach by using the two-step system-GMM. The results of the autoregressive (AR) errors test indicate that all p-values of the AR(2) errors test were significantly larger than 0.1, therefore, the null hypothesis of no autocorrelation of order 2 was not rejected; the results of the Hansen test indicate that all p-values were significantly larger than 0.1, therefore, the null hypothesis that the tool variables are exogenous as a group was not rejected; the results of the Sargan test indicate that all p-values were significantly larger than 0.1, therefore, the null hypothesis of over-identifying restrictions was not rejected. Therefore, the specification for DPD analysis with the GMM approach in this study is reasonable and valid and can be estimated by using the two-step system-GMM. It is shown that effect direction and statistical significance of most independent variables' estimated coefficients in Table 7 were totally the same as the panel data fixed effect model's estimation results in Table 5, which implies that the DPD analysis with the GMM approach supports the robustness of the baseline model estimators. At the same time, all the lagged dependent variables were positive with statistically significant, showing that food security's four pillars change tardily over time and are determined by their previous conditions, which justifies the specification of the dynamic panel model and the adoption of the two-step system-GMM for estimating DPD analysis with the GMM approach.

Table 7. Estimation results of DPD analysis with GMM approach.

Variables	(9): *DES*	(10): *RLD*	(11): *PCF*	(12): *PUS*
DES(-1)	0.815 *** (3.623)			
RLD(-1)		0.794 *** (4.189)		
PCF(-1)			0.802 *** (4.351)	
PUS(-1)				0.739 *** (5.026)
TO	−1.518 *** (−4.129)	−0.002 *** (−3.871)	−1.109 ** (−2.117)	−0.109 *** (−3.214)
TO^2	0.029 *** (2.918)	0.00001 ** (2.252)	0.010 ** (2.183)	0.001 *** (3.826)
GDPC	0.076 * (1.923)	0.0002 ** (2.497)	0.001 * (1.918)	0.003 *** (4.724)
EA	−1.674 *** (−4.918)	−0.003 ** (−2.286)	−0.895 * (−1.903)	−0.386 *** (−3.951)
AL	−8.752 * (−1.909)	−2.742 (−1.158)	−5.652 (−0.970)	−3.283 ** (−2.160)
FWA	−2.257 *** (−9.184)	−0.013 (−0.635)	−2.034 (−1.562)	−1.232 * (−1.921)
AP	0.095 *** (3.706)	0.001 (0.548)	0.009 *** (2.781)	0.021 *** (3.056)
GDPG	1.652 (0.524)	0.013 ** (2.124)	0.643 (0.571)	0.104 (0.634)
PG	−4.673 ** (2.108)	−0.205 (−0.817)	−2.428 ** (−2.061)	−0.085 (−1.236)
ND	−0.094 (−1.076)	−0.0001 (−0.443)	−0.373 (−0.692)	−0.012 (−0.784)
IR	−0.011 (−0.593)	−0.003 ** (−1.982)	0.304 (0.758)	−0.026 ** (−2.205)
p-value of AR(1) Errors Test	0.056	0.017	0.024	0.019
p-value of AR(2) Errors Test	0.439	0.452	0.519	0.523
p-value of Hansen Test	0.723	0.845	0.906	0.894
p-value of Sargan Test	0.384	0.372	0.268	0.291

Note: Numbers in parentheses are t-statistics values; *p*-values—values of probability; AR—autoregressive; ***—1% significance level; **—5% significance level; *—10% significance level.

5. Conclusions and Policy Implications

Using the 2001–2018 panel data of Central Asian countries, this study is the first to quantitatively estimate trade openness and other factors' impact on food security's four pillars (including availability, access, stability, and utilization) and traces a U-shaped (or inverted U-shaped) relationship between trade openness and food security by adopting a panel data fixed effect model as the baseline model, and then conducts a robustness test by using an LS procedure for the pooled data and a DPD analysis with the GMM approach, simultaneously.

Estimates from the panel data fixed effect model indicated a U-shaped relationship between trade openness and food security's four pillars in Central Asian countries, which means that the initial stages of trade openness negatively impact food security; furthermore, beyond a certain threshold of trade openness, food security status tends to improve, therefore, the participation in the world markets through international trade can finally improve food security in Central Asian countries. The empirical findings also confirm that other economic factors and non-economic factors chosen as control variables in this study can be important determinants of food security. GDP per capita, GDP growth, and agricultural productivity have contributed to the improvement of food security; employment in agriculture, arable land, freshwater withdrawals in agriculture, population growth, natural disasters, and inflation rate have negative impacts on food security. Estimates from the LS procedure for the pooled data and the DPD analysis with the GMM approach indicated that estimates from the panel data fixed effect model were robust and valid.

According to the above research conclusions, we can develop the following policy enlightenments to improve food security status for Central Asian countries: first, the countries need to build more robust trade policies and food safety technical rules that adhere to relevant rules of the WTO, such as decreasing burdensome and intricate trade regulations, increasing awareness about food safety standards, and improving the infrastructure of food safety and sanitary and phytosanitary, in order to maximize the positive effects of trade openness on food security. Second, the countries need to further boost government governance reform, to ensure sustained and steady economic growth and avoid large fluctuations in economic growth and exchange rate. Great importance should also be attached to social protection and safety nets policies, such as cash transfer programs, and transfers and subsidization of food [70], which are very important for low-income classes in these countries when facing various shocks and crises, including the ongoing COVID-19 pandemic. Third, the countries need to improve domestic agricultural support policies, especially producer support for agricultural inputs [27], research and extension of agricultural technology, and construction of agriculture-related infrastructure, to ensure reasonable food self-sufficiency by improving the utilization efficiency of agricultural resources, increasing agricultural productivity, and enhancing the capacity of agricultural producers especially small-scale farmers to better respond to natural disasters and climate change. Forth, the countries need to be more proactive in integrating into main global and regional cooperative platforms to obtain more external support for improving their food security status, such as China's Belt and Road Initiative (BRI) which covers resources of urgent need in these countries, including agricultural investment, agricultural technology transfer, and massive investment in infrastructure [94,95], and promote the interface between these countries' national development strategies and policies and main cooperation platforms based on mutual benefit and win-win cooperation.

This study has important scientific value and practical significance. Our research represents an attempt to provide further empirical evidence on the vital unresolved issue at the country level of typical transition countries, which will contribute to the research progress on the relationship between reform of trade policy and improvement of food security, and by extending research to food security's four pillars, our research is a significant improvement over past studies mainly using poverty indicators for food security. Our research findings provide scientific support for sustainable food system strategies in Central Asian countries and also provide valuable implications for other transition countries and developing countries facing the common difficulty of realizing the Zero Hunger goal by 2030.

Further research should address some limitations of this study. First, limited by the data availability, our analysis selected one indicator for each of food security's four pillars in Central Asian countries. However, there are a series of indicators for denoting and interpreting each of food security's four pillars in the FAOSTAT. Hence, future studies should build an indicator system to evaluate each of the four pillars of food security, and based on which, more comprehensive evaluation results can be obtained. Second, based on comparison and by referring to the practice of previous studies [14,36,43,47], the share of trade value (total trade volume in goods and services) over GDP has been employed to measure trade openness in this study. Tariff [96], and free trade agreements (FTAs) [97] are also used as proxies for trade openness in some literature and deserve further analysis of these alternative variables' impact on food security.

Author Contributions: Z.S. contributed towards the conceptualization, methodology, formal analysis, data collection and curation, writing—original draft preparation, writing—review and editing, project administration, and funding acquisition; D.Z. contributed towards the data collection and curation, and writing—original draft preparation. All authors have read and agreed to the published version of the manuscript.

Funding: This research was funded by the National Natural Science Foundation of China (71961147001; 71703157), and the Agricultural Science and Technology Innovation Project of Chinese Academy of Agricultural Sciences (ASTIP-IAED-2021-06; ASTIP-IAED-2021-ZD-02).

Data Availability Statement: The data presented in this study are available on request to authors.

Conflicts of Interest: The authors declare no conflict of interest. The funders had no role in the design of the study; in the collection, analyses, or interpretation of data; in the writing of the manuscript, or in the decision to publish the results.

References

1. International Food Policy Research Institute. *2021 Global Food Policy Report: Transforming Food Systems after COVID-19*; International Food Policy Research Institute: Washington, DC, USA, 2021.
2. Food and Agriculture Organization of the United Nations; International Fund for Agricultural Development; United Nations Children's Fund; World Food Programme; World Health Organization. *The State of Food Security and Nutrition in the World 2021: Transforming Food Systems for Food Security, Improved Nutrition and Affordable Healthy Diets for All*; Food and Agriculture Organization of the United Nations: Rome, Italy, 2021.
3. Swinnen, J.; McDermott, J. Covid-19 and global food security. *EuroChoices* **2020**, *19*, 26–33. [CrossRef]
4. Galanakis, C.M. The food systems in the era of the coronavirus (COVID-19) pandemic crisis. *Foods* **2020**, *9*, 523. [CrossRef] [PubMed]
5. Fan, S.G.; Teng, P.; Chew, P.; Smith, G.; Copeland, L. Food system resilience and COVID-19—Lessons from the Asian experience. *Glob. Food Secur.* **2021**, *28*, 100501. [CrossRef]
6. Laborde, D.; Martin, W.; Vos, R. Impacts of COVID-19 on global poverty, food security, and diets: Insights from global model scenario analysis. *Agric. Econ.* **2021**, *52*, 375–390. [CrossRef]
7. WFP at a Glance: A Regular Lowdown on the Facts, Figures and Frontline Work of the World Food Programme. Available online: https://www.wfp.org/stories/wfp-glance (accessed on 10 June 2021).
8. United Nations. *United Nations Millennium Declaration*; United Nations: New York, NY, USA, 2000.
9. United Nations. *Transforming Our World: The 2030 Agenda for Sustainable Development*; United Nations: New York, NY, USA, 2015.
10. Food and Agriculture Organization of the United Nations. *Rome Declaration on World Food Security*; Food and Agriculture Organization of the United Nations: Rome, Italy, 1996.
11. Food and Agriculture Organization of the United Nations. *Trade Reforms and Food Security: Country Case Studies and Synthesis*; Food and Agriculture Organization of the United Nations: Rome, Italy, 2006.
12. Food and Agriculture Organization of the United Nations. *Food Security and International Trade: Unpacking Disputed Narratives*; Food and Agriculture Organization of the United Nations: Rome, Italy, 2015.
13. Gibson, M. Food security—A commentary: What is it and why is it so complicated. *Foods* **2012**, *1*, 18–27. [CrossRef]
14. Fusco, G.; Coluccia, B.; Leo, F.D. Effect of trade openness on food security in the EU: A dynamic panel analysis. *Int. J. Environ. Res. Public Health* **2020**, *17*, 4311. [CrossRef]
15. Food and Agriculture Organization of the United Nations. *The State of Food Insecurity in the World 2001*; Food and Agriculture Organization of the United Nations: Rome, Italy, 2001.
16. Food and Agriculture Organization of the United Nations; International Fund for Agricultural Development; United Nations Children's Fund; World Food Programme; World Health Organization. *The State of Food Security and Nutrition in the World 2020: Transforming Food Systems for Affordable Healthy Diets*; Food and Agriculture Organization of the United Nations: Rome, Italy, 2020.
17. Babu, S.C.; Tashmatov, A. Attaining food security in Central Asia—Emerging issues and challenges for policy research. *Food Policy* **1999**, *24*, 357–362. [CrossRef]
18. Babu, S.C.; Pinstrup-Andersen, P. Achieving food security in Central Asia—Current challenges and policy research needs. *Food Policy* **2000**, *25*, 629–635. [CrossRef]
19. Pandya-Lorch, R.; Rosegrant, M.W. Prospects for food demand and supply in Central Asia. *Food Policy* **2000**, *25*, 637–646. [CrossRef]
20. Babu, S.C.; Reidhead, W. Poverty, food security, and nutrition in Central Asia: A case study of the Kyrgyz Republic. *Food Policy* **2000**, *25*, 647–660. [CrossRef]
21. Aigarinova, G.T.; Akshatayeva, Z.; Alimzhanova, M.G. Ensuring food security of the Republic of Kazakhstan as a fundamental of modern agricultural policy. *Procedia-Soc. Behav. Sci.* **2014**, *143*, 884–891. [CrossRef]
22. World Bank. *Collaboration Is Key to Food Security in Central Asia*; World Bank: Washington, DC, USA, 2016.
23. Lioubimtseva, E. Food security factors and trends in Central Asia. *Encycl. Food Secur. Sustain.* **2019**, *3*, 134–141.
24. Babu, S.C.; Akramov, K.T. Emerging food and nutrition security challenges in Central Asia: Implications for BRICS and emerging economies. In *Handbook of BRICS and Emerging Economies*, 1st ed.; Anand, P.B., Fennell, S., Comim, F., Eds.; Oxford University Press: Oxford, UK, 2021; ISBN 9780198827535.
25. Swinnen, J.; Burkitbayeva, S.; Schierhorn, F.; Prishchepov, A.V.; Müller, D. Production potential in the "bread baskets" of Eastern Europe and Central Asia. *Glob. Food Secur.* **2017**, *14*, 38–53. [CrossRef]
26. Samad, G.; Abbas, Q. Infrastructure in Central Asia and the Caucasus. In *Developing Infrastructure in Central Asia: Impacts and Financing Mechanisms*, 1st ed.; Yoshino, N., Huang, B., Azhgaliyeva, D., Abbas, Q., Eds.; Asian Development Bank Institute: Tokyo, Japan, 2021; ISBN 9784899742319.
27. Burkitbayeva, S.; Liefert, W.; Swinnen, J. Agricultural development and food security in Eastern Europe and Central Asia. In *Agricultural Development: New Perspectives in a Changing World*, 1st ed.; Otsuka, K., Fan, S., Eds.; International Food Policy Research Institute: Washington, DC, USA, 2021; ISBN 9780896293830.
28. World Food Programme. *WFP Tajikistan Country Brief in June 2021*; World Food Programme: Rome, Italy, 2021.
29. World Food Programme. *WFP Kyrgyz Republic Country Brief in June 2021*; World Food Programme: Rome, Italy, 2021.
30. Chikhuri, K. Impact of alternative agricultural trade liberalization strategies on food security in the Sub-Saharan Africa region. *Int. J. Soc. Econ.* **2013**, *40*, 188–206. [CrossRef]

31. Luo, P.; Tanaka, T. Food import dependency and national food security: A price transmission analysis for the wheat sector. *Foods* **2021**, *10*, 1715. [CrossRef]
32. Zhang, C.; Yang, Y.Z.; Feng, Z.M.; Xiao, C.W.; Lang, T.T.; Du, W.P.; Liu, Y. Risk of global external cereals supply under the background of the COVID-19 pandemic: Based on the perspective of trade network. *Foods* **2021**, *10*, 1168. [CrossRef]
33. DG Okonjo-Iweala Highlights Vital Role of Trade for Global Food Security. Available online: https://www.wto.org/english/news_e/news21_e/dgno_06jul21_e.htm (accessed on 10 July 2021).
34. Tanaka, T.; Hosoe, N. Does agricultural trade liberalization increase risks of supply-side uncertainty? Effects of productivity shocks and export restrictions on welfare and food supply in Japan. *Food Policy* **2011**, *36*, 368–377. [CrossRef]
35. Hussain, M.A.; Dawson, C.O. Economic impact of food safety outbreaks on food businesses. *Foods* **2013**, *2*, 585–589. [CrossRef]
36. Dithmer, J.; Abdulai, A. Does trade openness contribute to food security? A dynamic panel analysis. *Food Policy* **2017**, *69*, 218–230. [CrossRef]
37. Schillhorn van Veen, T.W. International trade and food safety in developing countries. *Food Control* **2005**, *16*, 491–496. [CrossRef]
38. Food and Agriculture Organization of the United Nations. *Trade Reforms and Food Security: Conceptualizing the Linkages*; Food and Agriculture Organization of the United Nations: Rome, Italy, 2003.
39. Sanogo, I.; Amadou, M.M. Rice market integration and food security in Nepal: The role of cross-border trade with India. *Food Policy* **2010**, *35*, 312–322. [CrossRef]
40. Paarlberg, R. The weak link between world food markets and world food security. *Food Policy* **2000**, *25*, 317–335. [CrossRef]
41. Moon, W. Is agriculture compatible with free trade? *Ecol. Econ.* **2011**, *71*, 13–24. [CrossRef]
42. Dang, Q.; Konar, M. Trade openness and domestic water use. *Water Resour. Res.* **2018**, *54*, 4–18. [CrossRef]
43. Kang, H. A study on the relationship between international trade and food security: Evidence from less developed countries (LDCs). *Agric. Econ.* **2015**, *61*, 475–483. [CrossRef]
44. Zhu, Y. International trade and food security: Conceptual discussion, WTO and the case of China. *China Agric. Econ. Rev.* **2016**, *8*, 399–411. [CrossRef]
45. Diaz-Bonilla, E.; Ron, J.F. *Food Security, Price Volatility and Trade: Some Reflections for Developing Countries*; International Centre for Trade and Sustainable Development (ICTSD) Issue Paper 27; ICTSD: Geneva, Switzerland, 2011.
46. McKay, F.H.; Haines, B.C.; Dunn, M. Measuring and understanding food insecurity in Australia: A systematic review. *Int. J. Environ. Res. Public Health* **2019**, *16*, 476. [CrossRef] [PubMed]
47. Nguyen, M.T.; Bui, T.N. Trade openness and economic growth: A study on Asean-6. *Economies* **2021**, *9*, 113. [CrossRef]
48. Nickell, S. Biases in dynamic models with fixed effects. *Econometrica* **1981**, *49*, 1417–1426. [CrossRef]
49. Holtz-Eakin, D.; Newey, W.; Rosen, H.S. Estimating vector autoregressions with panel data. *Econometrica* **1988**, *56*, 1371–1395. [CrossRef]
50. Arellano, M.; Bond, S. Some tests of specification for panel data: Monte Carlo evidence and an application to the employment equations. *Rev. Econ. Stud.* **1991**, *58*, 277–297. [CrossRef]
51. Blum, W.E.H. Soil and land resources for agricultural production: General trends and future scenarios-A worldwide perspective. *Int. Soil Water Conserv.* **2013**, *1*, 1–14. [CrossRef]
52. Gordon, L.J.; Finlayson, C.M.; Falkenmark, M. Managing water in agriculture for food production and other ecosystem services. *Agric. Water Manag.* **2010**, *97*, 512–519. [CrossRef]
53. Kirby, M.; Ahmad, M.; Mainuddin, M.; Khaliq, T.; Cheema, M.J.M. Agricultural production, water use and food availability in Pakistan: Historical trends, and projections to 2050. *Agric. Water Manag.* **2017**, *179*, 34–46. [CrossRef]
54. Funk, C.C.; Brown, M.E. Declining global per capita agricultural production and warming oceans threaten food security. *Food Secur.* **2009**, *1*, 271–289. [CrossRef]
55. Baldos, U.L.; Hertel, T.W. Global food security in 2050: The role of agricultural productivity and climate change. *Aust. J. Agric. Resour. Econ.* **2014**, *58*, 554–570. [CrossRef]
56. Ayodeji, O.; Koye, B.; Chidozie, O.; Jens-Peter, J. *Agricultural Productivity and Food Supply Stability in Sub-Saharan Africa: LSDV and SYS-GMM Approach*; Munich Personal RePEc Archive (MPRA) Paper No. 90204; University Library of Munich: Munich, Germany, 2018.
57. Kavallari, A.; Fellmann, T.; Gay, S.H. Shocks in economic growth = shocking effects for food security? *Food Secur.* **2014**, *6*, 567–583. [CrossRef]
58. Food and Agriculture Organization of the United Nations. *The Impact of Disasters on Agriculture and Food Security*; Food and Agriculture Organization of the United Nations: Rome, Italy, 2015.
59. Hawkins, R.J. Macroeconomic susceptibility, inflation, and aggregate supply. *Physica A* **2017**, *469*, 15–22. [CrossRef]
60. Dellas, H.; Gibson, H.D.; Hallb, S.G.; Tavlas, G.S. The macroeconomic and fiscal implications of inflation forecast errors. *J. Econ. Dyn. Control* **2018**, *93*, 203–217. [CrossRef]
61. Akbar, M.; Jabbar, A. Impact of macroeconomic policies on national food security in Pakistan: Simulation analyses under a simultaneous equations framework. *Agric. Econ.* **2017**, *63*, 471–484.
62. Manap, N.M.A.; Ismail, N.W. Food security and economic growth. *Int. J. Mod. Trends Soc. Sci.* **2019**, *2*, 108–118.
63. David, J.G.; Bauer, A. The costs of transition in Central Asia. *J. Asian Econ.* **1998**, *9*, 345–364.
64. Food and Agriculture Organization of the United Nations. *Regional Overview of Food Insecurity in Europe and Central Asia: Focus on Healthy and Balanced Nutrition*; Food and Agriculture Organization of the United Nations: Rome, Italy, 2015.

65. Baydildina, A.; Akshinbay, A.; Bayetova, M.; Mkrytichyan, L.; Haliepesova, A.; Ataev, D. Agricultural policy reforms and food security in Kazakhstan and Turkmenistan. *Food Policy* **2000**, *25*, 733–747. [CrossRef]
66. Kassam, S.N.; Akramkhanov, A.; Nishanov, N.; Nurbekov, A. *From Destatization to Land Privatization to Agrarian Reform in Post-Soviet Uzbekistan: How to Support Paradigm Shifts to Sustainable Land Use Management Practices?* International Centre for Agricultural Research in the Dry Areas (ICARDA) Policy Brief: Beirut, Lebanon, 2016.
67. Babu, S.C.; Sengupta, D. Policy reforms and agricultural development in Central Asia: An overview of issues and challenges. In *Policy Reformsand Agriculture Development in Central Asia*, 1st ed.; Djalalov, S., Babu, S.C., Eds.; Springer: New York, NY, USA, 2006; ISBN 9780387297798.
68. Food and Agriculture Organization of the United Nations. *Regional Overview of Food Insecurity in Europe and Central Asia 2016: The Food Insecurity Transition*; Food and Agriculture Organization of the United Nations: Budapest, Hungary, 2017.
69. Food and Agriculture Organization of the United Nations. *Regional Overview of Food Insecurity in Europe and Central Asia 2019: Structural Transformations of Agriculture for Improved Food Security, Nutrition and Environment*; Food and Agriculture Organization of the United Nations: Budapest, Hungary, 2019.
70. Food and Agriculture Organization of the United Nations; World Food Programme; United Nations Economic Commission for Europe; United Nations Children's Fund; World Health Organization; World Meteorological Organization. *Regional Overview of Food Insecurity in Europe and Central Asia 2020: Affordable Healthy Diets to Address All Forms of Malnutrition for Better Health*; Food and Agriculture Organization of the United Nations: Budapest, Hungary, 2021.
71. Profillidis, V.A.; Botzoris, G.N. Statistical Methods for Transport Demand Modeling. In *Modeling of Transport Demand*, 1st ed.; Profillidis, V.A., Botzoris, G.N., Eds.; Elsevier Inc.: San Diego, CA, USA, 2019; ISBN 9780128115138.
72. Arsova, A.; Örsalcd, D.D.K. A panel cointegrating rank test with structural breaks and cross-sectional dependence. *Econom. Stat.* **2021**, *17*, 107–129. [CrossRef]
73. Örsalcd, D.D.K.; Droge, B. Panel cointegration testing in the presence of a time trend. *Comput. Stat. Data Anal.* **2014**, *76*, 377–390.
74. Pedroni, P. Critical values for cointegration tests in heterogeneous panels with multiple regressors. *Oxf. Bull. Econ. Stat.* **1999**, *61*, 653–670. [CrossRef]
75. Pedroni, P. Panel cointegration: Asymptotic and finite sample properties of pooled time series tests with an application to the PPP hypothesis. *Econ. Theory* **2004**, *20*, 597–625. [CrossRef]
76. Kao, C. Spurious regression and residual-based tests for cointegration in panel data. *J. Econ.* **1999**, *90*, 1–44. [CrossRef]
77. Maddala, G.; Wu, S. A comparative study of unit root tests and a new simple test. *Oxf. Bull. Econ. Stat.* **1999**, *61*, 631–652. [CrossRef]
78. Hausman, J.A. Specification tests in econometrics. *Econometrica* **1978**, *46*, 1251–1271. [CrossRef]
79. Teignier, M. The role of trade in structural transformation. *J. Dev. Econ.* **2018**, *130*, 45–65. [CrossRef]
80. Otero, G.; Pechlaner, G.; Gürcan, E.C. The political economy of "food security" and trade: Uneven and combined dependency. *Rural Sociol.* **2013**, *78*, 263–289. [CrossRef]
81. Batsaikh, U.; Dabrowski, M. Central Asia—Twenty-five years after the breakup of the USSR. *Russ. J. Econ.* **2017**, *3*, 296–320. [CrossRef]
82. Tamea, S.; Laio, F.; Ridolfi, L. Global effects of local food-production crises: A virtual water perspective. *Sci. Rep.* **2016**, *6*, 18803. [CrossRef]
83. Tu, C.; Suweis, S.; D'Odorico, P. Impact of globalization on the resilience and sustainability of natural resources. *Nat. Sustain.* **2019**, *2*, 283–289. [CrossRef]
84. Dupas, M.C.; Halloy, J.; Chatzimpiros, P. Time dynamics and invariant sub-network structures in the world cereals trade network. *PLoS ONE* **2019**, *14*, e0216318. [CrossRef] [PubMed]
85. Raymond, A.B.; Alpha, A.; Ben-Ari, T.; Daviron, B.; Nesme, T.; Tétart, G. Systemic risk and food security: Emerging trends and future avenues for research. *Glob. Food Secur.* **2021**, *29*, 100547. [CrossRef]
86. Chen, C.; Chen, G.; Yao, S. Do imports crowd out domestic consumption? A comparative study of China, Japan and Korea. *China Econ. Rev.* **2012**, *23*, 1036–1050. [CrossRef]
87. Ksenofontov, M.Y.; Polzikov, D.A.; Urus, A.V. Scenarios of the EAEU agricultural market development in the long term. *Stud. Russ. Econ. Dev.* **2020**, *31*, 700–716. [CrossRef]
88. Zhang, J.; Chen, Y.; Li, Z. Assessment of efficiency and potentiality of agricultural resources in Central Asia. *J. Geogr. Sci.* **2018**, *28*, 1329–1340. [CrossRef]
89. Felbermayr, G.; Gröschl, J. Naturally negative: The growth effects of natural disasters. *J. Dev. Econ.* **2014**, *111*, 92–106. [CrossRef]
90. Coulibaly, T.; Islam, M.; Managi, S. The impacts of climate change and natural disasters on agriculture in African countries. *Econ. Disaters Clim. Chang.* **2020**, *4*, 347–364. [CrossRef]
91. Naceur, S.B.; Hosny, A.; Hadjian, G. How to de-dollarize financial systems in the Caucasus and Central Asia? *Empir. Econ.* **2019**, *56*, 1979–1999. [CrossRef]
92. Sun, Z.L.; Li, X.D. The trade margins of Chinese agricultural exports to ASEAN and their determinants. *J. Integr. Agric.* **2018**, *17*, 2356–2367. [CrossRef]
93. Beck, N.; Katz, J.N. What to do (and not to do) with time series-cross-section data in comparative politics. *Am. Polit. Sci. Rev.* **1995**, *89*, 634–647. [CrossRef]

94. Sternberg, T.; McCarthy, C.; Hoshino, B. Does China's Belt and Road Initiative threaten food security in Central Asia? *Water* **2020**, *12*, 2690. [CrossRef]
95. Tortajada, C.; Zhang, H. When food meets BRI: China's emerging Food Silk Road. *Glob. Food Secur.* **2021**, *29*, 100518. [CrossRef]
96. Fathelrahman, E.; Davies, S.; Muhammad, S. Food trade openness and enhancement of food security—Partial equilibrium model simulations for selected countries. *Sustainability* **2021**, *13*, 4107. [CrossRef]
97. Zolin, M.B.; Cavapozz, D.; Mazzarolo, M. Food security and trade policies: Evidence from the milk sector case study. *Br. Food J.* **2021**, *123*, 59–72. [CrossRef]

Article

Let Them Eat Fish!—Exploring the Possibility of Utilising Unwanted Catch in Food Bank Parcels in The Netherlands

Madhura Rao [1,*], Lea Bilić [2], Joanna Duwel [2], Charlotte Herentrey [2], Essi Lehtinen [2], Malin Lee [2], María Alejandra Díaz Calixto [2], Aalt Bast [2,3] and Alie de Boer [1]

1   Campus Venlo, Food Claims Centre Venlo, Faculty of Science and Engineering, Maastricht University, 5900 AA Venlo, The Netherlands; a.deboer@maastrichtuniversity.nl

2   Campus Venlo, University College Venlo, Maastricht University, 5900 AA Venlo, The Netherlands; l.bilic@student.maastrichtuniversity.nl (L.B.); j.duwel@student.maastrichtuniversity.nl (J.D.); c.herentrey@student.maastrichtuniversity.nl (C.H.); e.lehtinen@student.maastrichtuniversity.nl (E.L.); malin.lee@student.maastrichtuniversity.nl (M.L.); m.diazcalixto@student.maastrichtuniversity.nl (M.A.D.C.); a.bast@maastrichtuniversity.nl (A.B.)

3   Department of Pharmacology & Toxicology, Faculty of Health, Medicine and Life Sciences, Maastricht University, 6229 ER Maastricht, The Netherlands

*   Correspondence: m.rao@maastrichtuniversity.nl; Tel.: +31-433883666

Citation: Rao, M.; Bilić, L.; Duwel, J.; Herentrey, C.; Lehtinen, E.; Lee, M.; Díaz Calixto, M.A.; Bast, A.; de Boer, A. Let Them Eat Fish!—Exploring the Possibility of Utilising Unwanted Catch in Food Bank Parcels in The Netherlands. *Foods* 2021, 10, 2775. https://doi.org/10.3390/foods10112775

Academic Editors: António Raposo, Renata Puppin Zandonadi, Raquel Braz Assunção Botelho and Theodoros Varzakas

Received: 9 September 2021
Accepted: 8 November 2021
Published: 11 November 2021

Publisher's Note: MDPI stays neutral with regard to jurisdictional claims in published maps and institutional affiliations.

Abstract: The Common Fisheries Policy of the European Union was reformed in 2013 with the aim of improving the sustainability of the fishing sector. The Landing Obligation, a cornerstone of this reform, requires fishers to land their unwanted catch instead of discarding it at sea. Existing literature pays little attention to what becomes of this unwanted catch once it is landed. To further the discourse on the sustainable valorisation of unwanted catch, this study explores whether unwanted catch that is safe for human consumption could be used for improving food security. The paper focuses on Dutch food banks, which deliver critical food aid to over 160,000 individuals yearly but struggle to provide all dependant recipients with nutritionally balanced food parcels. The research question is addressed in two ways. The food bank recipients' willingness to consume UWC is evaluated quantitatively through a survey. Next to this, data from interviews with relevant stakeholders are analysed qualitatively. Results indicate that the Food Bank Foundation and its recipients are willing to receive this fish if it is safe to consume and accessible. However, various factors such as existing infrastructure, lack of economic incentive to donate, competition from non-food and black markets, and the fishing industry's conflict with the landing obligation might pose barriers to this kind of valorisation. The dissonance between fisheries, food, and sustainability policies is discussed and identified as a key limiting factor. To bridge the differences between these policy areas, we propose public-private partnerships and voluntary agreements among involved stakeholders.

Keywords: landing obligation; Common Fisheries Policy; food waste; food security; sustainability

1. Introduction

As part of the 2013 Common Fisheries Policy (CFP) reform, the European Commission introduced a Landing Obligation (LO) which requires all catches of species subject to catch quotas and minimum conservation reference size to be landed and counted against quota. The objective of this requirement is to promote selective fishing and significantly reduce the incidence of unwanted catches (UWC). Despite implementing selective fishing strategies such as the use of specialised gear and locating species-specific hotspots, some untargeted fish are still likely to end up being caught and landed [1]. Considering that fishers have traditionally discarded such fish at sea, routes to fully utilise the landed UWC are currently underdeveloped. The 2015–2019 implementation period for the LO saw the sector focus on negotiating requirements and engineering ways to fish more selectively. However, little attention was paid to developing sustainable ways to utilise UWC that would be landed as a consequence of the LO.

Next to conserving marine biological resources, reducing waste was a key incentive in bringing about this reform [2]. When UWC with low survival rate is discarded at sea, it contributes neither to the health of the fish stock nor towards meeting the dietary needs of the growing population. Therefore, the practice of discarding UWC at sea can be seen as wasteful and undesirable [2]. However, not utilising this UWC once it is landed is as wasteful and undesirable, if not more. Given that meeting current and future protein needs of the population is a significant challenge for the food system [3], not making appropriate use of the landed UWC to supplement the human diet when feasible is a missed opportunity. This is true also for developed countries like the Netherlands, where over 1 million people out of the 17.4 million population are known to be part of low-income households [4]. Of these, several individuals require social assistance to procure sufficient food [4]. The national Food Bank Foundation, Voedselbanken Nederland, aids these individuals by providing weekly food parcels through its network of 171 food banks [5]. Based on existing literature, it is known that Dutch food banks, like their counterparts around the world, struggle to provide recipients with nutritionally balanced parcels [6,7]. In the Netherlands, fish and fruit are particularly scarce in food bank parcels [7]. Neter et al. (2018) found that 99% of the food bank recipients did not consume fish in amounts considered sufficient by the national dietary guidelines [7]. This issue cannot be ignored, as the number of food bank recipients continues to grow [5]. Although it is known that there is a lack of fish in food bank parcels, no previous studies have enquired as to whether food bank recipients are keen on consuming more fish if it is made available to them. Our study provides a first foray into this topic.

Together, the issues of UWC underutilisation and fish deficiency in food bank parcels concern several of the United Nations Sustainable Development Goals (UN SDGs), such as goal 2-zero hunger, goal 3-good health and well-being, goal 12-responsible consumption and production, and goal 14-life below water. The goals stand for eliminating hunger and food insecurity, ensuring a healthy life, making consumption and production more sustainable, and ensuring the sustainable use of marine resources. The SDGs give UN Member States, including the Netherlands, a blueprint for a more sustainable future. Although many actions have already been taken, most Member States are not on track to reach the goals by 2030 [8]. By sustainably utilising UWC and increasing the protein content of food bank parcels, the Netherlands would come closer to realising the SDGs. However, with existing EU policies, market conditions, and stakeholder relations, addressing the issue is not straightforward, and may require several large-scale changes.

Currently under the LO, UWC that is under the minimum conservation reference size (MRCS) cannot be sold for direct human consumption in the EU. Instead, it ends up as pet food, fish meal, or other products. According to Cashion et al. (2017), on a global scale as much as 90% of fish which are safe for human consumption end up being used for fishmeal, fish oil, and other non-food applications [9]. This does not align with the food waste hierarchy developed by Papargyropoulou et al. (2014), which states that before recycling food into animal feed, redistribution for human consumption must be considered, if feasible [10]. Iñarra et al. (2019) have adapted the hierarchy specifically for fish waste management, starting again with (1) prevention and reduction, and continuing to (2) human consumption, (3) bio-products, (4) animal feed, (5) industrial uses, (6) production of energy, (7) agronomic purposes, and, finally (8) disposal [11]. In line with this, the LO promotes prevention and reduction of fish waste as a top priority. However, it forfeits the second step in the hierarchy to avoid creating a market for <MRCS fish. In the case of UWC > MRCS, the LO allows for the fish to be utilised for human consumption. Since substantial efforts to create a market for these fish have not been undertaken, any landed UWC that does not have a high market value as a food product is likely to be used for animal feed production, non-food applications, or even disposed of.

At the time of writing this paper, literature on the successful utilisation of UWC in Europe was found to be scarce. This is not surprising considering that the LO was implemented only recently and the issue of UWC valorisation is rather new. DiscardLess, an EU Horizon2020 project with a focus on reducing fishing discards in the region, collated information about some ongoing initiatives. A 2019 report from the project indicated that unavoidable UWC in parts of Denmark is either discarded after landing or sold to fish meal or fish oil producers [12]. In the Netherlands and Germany, the pet food industry is the largest buyer of UWC. In Boulogne-Sur-Mer, the biggest harbour in France, UWC was largely found to be sold to a local company that specialises in the production of cosmetics, health products, and fish meal. Pescanova, the biggest fishing company in Spain, was reported to be selling UWC outside of the EU [12]. Some pilot projects focusing on valorising UWC for human consumption were found in the literature as well. Two studies from Spain showed the successful valorisation of UWC into fish mince products such as fish finger and burger patties [13,14]. A project based in northern Finland was also reported to be able to set up a fish mincing facility to utilise UWC for human consumption [15]. No literature focusing on utilising UWC for food charities in Europe could be found.

The idea of donating UWC to food banks, however, is not entirely novel. SeaShare is a US-based organisation that enables the donation of similar type of UWC to hunger relief organisations. SeaShare, as described by Watson et al. (2020), partly relies on what the US legislation calls a prohibited species donation (PSD) programme [16]. During its pilot run in 1993, the PSD program authorised the donation of salmon that was classified as prohibited species in order to prevent valuable fish protein from leaving the food supply chain [16]. Later in 1996, Amendments 26 and 29 to the Fishery Management Plans for Groundfish in the North Pacific (61 FR 38358) allowed a full-fledged programme to be established [16]. SeaShare, which is a non-profit organisation, supported the PSD programme by taking on the logistical and financial burden of processing and distributing this fish [16]. While this Prohibited Species Donation Program serves as inspiration for what could be done in the Netherlands, due to the differences in intra-industrial relationships, as well as in fishing-related laws and regulations in the US and EU, the possibilities within the Netherlands warrant a separate evaluation.

Therefore, this study explores the possibility of including more fish in Dutch food bank parcels by utilising UWC. It does this by assessing whether food bank recipients would be interested in such fish. This is followed by an analysis of relevant stakeholders' views on valorising UWC in this manner.

2. Methods

2.1. Study Design

This mixed methods study was designed to investigate the possibility of utilising unwanted catch in Dutch food bank parcels from two perspectives. The first perspective focuses on whether food bank recipients are interested in receiving more fish in their parcels and if they are keen on consuming UWC. It was important to ascertain this because without an interest from the food bank recipients, further investigations would not be justified. Therefore, to understand whether food bank recipients want more fish in their parcels and whether they are willing to eat UWC, a quantitative approach was deemed appropriate. The goal with this approach was to survey the terrain before proceeding with further data collection. A detailed description of the method employed is provided in Section 2.2.

The second perspective looks at the fisheries supply chain and seeks to identify whether it is economically, logistically, and legally possible to utilise UWC in this manner. To answer this question, it was necessary to gather insights from stakeholders and experts working directly with this supply chain. A qualitative approach using semi-structured interviews was seen as the best fit for collecting and analysing such data since it seeks to contribute to an improved understanding of social realities [17]. It does this by drawing attention to processes, meanings, patterns, and structural features [17]. The purpose this perspective serves is to mine the surveyed terrain. A detailed description of the method employed is provided in Section 2.3.

2.2. Questionnaires—Statistical Analysis

Questionnaires were used to determine whether there was a demand for more fish in the food banks and if food bank recipients were willing to consume UWC. Printed questionnaires were distributed to the recipients of Food Bank Venlo during May and June 2021. Two hundred and ninety-eight paper questionnaires were packed in individual envelopes and were transported to the food bank for distribution. In the food bank, volunteers placed the envelopes in the recipients' food parcels.

Food bank recipients had two weeks to participate in the research and involvement was entirely voluntary. The self-administered questionnaires were in Dutch and consisted of six questions, five of which were multiple-choice. The questions were asked in the following order: (1) age, (2) gender, (3) food bank visits per month, (4) frequency of receiving fish from the food bank, (5) willingness to receive more fish, and (6) willingness to eat UWC. The term UWC was explained to the participants in the introduction and repeated in question 6. The first two questions were asked to determine whether the demographic of the sample matched with that of food bank recipients from across the country. This data was also collected to examine whether age and gender were associated with desire for more fish and willingness to consume UWC. The third and fourth questions were asked for the purpose of determining whether the frequency of visiting the food bank was associated with the quantity of fish received. Finally, respondents were asked if they would be willing to eat UWC. The definition of UWC was provided along with this question but description of the fish (species, size, state) was not included. This was done in order to understand whether respondents' interest in fish was associated with the fish's status as UWC solely. The number of questions were limited so as to require as little of the respondents' time as possible. The food bank recipients could either return the filled-in questionnaires during their next visit to the food bank or scan a QR-code for the online version created with Qualtrics. Returned paper questionnaires were collected from the food bank and online responses were retrieved from Qualtrics. The relationship between gender and willingness to consume UWC and age and willingness to consume UWC were tested using a chi-square test of independence.

2.3. Interviews

Qualitative data were collected through in-depth semi-structured interviews with experts working with the Dutch Food Bank Foundation and the fisheries sector. The theoretical sampling strategy of Glaser & Strauss (1967) was used to create a suitable sample [18]. This strategy entails collecting, coding, and analysing data simultaneously to inform decisions regarding what data to collect next and where to find them. In this approach, data collection, analysis and theory emergence take place in a parallel manner [18]. All participants were contacted via email with a brief explanation about the purpose of the study and the relevance of their participation. Ten participants were interviewed, following which recruitment was stopped because data from the interviews were no longer contributing to the further development of existing codes. This phenomenon is described by Glaser & Strauss (1967) as theoretical saturation and marks the end of data collection [18]. Table 1 provides an overview of the participants' profiles and expertise. The interviews were conducted in English and lasted for 45 to 75 min. With the participants' consent, interviews were recorded and transcribed *verbatim*. All participants were invited to review their interview transcripts for any inaccuracies.

As described above, the process of analysing the transcripts to identify recurring themes and ideas took place continually alongside data collection. Relevant phrases, sentences, or even entire sections were coded inductively using the in vivo technique described by Saldaña (2021) [19]. At the end of the first coding cycle, 30 codes were identified. In the next round, codes that were related to each other were grouped together into 11 categories. Subsequently, the 11 categories were further grouped together into four themes informed by the analysed data as well as previously read literature on the topic.

Table 1. Overview of participants and their expertise.

Participant	Expertise	Description
P1	Supply chain	Procurement and supply chain specialist connected with the Netherlands Foodbank Association.
P2	Market research	Market researcher connected with the Netherlands Foodbank Association.
P3	Food safety	Food safety specialist connected with the Netherlands Foodbank Association.
P4	Marine ecology	Research scientist based in the US, specialising in quantitative marine ecology.
P5	Marine ecology	Research scientist based in the US, specialising in marine conservation ecology.
P6	Fisheries governance	Independent consultant having extensive experience with NGOs focused on sustainable fishing.
P7	Fisheries governance	Project manager for sustainable fisheries governance programmes in the Netherlands.
P8	Fish trading and processing	Commercial head for a fish trading and processing business in the Netherlands.
P9	Fish trading and management	Director for a national fish auction in the Netherlands.
P10	Fisheries political affairs	Director for various association advocating for the interests of fishers in the Netherlands.

ATLAS.ti version 8.4 was used for all rounds of coding. The first and third authors undertook data analysis independently following the method described above and later compared and consolidated their findings. As an additional validation strategy, the last author oversaw the comparison and merging of the two independent analyses. Wherever necessary, excerpts, quoted *verbatim* unless modified to improve readability or to ensure anonymity, have been used in Section 3 to underpin the findings.

2.4. Ethical Considerations

Food bank recipients received the questionnaires as part of their regular food parcels. They were informed in writing about the aim of the study and that filling out and returning the questionnaires was voluntary. This approach was chosen over the researchers personally handing out questionnaires to ensure complete anonymity and so that the food bank recipients would experience no discomfort or pressure. Each envelope contained a piece of candy as acknowledgment for participation.

All participants who were interviewed were informed about the aim of the study, data storage, and privacy. Participants were given the opportunity to ask questions about these topics. They were informed that they could choose not to answer any questions asked by the researchers and to stop the interview at any given point. All participants gave their written informed consent.

3. Results

3.1. Demand and Supply

3.1.1. Demand for More Fish in Food Bank Parcels

Out of the 298 distributed questionnaires, 46% were returned, and the total number of responses was 138. Gender representation was evenly split with 69 female and 68 male participants. One participant did not report their gender. The age distribution ranged from 21 to 76 years old, and the respondents' mean age was found to be 45.23 years, with a standard deviation of 12 years. Thirteen individuals did not report their age. It was found that the study sample's demographic closely matches that of food bank recipients throughout the country (Voedselbanken Nederland, 2020).

Out of the 138 respondents, 61 reported never receiving fish in their food parcels and 52 reported receiving fish once a month. The rest of the respondents reported receiving fish more often. However, visiting the food bank more frequently did not increase the quantity of fish received. Of the 72.46% of respondents who visited the food bank once per week, 38.00% reported that they received fish once per month. Of those who reported visiting more frequently, 37.04% reported receiving fish once per month.

Regarding food bank recipients' willingness to eat more fish, 106 (72.45%) respondents indicated that they wanted more fish in their diet. Of these, 98.11% were willing to eat UWC. Results indicate that 81% of the female respondents wanted more fish in their diet, while 72% of the male respondents reported the same. This difference between genders with regard to desire for more fish is in line with the larger Dutch population, where women were observed to consume fish more frequently than men [20]. However, with regard to UWC, results from the chi-square test indicated that the relationship between gender and willingness to consume UWC was not significant at baseline (X^2 (2, N = 136) = 4.2843, p = 0.117). Thus, female participants were not less or more likely than male participants to not be willing to eat UWC. Similarly, age did not show a significant association with willingness to consume UWC (X^2 (4, N = 124) = 1.2278, p = 0.881). None of the age groups in the sample were therefore more willing to consume UWC than others. This contradicts previous studies which indicate older consumers to be more accepting of food that is traditionally seen as surplus or waste [3,21]. However, the oldest age group (50–76) is underrepresented in our sample as compared to the national food bank recipient population, and different results are possible if more individuals in the age group were to be included in the study. Tables 2 and 3 provide an overview of gender in relation to willingness to consume UWC and age in relation to willingness to consume UWC. Missing data were not included while computing the chi-square test.

Table 2. Gender in relation to the willingness to consume UWC.

	Total Number of Participants (% by Gender)	Willing * (% m/f)	Not Willing * (% m/f)	Unsure * (%m/f)	Missing Values	*p*-Value
Gender						0.117
Male	68 (49.3%)	52 (47.7%)	10 (76.9%)	6 (42.9%)		
Female	69 (50.0%)	57 (52.3%)	3 (23.1%)	8 (57.1%)	1	
Missing	1	1	0	0		

* = Reported willingness to consume UWC.

Table 3. Age groups in relation to the willingness to consume UWC.

	Total Number of Participants (% by Age Group)	Willing * (% by Age Groups)	Not Willing * (% by Age Groups)	Unsure * (% by Age Groups)	Missing Values	*p*-Value
Age group						0.882
18–29	15 (10.9%)	12 (12.1%)	2 (16.7%)	1 (7.1%)		
30–49	67 (48.5%)	53 (53.5%)	7 (58.3%)	7 (50.0%)		
50–76	43 (31.2%)	34 (34.3%)	3 (25.0%)	6 (42.9%)		
Missing	13	11	1	0	1	

* = Reported willingness to consume UWC.

The board of the Food Bank Foundation seems to be aware of the lack of fish in the parcels. During the interviews, relevant participants acknowledged that food parcels rarely contained fish. It was mentioned that several food bank recipients could not consume various meats for religious reasons. Therefore, as per interviewees, adding more fish to the parcels would be seen as favourable by the recipients because it would improve the nutritional content of the parcels and religious or social restrictions would not apply. It was mentioned that the Food Bank Foundation was already working to identify actors in the fish supply chain who were willing to donate their surplus fish.

3.1.2. Availability of Edible UWC in The Netherlands

From the results presented above, it is evident that the Food Bank Foundation and its recipients are interested in UWC. However, our inquiry regarding the availability of edible UWC did not yield as concordant a response. Neither the interviewed study participants nor existing literature could provide definitive information regarding how much edible UWC was available or discarded in the Netherlands. One of the interviewees was reliably informed that up to half of the demersal (groundfish) catch was unwanted. Other interviewees working with the fishing industry were also aware that a large quantity of fish was caught unintentionally by demersal fleets but did not provide an estimate regarding the numbers. In 2018, Stichting De Noordzee, a Dutch non-profit organisation focused on sustainability in the North Sea, reported that the Dutch demersal fisheries discarded more than 70,000 tons of fish at sea every year [22]. The Common Dab and the European Plaice were reported to be the most discarded fish [22]. The same article mentions that the LO could help reduce these numbers by stimulating the fishing industry to engineer creative ways to avoid UWC instead of landing it [22]. The article, however, does not mention what happens to this UWC when it is landed.

Based on the data collected from our interviews, it appears that UWC which is landed has three possible destinations. UWC that has demand as food for human consumption enters the food supply chain through mainstream fish auctions. UWC that does not have market value as food, including <MCRS catch that cannot be used for human consumption as per Regulation (EU) No 1380/2013, is sold for other applications such as animal feed, pet food, fish oil, food additives, pharmaceuticals, and cosmetics [23]. Lastly, unwanted catch that has no demand at all or cannot be used due to reasons such as spoilage or damage is discarded. Finding uses for all of the landed UWC is anticipated to be one of the most challenging impacts of the landing obligation [24]. There are some existing market opportunities for these fish, but as reported by Hedley et al. (2015), it is clear that new markets will need to be developed if the incoming stream of UWC is to be fully utilised [24].

If the food waste hierarchy is to be applied in this context, safe to consume UWC should stay in the human food supply chain to ensure sustainable utilisation [10]. Donating part of the UWC to the food banks could be a way to do this without creating a market for these fish. However, there are several economic barriers to this, as further discussed in Section 3.2. When asked whether it would be feasible to use UWC < MCRS in food bank parcels, none of the interviewees working with the fisheries responded enthusiastically. Some interviewees believed that if the <MCRS fish were to be utilised for human consumption, it would take away the fishing industry's incentive to fish more selectively. Most interviewees did not think that the situation in the Netherlands was unfavourable. P10 described it as: *"The fishermen sell all their fish at the auction—target species as well as non-target species. All the fish which is marketable is being sold. There is no fish which is not being sold"*. This excludes under MCRS catch, but interviewees did not seem to view this fish as food. This can be extended to UWC without a market value in general.

Those working on the ground might have a different perspective on the issue. In a paper published by de Vos et al. (2016), all Dutch fishers interviewed for the study expressed their aversion of being obliged to land < MCRS catch due to reasons related to principle, profitability, and environmental protection [25]. An episode of the Dutch television show Keuringsdienst van Waarde that documented this issue also indicated that fishers saw this legislative requirement as unreasonable, wasteful, and unsustainable [26]. However, this does not imply that the fishers would be willing to donate this fish. As per results from Maynou et al. (2018), European fishers saw donating UWC as the least favourable valorisation route [27]. In comparison, other stakeholders like NGOs, researchers, and industry representatives expressed moderate interest. The same study compared the opinions of stakeholders by fishing regions. Stakeholders involved in North Sea fisheries considered charity to be the least favourable option [27].

In the US, a programme that facilitates the donation of UWC that cannot be used for any other purposes has been successfully established. P4, who studied the impact of this programme, emphasized the importance of stakeholder cooperation, community support, and goodwill in establishing it. Based on the responses of participants working with the Dutch fisheries, the fishing sector in the Netherlands does not seem to have any prior experience with working with food banks. As a result, cooperation or goodwill cannot be expected from stakeholders yet. NGOs or civil society organisations could play a role in brokering such an understanding between the food banks and the fishing sector. However, based on interviewees' responses, the fisheries in the Netherlands do not currently share a cordial relationship with such organisations.

Several interviewees working with the fishing industry believed that the Food Bank Foundation should try to procure fish from a later stage of the supply chain as opposed to the pre-auction and auction stages. Frozen fish close to its expiration date, procured from various stages of the supply chain was suggested as a possible option. However, given that this study focuses on UWC, this option will not be explored further.

3.2. Economic Feasibility

3.2.1. Paying for Unwanted Catch

Bringing UWC that is currently used for non-food applications back into the food supply chain is a sustainable way of utilising such fish. Donating it to the food bank is an attractive way to do this from the perspectives of food security and public health, but it raises concerns regarding economic viability. The food banks in the Netherlands do not pay for procuring food. Next to providing food aid, the Food Bank Foundation aims to reduce food waste by utilising surplus food that would have ended up as waste otherwise [28]. Therefore, it relies on food businesses seeing the merit of donating their surplus. In this case, however, fishers may not see the need to donate their UWC because it can be sold to other destinations and used for other purposes, such as animal feed or pet food. Income from such sale itself is unlikely to offset the costs incurred from keeping UWC on board [29]. However, donating UWC instead would further reduce profits. Interviewees working with the fishing industry expressed their understanding regarding donation being the more sustainable option but also regarded it as an unlikely scenario due to its impact on profits. Considering that implementing the LO is likely to cost demersal cutter fleets between EUR 5.6 and 12.3 million in transition costs [29], the sector's worries regarding profits and economic viability are well founded.

Additionally, the market value of UWC is dynamic. One of the interviewees described the example of the octopus to demonstrate this: *"Five years ago, octopus was a low-value side catch. It was sold at 2 euros per kilogram. But these days, it has a value of around 18 euros per kilogram. Nowadays, we see that the demand for fish is healthy and even the side-catch is expensive"*. In hopes of receiving a higher price for their UWC in the future, fishers may not want to donate their UWC to the food bank and classify it as low value fish.

When asked whether paying a small amount of money for procuring UWC was something the Food Bank Foundation would consider, interviewees working with the food banks conveyed that this would not be feasible. Interviewees indicated that over the last years, the food banks' supply has decreased while the number of people signing up to receive food aid has increased. Additionally, they believed that if the foundation started paying for one category of products, all their procurement partners would expect to be paid.

3.2.2. Processing Costs

Economic barriers to donating UWC do not end at the procurement stage. If we were to consider the hypothetical scenario of fishers being willing to donate their low value UWC to the food banks, processing costs would still be a concern. Processing the fish would be necessary because providing whole, unprocessed fish to the food bank recipients might lead to them wasting it as a result of not knowing how to handle it. A cross-cultural study by Olsen et al. (2007) found the Dutch population to have the least positive outlook towards preparing fish at home compared to the other four European countries considered in the sample [30]. Given that the food bank recipients are known to consume particularly low quantities of fish [7], interviewees working with the food bank agreed that it would be important to introduce UWC in an accessible manner. Therefore, processing costs must be taken into consideration. This challenge might be easier to overcome than paying for procuring fish. Although the Food Bank Foundation does not spend on procuring food, interviewees stated that it is willing to do so for setting up infrastructure to process the food that it procures. Currently, food banks source most of their products from the retail stage of the supply chain. However, P1 mentioned that the board was looking to procure at least a part of its supply from earlier stages of the supply chain because the retail sector was becoming increasingly efficient at managing its surplus. Procuring from earlier stages of the supply chain would entail some degree of processing and as per the information provided by relevant participants, the board of the Food Bank Foundation is looking into the possibility of creating a separate foundation that handles the procurement and processing of such food. However, processes such as filleting fish are highly specialised and relatively expensive. It is therefore important to undertake a thorough cost-benefit analysis before investing in such facilities. It is important to note, however, that utilising fishing by-products is not completely new to the fishing industry. Operations to valorise fish by-products including cutting, boiling, drying, and ensiling exist across European harbours but limited attempts to introduce UWC in existing manufacturing operations have been made [11,27].

If UWC < MCRS is to be utilised in food bank parcels, processing costs might be an important factor to consider given the small size and bony structure of such fish. However, interviewees working with the Dutch fishing industry were far less willing to discuss the possibility of donating under MCRS catch to the food banks. Some interviewees mentioned that utilising < MCRS catch in other sustainable ways was not economically viable either. Regarding such trials, they elaborated: *"Fishers are not allowed to sell certain unwanted catch for human consumption, so it must go for products like pet food, fish meal, and fish oil. For demersal species, this is just not economically viable. We did look at whether you could extract high level proteins or turn it into fish oil instead of using it for pet food for instance. But the problem is that they would have to treat the unwanted catch in accordance with the same quality standards they use for their commercial fish. That is impossible"*.

Upon enquiring if they considered this fish to be safe for consumption, all answered favourably. However, they stated the LO requirement to not use this fish for human consumption as the restricting factor. When we enquired about the same issue but in a hypothetical situation where the LO allowed the donation of UWC < MCRS to the food bank, interviewees began to propose other profitable ways of utilising this fish. Another hypothetical scenario suggested by the researchers focused on the government providing monetary compensation to fishers for donating UWC or UWC < MCRS to the food banks. This option was not seen as favourable due to the possibility of it turning into a perverse incentive. However, targeted benefits such as tax deductions or subsidies related to sustainability were viewed more favourably. In the US context, P4 described similar incentives to have worked positively for the fish donation programme.

3.2.3. Competition from Non-Food and Black Markets

UWC that does not have a high market value as food is redirected to animal feed, pet food, or other technical uses. These supply chains are well established in the Netherlands and would therefore pose as a barrier to entry for the food bank. Some interviewee responses indicated that selling UWC for non-food applications is not favoured by the fishers, but given that they are able to recover a part of their costs through these transactions, they cannot refuse them. Despite not bringing in as much value as the food supply chain, non-food chains are important to the fishing industry. For instance, several participants mentioned that the recent shutting down of Dutch mink farms had negatively impacted the fishing industry because mink farmers no longer purchased fish as food for their minks. One participant expressed the situation as follows: *"Feed producers pay maybe seven to eight cents per kilo of this fish and that is not the best way to bring these proteins back into the chain. Seven cents do not make the fishers happy anyway. One cent more than the feed producers and they would rather give it to the food bank"*. However, as discussed above, the Food Bank Foundation is unwilling to pay for procuring this fish. Selling UWC to fish meal and oil producers is also seen as more advantageous than donating it to charity because these well-established supply chains are known to be able to cope with uncertainties like fluctuations in composition and quality [27]. Fish meal is also likely to be seen as an economically advantageous choice because Europe is a net importer of fish for non-food use [31].

Some interviewees mentioned that the existence of black markets for under MCRS catch might pose as a barrier for donating it to the food banks. It was suggested that legalising the use of UWC < MCRS in food banks may help these shadow markets proliferate and make it more challenging for national authorities to implement the LO. One of the participants stated, *"if a fishing vessel is found with undersized fish being stored in a frozen state for sale as food, the fisherman could simply say that it is all for the food bank and then sell it illegally"*. Bellido et al. (2017) discuss this in the context of the Mediterranean region wherein they note that implementing the LO may lead to the expansion of illegal markets for fish below the minimum size [32]. However, this is discussed in the context of simply keeping the fish onboard to land it and not donating it to food banks.

3.3. Logistics and Infrastructure

3.3.1. Fishing Industry's Perspective

Prior to the LO being implemented, fishers were able to select what fish to keep on board and what to discard or release at sea. As a result, fishing vessels have been designed to accommodate catch that can be brought ashore and sold at an attractive price [33]. Interviewees mentioned that vessel infrastructure and labour costs would be a cause for concern if fishers were expected to treat UWC in the same way they did regular catch. If UWC is to be used by the food banks, a high level of food safety will have to be ensured by applying standard bleeding, cleaning, sorting, and cold storage procedures. This, in turn, will require labour, space, and machine capacity on board. Without any monetary returns, fishers are unlikely to be willing to undertake these tasks. Even in the current situation where UWC can be sold at a low price for non-food purposes, fishers are not in favour of processing and accommodating this fish onboard [33].

When it comes to infrastructure related investments, interviewees working with the fishing industry in the Netherlands stated that the sector is heavily focused on developing fishing gear that will enable more selective fishing, thereby eliminating UWC. Based on interviewee responses, developing vessel capacity to handle UWC can be seen as a contradiction to improving fishing gear and, as a result, fishers might be unwilling to invest in it. The work of Viðarsson et al. (2019) confirms that vessel owners are reluctant to invest in technology to process and preserve UWC onboard [33]. The same paper suggests silage production from <MCRS catches onboard the vessels [33]. This silage could be used for producing fish meal for animal feed and other uses but render it unusable for food bank use.

It was not possible to access information regarding how UWC is currently handled aboard Dutch vessels. Generally, the main challenges for onboard management are associated with catches under MCRS [33]. UWC > MCRS is often destined for human consumption and can therefore be managed as per the traditional onboard handling processes.

3.3.2. Food Donation Logistics

If we were to assume that the fishing industry is willing to cooperate with the food banks for the donation of UWC < MCRS, the Food Bank Foundation would need to invest in suitable logistics and infrastructure to be able to use this fish. Although fish is rarely provided in food bank parcels, the cold chain for frozen meat products is already established. As per relevant interviewees, the Food Bank Foundation would need to expand its current cold storage and transportation facilities if it was to receive more fish.

This investment was seen as viable if a regular flow of processed, ready-to-cook fish could be ensured. However, given that the fishing industry's objective is to fish more selectively and reduce UWC, ensuring a steady flow of fish for the food bank could be challenging. Considering these uncertainties, it would be seen as risky for the Food Bank Foundation to invest in expanding its cold chain specifically for fish. A possible solution to this could be renting such logistics facilities. As per the information provided by interviewees, the Food Bank Foundation already rents such a warehouse to manage its current flow of frozen products.

Regarding food safety, relevant interviewees were confident about the volunteers being able to handle fish if the food banks were to receive it processed and then frozen. The food banks work with an in-house private standard for food safety which already includes provisions for handling fish. Therefore, it can be expected that ensuring the safety of UWC in food bank parcels is feasible if it is handed over to the food banks in good condition.

Dutch food banks are likely to be able to incorporate more fish in its inventory more easily compared to the US fish donation programme described by P4 and P5. The programme described by them is based in Alaska and required significant infrastructural investment to make it operational. The Dutch food banks, in comparison, have a well-established system for procuring, transporting, storing, and redistributing surplus food, including products that need cold storge facilities. However, based on interviewee responses, the fishing industry may not feel confident about food banks' capacity to handle UWC. One of the interviewees described that the industry itself engages in charitable acts but the food banks may not be able to replicate this: *"Sometimes fishers donate fish for social causes. One or two times a year, they participate in a project for homeless people in Rotterdam where they donate fish, fry it, and give it to the homeless people. But if you want to incorporate fresh fish into the parcels of the food banks, that's going to cause a bit of a logistical problem because the food banks are absolutely not equipped to handle whole fish that is not processed"*.

3.4. Fisheries Policy and Legislation

The Common Fisheries Policy and the Landing Obligation and their impact on UWC valorisation came up as a prominent theme in our analysis. The LO, and by extension the CFP, share a rather paradoxical relationship with the research question this study seeks to answer. The possibility of donating UWC to the food banks arises largely because fishers are now obliged to land fish that they unintentionally catch. Prior to the implementation of Regulation (EU) No 1380/2013, fishers could simply discard UWC at sea. It is only due to the Landing Obligation that this fish will need to be brought ashore, thus giving rise to surplus fish that needs to be utilised. At the same time, the LO also creates barriers for donating this fish to the food banks.

Firstly, the fishing industry's discontentment towards the LO may negatively impact its willingness to donate UWC. One of the interviewees working in close cooperation with the fishers described the LO as following: *"This landing obligation is not workable. It's not doable. It's not enforceable. We need to turn away from this non-workable regulation and move towards a workable regulation. But everyone is so politicised. And everybody is so concerned with the issue of discarding at sea. Nobody on the policy side wants to give in"*. This sentiment of the fishing industry towards the LO has been discussed in existing literature as well [25,34–36]. To prepare for the implementation of the LO, the Dutch government launched a working group (werkgroep Aanlandplicht) in autumn 2012 [36]. Scientists, NGOs, industry representatives, and civil servants from the ministry and the control agency were invited to be part of this working group [35]. However, the Dutch fishing industry reacted negatively to the possibility of having to land UWC and refused to participate in the working group [35]. Based on interviewee responses, it is evident that the fishing industry continues to hope that legislators will take the fishers' dissatisfaction into consideration and repeal the requirement to land UWC. Establishing a system where UWC is donated to the food banks might create dependency on landing UWC and therefore make it difficult to reverse LO. Regarding the push to make amends to the LO, an interviewee working on fisheries governance in the Netherlands elaborated: *"There is a strong call to re-evaluate the landing obligation. Everybody in the industry says it's not working. They say it's actually contributing more to illegal fisheries. The fishers are so worried about fisheries being closed and therefore, about their survival. We are very lucky to still be allowed to go on board and see the actual catch composition for research. But we also know from colleagues in other countries that fishermen tell them: 'Sorry, we cannot take you anymore, because you would see that we are actually discarding unwanted fish at sea'"*.

Secondly, the restriction on using UWC < MCRS for human consumption impedes the possibility to donate this fish. Interviewees frequently stated that donating < MCRS fish to the food banks would not be possible because (Article 15(11) of) Regulation (EU) No 1380/2013 does not permit the use of such fish for human consumption [23]. They viewed this requirement as non-negotiable and central to the LO. However, during the formative stages of this legislation, the sale of UWC < MCRS for direct human consumption was hotly debated [37]. Some Member States, especially those from the Baltic basin, proposed that once the undersized fish are caught and counted against quota, they should be given appropriate value [37]. They expected that this value would be low enough to deter fishers from catching too many undersized fish [37]. However, other Member States, particularly from the Mediterranean basin, were of the opinion that this could encourage fishers to catch undersized fish because, unlike the Baltic, consumer demand for small fish is high in the Mediterranean region [37]. As a compromise, legislators concluded that UWC < MCRS should be landed, counted against quota, and then sold only for non-human consumption [37]. These diverging views on the use of <MCRS coupled with the fact that consumer demand for small fish varies significantly across the continent may indicate that one policy may not fit all Member States. It is important to note that the decision to not allow the direct human consumption of <MCRS was based on general knowledge and is not supported by empirical evidence [37].

This reluctance to consider that undersized fish could be used for human consumption can be seen in other pieces of European legislation too. For instance, the 2013 proposal to regulate fishing in the Skagerrak reflects this [38]. In the draft version, Article 5 initially stipulated the following: *'(. . .) the sale of catches of that stock below the minimum conservation reference size shall be restricted to reduction to fish meal, pet food or other non-human consumption products only, or for charitable purposes.'* The final version of the proposal omitted *'or for charitable purposes'*. A report recording the decision-making process explained that this was done because *'it is not appropriate that juveniles be sold for charitable purposes'* [39]. Further, it stated *'it would have been possible to amend the provision so that juvenile fish may be given to charitable purposes, but as there is no such tradition in the Member States surrounding the Skagerrak, such a provision would have no place in the Regulation'* [39]. This goes on to show that the aversion to donate UWC for use as food or for charity is based on arbitrary factors rather than scientific evidence.

4. Discussion

The aim of this research was to explore the possibility of including more fish in Dutch food bank parcels by utilising unwanted catch. This study is socially and environmentally relevant because donating UWC to the food banks could not only improve the nutritional quality of the food parcels but also reduce food waste from the fishing industry. Based on the results discussed above, economic and legislative barriers, stakeholder relations, and the state of logistics and infrastructure would currently make the donation of UWC to food banks challenging. In this section, we discuss the possibility of overcoming these barriers.

The push to ban discards at sea came from a place of concern regarding food waste. Discussions about high volumes of discards in European fisheries were ongoing at the Commission when British celebrity chef Hugh Fearnley-Whittingstall's 2010 public campaign gained popularity across the Member States. Celebrities, influential retailers, environmental NGOs, and the general public expressed solidarity with Fearnley-Whittingstall's demand to end the wasteful practice of discarding UWC at sea [36]. The campaign, backed by over 650,000 petitioners, played an important role in influencing legislators to pass a law requiring fishers to land their unwanted catch instead of discarding it at sea [25,34,36]. The Commission hoped that the LO would encourage the fishing industry to focus its attention on developing selective fishing practices and eventually, significantly reducing the existence of UWC. However, developing selective fishing techniques that reduce discard rates to the 5% benchmark set by the CFP cannot be realised instantaneously. This is especially true for fisheries such as the demersal fleets that fish in the North Sea where the discard rate has historically been as high as 40% [40]. Until such tools and technologies are developed and successfully implemented, UWC will continue to be landed. The initial emphasis on food waste that led to the realisation of LO was eventually replaced by concerns regarding the health of fish stocks and marine ecology. Neither the CFP nor connected policy areas provide Member States with guidelines regarding how surplus fish that would be landed as a consequence of the LO should be sustainably utilised. The policy formulation process can be considered top-down due to the fishing industry's minimal involvement in it. The fishing sector's aversion to this approach is evident in the results. De Vos et al. (2016) indicate that fishers feel that their professional knowledge was disregarded by policymakers and scientists who lack a pragmatic understanding of how the industry is organised [25]. In their paper on the US-based seafood donation programme SeaShare, Watson et al. (2020) describe that a flexible, bottom-up approach coupled with regulatory changes enabled the programme's success [16].

In the European context, literature recognising the food waste problem associated with the Landing Obligation is scant. The European Court of Auditors' 2016 report on food waste is one of the few public documents that acknowledges that UWC landed as a result of the LO could end up as food waste [41]. The report describes the lack of legal provision to donate UWC in the new CFP as a missed opportunity. Furthermore, it encourages the Commission to assess the possibility of including legal provisions in the CFP to donate surplus fish [41]. The paper published by Vaqué (2017) makes an identical appeal [42]. The EU FUSIONS research project (2012–2016) also identified the risk of the LO turning fish waste at the sea into food waste on land [43]. This paper adds to the limited body of literature focusing on this issue and recommends better cohesion between the CFP and the EU's food waste reduction ambitions. Including a legal provision to donate surplus UWC would be a first step in facilitating this. It is unlikely that such provisions would discourage selective fishing. Landing unwanted catch would still be seen as unfavourable by fishers due to limited onboard storage and low value fish getting counted against quota. The incentive to fish selectively would considerably outweigh the incentive to land UWC and donate it to charity.

If donations are to be operationalised in the EU, legal provisions will have to be supported by suitable infrastructure. One way to facilitate this could be through the European Maritime Fisheries and Aquaculture Fund (EMFAF) which entered into force in July 2021. Part of its EUR 6,108,000,000 budget is allocated to projects focused on ensuring food security through the supply of seafood products in the Member States [44]. Additionally, the fund was set up to help fulfil the objectives of the EU Green Deal, which explicitly states reducing food waste as one of its goals [45,46]. Donating UWC to food banks not only improves food security through seafood products but also reduces food waste and ensures that UWC is utilised sustainably. To safeguard fishers' economic interests, such projects could be supported by public-private partnerships. Iñarra et al. (2020) show that turning UWC > MRCS into fish burgers or other minced fish derived products is an economically viable way to utilise UWC while also increasing fish consumption [14]. If the fish is delivered to consumers, including food bank recipients, in an affordable and easy to cook format, it is likely to be well received. The paper published by Iñarra et al. (2020) describes results from one of the few case studies that explore the possibility of utilising UWC for human consumption and will need to be replicated in different contexts. However, it considers UWC < MRCS to be ineligible for such operations largely due to the legal restriction on using such fish for direct consumption [14]. If the EMFAF is utilised to set up operations to produce new fish products from UWC and if direct human consumption of UWC < MRCS is permitted for charitable purposes, part of the fish products can be donated to the food banks. This could help ease the fishing industry's concerns regarding the economic consequences of donating UWC. At the same time, such funding and state support could help the food banks set up infrastructure and logistics to safely handle the fish.

Voluntary agreements as a form of private governance could also help advance the sustainable utilisation of unwanted catch. EU REFRESH (2015–2019), a project focused on reducing food waste across the EU, recommended voluntary agreements (VAs) as a tool for private actors to fill legislative gaps with regard to food waste valorisation [47]. VAs can be used as an alternative course of action to traditional legislation and can be directed by government officials, businesses, or other actors [47]. In the context of UWC utilisation, the VA could focus on developing new markets, setting up infrastructure, and discouraging the illegal trade of UWC. Donating a share of the UWC to improve food security could be a part of the agreement. Relevant actors who could collaborate include fishing industry associations, fish processors, NGOs, and the Food Bank Foundation. The EMFAF or other national funds set up for the purpose of improving food security could help finance such programmes.

Some limitations faced by existing food waste-focussed VAs in the Netherlands are already known. It is important to take these into account while developing any new agreements on UWC valorisation. For instance, Piras et al. (2018) point out that besides SDG 12.3, the Netherlands does not have a specific national food waste reduction target. Due to the lack of dedicated policy measures, government support for voluntary actions is limited [47]. The same report also indicated that a paucity of food waste data makes it challenging to arrive at objective targets. Current food waste VAs in the Netherlands also lack built-in sanctions for non-compliance [47]. This makes it possible for free riders to take undue advantage of such agreements by joining them only to improve their public image without taking concrete action to reduce food waste [47]. Next to this, prior to setting up any programmes, taking demand-side barriers into consideration is critical. To address such barriers, van Putten et al. (2019) recommend selling unwanted catch at an affordable price and educating consumers about preparing such fish [48]. Efforts to increase consumer acceptance should be directed towards the general public and not only food bank recipients because several types of UWC can be used for direct human consumption.

Strengths and Limitations

This paper makes a first attempt to analyse whether unwanted catch that is landed as a consequence of the new EU Common Fisheries Policy can be utilised to improve food security. It adds valuable insight to a limited body of literature that discusses the issue of unwanted catch through the lens of food waste valorisation. It is, however, not without limitations. Not all relevant stakeholders could be interviewed for this research. Some stakeholders ($n = 7$) invited to take part in this study did not agree to participate. Additionally, fishers, who would be the ultimate decision makers regarding the donation of UWC, were not included in the sample. This is due to the scope of our enquiry being limited to understanding this previously unexplored issue from a bird's-eye view as opposed to mapping the reality on the ground. Future research on this topic could highlight the viewpoint of legislators, policymakers, and fishers, thereby filling the gap in the literature. Next to this, despite employing validation strategies such as peer reviews and multi-author coding while analysing data from the interviews, a certain degree of bias could be present in this study due to its qualitative nature.

Lastly, quantitative data was collected by recruiting participants from only one out of the 171 food banks in the Netherlands. Whether food bank recipients across the country share the same inclination towards eating UWC remains unknown. The respondents were only provided with the definition of unwanted catch. Additional information such as fish species and condition (processed, unprocessed), and taste tests might yield different results. However, prior to this research, it was not known whether the food banks and their recipients were inclined to receive more fish. Their attitude concerning unwanted catch had never been studied either. Results presented in this study open the doors to further enquiry on this issue.

5. Conclusions

This study explored the possibility of including more fish in Dutch food bank parcels by utilising unwanted catch. It did this, firstly, by gauging whether food bank recipients would be interested in such fish. This was followed by an analysis of relevant stakeholders' opinions on such an initiative. By considering unwanted catch utilisation, fish shortage in Dutch food bank parcels, and food waste valorisation together, this paper paves the way for a better understanding of all three issues and their possible interconnectedness.

Currently, several economic, legislative, social, and logistical barriers stand in the way of donating unwanted catch to food banks. However, given the European Commission's, and in turn the Dutch government's, strong focus on achieving the SDGs, change is possible. As highlighted in this paper, sustainably utilising unwanted catch and reducing food waste are often not viewed by policymakers as interconnected issues. Through this study, we aim to challenge this narrative. Legislative change, sufficient funding, and industry-led initiatives such as voluntary agreements can make donating safe-to-consume surplus fish to food banks a reality in the near future. Such an initiative could potentially improve health, accommodate the increasing demand for sustainable protein, and prevent wastage of valuable marine resources.

Author Contributions: Data curation: C.H., E.L., J.D., L.B., M.L., M.A.D.C., M.R.; Formal analysis: A.d.B., J.D., L.B., M.R.; Investigation: C.H., E.L., L.B., M.L., M.A.D.C., M.R.; Methodology: M.R.; Supervision: A.d.B., A.B.; Writing—original draft: M.R.; Writing—review & editing: A.d.B., A.B., J.D., L.B. All authors have read and agreed to the published version of the manuscript.

Funding: This research did not receive any specific grant from funding agencies in the public, commercial, or not-for-profit sectors.

Institutional Review Board Statement: The study was conducted according to the guidelines of the Declaration of Helsinki. The protocol was reviewed and approved by an internal ethics committee of Maastricht University, University College Venlo.

Informed Consent Statement: Participants involved in this study consented to their involvement by signing a declaration of informed consent. The questionnaires, both digital and physical, asked for survey participants' consent.

Data Availability Statement: The corresponding author can be contacted to request a copy of the code book.

Acknowledgments: The authors would like to thank all study participants and the volunteers of Food Bank Venlo for their cooperation. The feedback provided by Frits van Osch is greatly appreciated by all authors. This research has been made possible with the support of the Dutch Province of Limburg.

Conflicts of Interest: The authors declare that there are no conflict of interest.

References

1. Reid, D.G.; Calderwood, J.; Afonso, P.; Bourdaud, P.; Fauconnet, L.; González-Irusta, J.M.; Mortensen, L.O.; Ordines, F.; Lehuta, S.; Pawlowski, L.; et al. *The Best Way to Reduce Discards Is by Not Catching Them! In The European Landing Obligation: Reducing Discards in Complex, Multi-Species and Multi-Jurisdictional Fisheries*; Uhlmann, S.S., Ulrich, C., Kennelly, S.J., Eds.; Springer International Publishing: Cham, Switzerland, 2019; pp. 257–278.
2. Karp, W.A.; Breen, M.; Borges, L.; Fitzpatrick, M.; Kennelly, S.J.; Kolding, J.; Nielsen, K.N.; Viðarsson, J.R.; Cocas, L.; Leadbitter, D. Strategies used throughout the world to manage fisheries discards–Lessons for implementation of the EU Landing Obligation. In *The European Landing Obligation: Reducing Discards in Complex, Multi-Species and Multi-Jurisdictional Fisheries*; Uhlmann, S.S., Ulrich, C., Kennelly, S.J., Eds.; Springer International Publishing: Cham, Switzerland, 2019; pp. 3–26. ISBN 978-3-030-03307-1.
3. Henchion, M.; Hayes, M.; Mullen, A.M.; Fenelon, M.; Tiwari, B. Future Protein Supply and Demand: Strategies and Factors Influencing a Sustainable Equilibrium. *Foods* **2017**, *6*, 53. [CrossRef] [PubMed]

4.   Centraal Bureau voor de Statistiek. Coronacrisis Leidt Tot Ongekende Daling Aantal Banen. Available online: https://www.cbs. nl/nl-nl/nieuws/2020/33/coronacrisis-leidt-tot-ongekende-daling-aantal-banen (accessed on 22 March 2021).
5.   Voedselbanken Nederland. Feiten En Cijfers Voedselbanken Nederland 2020. Available online: https://voedselbankennederland. nl/wp-content/uploads/2020/03/DEF-Feiten-en-Cijfersper31-12-2019-.pdf (accessed on 19 October 2021).
6.   Bazerghi, C.; McKay, F.H.; Dunn, M. The Role of Food Banks in Addressing Food Insecurity: A Systematic Review. *J. Community Health* **2016**, *41*, 732–740. [CrossRef] [PubMed]
7.   Neter, J.E.; Dijkstra, S.C.; Dekkers, A.L.M.; Ocké, M.C.; Visser, M.; Brouwer, I.A. Dutch food bank recipients have poorer dietary intakes than the general and low-socioeconomic status Dutch adult population. *Eur. J. Nutr.* **2017**, *57*, 2747–2758. [CrossRef] [PubMed]
8.   Eurostat. Sustainable Development in the European Union: Monitoring Report on Progress towards the SDGs in an EU Context; Luxembourg. 2021. Available online: https://ec.europa.eu/eurostat/documents/3217494/12878705/KS-03-21-096-EN-N.pdf/ 8f9812e6-1aaa-7823-928f-03d8dd74df4f?t=1623741433852 (accessed on 19 October 2021).
9.   Cashion, T.; Le Manach, F.; Zeller, D.; Pauly, D. Most fish destined for fishmeal production are food-grade fish. *Fish Fish.* **2017**, *18*, 837–844. [CrossRef]
10.   Papargyropoulou, E.; Lozano, R.; Steinberger, J.; Wright, N.; bin Ujang, Z. The food waste hierarchy as a framework for the management of food surplus and food waste. *J. Clean. Prod.* **2014**, *76*, 106–115. [CrossRef]
11.   Iñarra, B.; Bald, C.; Cebrián, M.; Antelo, L.T.; Franco-Uría, A.; Vázquez, J.A.; Pérez-Martín, R.I.; Zufía, J. *What to Do with Unwanted Catches: Valorisation Options and Selection Strategies*; Springer: Cham, Switzerland, 2018; pp. 333–359. [CrossRef]
12.   Larsen, E.; Iñarra, B.; Peral, I. Economic Feasibility Study for the Best Use of Unavoidable Unwanted Catches, Avoiding Creating Incentives to the Fisheries; DiscardLess. 2019. Available online: https://docs.google.com/viewerng/viewer?url=http: //www.discardless.eu/media/results/DiscardLess_Deliverable_-_D6-4_27February2019.pdf (accessed on 19 October 2021).
13.   Blanco, M.; Domínguez-Timón, F.; Perez-Martin, R.; Fraguas, J.; Ramos-Ariza, P.; Vázquez, J.A.; Borderías, A.J.; Moreno, H.M. Valorization of recurrently discarded fish species in trawler fisheries in North-West Spain. *J. Food Sci. Technol.* **2018**, *55*, 4477–4484. [CrossRef] [PubMed]
14.   Iñarra, B.; Bald, C.; Cebrián, M.; Peral, I.; Llorente, R.; Zufía, J. Evaluation of unavoidable unwanted catches valorisation options: The Bay of Biscay case study. *Mar. Policy* **2019**, *116*, 103680. [CrossRef]
15.   Minced Fish from Unwanted By-Catch. Available online: https://webgate.ec.europa.eu/fpfis/cms/farnet2/on-the-ground/ good-practice/short-stories/minced-fish-unwanted-catch_en (accessed on 19 October 2021).
16.   Watson, J.T.; Stram, D.L.; Harmon, J. Mitigating Seafood Waste Through a Bycatch Donation Program. *Front. Mar. Sci.* **2020**, *7*, 923. [CrossRef]
17.   Flick, U.; von Kardoff, E.; Steinke, I. *A Companion to Qualitative Research*; SAGE: Thousand Oaks, CA, USA, 2004; ISBN 978-1-84860-523-7.
18.   Glaser, B.G.; Strauss, A.L. *Discovery of Grounded Theory: Strategies for Qualitative Research*; Routledge: New York, NY, USA, 1967; ISBN 978-1-351-52216-8.
19.   Saldana, J. *The Coding Manual for Qualitative Researchers*; SAGE: Thouand Oaks, CA, USA, 2021; ISBN 978-1-5297-5599-2.
20.   van Rossum, C.; Buurma-Rethans, E.; Vennemann, F.; Beukers, M.; Brants, H.; de Boer, E.; Ocké, M. The Diet of the Dutch: Results of the First Two Years of the Dutch National Food Consumption Survey 2012–2016; Rijksinstituut voor Volksgezondheid en Milieu RIVM. 2016. Available online: https://www.rivm.nl/bibliotheek/rapporten/2016-0082.pdf (accessed on 19 October 2021).
21.   Djamel, R.; Jose Maria, G. Valorisation of Food Surpluses and Side-Flows and Citizens' Understanding: REFRESH Deliverable D1.7. 2018. Available online: https://eu-refresh.org/sites/default/files/Food%20waste%20valorisation%20and%20citizens% 20understanding%20_Final%20report.pdf (accessed on 19 October 2021).
22.   Stichting De Noordzee. Aandlandplicht en Ongewenste Bijvangst: De Meest Gestelde Vragen. Available online: https://www. noordzee.nl/aandlandplicht-en-ongewenste-bijvangst-de-meestgestelde-vragen (accessed on 29 July 2021).
23.   Regulation (EU) No 1380/2013 of the European Parliament and of the Council of 11 December 2013 on the Common Fisheries Policy, Amending Council Regulations (EC) No 1954/2003 and (EC) No 1224/2009 and Repealing Council Regulations (EC) No 2371/2002 and (EC) No 639/2004 and Council Decision 2004/585/EC. 2013. Available online: https://eur-lex.europa.eu/ LexUriServ/LexUriServ.do?uri=OJ:L:2013:354:0022:0061:EN:PDF (accessed on 19 October 2021).
24.   Hedley, C.; Catchpole, T.; Santos, A.R. *The Landing Obligation and Its Implications on the Control of Fisheries*; The European Parliament: Brussels, Belgium, 2015; p. 122. Available online: https://www.europarl.europa.eu/RegData/etudes/STUD/2015/563381/ IPOL_STU(2015)563381_EN.pdf (accessed on 19 October 2021).
25.   De Vos, B.; Döring, R.; Aranda, M.; Buisman, F.; Frangoudes, K.; Goti, L.; Macher, C.; Maravelias, C.; Murillas-Maza, A.; van der Valk, O.; et al. New modes of fisheries governance: Implementation of the landing obligation in four European countries. *Mar. Policy* **2016**, *64*, 1–8. [CrossRef]
26.   Keuringsdienst van Waarde: Bijvangst. Available online: https://www.npo3.nl/keuringsdienst-van-waarde/08-02-2018/KN_16 96748 (accessed on 19 October 2021).
27.   Maynou, F.; Kraus, G.; Pinello, D.; Accadia, P.; Sabatella, E.; Spinadin, M. Handling, storage, transport and utilization of unwanted catches: MINOU Deliverable 2.19. Available online: http://minouw-project.eu/wp-content/uploads/2018/11/D2.19-Handling-Storage-Transport.pdf (accessed on 19 October 2021).

28. Voedselbank Nederland. Missie en Visie Voedselbanken. Available online: http://voedselbankennederland.nl/wat-we-doen/missie-en-visie-voedselbanken (accessed on 2 August 2021).
29. Buisman, E.; van Oostenbrugge, H.; Beukers, R. Economische Effecten van Een Aanlandplicht Voor de Nederlandse Visserij; LEI Wageningen UR: Den Haag. 2013. Available online: https://edepot.wur.nl/283011 (accessed on 19 October 2021).
30. Olsen, S.O.; Scholderer, J.; Brunsø, K.; Verbeke, W. Exploring the relationship between convenience and fish consumption: A cross-cultural study. *Appetite* **2007**, *49*, 84–91. [CrossRef] [PubMed]
31. EUMOFA. The EU Fish Market: 2020 Edition; Luxembourg. 2020. Available online: https://www.eumofa.eu/documents/20178/415635/EN_The+EU+fish+market_2020.pdf (accessed on 19 October 2021).
32. Bellido, J.M.; García-Rodriguez, M.; García-Jiménez, T.; González-Aguilar, M.; Carbonell-Quetglas, A. Could the obligation to land undersized individuals increase the black market for juveniles: Evidence from the Mediterranean? *Fish Fish.* **2016**, *18*, 185–194. [CrossRef]
33. Viðarsson, J.R.; Einarsson, M.I.; Larsen, E.P.; Valeiras, J.; Ragnarsson, S.Ö. Onboard and Vessel Layout Modifications. In *The European Landing Obligation: Reducing Discards in Complex, Multi-Species and Multi-Jurisdictional Fisheries*; Uhlmann, S.S., Ulrich, C., Kennelly, S.J., Eds.; Springer International Publishing: Cham, Switzerland, 2019; pp. 319–331.
34. Kraan, M.; Verweij, M. Implementing the Landing Obligation. An Analysis of the Gap Between Fishers and Policy Makers in the Netherlands. In *Collaborative Research in Fisheries: Co-creating Knowledge for Fisheries Governance in Europe*; Holm, P., Hadjimichael, M., Linke, S., Mackinson, S., Eds.; MARE Publication Series; Springer International Publishing: Cham, Switzerland, 2020; pp. 231–248. ISBN 978-3-030-26784-1.
35. Pastoors, M.A.; Buisman, E.; van Oostenbrugge, H.; Kraan, M.L.; van Beek, F.A.; Uhlmann, S.S.; van Overzee, H.M.J. Fasering Discard Ban: IMARES Wageningen UR. 2014. Available online: https://edepot.wur.nl/301924 (accessed on 19 October 2021).
36. Van Hoof, L.; Kraan, M.; Visser, N.M.; Avoyan, E.; Batsleer, J.; Trapman, B. Muddying the Waters of the Landing Obligation: How Multi-level Governance Structures Can Obscure Policy Implementation. In *The European Landing Obligation: Reducing Discards in Complex, Multi-Species and Multi-Jurisdictional Fisheries*; Uhlmann, S.S., Ulrich, C., Kennelly, S.J., Eds.; Springer International Publishing: Cham, Switzerland, 2019; pp. 179–196.
37. Borges, L. The unintended impact of the European discard ban. *ICES J. Mar. Sci.* **2020**, *78*, 134–141. [CrossRef]
38. European Parliament. P7_TA (2013)0117 Technical and Control Measures in the Skagerrak. 2013. Available online: https://eur-lex.europa.eu/legal-content/EN/TXT/PDF/?uri=CELEX:52013AP0117&from=EN (accessed on 19 October 2021).
39. European Committee on Fisheries. Report on the Proposal for a Regulation of the European Parliament and of the Council on Certain Technical and Control Measures in the Skagerrak and Amending Regulation (EC) No 850/98 and Regulation (EC) No 1342/2008. 2013. Available online: https://www.europarl.europa.eu/doceo/document/A-7-2013-0051_EN.html (accessed on 19 October 2021).
40. IMARES Wageningen UR. Discard Atlas of North Sea Fisheries; Wageningen. 2014. Available online: https://edepot.wur.nl/315708 (accessed on 19 October 2021).
41. European Court of Auditors Combating Food Waste: An Opportunity for the EU to Improve the Resource-Efficiency of the Food Supply Chain; Luxembourg. 2016, p. 74. Available online: https://www.eca.europa.eu/Lists/ECADocuments/SR16_34/SR_FOOD_WASTE_EN.pdf (accessed on 19 October 2021).
42. Vaqué, L.G. French and Italian Food Waste Legislation: An Example for Other EU Member States to Follow? *Eur. Food Feed Law Rev.* **2017**, *12*, 224–233.
43. Vittuari, M.; Politano, A.; Gaiani, S.; Canali, M.; Elander, M.; Aramyan, L.; Gheoldus, M.; Easteal, S.; Timmermans, T.; Bos-Brouwers, H. Review of EU Legislation and Policies with Implications on Food Waste: EU FUSIONS Final Report. 2015. Available online: https://www.eu-fusions.org/index.php/download?download=161:review-of-eu-legislation-and-policies-with-implications-on-food-waste (accessed on 19 October 2021).
44. Regulation (EU) 2021/1139 of the European Parliament and of the Council of 7 July 2021 Establishing the European Maritime, Fisheries and Aquaculture Fund and Amending Regulation (EU) 2017/1004. 2021. Available online: https://eur-lex.europa.eu/legal-content/EN/TXT/PDF/?uri=CELEX:32021R1139&from=EN (accessed on 19 October 2021).
45. European Commission. The European Green Deal; Brussels, Belgium. 2019. Available online: https://ec.europa.eu/info/sites/default/files/european-green-deal-communication_en.pdf (accessed on 19 October 2021).
46. European Commission European Maritime Fisheries and Aquaculture Fund (EMFAF). Available online: https://ec.europa.eu/oceans-and-fisheries/funding/emfaf_en (accessed on 12 August 2021).
47. Piras, S.; García Herrero, L.; Burgos, S.; Colin, F.; Gheoldus, M.; Ledoux, C.; Parfitt, J.; Jarosz, D.; Vittuari, M. Unfair Trading Practice Regulation and Voluntary Agreements Targeting Food Waste: EU REFRESH Deliverable 3.2. 2018. Available online: https://eu-refresh.org/sites/default/files/REFRESH_D3.2_UTPs%20and%20VAs%20targeting%20food%20waste_07.2018.pdf (accessed on 19 October 2021).
48. Van Putten, I.; Koopman, M.; Fleming, A.; Hobday, A.J.; Knuckey, I.; Zhou, S. Fresh eyes on an old issue: Demand-side barriers to a discard problem. *Fish. Res.* **2018**, *209*, 14–23. [CrossRef]

Article

Food Myths or Food Facts? Study about Perceptions and Knowledge in a Portuguese Sample

Sofia G. Florença [1,2,*] , Manuela Ferreira [2], Inês Lacerda [3] and Aline Maia [3]

[1] Faculty of Food and Nutrition Sciences, University of Porto, 4150-180 Porto, Portugal
[2] UICISA:E, Polytechnic Institute of Viseu, 3504-510 Viseu, Portugal; manuelaferreira@sc.ipv.pt
[3] ACES Dão-Lafões, Grouping of Health Centres of Dão-Lafões, 3514-511 Viseu, Portugal; IMLacerda@arscentro.min-saude.pt (I.L.); ADMaia@arscentro.min-saude.pt (A.M.)
* Correspondence: sofiaflorenca@outlook.com

Abstract: Food myths are nutritional concepts poorly justified or even contradict existing scientific evidence that individuals take as the truth. Knowledge in nutrition is an important tool in tackling misinformation and in the promotion of adequate food choices. This study aimed to investigate the beliefs and perceptions of a sample of the Portuguese population regarding a series of food myths and facts, evaluating, consequently, the level of knowledge and the main sources of information. The research was conducted on a sample of 503 participants, using a questionnaire disclosed online, by email, and social networks, between May and June of 2021. Thirty statements, some true and others false, were analyzed to assess people's perceptions. Based on the respondents' answers, a score was calculated for each statement, allowing to differentiate the correct (positive score) from incorrect (negative score) perceptions. The results showed that most statements obtained positive scores, corresponding to correct perceptions. Moreover, the level of knowledge was measured, being very high for 21.7% of the participants and high for 42.1%. The main sources where the participants acquire nutritional information are scientific journals (43.3%), website of the Portuguese General Health Office (DGS) (31.4%), and technical books (31.0%), which is concordant with the level of trust in these sources. Hence, it was concluded that, despite the levels of nutritional knowledge, there are still several food myths that need to be debunked, through the proper channels, in order to promote healthy, balanced, and adequate eating behaviors.

Keywords: food myths; sources of information; nutritional knowledge; health

Citation: Florença, S.G.; Ferreira, M.; Lacerda, I.; Maia, A. Food Myths or Food Facts? Study about Perceptions and Knowledge in a Portuguese Sample. *Foods* **2021**, *10*, 2746. https://doi.org/10.3390/foods10112746

Academic Editors: Theodoros Varzakas, António Raposo, Raquel Braz Assunção Botelho and Renata Puppin Zandonadi

Received: 7 October 2021
Accepted: 5 November 2021
Published: 9 November 2021

Publisher's Note: MDPI stays neutral with regard to jurisdictional claims in published maps and institutional affiliations.

1. Introduction

Food is one of the basic necessities of humans, being fundamental in the promotion of health and well-being, as well as the prevention of diseases [1]. However, even though food consumption has the main goal of satisfying a physiological need, hunger, other factors also determine what is the motivation, quantity, frequency, and food choice. Some of these factors are related to appetite, cost, accessibility, emotions, culture, and social interactions [2,3]. Food choices are, therefore, an overlap of several domains, from exact sciences, for example nutrition, to social policies and the individual behavior [1]. A work by Milošević et al. [4] carried out in six Balkan countries, based on the Food Choice Questionnaire, allowed identifying eight factors underlying food choice and five groups of consumers according to their motivational profiles. Hence, the authors found the participants distributed by groups depending on their food purchasing behavior and socioeconomic characteristics. Cunha et al. [5] published a review about the application of the Food Choice Questionnaire in different countries, and they concluded that different contexts influence people's responses and, therefore, even using the same instruments, the results might be differentiated among countries. Guiné et al. [6] also investigated the eating motivations in 16 countries and found them related to different types of factors:

health, emotions, price and availability, society and culture, environment and politics, and marketing and commercials.

Beliefs are one of the many factors that influence eating behaviors, arising from a set of concepts adopted by an individual as the truth. Food myths are beliefs; that is, nutritional concepts poorly justified or even contradicting existing scientific evidence that are, however, taken as the truth by the individual [2,7]. Many of the existing myths are born from the current misinformation, and it is this lack of trustworthy information and knowledge about nutrition that results, most of the time, in diseases such as the metabolic syndrome, hypertension, diabetes, atherosclerosis, and cardiovascular diseases [2]. However, can eating behavior really be influenced by nutritional knowledge? The main goal of numerous food education programs is to improve nutritional knowledge, contributing to the establishment and maintenance of healthy, balanced, and varied eating habits. A review by Worsley [7] highlighted the implications of the nutritional knowledge for public health, as follows: develop more proactive educational programs, attribute greater importance to the education of children and adults, consider people's personal food motivations, and continue developing research that allows better understanding consumer trends, and design appropriate interventions. According to another systematic review by Spronk et al. [8], nutritional knowledge is positively correlated, even if weakly, to food consumption, and so, the promotion of nutritional literacy should be considered a relevant nutritional policy to the modification of eating behaviors.

The lack of knowledge about certain food items can be related to their own perception and misperception of the information available. Hence, food items could be perceived as good when in fact, they are not, mainly due to cognitive errors. Marsola et al. [9] evaluated the perceptions of Brazilian consumers about the risks and benefits of different foods and how they perceived their related health effects. They concluded that income, education, and the absence of children are factors that contribute to increased perception of the health benefits as compared to the risks. Additionally, nutritional knowledge, is, many times, obtained through unreliable information sources [10]. A Canadian study has shown that, between 2004 and 2008, the primary sources of information about nutrition and food were food labels, followed by magazines/journals/books, colleagues/friends/family, internet, and health care workers. When evaluating the level of credibility provided by the sources, nutritionists and health professionals were the ones who obtained a better level of credibility [10]. Nutrition knowledge is an important tool in tackling misinformation and in the promotion of adequate eating behaviors. However, the information sources must be reliable and trustworthy, hence the primary role of health professionals in educating and monitoring the food habits of the population in a country, thus allowing the control of various chronic pathologies influenced by lifestyle [10]. According to the Global Burden of Disease report [11], from 2017, inadequate food habits are responsible for 11 million deaths every year, from which the main cause of death are diseases intrinsically linked to maladjusted eating behaviors, such as cardiovascular diseases, cancer, and type 2 diabetes. These data corroborates the existence of a new global pandemic of chronic diseases that are directly connected with food [11,12]. Hence, nutrition has assumed a prominent role in the prevention, control, treatment, and even regression of several pathologies that are influenced by eating habits and behaviors, being important as a part of a healthy and balanced lifestyle [12,13].

In the scientific literature studies, it is possible to find the importance of food or nutritional literacy for improved health [14–16], how this literacy can be achieved [17–19], and what factors contribute to increase food literacy in the general population or among specific groups [20–22]. However, the scientific literature is lacking research studies focused on the myths related to human food consumption. Hence, there is a need to deepen the research into the wrong perceptions people sometimes have about some aspects related with food and eating, as well as the nutrients and their roles for human health. In this context, it is proposed to contribute to bringing some highlight into the matter of what are food myths and what are facts scientifically supported. The aim of this research was to investigate the beliefs and perceptions of a sample of the Portuguese population regarding a series of food myths and facts, evaluating, consequently, the level of knowledge. The main sources of information concerning nutritional knowledge were also analyzed, as well as their level of trust.

2. Materials and Methods

2.1. Instrument and Data Collection

The instrument used in this study was a questionnaire developed for this research. After preparing the questionnaire, and before its application, it was approved by the Ethical Commission at the Faculty of Food and Nutrition Sciences of University of Porto (reference N° 28/2021/CEFCNAUP/2021).

Since there were no previous studies related to this particular type of work to identify food myths in the literature, the questionnaire was based on a search previously made on the internet and blogs about nutrition to identify the main problematic topics. Additionally, the input from patients at appointments with nutritionists was also considered to design the questions. The questionnaire consisted of four parts. The first section collected data on sociodemographic data: age, sex, highest level of education completed, living environment, marital status, and area of professional activity. In the second section, some questions were included to obtain information regarding anthropometric data and lifestyle choices: height and weight (auto reported data), whether the participant is responsible for buying the food he eats, practicing a balanced diet, or the practice of specific food regimes. In the third section, thirty sentences were included, some of them true and others false (Table 1), to evaluate the level of concordance concerning some food myths and facts, assessing the sample's perceptions on the subjects on a 5 point Likert scale: 1—strongly disagree, 2—disagree, 3—neutral, 4—agree, 5—strongly agree). Finally, the fourth section of the questionnaire consisted in the evaluation of the main sources of information and their level of trust: YouTube channels of influencers, blogs from nutritionists, website of the Portuguese General Health Office (DGS), website of the World Health Organization (WHO), shops of natural products, scientific journals, webpages, social networks, television, technical books, pharmacies, health centers, or appointments with medical doctors or nutritionists. The frequency of use of the different sources of information about nutritional facts was assessed in the scale: never, few times, many times. The scale for assessing the level of trust was: 1—no trust at all, 2—some trust, 3—much trust, 4—full trust.

The data were collected between May and June of 2021, and applied to a sample, chosen according to the facility of recruitment and willingness to participate. This survey was applied online through the Google Forms software and disclosed to people by different tools, such as email and social networks. Participation in the study was voluntary and only allowed for adults aged 18 years or older, but no upper age limit was established. All respondents gave their informed consent to participate and the confidentiality of the answers provided was guaranteed. The inclusion criteria to partake in the study, apart from age, was the access to internet, email, social networks, and a computer, as well as the skills necessary to be able to access the questionnaire. It guaranteed confidentiality of the responses by all participants, and these consented freely in their participation. Furthermore, they could abandon the questionnaire at any time without submitting the form. The design and application of the questionnaire has respected all ethical issues.

Considering the strategy used to collect the data, which occurred under COVID-19 limitations, the questionnaire was applied to a convenience sample, recruited according to facility of contact and disposition to take part in the research. Although convenience samples have some limitations, such as not allowing direct generalization of the conclusions from the sample to the whole population, they can also have some benefits, namely easiness of recruitment. Therefore they have been reported as a good tool to carry out research with exploratory nature [2,3]. Despite being a convenience sample, adequacy of the sample size was calculated, to serve as an indicator during data collection. For this, some assumptions were considered: confidence interval = 90%; Z score = 1.645; power of the test = 95% (minimum acceptable probability of preventing type II error = 0.05) [4,5]; Portuguese population in 2019 (the latest year available when the data collection started) = 10,283,822 people [6]: assumed that ~7.5 million were adults and the target population was 25% = 1875 thousand. Considering these conditions, calculation of the minimum sample size resulted in 203 adults [7,8]. The number of valid questionnaires obtained in this survey was 503, over the minimum number previously calculated.

Table 1. Myths and facts used in the questionnaire.

Code	Statement	Nature
S1	Drinking water during meals, contributes to weight gain.	Myth
S2	The digestion process begins in the mouth.	Fact
S3	Fruit should be eaten before meals.	Myth
S4	Egg consumption increases blood cholesterol.	Myth
S5	Drinking milk is bad for health.	Myth
S6	Eating carbohydrates at night leads to an increase in weight gain.	Myth
S7	Fat is important to the human body.	Fact
S8	Fruit should be eaten after meals.	Myth
S9	Fiber intake is important for normal bowel function.	Fact
S10	Gluten-free foods are better for health and should, there-fore, be adopted by all.	Myth
S11	Cheese consumption is bad for memory.	Myth
S12	Coconut oil is healthier than olive oil.	Myth
S13	Lactose-free foods are better for health and should, there-fore, be adopted by all.	Myth
S14	Children have different nutritional needs than those for adults.	Fact
S15	Fruits and vegetables do not contribute to weight gain.	Myth
S16	Normal potatoes are more caloric than sweet potatoes.	Myth
S17	Diet should be adapted to a person's blood group.	Myth
S18	Not having a balanced and varied diet can lead to the development of multiple diseases.	Fact
S19	The alkaline diet allows balancing the acidity in the blood.	Myth
S20	Drinking, while fasting, a glass of water with lemon helps in weight loss.	Myth
S21	Inadequate eating habits are the third risk factor for the loss of years of healthy life.	Fact
S22	Ingesting high amounts of protein helps in the faster formation of muscles.	Myth
S23	Pregnant women should be eating for two.	Myth
S24	Cold water should not be drunk.	Myth
S25	The day should always start with breakfast.	Fact
S26	Water is essential to the normal function of all organs.	Fact
S27	Soy milk is healthier than cow's milk.	Myth
S28	Orange should not be eaten at the same time as milk or yogurt.	Myth
S29	Dairy products should be consumed in between two and three portions per day.	Fact
S30	All food additives (E's) are harmful to health.	Myth

2.2. Data Analysis

The statistical analysis was conducted using Excel 2016 and SPSS software V25 (IBM, Inc., Armonk, NY, USA). For treatment of data, basic statistical tools were used, such as frequencies, percentiles, mean values, and standard deviation.

For the data treatment, variable age was categorized into the following groups: young adults (18–30 years), middle-aged adults (31–50 years), senior adults (51–65 years), and elderly (aged 66 years or more). The body mass index (BMI) was calculated from the collected data as weight divided by squared height. The BMI was then classified according to the following categories of the WHO: underweight (<18.5 kg/m^2), normal weight (18.5–24.9 kg/m^2), overweight (25–29.9 kg/m^2), and obesity (≥ 30 kg/m^2). For those with BMI over 30 kg/m^2, there are still differentiated classes of obesity; however, for the objectives of this study, these four classes were considered satisfactory enough [9–11]. The normality of the distribution was evaluated for some variables (age and BMI), testing the null hypothesis (H0: The data follow a normal distribution).

Spearman correlations (r_s) were used to measure the association between some variables. The values of r_s varied between -1 and 1, with negative values corresponding to inverse associations between the variables. The reference absolute values considered were: $r_s = 0$—no correlation, $r_s \in {]}0.0, 0.2{[}$—very weak correlation, $r_s \in [0.2, 0.4{[}$—weak correlation, $r_s \in [0.4, 0.6{[}$—moderate correlation, $r_s \in [0.6, 0.8{[}$—strong correlation, $r_s \in [0.8, 1.0{[}$—very strong correlation, $r_s = 1$—perfect correlation [12–14].

The agreement with the statements included in the questionnaire was used to calculate the scores for each item. However, before this, all of the incorrect statements (myths) were reversed to obtain the 30 items in the same scale of correct information. The variables were recoded from the 5-point Likert scale into the following scale: -1 = fully incorrect perception, -0.5 = partially incorrect perception, 0 = no opinion, 0.5 = partially correct perception, and 1 = fully correct perception. Then the scores for each of the 30 items were calculated as the sum of the values for each of the 503 participants. Hence, for each item, the sum score for perceptions varied between -503 to 503.

Additionally, the level of knowledge was calculated for each participant, as the sum of scores obtained for all items, after the previous operations of reversing the incorrect items and recoding into the scale from -1 to 1. In this way, for each participant, the sum varied between -30 and 30, and the values were then recoded into a categorical variable as class of knowledge: very low $[-30;-15]$, low $]-15;0]$, medium $]0;10]$, high $]10;20]$, and very high knowledge $]20;30]$. This variable was submitted to tree classification analysis for evaluation of the relative importance of the possible influential variables: sociodemographic variables (sex, age group, education, living environment, and marital status), other variables (job-nutrition, job-food, job-sport, job-health, BMI class, practice of balanced diet). A classification and regression tree (CRT) algorithm, with cross-validation and with minimum change in improvement of 0.0001, was used for treatment of the data, considering a limit of 5 levels and with a minimum number of cases in parent and child nodes equal to 30 and 15, respectively [15].

The variables frequency of use, of the different sources of information on nutritional facts and level of trust in those sources, were correlated by Spearman correlations.

The level of significance considered in the analyses was 5% ($p < 0.05$).

3. Results

3.1. Sample Characterization

In this survey, half of the participants were less than 44 years, with only 5% of the respondents under 21 years old, and 5% over 65 years old. It was further observed that 25% of participants were under 29 years old and 75% were over 52 years old. It was observed that age did not follow a normal distribution ($p < 0.0005$).

Regarding the sociodemographic characteristics (Table 2), the vast majority were middle-aged adults (31–50 years) (42.9%), women (77.8%), with university education (46.1%), living in urban environment (68.4%), and living in union or co-habiting (57.1%).

3.2. Lifestyle and Anthropometric Data

In this sample, most participants worked in professional areas related to nutrition (28.2%), health (33.4%), and food (34.8%) (Table 3). Of the people who answered the survey, 35.4% practiced caloric restriction and 8.5% flexitarianism (Table 3).

The acquisition of food was also questioned in this survey, with 88.3% of the participants being responsible for the purchase of the foods they consume.

The frequency of how often the respondents believe to be practicing a balanced and healthy diet was investigated based on the scale: always, frequently, sometimes, rarely, and never. The results showed that the majority, 52.5%, answered frequently, followed by 37% who answered sometimes, with only 5.6% acknowledging always, 4.4% rarely, and 0.6% never.

Table 2. Sociodemographic characterization of the sample at study.

	Variable	N	%
Age group	Young adults (18–30 years)	137	27.2
	Middle aged adults (31–50 years)	216	42.9
	Senior adults (51–65 years)	126	25.0
	Elderly ($\geq$66 years)	24	4.8
Sex	Women	390	77.8
	Men	111	22.2
Education level	Basic	10	2.0
	Secondary	87	17.3
	University	232	46.1
	Post-graduation	174	34.6
Living environment	Urban	287	68.4
	Suburban	57	11.3
	Rural	102	20.3
Marital status	Single	172	34.2
	Married/Union	287	57.1
	Divorced/Separate	38	7.6
	Widowed	6	1.2
Total		503	100

Table 3. Anthropometric and behavioral characterization of the sample at study.

Profession	N	%	Diet	N	%
Other	197	39.2	Calorie Restriction	178	35.4
Food	175	34.8	Flexitarianism	43	8.5
Health	168	33.4	Vegetarianism	22	4.4
Nutrition	142	28.2	Fruitarianism	21	4.2
Biology	94	18.7	Crudism	10	2.0
Agriculture	81	16.1	Veganism	8	1.6
Environment	77	15.3	Religious Restriction	7	1.4
Sports	40	8.0			

BMI percentile	BMI (kg/m^2)	BMI Class [1]	%
5%	19.1	Underweight	3.4
10%	20.0	Normal weight	57.5
25%	21.4	Overweight	32.6
50%	23.7	Obesity	6.6
75%	26.3		
90%	28.9		
95%	31.3		

[1] Underweight (BMI $<$ 18.5 kg/m^2), normal weight (18.5 $\leq$ BMI $<$ 25.0 kg/m^2), overweight (25.0 $\leq$ BMI $<$ 30.0 kg/m^2), obesity (BMI $\geq$ 30 kg/m^2).

Some data were collected regarding the height and weight of those who partook in the study, in order to calculate BMI. It was found that BMI did not follow a normal distribution ($p < 0.0005$). There were 5% of the participants with a BMI under 19.1 kg/m^2 and 5% with a BMI over 31.3 kg/m^2 (Table 3). For this sample, 57.5% had normal weight (18.5–24.9 kg/m^2) and 32.6% were overweight (25–29.9 kg/m^2) (Table 3). A significant ($p < 0.01$) negative correlation was found between BMI and the self-reported practice of a balanced and healthy diet; however, the value was low ($r_s = -0.271$), meaning that the correlation was weak.

3.3. Perceptions Regarding Food Myths and Facts

Table 4 shows the items included in the survey related to different food myths and facts, and the corresponding sum score, for which negative values indicate wrong perception and positive values indicate correct perception. Items S26, S9, S11, and S18 exhibit the highest scores (484.5, 445.0, 441.5, and 437.0, respectively), demonstrating that these statements are known to be true by the participants. However, some items did not have such a high score, meaning that people did not know about these topics or had false beliefs about them. The items with the lowest scores were S6 (-10), S8 (25.5), S30 (37.0), and S3 (38.0).

Table 4. Item scores regarding the perception for each myth/fact, in descending order.

Items	Mean $\pm$ SD [1]	Sum Score [2]
S26. Water is essential to the normal function of all organs.	0.96 ± 0.18	484.5
S9. Fiber intake is important for normal bowel function.	0.89 ± 0.25	445.0
S11. Cheese consumption is bad for memory.	0.88 ± 0.31	441.5
S18. Not having a balanced and varied diet can lead to the development of multiple diseases.	0.87 ± 0.30	437.0
S23. Pregnant women should be eating for two.	0.83 ± 0.33	418.0
S14. Children have different nutritional needs than those for adults.	0.82 ± 0.33	410.5
S2. The digestion process begins in the mouth.	0.77 ± 0.49	385.0
S1. Drinking water during meals, contributes to weight gain.	0.70 ± 0.50	351.5
S11. Cheese consumption is bad for memory.	0.88 ± 0.31	441.5
S18. Not having a balanced and varied diet can lead to the development of multiple diseases.	0.87 ± 0.30	437.0
S23. Pregnant women should be eating for two.	0.83 ± 0.33	418.0
S14. Children have different nutritional needs than those for adults.	0.82 ± 0.33	410.5
S2. The digestion process begins in the mouth.	0.77 ± 0.49	385.0
S1. Drinking water during meals, contributes to weight gain.	0.70 ± 0.50	351.5
S25. The day should always start with breakfast.	0.68 ± 0.52	341.5
S21. Inadequate eating habits are the third risk factor for the loss of years of healthy life.	0.66 ± 0.43	330.5
S7. Fat is important to the human body.	0.54 ± 0.53	269.5
S17. Diet should be adapted to a person's blood group.	0.52 ± 0.58	259.0
S12. Coconut oil is healthier than olive oil.	0.51 ± 0.56	257.0
S13. Lactose-free foods are better for health and should, therefore, be adopted by all.	0.50 ± 0.59	250.5
S5. Drinking milk is bad for health.	0.47 ± 0.59	237.5
S4. Egg consumption increases blood cholesterol.	0.43 ± 0.54	217.5
S10. Gluten-free foods are better for health and should, therefore, be adopted by all.	0.41 ± 0.60	206.5
S27. Soy milk is healthier than cow's milk.	0.35 ± 0.60	175.0
S20. Drinking, while fasting, a glass of water with lemon helps in weight loss.	0.34 ± 0.61	172.5
S28. Orange should not be eaten at the same time as milk or yogurt.	0.32 ± 0.62	160.5
S19. The alkaline diet allows balancing the acidity in the blood.	0.24 ± 0.59	118.5
S15. Fruits and vegetables do not contribute to weight gain.	0.24 ± 0.59	118.0
S22. Ingesting high amounts of protein helps in the faster formation of muscles.	0.14 ± 0.57	68.5
S29. Dairy products should be consumed in between two and three portions per day.	0.13 ± 0.58	66.5
S24. Cold water should not be drunk.	0.11 ± 0.66	57.0
S16. Normal potatoes are more caloric than sweet potatoes.	0.10 ± 0.68	50.0
S3. Fruit should be eaten before meals.	0.08 ± 0.62	38.0
S30. All food additives (E's) are harmful to health.	0.07 ± 0.65	37.0
S8. Fruit should be eaten after meals.	0.05 ± 0.58	25.5
S6. Eating carbohydrates at night leads to an increase in weight gain.	-0.02 ± 0.71	-10.0

[1] Values given as mean and standard deviation, on the scale from -1 (fully incorrect perception) to 1 (fully correct perception). [2] The sum score ranges from -503 (fully incorrect perception) to 503 (fully correct perception).

Table A1 in Appendix A presents the correlations between perceptions regarding food myths/facts and the items (variables S1 to S30). Concerning the relation between age and perception of food myths/facts, some weak positive correlations were found to be significant ($p < 0.01$) for items S6 ($r_s = 0.309$), S3 ($r_s = 0.361$) and S1 ($r_s = 0.286$). For variable sex, there were fewer significant correlations ($p < 0.01$), being that the highest value found was for item S23 ($r_s = 0.208$), demonstrating that the correlation is weak. Variables education and environment have very weak influence on the perceptions about food myths and facts. Considering variable marital status, there are some positive significant ($p < 0.01$)

correlations with items S6 (r_s = 0.266) and S3 (r_s = 0.259), even though the associations are weak due to the low values. Being professionally connected to areas, such as nutrition, food, sports, and health also influences the perception cornering food myths and facts. Some significant ($p < 0.01$) weak positive correlations were found between nutrition and items S6 (r_s = 0.464), S19 (r_s = 0.410), and S20 (r_s = 0.396). On the other hand, some significant ($p < 0.01$) weak negative correlations were found between nutrition as items such as S29 (r_s = −0.389). As regards to variable food, significant ($p < 0.01$) weak positive correlations were found for items S30 (r_s = 0.354), S13 (r_s = 0.331), and S6 (r_s = 0.326). With respect to sports, a significant ($p < 0.01$) weak positive correlation was found for item S26 (r_s = 0.213). Respecting health, some significant ($p < 0.01$) weak positive correlations were found for items S6 (r_s = 0.296), S17 (r_s = 0.282) and S13 (r_s = 0.279). For variable balance diet, there were few significant correlations with perception, and the ones that exist are very weak. Regarding the relation between BMI class and perception of food myths/facts, some significant ($p < 0.01$) weak positive correlations were found such as for item S20 (r_s = 0.201).

3.4. Level of Knowledge

Concerning the level of knowledge about the 30 questions asked related to food myths and facts, a vast majority of the participants had a high (42.1%) or medium (35.6%) level of knowledge, with some showing a very high level of knowledge (21.7%), a few showing a low level (0.6%), and no respondents showing a very low level.

Figure 1 presents the classification tree for level of knowledge as a function of the sociodemographic variables. The obtained tree is four levels deep, with 13 nodes, from which, 7 are terminal. The risk estimate was 0.535 for re-substitution and 0.586 for cross-validation, with standard errors of 0.022 in both cases. According to the results obtained, the first discriminant variable was age, indicating that young adults have higher level of knowledge (30.7% of very high knowledge) in comparison with middle aged adults, senior adults and elderly (18.3% of very high knowledge). For young adults, the next discriminating variable was sex, showing that female participants are better-informed (35.0% of very high knowledge) than male participants (5.0% of very high knowledge). The education level is the discriminating variable following gender, in which respondents with basic, secondary and university education have superior level of knowledge (38.2% of very high knowledge) than postgraduate respondents (13.3% of very high knowledge).

The tree in Figure 2 shows the influence of other variables that characterize the sample on the level of knowledge. This tree also has 4 levels and 15 nodes—8 of them terminal. The risk estimate was 0.507 for re-substitution and 0.583 for cross-validation, with standard errors of 0.022 in both cases. The results show that, for the first discriminant variable, participants whose jobs were related to nutrition had very high levels of knowledge (42.2%). The following discriminant was BMI class, in which respondents with overweight or obesity had lower levels of knowledge (27.0% very high) than participants underweight or with normal weight (49.3%). For underweight and normal weight people, the next discriminating variable was professional area of health, showing that those who worked in the health sector had higher levels of knowledge (52.5% very high). A balanced diet was the next discriminant variable, indicating that those participants who always or frequently practiced it had very high levels of knowledge (56.8%) when compared to those who never, rarely, or only sometimes practiced it.

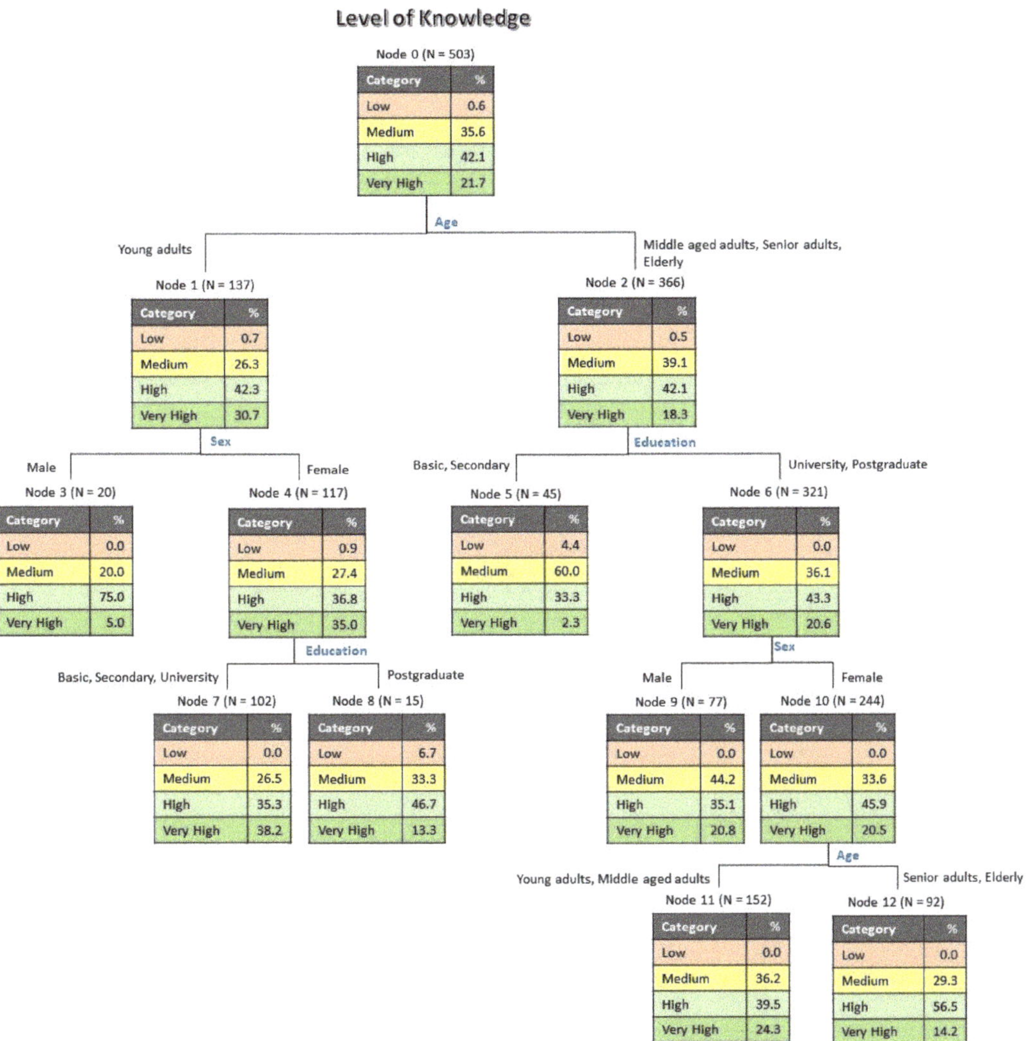

Figure 1. Tree classification for level of knowledge as influenced by sociodemographic variables.

3.5. Sources of Information

The main sources of information about nutritional facts were also evaluated. The most frequently used sources were scientific journals (34.4%), DGS website (31.4%), technical books (31.0%), appointments with professionals (medical doctors or nutritionists) (29.2%), and the WHO website (27.2%) (Table 5). Among the least used sources were YouTube channels of influencers (3.0%), natural food stores (7.4%), and pharmacies (8.3%).

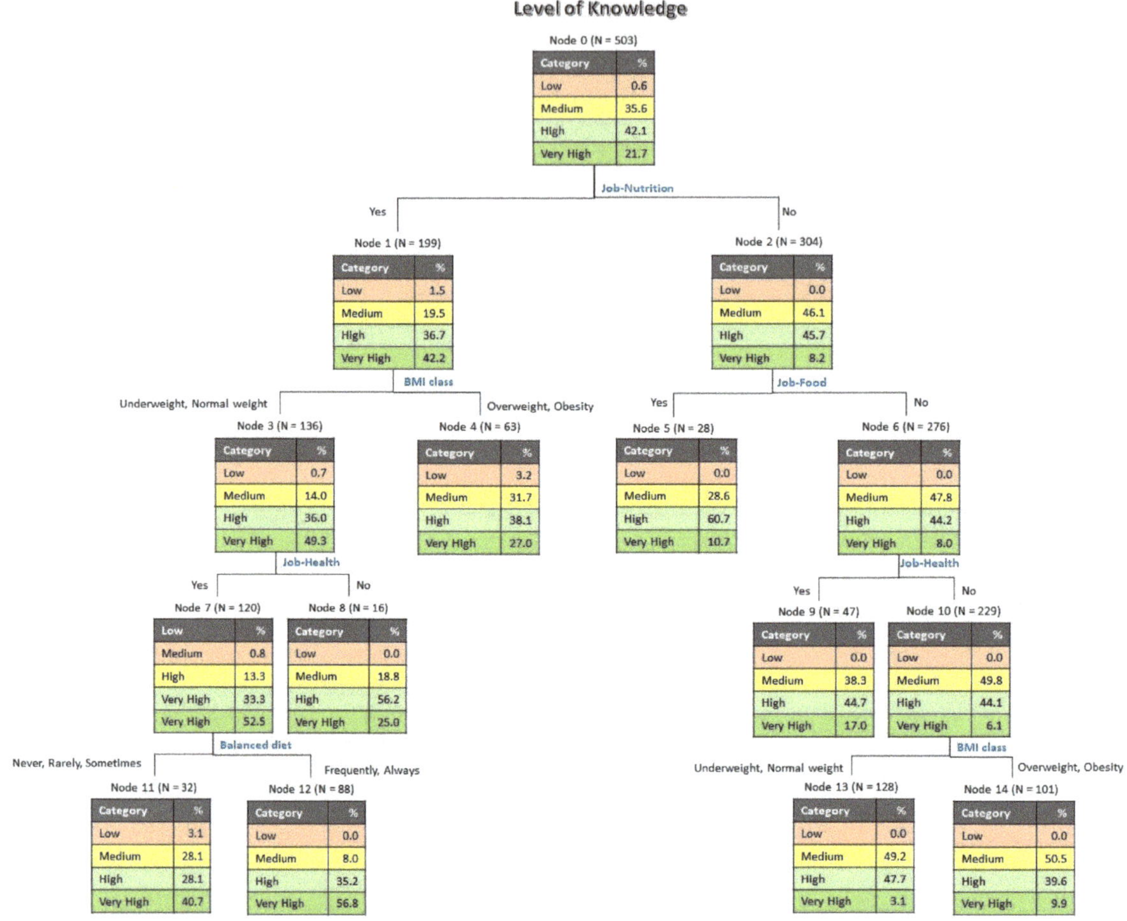

Figure 2. Tree classification for level of knowledge as influenced by other variables, professional areas, BMI, and practice of balanced diet.

The sources with the highest levels of trust were doctor/nutritionist appointments (44.7%), WHO website (42.5%), and DGS website (36.2%). Significant ($p < 0.01$) positive correlations were found between the utilization of information sources and the level of trust, such as, for example, social media ($r_s = 0.570$), YouTube channels of influencers ($r_s = 0.568$), and natural food stores ($r_s = 0.539$) (Table 5).

Some significant ($p < 0.01$) negative correlations were found between knowledge and the utilization of some sources of information, such as natural food stores ($r_s = -0.355$), television ($r_s = -0.285$), and influencers' YouTube channels ($r_s = -0.240$). Nevertheless, the values are low, which means that the correlations are weak (Table 6). On the other hand, significant ($p < 0.01$) positive correlations were found between knowledge and the use of some sources of information, such as DGS website ($r_s = 0.320$), WHO website ($r_s = 0.357$), and scientific journals ($r_s = 0.385$), although, again, with low values corresponding to weak correlations (Table 6).

Table 5. Expressed opinions and association between the frequency of utilization and level of trust in the information sources.

Information Sources	Frequency of Use (% of Respondents)			Level of Trust (% of Respondents)				Spearman Correlation [1]
	Never	Few Times	Many Times	No Trust	Some Trust	Much Trust	Full Trust	
YouTube channels of influencers	77.3	19.7	3.0	81.3	18.3	0.4	0	0.568 **
Nutritionist blogs	32.6	46.5	20.9	10.9	60.0	25.6	3.4	0.383 **
DGS website	25.3	43.3	31.4	2.0	22.7	39.2	36.2	0.335 **
Natural food stores	55.9	36.8	7.4	35.0	50.3	11.9	2.8	0.539 **
Scientific journals	27.0	38.6	34.4	2.4	23.7	39.8	34.2	0.362 **
Internet pages	28.2	53.3	18.5	39.0	56.5	4.0	0.6	0.483 **
Social media	54.5	36.6	8.9	67.0	30.8	2.0	0.2	0.570 **
Television	36.2	54.1	9.7	28.8	56.1	13.5	1.6	0.649 **
Technical books	28.6	40.4	31.0	3.0	25.0	38.2	33.8	0.304 **
WHO website	34.2	38.6	27.2	2.4	20.7	34.4	42.5	0.337 **
Pharmacies	58.3	33.4	8.3	13.7	43.7	29.0	3.5	0.341 **
Health centers	38.6	44.9	16.5	4.8	30.2	40.2	24.9	0.301 **
Appointments with Medical doctors/ nutritionists	27.6	43.1	29.2	2.0	18.7	34.6	44.7	0.263 **

[1] Spearman correlation between frequency of utilization and level of trust. ** Correlation is significant at the 0.01 level.

Table 6. Associations between level of knowledge and the frequency of utilization and level of trust in the information sources.

Information Sources	Spearman Correlation between Level of Knowledge and Frequency of Utilization of Information Sources	Spearman Correlation between Level of Knowledge and Level of Trust in the Information Sources
YouTube channels of influencers	−0.240 **	−0.234 **
Nutritionist blogs	−0.063	−0.021
DGS Website	0.320 **	0.357 **
Natural food stores	−0.355 **	−0.416 **
Scientific journals	0.385 **	0.380 **
Internet pages	−0.121 **	0.006
Social media	−0.209 **	−0.161 **
Television	−0.285 **	−0.266 **
Technical books	0.283 **	0.263 **
WHO website	0.357 **	0.342 **
Pharmacies	−0.124 **	−0.179 **
Health centers	0.083	0.086
Appointments with Medical doctors/nutritionists	0.072	0.213 **

** Correlation is significant at the 0.01 level.

Regarding the relation between knowledge and the level of trust in information sources, some negative correlations were found significant ($p < 0.01$) for natural food stores ($r_s = -0.416$), television ($r_s = -0.266$), and influencers' YouTube channels ($r_s = -0.234$). Again, the correlations are weak because of the low values (Table 6).

4. Discussion

Healthy and well-balanced eating habits are important for the maintenance of general health and well-being, prevention of several diseases, and the increase of life expectancy. However, in reality, there are several dietary myths that compromise people's nutritional knowledge, making it difficult to follow adequate dietary patterns [13]. Hence the relevance of credible and trustworthy sources of nutritional information, in order to optimize the levels of knowledge and, consequently, improve the nutritional behaviors [23]. This study has shown (as previously seen in Table 5) that the main sources used to obtain information about nutrition are scientific journals, DGS website, and technical books. In a study regarding the use of nutritional information, Goodman et al. [10] determined that magazines, newspapers, and books were the third most common source of information, surpassed only by the internet. Even though these are the most utilized sources, doctors and nutritionists are, in this research, the most trusted and reliable. Goodman et al. [10] and Quaidoo et al. [23], in their studies, have also confirmed that healthcare professionals were perceived as the most credible sources of nutritional information.

The grand majority of the participants in this study recognize the importance of water (corresponding to the highest score in Table 4) and adequate hydration, for general health status, since it is involved in the functioning of all cells in the human body. A narrative review has evidenced that dehydration impairs cognitive functions such as attention, motor coordination, and executive functions. Moreover, high fluid intake has been linked to a lower risk of incidence of kidney stones, as well as to maintaining a healthy and functional gastrointestinal system [24,25].

Regarding the statement about drinking milk being bad for health, our results showed that the score was intermediate, revealing that still a lot of participants believe milk to have a negative impact in health. To the matter of milk intake, there is some ambiguity as to which effects dairy products have on health. Milk is an important source of high-quality protein, vitamins and minerals such as Ca, P, Mg, Zn, Se, vitamins A, D, E, and B complex, and bioactive peptides with antimicrobial, antiviral, antifungal, antioxidant, and antibacterial properties [26]. According to a review [27], milk and dairy products have proven to reduce the risk of obesity in children and improve body composition in adults. Additionally, dairy product intake contributes to a reduced risk of type 2 diabetes, bone health maintenance in children and adolescents, and consuming 200 to 300 mL per day does not increase the risk of cardiovascular disease. Milk and dairy probably protect, as well, against colorectal cancer, bladder cancer, gastric cancer, and breast cancer. There is, however, a potential risk for prostate cancer, but the evidence is scarce and inconsistent [27].

One of the items investigated in this survey, which has had a score below the average, is the adoption of a gluten-free diet for better health. This shows that people still have incorrect perceptions about gluten-free diets and think they should be adopted generally. A similar result was reported by Jones [28], according to which 65% of Americans think that gluten-free foods are healthier. In recent years, more people have adopted a gluten-free diet, even though without being diagnosed for celiac disease or non-celiac gluten sensitivity, in which cases, a gluten-free diet is a necessity. There are significant disadvantages in embracing this type of diet, mainly, the lower levels of fiber, iron, zinc, and potassium that products with grains, such as breads and cereals, have, as well as an increased risk of nutritional deficiencies of B vitamins, iron, and other minerals [28].

The statement S30 "All food additives (E's) are harmful to health" was the third item with the lowest score, corresponding to a generalized wrong perception among the participants in this study. Food additives are substances that are added to products with a specific purpose, such as preservation, coloring, sweetening, and other reasons. All food additives authorized for use are considered safe by the Scientific Committee on Food and/or the European Food Safety Authority. Many of these additives are, in fact, naturally present in foods. As an example tomatoes are rich in lycopene (E160c), oranges have ascorbic acid, or vitamin C (E300), and apples have riboflavin, or vitamin B12 (E101) [29,30].

In general, it was observed that the participants in this study have a wrong perception about the adequacy of eating the fruit before or after a meal (statements S3 and S8). In reality, fruits are an important source of fiber, several vitamins, minerals and phytochemicals, having numerous studies associated the intake of fruits and vegetables with a reduced risk of all-cause mortality, cardiovascular diseases, diabetes, hypertension, and several types of cancer [31,32]. Due to all of these benefits, the World Health Organization recommends the intake of at least 400 g of fruit and vegetables per day [31,32]. In general, the consumption of fresh fruit, prior to or as part of a meal can promote satiety, and reduce hunger and the total energy intake during meals or throughout the day. These properties come, majorly, from the high fiber content, which stimulates the release of satiety hormones [32]. Thus, eating fruit right before or after the meal does not bring any benefit apart from those of eating fruit. Only one clinical study has shown that, if fruit is eaten 30 to 120 min prior to the meals, this might have some advantages on maximizing satiety and reduction of energy intake [33].

The item with the lowest score is S6 "Eating carbohydrates at night leads to an increase in weight gain", demonstrating the incorrect perceptions of the participants about this. Firstly, carbohydrates are not all the same. Some carbohydrates present in vegetables, fruits, and whole-grains are mostly fiber, contrarily to others, such as those in rice, potato, and pasta, which are poorer in dietary fiber. Fiber has demonstrated the ability to facilitate the weight lost, since it helps stimulate satiety hormones, slowing gastric emptying, and reducing appetite, and energy intake [32,34,35]. Secondly, carbohydrates only contribute to weight gain when they are eaten in excessive amounts. When this happens, glucose is transformed into triglycerides, which are then stored in the adipose tissue (fat). Obesity is a disorder in which there is an imbalance between energy intake and expenditure, therefore, the energy present in foods, from all macronutrients, is of extreme importance. Overall, carbohydrates are relevant in the human body, and at night, the most important is to prioritize low density and high fiber foods as well as the amounts consumed. Several studies defend that, for weight loss, the largest meal should be consumed at breakfast or lunch and not in the evening [36,37].

This study has also shown that age is the main discriminant for the level of knowledge exhibited by the participants, with young adults having higher levels of knowledge when compared to other age classes. This might be because in the past years there were attempts to better educate people to eat healthy, as a way to reduce the expenses with health care for patients suffering from non-communicable food-related pathologies. One's sex is another important discriminant, where women have been shown to have superior levels of knowledge than men. The next discriminant that influences the knowledge is professional area connected with nutrition. Participants with no relation to this field have been shown to have lower levels of knowledge. Furthermore, being underweight and having a normal weight have determined higher levels of knowledge. In work by Guiné et al. [38], also evaluating the sociodemographic factors affecting knowledge about nuts and their health effects, the authors found, as the main discriminants, the level of education, sex, age, exercise, and BMI class. Some of these variables were also found to influence the knowledge about food facts and myths in the present work.

By providing tools to understand what people should eat, this type of work can contribute to the higher literacy of the population if appropriate educational programs and public health strategies are implemented. However, people's food choices are conditioned by many other factors, some related to personal traits, preferences, or cues, and others associated with social and cultural influences or even economic constraints [39–41].

Sociodemographic, anthropometric, and behavioral variables not only influenced the level of knowledge, but also the way participants responded to the questions on myths and facts during this research. The variable age influenced how respondents answered to certain myths/facts such as for example item S1 (Drinking water during meals, contributes to weight gain), item S3 (fruit should be eaten before meals), and item S6 (eating carbohydrates at night leads to an increase in weight gain). As for the professional area, variable nutrition has influenced how participants responded to items such as S6 (eating carbohydrates at night leads to an increase in weight gain), S19 (the alkaline diet allows for balancing the acidity in the blood), and S20 (drinking, while fasting, a glass of water with lemon helps in weight loss). Finally, variable health has influenced items S6 (eating carbohydrates at night leads to an increase in weight gain), S13 (lactose-free foods are better for health and should, therefore, be adopted by all), and S17 (diet should be adapted to a person's blood group). Age has been reported as a factor affecting people's care about diet and motivations for food choice. In fact, with increasing age, people tend to think more about health aspects, as a natural way to prevent age-related pathologies or to improve the global health status [17,18].

5. Conclusions

This research has shown that, for this sample, the level of knowledge about nutritional facts is quite high. However, this study has also demonstrated that there are still several food myths that need to be debunked through trustworthy information sources, in order to promote healthy, balanced, and appropriate eating behaviors. Aspects that need more attention from public health authorities and educational programs were identified, and will enable the design of more adequate strategies, to improve the level of knowledge of the Portuguese on those aspects in particular. In this way, people could be better educated to make more appropriate food choices and improve their health. This will result, not only in gains, in terms of individual health, but also in public health, and a reduction in the burden associated with non-communicable food-related diseases, such as diabetes, obesity, or heart diseases.

Furthermore, this study concluded that sociodemographic, anthropometric, and behavioral variables not only influenced the level of knowledge, but also the way participants responded to the questions on myths and facts. Hence, despite its limitations, this research provides a scientific perspective on the perception about several food myths and facts. In particular, by highlighting the information that people already have and those aspects that need improvement, the focus of the information campaigns can be directed to target specific myths. Additionally, the sociodemographic characteristics must also be considered when planning information techniques, as the results showed that different sociodemographic groups have different levels of knowledge.

Although providing very useful and new insights into the perceptions about food myths and facts among the Portuguese, this study has some limitations that need to be highlighted. One limitation is related to the nature of the sample, since, as a convenience sample, the results are not statistically representative to be generalized. One other limitation concerns the high levels of education of the participants, which might somehow bias the results. Moreover, the professional areas of a great number of participants are related to food or nutrition; therefore, this could influence their responses.

Author Contributions: Conceptualization, S.G.F., I.L. and A.M.; methodology, S.G.F.; software, S.G.F.; validation, A.M. and I.L.; formal analysis, S.G.F.; investigation, S.G.F., I.L. and A.M.; resources, S.G.F., I.L., A.M. and M.F.; data curation, S.G.F.; writing—original draft preparation, S.G.F.; writing—review and editing, S.G.F., I.L., A.M. and M.F.; visualization, S.G.F.; supervision, A.M. and I.L.; project administration, I.L. and A.M.; funding acquisition, M.F. All authors have read and agreed to the published version of the manuscript.

Funding: This research received no external funding. The APC was funded by the Portuguese Foundation for Science and Technology (FCT) project refª UIDB/00742/2020.

Institutional Review Board Statement: The study was conducted according to the guidelines of the Declaration of Helsinki, and approved by the Ethical Commission at the Faculty of Food and Nutrition Sciences of University of Porto (reference N° 28/2021/CEFCNAUP/2021).

Informed Consent Statement: Informed consent was obtained from all subjects involved in the study.

Data Availability Statement: Data are available from the corresponding author upon reasonable request.

Acknowledgments: This work results from the final graduation work of the first author, Sofia Florença, in Nutrition Sciences from the FCNAUP, finalized in July 2021. This research was supported by national funds through the FCT—Foundation for Science and Technology, I.P., within the scope of the project refª UIDB/00742/2020. Furthermore, we would like to thank the UICISA:E Research Centre and the Polytechnic Institute of Viseu for their support.

Conflicts of Interest: The authors declare no conflict of interest.

Appendix A

Table A1 presents the Spearman correlations between perception about the food myths/facts and some sociodemographic, anthropometric and behavioral variables.

Table A1. Values of Spearman correlation between perception about the food myths/facts and some sociodemographic, anthropometric, and behavioral variables.

Code	Item	Age	Sex	Education	Environment	Marital	Nutrition	Food	Sport	Health	Bal. Diet	BMI Class
S1	Drink water at meals	0.286 **	0.037	0.049	−0.039	0.172 **	0.286 **	0.281 **	0.045	0.183 **	−0.106 *	0.113 *
S2	Beginning of digestion	0.033	−0.002	0.089 *	0.032	−0.028	−0.202 **	−0.158 **	−0.024	−0.167 **	0.130 **	0.001
S3	Fruit before meals	0.361 **	−0.013	0.083	0.038	0.259 **	0.307 **	0.186 **	0.046	0.113 *	−0.042	0.134 **
S4	Eggs and cholesterol	0.041	0.089 *	−0.035	0.143 **	0.042	0.095	0.064	−0.084	0.018	−0.128 **	−0.060
S5	Milk is bad for health	0.143 **	0.044	−0.083	0.012	0.104 *	0.282 **	0.326 **	−0.026	0.214 **	−0.046	0.101 *
S6	Carbohydrates at night and weight	0.309 **	−0.002	0.061	0.011	0.266 **	0.464 **	0.299 **	0.087	0.296 **	−0.144 **	0.158 **
S7	Importance of fat	−0.195 **	−0.042	0.049	0.043	−0.174 **	−0.369 **	−0.331 **	−0.075	−0.225 **	0.132 **	−0.117 **
S8	Fruit after meals	−0.138 **	0.130 **	0.008	0.047	−0.141 **	−0.008	−0.042	−0.011	0.053	−0.012	−0.024
S9	Fiber and bowel function.	0.164 **	−0.069	0.108 *	−0.006	0.103 *	−0.226 **	−0.188 **	−0.043	−0.114 *	0.136 **	0.008
S10	Gluten-free foods are better for all	0.148 **	0.099 *	−0.073	0.002	0.107 *	0.275 **	0.203 **	−0.052	0.177 **	−0.086	0.133 **
S11	Cheese is bad for memory	−0.071	0.011	−0.086	0.007	−0.049	0.040	0.015	−0.028	0.112 *	0.006	−0.001
S12	Coconut oil versus olive oil	−0.066	0.072	−0.023	−0.091 *	−0.001	0.306 **	0.261 **	0.033	0.182 **	−0.117 **	0.084
S13	Lactose-free foods are better for all	0.120 **	0.059	−0.144 **	−0.029	0.070	0.363 **	0.331 **	−0.007	0.279 **	−0.056	0.098 *
S14	Children's nutritional needs	−0.066	−0.109 *	−0.007	0.001	−0.045	−0.147 **	−0.118 *	−0.004	−0.103 *	0.026	0.004
S15	Fruits/vegetables and weight gain	0.061	0.052	−0.090 *	−0.017	0.014	0.133 **	0.058	0.041	0.085	−0.071	0.107 *
S16	Sweet potatoes have less calories	0.180 **	0.042	0.060	0.008	0.149 **	0.376 **	0.295 **	0.119 *	0.220 **	−0.128 **	0.039
S17	Diet and blood group	0.070	0.072	0.023	−0.030	0.049	0.335 **	0.234 **	0.038	0.282 **	−0.101*	0.110 *
S18	Balanced varied diet and diseases	0.047	−0.130 **	0.102 *	0.028	0.061	−0.133 **	−0.099 *	0.107 *	−0.026	0.096 *	−0.009
S19	Alkaline diet and blood acidity	0.211 **	0.129 **	−0.043	−0.014	0.175 **	0.410 **	0.321 **	0.028	0.258 **	−0.058	0.145 **
S20	Water with lemon and weight loss	0.238 **	0.051	−0.030	0.019	0.218 **	0.396 **	0.313 **	0.030	0.247 **	−0.149 **	0.201 **
S21	Eating habits and risk of disease	0.160 **	−0.064	0.026	−0.037	0.132 **	−0.079	−0.056	−0.096	−0.068	0.077	0.046
S22	Protein and muscle formation	−0.087	0.049	−0.023	−0.032	−0.099 *	0.128 *	0.076	−0.017	0.174 **	−0.016	0.044
S23	Pregnant women eating for two	−0.106 *	0.208 **	−0.036	0.070	−0.194 **	0.191 **	0.086	0.037	0.133 *	−0.073	0.029
S24	Cold water should not be drunk	0.153 **	0.022	−0.086	−0.035	0.058	0.271 **	0.184 **	−0.073	0.208 **	−0.016	0.022
S25	Importance of breakfast	0.191 **	0.087	−0.043	−0.021	0.158 **	0.223 **	0.163 **	0.083	0.090	−0.087	0.092 *
S26	Water is essential for organs	0.115 **	−0.065	0.051	0.007	0.046	0.125 *	0.062	0.213 **	0.090	0.082	−0.053
S27	Soy milk is healthier	0.051	0.058	−0.139 **	−0.012	0.045	0.359 **	0.313 **	−0.005	0.228 **	−0.135 **	0.117 **
S28	Orange and dairy products	0.040	−0.006	−0.034	−0.076	0.016	0.227 **	0.210 **	0.007	0.186 **	−0.010	0.067
S29	Portions of dairy products	−0.117 **	−0.152 **	−0.004	−0.020	−0.094 *	−0.389 **	−0.321 **	−0.104	−0.273 **	0.162 **	−0.057
S30	Food additives (E's) and health	0.135 **	−0.059	−0.043	−0.121 **	0.118 **	0.356 **	0.354 **	0.060	0.141 **	−0.090 *	0.057

* Correlation significant at 0.05 level. ** Correlation significant at 0.01 level.

References

1. Tucker, K.L. Human Nutrition. In *Reference Module in Biomedical Sciences*; Elsevier: Amsterdam, The Netherlands, 2014.
2. Verbanac, D.; Maleš, Ž.; Barišić, K. Nutrition-facts and myths. *Acta Pharm.* **2019**, *69*, 497–510. [CrossRef] [PubMed]
3. Jiang, Y.; King, J.M.; Prinyawiwatkul, W. A review of measurement and relationships between food, eating behavior and emotion. *Trends Food Sci. Technol.* **2014**, *36*, 15–28. [CrossRef]
4. Milošević, J.; Žeželj, I.; Gorton, M.; Barjolle, D. Understanding the motives for food choice in Western Balkan Countries. *Appetite* **2012**, *58*, 205–214. [CrossRef] [PubMed]
5. Cunha, L.M.; Cabral, D.; Moura, A.P.; de Almeida, M.D.V. Application of the Food Choice Questionnaire across cultures: Systematic review of cross-cultural and single country studies. *Food Qual. Prefer.* **2018**, *64*, 21–36. [CrossRef]
6. Guiné, R.P.F.; Bartkiene, E.; Szűcs, V.; Tarcea, M.; Ljubičić, M.; Černelič-Bizjak, M.; Isoldi, K.; El-Kenawy, A.; Ferreira, V.; Straumite, E.; et al. Study about Food Choice Determinants According to Six Types of Conditioning Motivations in a Sample of 11,960 Participants. *Foods* **2020**, *9*, 888. [CrossRef] [PubMed]
7. Worsley, A. Nutrition knowledge and food consumption: Can nutrition knowledge change food behaviour? *Asia Pac. J. Clin. Nutr.* **2002**, *11* (Suppl. 3), S579–S585. [CrossRef]
8. Spronk, I.; Kullen, C.; Burdon, C.; O'Connor, H. Relationship between nutrition knowledge and dietary intake. *Br. J. Nutr.* **2014**, *111*, 1713–1726. [CrossRef]
9. Marsola, C.d.M.; Carvalho-Ferreira, J.P.d.; Cunha, L.M.; Jaime, P.C.; da Cunha, D.T. Perceptions of risk and benefit of different foods consumed in Brazil and the optimism about chronic diseases. *Food Res. Int.* **2021**, *143*, 110227. [CrossRef]
10. Goodman, S.; Hammond, D.; Pillo-Blocka, F.; Glanville, T.; Jenkins, R. Use of Nutritional Information in Canada: National Trends between 2004 and 2008. *J. Nutr. Educ. Behav.* **2011**, *43*, 356–365. [CrossRef] [PubMed]
11. Afshin, A.; Sur, P.J.; Fay, K.A.; Cornaby, L.; Ferrara, G.; Salama, J.S.; Mullany, E.C.; Abate, K.H.; Abbafati, C.; Abebe, Z.; et al. Health effects of dietary risks in 195 countries, 1990-2017: A systematic analysis for the Global Burden of Disease Study 2017. *Lancet* **2019**, *393*, 1958–1972. [CrossRef]
12. Downer, S.; Berkowitz, S.A.; Harlan, T.S.; Olstad, D.L.; Mozaffarian, D. Food is medicine: Actions to integrate food and nutrition into healthcare. *BMJ* **2020**, *369*, m2482. [CrossRef]
13. Lesser, L.I.; Mazza, M.C.; Lucan, S.C. Nutrition myths and healthy dietary advice in clinical practice. *Am. Fam. Physician* **2015**, *91*, 634–638. [PubMed]
14. Hemmer, A.; Hitchcock, K.; Lim, Y.S.; Butsch Kovacic, M.; Lee, S.-Y. Development of Food Literacy Assessment Tool Targeting Low-Income Adults. *J. Nutr. Educ. Behav.* **2021**. [CrossRef] [PubMed]
15. Palumbo, R.; Adinolfi, P.; Annarumma, C.; Catinello, G.; Tonelli, M.; Troiano, E.; Vezzosi, S.; Manna, R. Unravelling the food literacy puzzle: Evidence from Italy. *Food Policy* **2019**, *83*, 104–115. [CrossRef]
16. Vidgen, H.A.; Gallegos, D. Defining food literacy and its components. *Appetite* **2014**, *76*, 50–59. [CrossRef] [PubMed]
17. Perez, A.; Brown, A.; Elmore, S.; Hartson, K.; O'Neal, C.; King, K. P123 Nourishing Food Literacy, Community Health, and Sense of Place in Louisville, KY. *J. Nutr. Educ. Behav.* **2020**, *52*, S74. [CrossRef]
18. Steils, N.; Obaidalahe, Z. "Social food": Food literacy co-construction and distortion on social media. *Food Policy* **2020**, *95*, 101932. [CrossRef]
19. Thomas, H.; Azevedo Perry, E.; Slack, J.; Samra, H.R.; Manowiec, E.; Petermann, L.; Manafò, E.; Kirkpatrick, S.I. Complexities in Conceptualizing and Measuring Food Literacy. *J. Acad. Nutr. Diet.* **2019**, *119*, 563–573. [CrossRef] [PubMed]
20. Gartaula, H.; Patel, K.; Shukla, S.; Devkota, R. Indigenous knowledge of traditional foods and food literacy among youth: Insights from rural Nepal. *J. Rural Stud.* **2020**, *73*, 77–86. [CrossRef]
21. Højer, R.; Wistoft, K.; Frøst, M.B. Yes I can cook a fish; effects of a five week sensory-based experiential theme course with fish on 11- to 13- year old children's food literacy and fish eating behaviour–A quasi-experimental study. *Food Qual. Prefer.* **2021**, *92*, 104232. [CrossRef]
22. Rhea, K.C.; Cater, M.W.; McCarter, K.; Tuuri, G. Psychometric Analyses of the Eating and Food Literacy Behaviors Questionnaire with University Students. *J. Nutr. Educ. Behav.* **2020**, *52*, 1008–1017. [CrossRef] [PubMed]
23. Quaidoo, E.Y.; Ohemeng, A.; Amankwah-Poku, M. Sources of nutrition information and level of nutrition knowledge among young adults in the Accra metropolis. *BMC Public Health* **2018**, *18*, 1323. [CrossRef] [PubMed]
24. Liska, D.; Mah, E.; Brisbois, T.; Barrios, P.L.; Baker, L.B.; Spriet, L.L. Narrative Review of Hydration and Selected Health Outcomes in the General Population. *Nutrients* **2019**, *11*, 70. [CrossRef]
25. Popkin, B.M.; D'Anci, K.E.; Rosenberg, I.H. Water, hydration, and health. *Nutr. Rev.* **2010**, *68*, 439–458. [CrossRef] [PubMed]
26. Pereira, P.C. Milk nutritional composition and its role in human health. *Nutrition* **2014**, *30*, 619–627. [CrossRef] [PubMed]
27. Thorning, T.K.; Raben, A.; Tholstrup, T.; Soedamah-Muthu, S.S.; Givens, I.; Astrup, A. Milk and dairy products: Good or bad for human health? An assessment of the totality of scientific evidence. *Food Nutr. Res.* **2016**, *60*, 32527. [CrossRef]
28. Jones, A.L. The Gluten-Free Diet: Fad or Necessity? *Diabetes Spectr.* **2017**, *30*, 118–123. [CrossRef] [PubMed]
29. European Commission. *EU Food Additives: Making Our Food Safer*; Memo/13/480; European Commission: Brussels, Belgium, 2013; pp. 1–5.

30. Commision, E. COMMISSION REGULATION (EU) No 1129/2011 of 11 November 2011 amending Annex II to Regulation (EC) No 1333/2008 of the European Parliament and of the Council by establishing a Union list of food additives. *Off. J. Eur. Union* **2011**, *295*, 1–177.
31. Schwingshackl, L.; Hoffmann, G.; Kalle-Uhlmann, T.; Arregui, M.; Buijsse, B.; Boeing, H. Fruit and Vegetable Consumption and Changes in Anthropometric Variables in Adult Populations: A Systematic Review and Meta-Analysis of Prospective Cohort Studies. *PLoS ONE* **2015**, *10*, e0140846. [CrossRef]
32. Dreher, M.L. Whole Fruits and Fruit Fiber Emerging Health Effects. *Nutrients* **2018**, *10*, 1833. [CrossRef]
33. Abdul Hakim, B.N.; Yahya, H.M.; Shahar, S.; Abdul Manaf, Z.; Damanhuri, H. Effect of Sequence of Fruit Intake in a Meal on Satiety. *Int. J. Environ. Res. Public Health* **2019**, *16*, 4464. [CrossRef]
34. Anderson, J.W.; Baird, P.; Davis, R.H., Jr.; Ferreri, S.; Knudtson, M.; Koraym, A.; Waters, V.; Williams, C.L. Health benefits of dietary fiber. *Nutr. Rev.* **2009**, *67*, 188–205. [CrossRef]
35. Ramage, S.; Farmer, A.; Eccles, K.A.; McCargar, L. Healthy strategies for successful weight loss and weight maintenance: A systematic review. *Appl. Physiol. Nutr. Metab.* **2014**, *39*, 1–20. [CrossRef] [PubMed]
36. Madjd, A.; Taylor, M.A.; Delavari, A.; Malekzadeh, R.; Macdonald, I.A.; Farshchi, H.R. Beneficial effect of high energy intake at lunch rather than dinner on weight loss in healthy obese women in a weight-loss program: A randomized clinical trial. *Am. J. Clin. Nutr.* **2016**, *104*, 982–989. [CrossRef] [PubMed]
37. Raynor, H.A.; Li, F.; Cardoso, C. Daily pattern of energy distribution and weight loss. *Physiol. Behav.* **2018**, *192*, 167–172. [CrossRef] [PubMed]
38. Guiné, R.P.; Correia, P.; Fernandes, S.; Ramalhosa, E. Consumption of Nuts and Similar Dried Foods in Portugal and Level of Knowledge about their Chemical Composition and Health Effects. *J. Nuts* **2021**, *12*, 171–200. [CrossRef]
39. Leng, G.; Adan, R.A.H.; Belot, M.; Brunstrom, J.M.; de Graaf, K.; Dickson, S.L.; Hare, T.; Maier, S.; Menzies, J.; Preissl, H.; et al. The determinants of food choice. *Proc. Nutr. Soc.* **2017**, *76*, 316–327. [CrossRef] [PubMed]
40. Frank-Podlech, S.; Watson, P.; Verhoeven, A.A.C.; Stegmaier, S.; Preissl, H.; de Wit, S. Competing influences on healthy food choices: Mindsetting versus contextual food cues. *Appetite* **2021**, *166*, 105476. [CrossRef] [PubMed]
41. Djekic, I.; Bartkiene, E.; Szűcs, V.; Tarcea, M.; Klarin, I.; Černelić-Bizjak, M.; Isoldi, K.; El-Kenawy, A.; Ferreira, V.; Klava, D.; et al. Cultural dimensions associated with food choice: A survey based multi-country study. *Int. J. Gastron. Food Sci.* **2021**, *26*, 100414. [CrossRef]

 *foods*

Article

Food Security and Sustainability: Discussing the Four Pillars to Encompass Other Dimensions

Raquel de Pinho Ferreira Guiné, Maria Lúcia de Jesus Pato, Cristina Amaro da Costa, Daniela de Vasconcelos Teixeira Aguiar da Costa, Paulo Barracosa Correia da Silva and Vítor João Pereira Domingues Martinho *

Agricultural School (ESAV) and CERNAS-IPV Research Centre, Polytechnic Institute of Viseu (IPV), 3504-510 Viseu, Portugal; raquelguine@esav.ipv.pt (R.d.P.F.G.); mljesus@esav.ipv.pt (M.L.d.J.P.); amarocosta@esav.ipv.pt (C.A.d.C.); daniela@esav.ipv.pt (D.d.V.T.A.d.C.); pbarracosa@esav.ipv.pt (P.B.C.d.S.)
* Correspondence: vdmartinho@esav.ipv.pt

Abstract: The unadjusted intake of food constitutes a real challenge for the several sustainability dimensions. In this perspective, the main objectives of this research are to characterise the current contexts of food security, its relationship with sustainability, and identify proposals and actions that may support the design of more adjusted policies in the future. In addition, it is intended to assess if the food security pillars properly address the sustainability goals and if the evolution of undernutrition is accompanied by sustainable frameworks. In this way, statistical information from the FAOSTAT database was considered for the several dimensions of food security over the period 2000–2020. These data were analysed through factor-cluster approaches and panel data methodologies, namely those related to quantile regressions. As main insights, we may refer that undernutrition is more impacted by the availability of food and nutrients and political stability than by the level of GDP—Gross Domestic Product (except for the extreme cases). This means that the level of development is not the primary explanation for the problems of nutrition. The main focus of the national and international policies must be to improve the agrifood supply chains and to support political stability, in order to mitigate undernutrition worldwide and ensure a global access to sustainable and healthy diets. In addition, it is suggested to rethink the four pillars of food security (availability, access, utilisation and stability), in order to encompass other dimensions, such as climate change.

Keywords: FAOSTAT information; factor-cluster analyses; panel data approaches; quantile regressions

Citation: Guiné, R.d.P.F.; Pato, M.L.d.J.; Costa, C.A.d.; Costa, D.d.V.T.A.d.; Silva, P.B.C.d.; Martinho, V.J.P.D. Food Security and Sustainability: Discussing the Four Pillars to Encompass Other Dimensions. *Foods* **2021**, *10*, 2732. https://doi.org/10.3390/foods10112732

Academic Editors: António Raposo, Renata Puppin Zandonadi, Raquel Braz Assunção Botelho and Theodoros Varzakas

Received: 17 September 2021
Accepted: 5 November 2021
Published: 8 November 2021

Publisher's Note: MDPI stays neutral with regard to jurisdictional claims in published maps and institutional affiliations.

1. Introduction

There is a significant relationship between the level of development of the countries and food insecurity contexts [1]. Food insecurity is also impacted by political stability and climate change [2].

Food security is a multidimensional concept characterised by four pillars related to availability, access, utilisation, and stability [3]. These dimensions are interrelated with, for example, agricultural modernisation [4], social capital [5], kitchen equipment [6], and worldwide shocks [7].

In a broader perspective, food security is interrelated with sustainable food and agricultural sectors, where the needs of present and future generations are met in commitment with the environmental, social, and economic dimensions [8]. The concepts of food security and sustainability are interrelated. The sustainability concept was considered by the international community associated with sustainable developments, according to which the societies evolve without compromising the future generations [9].

In fact, food security is already considered by the Sustainable Development Goals (SDGs), namely in the goal 2 for zero hunger [10]. Nonetheless, the pandemic context brought additional challenges for these world objectives increasing the problems associated with hunger worldwide, and it is expected that it will worsen malnutrition, namely among children [11]. On the other hand, the question here is if the food security pillars also specifically address the several dimensions of sustainability. The scientific literature shows that sustainability could be better encompassed by the four pillars of food security [9,12–14].

From this perspective, the general and specific objectives are those presented in the following (summarised in Figure 1):

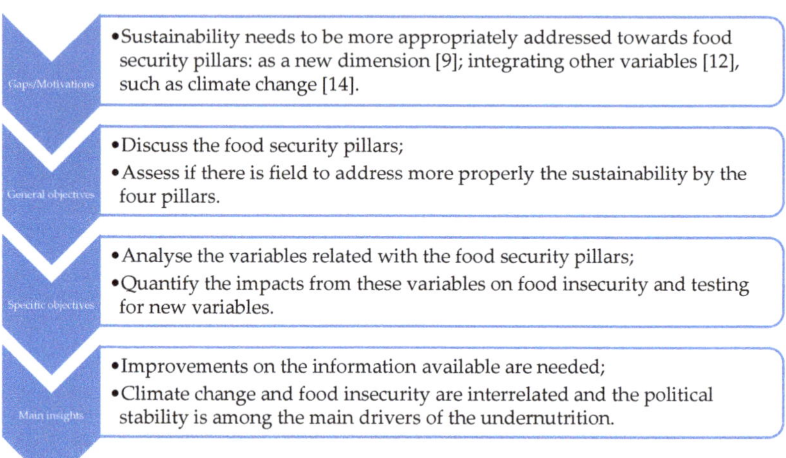

Figure 1. Gaps/motivations, general and specific objectives, and main insights.

General objectives: This research aims to characterize the current context of world food security and its relationships with sustainability, through statistical information from FAOSTAT and panel data regressions (with quantile approaches), in order to suggest rethinking the four pillars to encompass other variables, such as climate change, and propose new policy instruments. The main focus is to highlight if sustainability has been sufficiently addressed by the food security pillars (namely testing for the impact of new variables, such as those related to climate change), considering that food security is already a target for the SDGs. In addition, it is the objective of this research to assess if food security has been achieved with sustainable approaches.

Specific objectives: Specifically, it is intended to characterise the variables related to the food security pillars over the period of 2000–2020 and quantify the impacts from these variables on food insecurity, identifying potentialities to encompass other variables (related to climate change, for example) in these security dimensions. This assessment allowed us to identify weaknesses in the database (there is a lack of statistical information for several countries and years), weaknesses in the group of variables considered for each pillar (it would be important to consider other variables, such as those related to the temperature change) and the main impacts from the four pillars' variables on the food insecurity (with insights for the several stakeholders, namely for policymakers). This overview, considering the worldwide context and the approaches here carried out, has its novelty for the scientific community.

In summary, the literature shows that there are fields to be explored in these issues ("Sustainability needs to be more appropriately addressed towards food security pillars: as a new dimension [9]; integrating other variables [12], such as climate change [14]."). These fields were addressed through "—Analyse the variables related with the food security pillars; —Quantify the impacts from these variables on food insecurity and testing for new variables" that highlight "—Improvements on the information available are needed;

—Climate change and food insecurity are interrelated and the political stability is among the main drivers of the undernutrition".

The paper consists of six sections. After the introduction (Section 1), Section 2 contains a review of the concepts of food security and sustainability. The study design and materials are explained in Section 3, while Section 4 presents the main results of the study. To better assess the impacts of climate change on food security, Section 5 emphasises that the focus on food security should shift beyond the four pillars (availability, access, utilization, and stability) towards a vision that includes other sustainability variables. In Section 6, the discussion of the study is presented, as well as the main conclusions. In this last section, the study's many results are emphasized, the theoretical, political, and practical contributions are highlighted, and the limitations of the study as well as areas for future research are presented.

A preliminary literature survey shows that there are few (or none) studies about the topic of food security and sustainability considering the methodologies hereby described, namely the quantile regressions with the approaches addressed, highlighting the novelty of this research and the potential to analyse deeper these domains.

2. Literature Survey

Food security is an old concern [15] that motivated international organizations, such as the Food and Agriculture Organization, to define it with the following four dimensions (or pillars): availability, accessibility, utilization, and stability [16]. Food security is part of the Sustainable Development Goals [11]. Nonetheless, it is accepted that sustainability needs to be more appropriately addressed in the pillars of food security, maybe as a new dimension [9], or integrating other variables [12], where the food crises [13] and climate change [14] have its importance.

Food security is interrelated with sustainability. However, the associated contexts are complex and uncertain [17], showing that these interrelationships deserve continuous attention from several stakeholders and organizations [18], including the subnational governments [19], and multidisciplinary approaches [20] for particular cases [21].

Agricultural practices have a great impact on the interrelationships between food security and sustainability [22] around the world [23] and for the more diverse contexts [24]; namely, those related with chemical approaches [25], water use [26], soil salinity [27], pests and diseases control [28], concentration/diversification of farming productions [29], intensification/extensification [30], and associated with management strategies [31]. The same happens with the different pressures over the agricultural land associated with, for example, the urbanization and ever-growing population [32]. In any case, the farming sector and the associated dimensions are determinants for sustainable food security [33], including animal welfare [34], and the new technologies open new opportunities [35], including for monitoring [36] and assessments [37]. The population growth is particularly worrying in countries such as India, for example [38], but other specific contexts also deserve special attention [39], including those from Africa [40].

For more sustainable food security, alternative sources of food supply, as edible insects, may be interesting solutions [41,42] in a framework of protein transition [43] and farm to fork perspective [44]. Urban agriculture may be another solution; however, in this case, additional studies are needed for a more complete assessment [45]. Farming fish is another possibility to achieve more sustainable food security [46], as are the home food gardens [47], revitalising local food systems [48], using insects as bioconverters of organic waste [49], obtaining protein from food waste [50], food forestry [51], natural food [52], better governance and conception [53], hydroponic farm [54], organic farming [55], crops resilience assessments [56], wild edible greens [57], food loss, and waste management [58].

Food security and sustainability are worldwide concerns, specifically for the policymakers [59]. In fact, the policies are crucial for sustainable food security [60]. In any case, the paradigms of food security often change with international crises, as happens currently with the COVID−19 pandemic [61], and these frameworks bring new challenges for the different stakeholders [62] to promote balanced and healthy diets [63].

In these interrelationships between food security and sustainability, the circular economy approaches may play a relevant role, namely through insect larvae that bioconvert food waste in animal feed [64].

3. Material and Methods

To better clarify the several steps of this section for material and methods, Figure 2 presents a schematic description of the various phases.

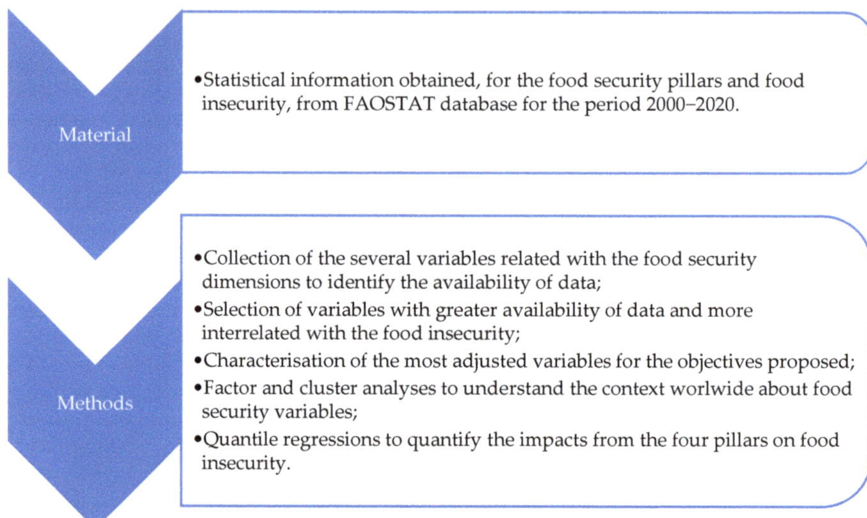

Figure 2. Schematic representation of the different steps under material and methods.

The quantile regressions are adjusted methodologies to deal with problems of outliers and non-normality in the statistical databases, and they were considered, for example, in the following studies related to the food security dimensions: factors affecting food diversity in Iran [65]; impacts on rice and wheat production in Pakistan [66]; implications from soil erosion on socio-economic domains in Malawi [67]; climate effects on the rice yield in India [68]; dietary deficiencies in the Indian context [69]; analysis of explanatory variables of calorie consumption in India [70]; food demand in Slovakia [71]; food insecurity in the population from the Afghanistan [72]; relationships between oil-export and food insecurity [73]; impacts from food insecurity on children [74]; food security in Nigerian urban households [75]; main drivers of healthy ageing in India [76]; effects from the internet use on the agricultural sustainability in China [77]; relationships among poverty and food security in Vietnam [78]; relations between religion and food security in India [79]; food security policies in India [80]; impacts from food prices on the nutrients intake in India [81]; interrelations between the efficiency of wheat and climate change in Pakistan [82]; crop diversity and welfare in Uganda [83]; drivers of food insecurity [84]; horticultural sector dynamics in Senegal [85]; and financial crisis and food supply in Mexico [86].

The statistical information was obtained from the FAOSTAT database [87] for the variables presented in Table 1 (and for the temperature change as an indicator for the climate changes) disaggregated across world countries and over the period of 2000–2020. The temperature change is correlated with the climate dimensions [88], being, in this way, an interesting indicator for the climate change contexts. Furthermore, it is one of the most straightforward quantitative indicators of climate change. These variables were grouped into five groups (level of development, nutritional and food availability, food insecurity, stability conditions, social and environmental conditions), following the suggestions from the statistical database. These five groups are relative to the four pillars of the food security concept (availability, access, utilization, stability) and a fifth set for food insecurity indicators (namely related to the number and prevalence of undernourishment and moderate and severe food insecurity). The four pillars were renamed as a suggestion to encompass other dimensions and to highlight the importance of rethinking these pillars and the food security dimensions. Some of these variables are presented in the database on average for periods of three years (Table 1). In these cases, to compare with other variables presented in annual values, it was considered the middle year as a reference. In addition, Table 1 shows that there is a lack of information for some countries and years. Due to this, the variables with lower observations were not considered in the analysis carried out in the following sections. The variables removed were: variable "Rail lines density (total route in km per 100 square km of land area)" from the group "level of development"; most variables related with the group "food insecurity"; and the variables "Percentage of children under 5 years affected by wasting (percent)" and "Prevalence of exclusive breastfeeding among infants 0–5 months of age" from the group "social and environmental conditions".

Table 1. Summary statistics for several variables related to food security and sustainability, worldwide over the period 2000–2020.

Groups of Variables	Observations	Mean	Standard Deviation	Min	Max
Level of development					
Gross domestic product per capita, PPP, dissemination (constant 2011 international $)	3883	18,905.540	21029.290	630.700	161,971.000
Rail lines density (total route in km per 100 square km of land area)	1452	2.753	2.903	0.000	12.100
Nutritional and food availability					
Average dietary energy supply adequacy (percent) (3-year average)	3238	118.464	15.244	37.000	160.000
Average protein supply (g/cap/day) (3-year average)	2802	78.804	20.510	23.200	143.300
Average supply of protein of animal origin (g/cap/day) (3-year average)	2802	35.464	20.350	3.000	103.000
Average value of food production (constant 2004–2006 I$/cap) (3-year average)	3116	259.279	241.337	1.000	2425.000
Dietary energy supply used in the estimation of prevalence of undernourishment (kcal/cap/day) (3-year average)	3238	2793.793	467.732	903.000	3901.000
Share of dietary energy supply derived from cereals, roots and tubers (kcal/cap/day) (3-year average)	2802	46.807	14.640	8.000	83.000
Food insecurity					
Number of moderately or severely food insecure people (million) (3-year average)	483	8.809	14.699	0.100	116.000
Number of people undernourished (million) (3-year average)	1708	6.833	23.229	0.100	249.600
Number of severely food insecure people (million) (3-year average)	391	3.117	4.897	0.100	43.000
Prevalence of moderate or severe food insecurity in the total population (percent) (3-year average)	520	25.865	21.698	2.000	88.300
Prevalence of severe food insecurity in the total population (percent) (3-year average)	502	9.187	11.469	0.500	61.800
Prevalence of undernourishment (percent) (3-year average)	2047	14.322	11.401	2.500	71.300

Table 1. *Cont.*

Groups of Variables	Observations	Mean	Standard Deviation	Min	Max
Stability conditions					
Cereal import dependency ratio (percent) (3-year average)	2645	27.914	57.439	−342.500	100.000
Per capita food production variability (constant 2004–2006 thousand int$ per capita)	3090	12.323	13.048	0.200	107.700
Per capita food supply variability (kcal/cap/day)	3361	42.027	28.035	1.000	259.000
Percent of arable land equipped for irrigation (percent) (3-year average)	2797	28.430	31.498	0.000	100.000
Political stability and absence of violence/terrorism (index)	3486	−0.032	0.962	−3.180	1.760
Value of food imports in total merchandise exports (percent) (3-year average)	3227	41.389	89.288	1.000	1380.000
Social and environmental conditions					
Percentage of children under 5 years affected by wasting (percent)	616	6.136	4.382	0.000	21.300
Percentage of children under 5 years of age who are overweight (modelled estimates) (percent)	3047	7.049	4.467	0.700	29.300
Percentage of children under 5 years of age who are stunted (modelled estimates) (percent)	3047	21.918	14.624	1.200	62.300
Percentage of population using at least basic drinking water services (percent)	3803	85.024	17.929	18.100	99.000
Percentage of population using at least basic sanitation services (percent)	3802	72.130	29.831	2.800	99.000
Percentage of population using safely managed drinking water services (Percent)	2372	67.400	31.055	2.300	99.000
Percentage of population using safely managed sanitation services (Percent)	2384	53.409	29.468	2.100	99.000
Prevalence of anaemia among women of reproductive age (15–49 years)	3607	27.772	13.305	7.300	62.900
Prevalence of exclusive breastfeeding among infants 0–5 months of age	425	35.945	19.595	0.100	88.400
Prevalence of low birthweight (percent)	2255	10.312	4.920	2.400	36.200
Prevalence of obesity in the adult population (18 years and older)	3044	16.407	10.246	0.600	61.000

Table 1 reveals that, on average, the world had, between 2000 and 2020, a Gross Domestic Product per capita (PPP, constant 2011) of 18905.540 I$, 2.753 km per 100 square km of land area for the rail density, 78.804 g/cap/day of average protein supply, 35.464 g/cap/day for average supply of protein of animal origin, 46.807 kcal/cap/day for the share of dietary supply derived from cereals, roots, and tubers, 14.322% of prevalence of undernourishment, 27.914% for cereal import dependency ratio, 28.430% for arable land equipped for irrigation, −0.032 for the political stability and absence of violence/terrorism (index) and several problems with social and environmental conditions. These results reveal that there is still much more to improve the nutritional conditions and the political stability, as well as the absence of violence/terrorism. In fact, Western Balkans and the European Union had, over the last years, values of 101.340 for average protein supply (g/cap/day), 57.300 average supply of protein of animal origin (g/cap/day) and 0.550 for political stability and absence of violence/terrorism (index), highlighting the way that must be run by some parts of the world in these domains [89].

In the different groups with the remaining variables, after the first selection identified before, the results presented in Table A1 (Appendix A) were considered to select the most representative variables in each group. Considering the great number of variables available in the database for four pillars, the correlation matrix brings insights that support the selection of the correlated variables. These are important findings to build the models for the regression approaches. These results were obtained following Stata procedures [90–95] for pairwise correlation matrices.

Table A1 (Appendix A) shows that there are positive and relatively strong correlations between the GDP (PPP, constant 2011 international $) and the supply of protein, political stability and access of the population to drinking water and sanitation services. In turn, there are negative and relatively strong relationships among the GDP and share of dietary energy supply from vegetables, percentage of children under 5 years of age who are stunted, and prevalence of anaemia among women of reproductive age (15–49 years). In addition, for the prevalence of undernourishment (percent), there are negative and relatively strong correlations with the GDP and the variables related to the "nutritional and food availability" group, with the exception of the variable "share of dietary energy supply derived from cereals, roots and tubers", for which there is a positive correlation. The prevalence of undernourishment is also negatively correlated with political stability and with the access of the population to drinking water and sanitation services.

Considering this assessment and the relevance of the variables for the objectives proposed, the following variables were selected for each one of the groups considered: gross domestic product per capita, PPP, dissemination (constant 2011 international $) for the "level of development", average protein supply (g/cap/day) (3-year average) for the group "nutritional and food availability", prevalence of undernourishment (percent) (3-year average) for the "food insecurity", political stability and absence of violence/terrorism (index) for "stability conditions" and percentage of population using safely managed drinking water services (percent) for "social and environmental conditions". It could be interesting to consider also the prevalence of severe and moderate food insecurity, however, the limited availability of data for these variables hampers its consideration in regressions with panel data. 1. In fact, it is not possible to consider these variables with around 500 observations (or less) as shows Table 1, when the "Prevalence of undernourishment (percent) (3-year average)", for example, has more than 2000 observations. In any case, for the cases (countries and years) where the observations are coincident the "Prevalence of undernourishment (percent) (3-year average)" (variable used by us) is strongly correlated (Table 2 obtained following Stata software procedures) with the severe and moderate undernourishment (showing that it is indifferent to use any of these variables and that our variable represents well the world context of undernutrition).

Table 2. Spearman's rank correlation matrix for the several forms of prevalence of undernourishment worldwide over the period of 2000–2020.

	Prevalence of Moderate or Severe Food Insecurity in the Total Population (Percent) (3-Year Average)	Prevalence of Severe Food Insecurity in the Total Population (Percent) (3-Year Average)	Prevalence of Undernourishment (Percent) (3-Year Average)
Prevalence of moderate or severe food insecurity in the total population (percent) (3-year average)	1.000		
Prevalence of severe food insecurity in the total population (percent) (3-year average)	0.9200 *	1.000	
	0.000		
Prevalence of undernourishment (percent) (3-year average)	0.7917 *	0.7421 *	1.000
	0.000	0.000	

Note: *, statistically significant at 1%.

To better analyse the evolution of these variables over the countries, averages over the period considered were calculated based on the summary statistics presented in Table 3, these being presented in Figures 3–7.

Figure 3 shows that the five higher averages for the gross domestic product per capita, PPP, dissemination (constant 2011 international $), over the period considered, were verified in Luxembourg, China (Macao SAR), Qatar, Bermuda, and Singapore. For the average protein supply (g/cap/day) (3-year average), the five countries with higher averages are Iceland, Israel, Lithuania, China (Hong Kong SAR), and Portugal (Figure 4). Luxembourg, Ireland, Norway, USA and Denmark are among the countries with higher average GDP and average protein supply. These results confirm the findings described before for the correlations between the level of development and the nutrients supply [96]. The prevalence of undernourishment (percent) (3-year average) has higher averages in countries such as Somalia, Haiti, Central African Republic, Democratic Republic of the Congo, Liberia, and Madagascar, among others (Figure 5). Some of the countries with higher averages for the GDP and nutrients supply are also those with higher averages for the political stability and absence of violence/terrorism index (Figure 6) and percentage of population using safely managed drinking water services (Figure 7).

Table 3. Summary statistics for selected variables related to food security and sustainability worldwide on average over the period of 2000–2020.

Variables	Observations	Mean	Standard Deviation	Min	Max
Gross domestic product per capita, PPP, dissemination (constant 2011 international $)	187.000	18,809.910	20,670.980	817.519	107,015.200
Average protein supply (g/cap/day) (3-year average)	167.000	78.724	19.928	36.547	133.135
Prevalence of undernourishment (percent) (3-year average)	123.000	13.446	11.047	2.600	61.529
Political stability and absence of violence/terrorism (index)	186.000	−0.036	0.920	−2.473	1.417
Percentage of population using safely managed drinking water services (Percent)	115.000	67.658	30.806	5.600	99.000

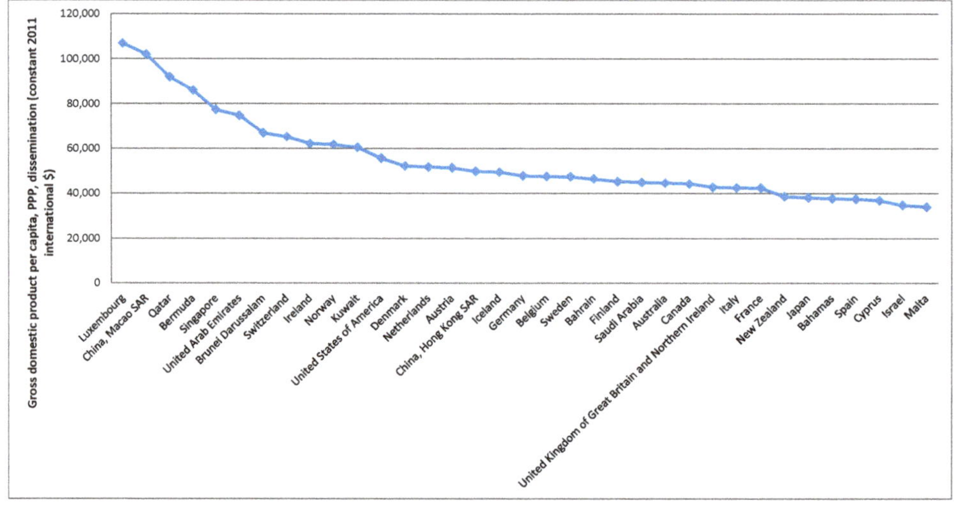

Figure 3. Top 35 countries for the level of development.

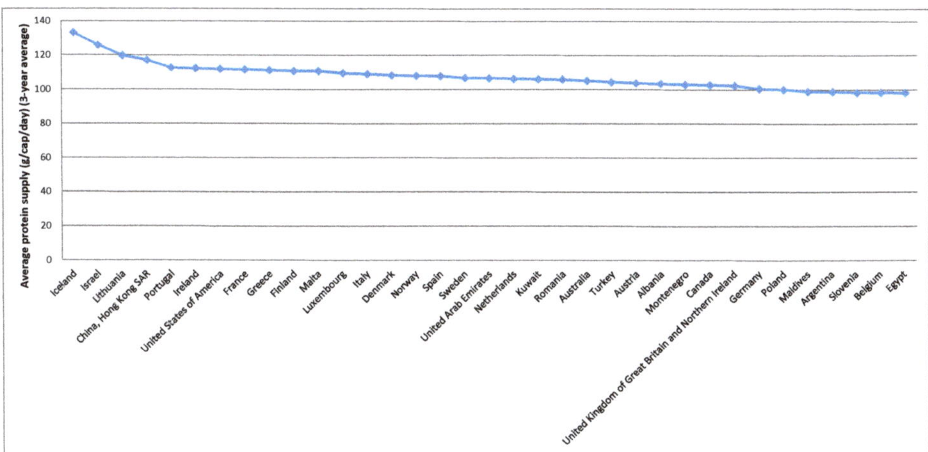

Figure 4. Top 35 countries for the nutritional and food availability.

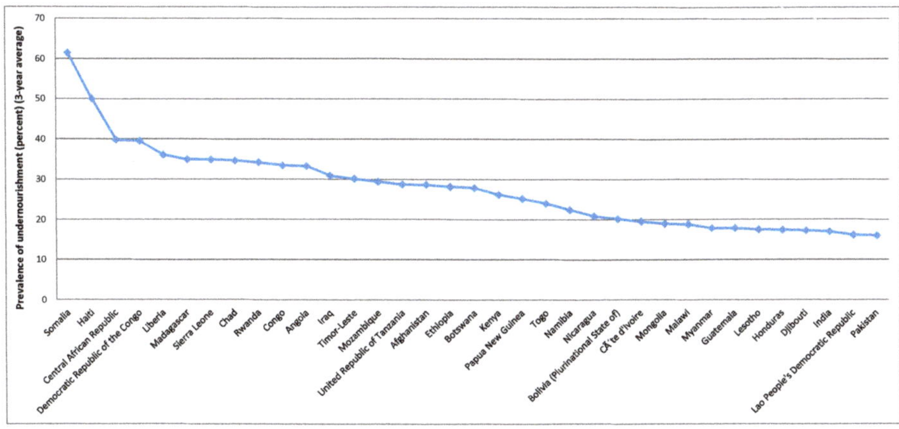

Figure 5. Top 35 countries for the level of food insecurity.

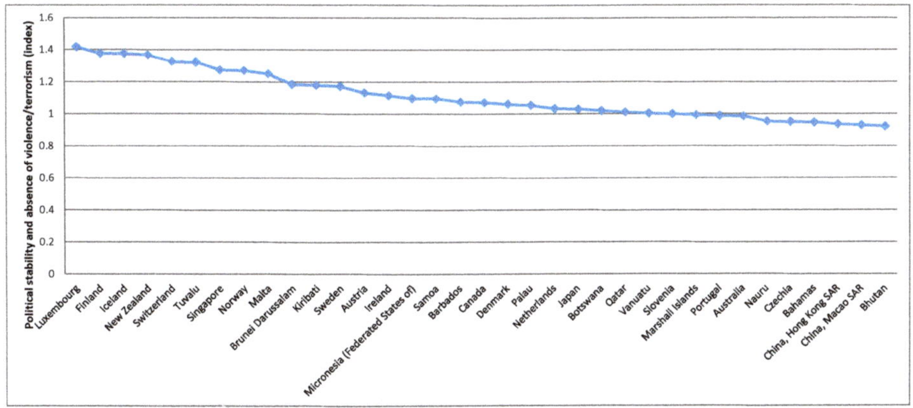

Figure 6. Top 35 countries for the stability conditions.

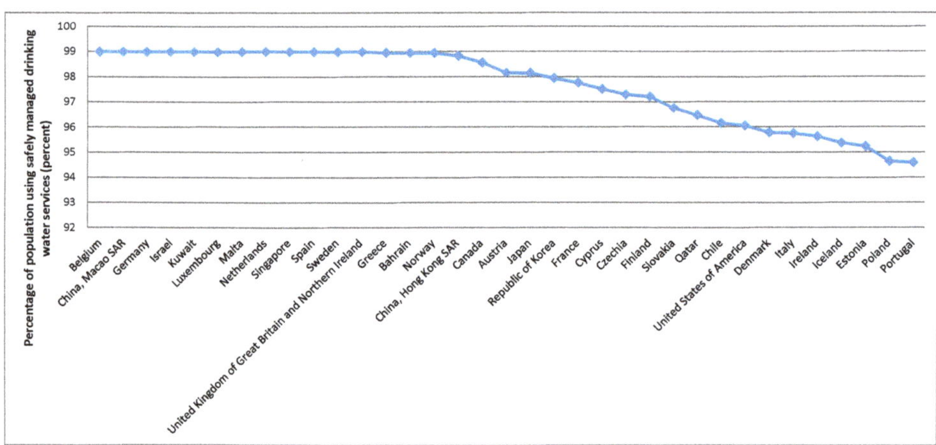

Figure 7. Top 35 countries for the social and environmental conditions.

To better explore these data, a factor-cluster analysis following Stata procedures [93–95,97–100] was performed in the next section, to identify world clusters for variables that might be considered better drivers to implement and design joint policies. In addition, quantile regressions in panel data [93,101–103] were carried out considering a model based on the Verdoorn–Kaldor laws [104–107] and taking into account, for example, Martinho [108]. In the developments associated with the Verdoorn–Kaldor laws, the output growth is exogenous and promotes productivity growth, benefiting all the economy and society. Thus, following these developments, in this research, a model considering the output growth as an explanatory variable was considered. Due to the existence of outliers and lack of normality for some data, the quantile regression, based on the median estimation, was considered.

A factor analysis was carried out to obtain uncorrelated factors and to deal with problems of collinearity among the variables in the cluster assessment [100]. For that, a principal-component factors methodology with rotation was considered.

4. Results

Figure 8 summarises the different phases of this section regarding the results obtained with the variables selected for the dimensions associated with food security.

The five variables considered are correlated with factor1, explaining 65.8% of the variability (Table 4). The less relevant variables in the definition of factor 1 are the gross domestic product per capita, PPP, dissemination (constant 2011 international $) and political stability and absence of violence/terrorism (index). Inversely, the percentage of population using safely managed drinking water services is the most relevant. The results for the KMO in Table 4 show the sampling adequacy.

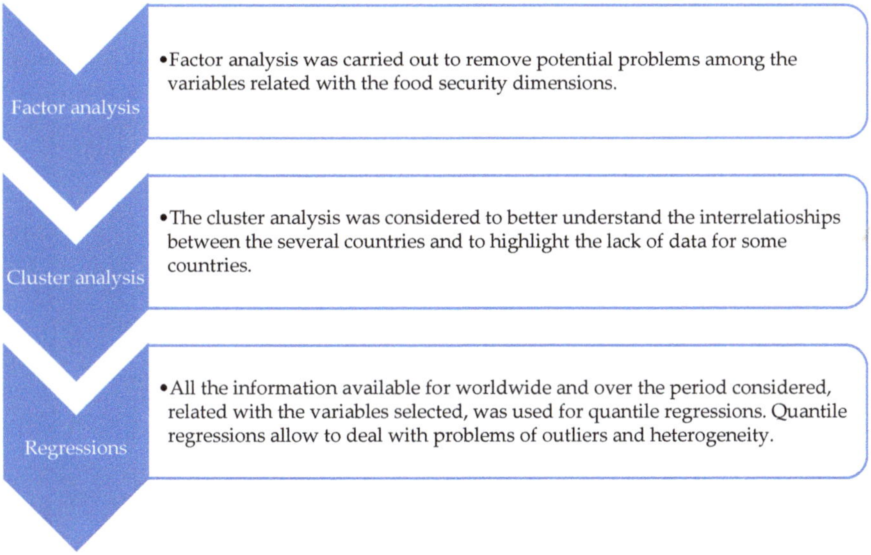

Figure 8. Summary of the several phases considered for the results.

Table 4. Results for rotated factor loadings (pattern matrix) and unique variances, worldwide on average over the period of 2000–2020.

Variables	Factor1	Uniqueness	KMO [1]
Gross domestic product per capita, PPP, dissemination (constant 2011 international $)	0.738	0.455	0.695
Average protein supply (g/cap/day) (3-year average)	0.882	0.222	0.832
Prevalence of undernourishment (percent) (3-year average)	−0.850	0.278	0.715
Political stability and absence of violence/terrorism (index)	0.666	0.556	0.815
Percentage of population using safely managed drinking water services (Percent)	0.896	0.198	0.768
Overall			0.762

Note: [1] Kaiser–Meyer–Olkin measure of sampling adequacy.

Considering the information from Figure 9, four clusters were used for the cluster analysis whose results are presented in Table 5. Combining the results from Table 5 with the data analysis carried out previously, the cluster 1 identifies a set of countries where it is crucial to solve problems of food insecurity, namely those related to the prevalence of undernourishment (percent) (3-year average). For the countries in cluster 2 and cluster 4, there are also problems of undernourishment, but they are not so expressive as in the countries of cluster1. Japan and Republic of Korea appear in cluster 3 due to relatively lower scores for the protein supply, below Turkmenistan and Kazakhstan, respectively. In addition, the Republic of Korea has relatively lower averages for political stability and the absence of violence/terrorism (index). Table 5 is missing other world countries due to the lack of information for some variables. The countries for which there is lack of data for some variables are removed from the analysis, because the observations do not match when different variables are interrelated.

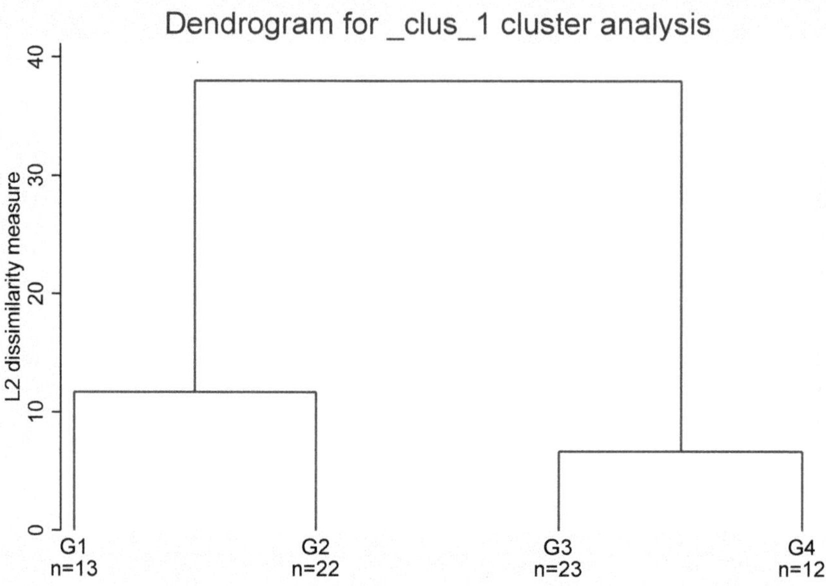

Figure 9. Results for cluster assessment, worldwide on average over the period of 2000–2020. Note: G1–G4 are branches (one for each cluster); and n is the number of observations in each branch.

Table 5. Identification of clusters, worldwide on average over the period of 2000–2020.

Clusters			
1	2	3	4
Afghanistan	Bangladesh	Albania	Algeria
Central African Republic	Cambodia	Belarus	Armenia
Chad	Colombia	Bosnia and Herzegovina	Azerbaijan
Congo	Ecuador	Brazil	Georgia
Côte d'Ivoire	Gambia	Bulgaria	Iran (Islamic Republic of)
Ethiopia	Ghana	Chile	Jordan
Iraq	Guatemala	China, Macao SAR	Mexico
Madagascar	Kiribati	Costa Rica	Morocco
Nigeria	Kyrgyzstan	Croatia	North Macedonia
Pakistan	Lao People's Democratic Republic	Cyprus	Samoa
Rwanda	Lebanon	Estonia	Serbia
Sierra Leone	Lesotho	Japan	Ukraine
Togo	Mongolia	Kazakhstan	
	Myanmar	Kuwait	
	Nepal	Latvia	
	Nicaragua	Malaysia	
	Paraguay	Montenegro	
	Peru	Oman	
	Philippines	Republic of Korea	
	Sao Tome and Principe	Russian Federation	
	Suriname	Slovakia	
	Uzbekistan	Tunisia	
		Turkmenistan	

Before the panel data regressions, considering the characteristics of the sample, the data used for the variables were tested for normality, following, for example, Alejo et al. [103]. The results shown in Table 6 confirm the lack of normality. To deal with these statistical problems, quantile regressions with panel data were considered [109]. This approach is also adjusted to deal with the heterogeneity of the countries in the sample. For that, it was considered a model based on the Verdoorn–Kaldor laws, where the food insecurity variable (prevalence of undernourishment (percent) (3-year average)) was regressed with variables related to several dimensions of sustainability (level of development, nutritional and food availability, stability conditions and social and environmental conditions).

Table 6. Normality analysis, worldwide over the period of 2000–2020.

	Coefficient	z	$p > |z|$
Skewness_e	−0.271	−0.760	0.447
Kurtosis_e	4.186 *	2.720	0.006
Skewness_u	−0.862	−1.280	0.201
Kurtosis_u	2.994 *	2.560	0.010
Joint test for Normality on e: chi2(2) = 8.000 *, Prob > chi2 = 0.018			
Joint test for Normality on u: chi2(2) = 8.200 *, Prob > chi2 = 0.017			

Note: *, statistically significant at 5%.

The results in Table 7 reveal that the undernourishment growth rates are mainly explained by the average protein supply (g/cap/day) (3-year average) growth rates and political stability, and absence of violence/terrorism (index) growth rates. Nonetheless, the political stability and absence of violence/terrorism improvement is not always synonymous of lower problems of undernutrition (as indicated by the sign found for the coefficient and the data analysis carried out before). On the other hand, the gross domestic product per capita, PPP, dissemination (constant 2011 international $) growth rates, and the percentage of the population using safely managed drinking water services (percent) growth rates are only statistically significant for the extreme cases (extreme quantiles). Furthermore, in the countries with lower growth rates for the prevalence of undernourishment (quantiles 0.10 and 0.20), the improvements in the GDP and drinking water services do not contribute to reducing the problems of undernutrition (the signals of the coefficients are positive).

Table 7. Results for panel data quantile regressions with the variables in growth rates, worldwide over the period of 2000–2020.

Prevalence of Undernourishment (Percent) (3-Year Average)	Coefficient	z	$p > |z|$
Quantile 0.10			
Gross domestic product per capita, PPP, dissemination (constant 2011 international $)	0.150 *	4.100	0.000
Average protein supply (g/cap/day) (3-year average)	−2.691 *	−49.820	0.000
Political stability and absence of violence/terrorism (index)	0.001 *	3.060	0.002
Percentage of population using safely managed drinking water services (Percent)	0.311 *	9.210	0.000
Quantile 0.20			
Gross domestic product per capita, PPP, dissemination (constant 2011 international $)	0.135 *	3.030	0.002
Average protein supply (g/cap/day) (3-year average)	−2.407 *	−14.720	0.000
Political stability and absence of violence/terrorism (index)	0.002 *	5.440	0.000
Percentage of population using safely managed drinking water services (Percent)	0.165 ***	1.670	0.094

Table 7. *Cont.*

Prevalence of Undernourishment (Percent) (3-Year Average)	Coefficient	z	$p > |z|$
Quantile 0.30			
Gross domestic product per capita, PPP, dissemination (constant 2011 international $)	−0.014	−0.440	0.657
Average protein supply (g/cap/day) (3-year average)	−2.275 *	−14.940	0.000
Political stability and absence of violence/terrorism (index)	0.003 *	19.480	0.000
Percentage of population using safely managed drinking water services (Percent)	−0.113 **	−2.400	0.017
Quantile 0.40			
Gross domestic product per capita, PPP, dissemination (constant 2011 international $)	−0.040	−1.140	0.256
Average protein supply (g/cap/day) (3-year average)	−2.259 *	−28.070	0.000
Political stability and absence of violence/terrorism (index)	0.003 *	15.510	0.000
Percentage of population using safely managed drinking water services (Percent)	−0.139 ***	−1.670	0.094
Quantile 0.50			
Gross domestic product per capita, PPP, dissemination (constant 2011 international $)	−0.005	−0.200	0.842
Average protein supply (g/cap/day) (3-year average)	−2.339 *	−26.550	0.000
Political stability and absence of violence/terrorism (index)	0.003 *	9.100	0.000
Percentage of population using safely managed drinking water services (Percent)	−0.063	−0.410	0.685
Quantile 0.60			
Gross domestic product per capita, PPP, dissemination (constant 2011 international $)	0.007	0.310	0.756
Average protein supply (g/cap/day) (3-year average)	−2.225 *	−23.670	0.000
Political stability and absence of violence/terrorism (index)	0.004 *	8.940	0.000
Percentage of population using safely managed drinking water services (Percent)	−0.100	−0.740	0.462
Quantile 0.70			
Gross domestic product per capita, PPP, dissemination (constant 2011 international $)	−0.005	−0.210	0.834
Average protein supply (g/cap/day) (3-year average)	−2.297 *	−16.500	0.000
Political stability and absence of violence/terrorism (index)	0.003 *	5.100	0.000
Percentage of population using safely managed drinking water services (Percent)	−0.021	−0.220	0.828
Quantile 0.80			
Gross domestic product per capita, PPP, dissemination (constant 2011 international $)	−0.045 **	−2.090	0.037
Average protein supply (g/cap/day) (3-year average)	−2.162 *	−24.610	0.000
Political stability and absence of violence/terrorism (index)	0.003 **	2.430	0.015
Percentage of population using safely managed drinking water services (Percent)	−0.225	−1.080	0.280
Quantile 0.90			
Gross domestic product per capita, PPP, dissemination (constant 2011 international $)	−0.119 *	−18.110	0.000
Average protein supply (g/cap/day) (3-year average)	−2.537 *	−168.980	0.000
Political stability and absence of violence/terrorism (index)	0.001 *	9.460	0.000
Percentage of population using safely managed drinking water services (Percent)	−0.094 *	−3.500	0.000
Mean VIF = 1.030			

Note: *, statistically significant at 1%; **, statistically significant at 5%; ***, statistically significant at 10%.

5. Encompassing the Climate Change among the Pillars for a More Sustainable Food Security

Figure 10 reveals a negative relationship between the prevalence of undernourishment and the temperature change over the period considered in the world countries. In other words, the increase in temperature change verified since 2000 was accompanied by a decrease in the prevalence of undernourishment.

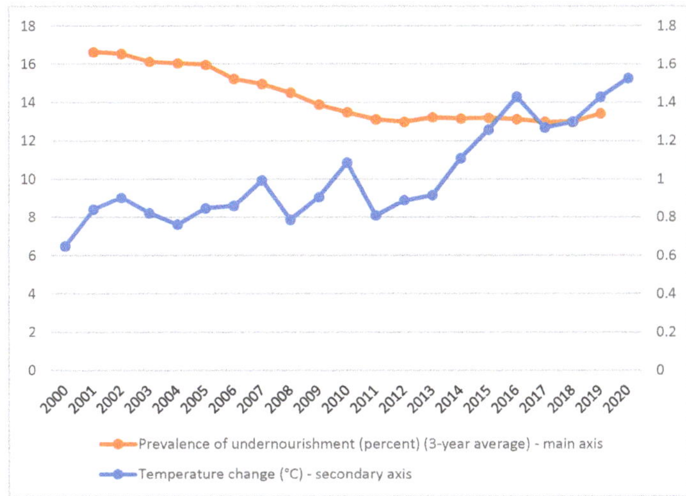

Figure 10. Benchmarking the prevalence of undernourishment and the temperature change, on average across the world over the period of 2000–2020.

To better assess the impacts of climate change on food insecurity, several regressions were repeated, considering now the temperature change as an indicator for climate change (Table 8). The granger causality test for balanced panel data, using the Stata software, supports the temperature change as an explanatory variable. For the period of 2000–2020, the results for the variables previously considered are similar. The findings for the temperature change growth rates are, in general, statistically significant since the quantile is 0.30 and consistently negative. This means that increases in the temperature change growth rates that promote decreases in the prevalence of undernourishment growth rates, nonetheless with coefficients around −0.001 (when the temperature change increases 1% point, the undernutrition decreases 0.001% points).

Table 8. Results for panel data quantile regressions with the variables in growth rates (encompassing climate change) worldwide over the period of 2000–2020.

Prevalence of Undernourishment (Percent) (3-Year Average)	Coefficient	z	$p > \lvert z \rvert$
Quantile 0.10			
Gross domestic product per capita, PPP, dissemination (constant 2011 international $)	0.077 **	2.090	0.037
Average protein supply (g/cap/day) (3-year average)	−2.674 *	−55.580	0.000
Political stability and absence of violence/terrorism (index)	0.001 *	4.080	0.000
Percentage of population using safely managed drinking water services (Percent)	0.337 *	3.740	0.000
Temperature change (°C)	0.000	0.160	0.875
Quantile 0.20			
Gross domestic product per capita, PPP, dissemination (constant 2011 international $)	0.077 *	3.080	0.002
Average protein supply (g/cap/day) (3-year average)	−2.718 *	−63.810	0.000
Political stability and absence of violence/terrorism (index)	0.002 *	5.370	0.000
Percentage of population using safely managed drinking water services (Percent)	0.152	1.640	0.101
Temperature change (°C)	0.000	0.810	0.420

Table 8. *Cont.*

Prevalence of Undernourishment (Percent) (3-Year Average)	Coefficient	z	$p > |z|$
Quantile 0.30			
Gross domestic product per capita, PPP, dissemination (constant 2011 international $)	−0.010	−0.450	0.652
Average protein supply (g/cap/day) (3-year average)	−2.280 *	−22.880	0.000
Political stability and absence of violence/terrorism (index)	0.003 *	12.740	0.000
Percentage of population using safely managed drinking water services (Percent)	−0.090 *	−2.760	0.006
Temperature change (°C)	−0.002 *	−6.060	0.000
Quantile 0.40			
Gross domestic product per capita, PPP, dissemination (constant 2011 international $)	−0.029	−0.800	0.424
Average protein supply (g/cap/day) (3-year average)	−2.243 *	−25.910	0.000
Political stability and absence of violence/terrorism (index)	0.003 *	13.550	0.000
Percentage of population using safely managed drinking water services (Percent)	−0.135 ***	−1.750	0.080
Temperature change (°C)	−0.001 *	−2.780	0.005
Quantile 0.50			
Gross domestic product per capita, PPP, dissemination (constant 2011 international $)	−0.021	−0.690	0.489
Average protein supply (g/cap/day) (3-year average)	−2.309 *	−28.610	0.000
Political stability and absence of violence/terrorism (index)	0.003 *	8.410	0.000
Percentage of population using safely managed drinking water services (Percent)	−0.165 ***	−1.710	0.087
Temperature change (°C)	−0.001 *	−3.670	0.000
Quantile 0.60			
Gross domestic product per capita, PPP, dissemination (constant 2011 international $)	−0.002	−0.090	0.927
Average protein supply (g/cap/day) (3-year average)	−2.302 *	−16.810	0.000
Political stability and absence of violence/terrorism (index)	0.003 *	3.070	0.002
Percentage of population using safely managed drinking water services (Percent)	0.117	0.880	0.379
Temperature change (°C)	−0.001 ***	−1.690	0.091
Quantile 0.70			
Gross domestic product per capita, PPP, dissemination (constant 2011 international $)	−0.034 *	−3.150	0.002
Average protein supply (g/cap/day) (3-year average)	−2.510 *	−24.620	0.000
Political stability and absence of violence/terrorism (index)	0.004 *	4.860	0.000
Percentage of population using safely managed drinking water services (Percent)	0.022	0.540	0.590
Temperature change (°C)	−0.001 *	−6.570	0.000
Quantile 0.80			
Gross domestic product per capita, PPP, dissemination (constant 2011 international $)	0.000	0.060	0.952
Average protein supply (g/cap/day) (3-year average)	−2.276 *	−31.300	0.000
Political stability and absence of violence/terrorism (index)	0.003 *	2.600	0.009
Percentage of population using safely managed drinking water services (Percent)	−0.398 *	−8.350	0.000
Temperature change (°C)	−0.001 **	−2.220	0.026
Quantile 0.90			
Gross domestic product per capita, PPP, dissemination (constant 2011 international $)	−0.131 *	−50.320	0.000
Average protein supply (g/cap/day) (3-year average)	−2.644 *	−195.770	0.000
Political stability and absence of violence/terrorism (index)	0.001 *	29.020	0.000
Percentage of population using safely managed drinking water services (Percent)	−0.160 *	−30.880	0.000
Temperature change (°C)	−0.000 *	−4.170	0.000
Mean VIF = 1.030			

Note: *, statistically significant at 1%; **, statistically significant at 5%; ***, statistically significant at 10%.

These findings highlight that, despite the lower marginal impacts, the improvements in the mitigation of malnutrition in the world have been supported by unsustainable approaches, showing that there are here fields that deserve special attention by the national and international organizations.

To benchmark, shorter periods (2000–2013 and 2014–2020) were considered to carry out the regressions and the results are those presented in Table 9. These findings highlight, again, the relevance of the protein supply to reduce the prevalence of undernourishment. In addition, the results show that there are differences between the two shorter periods. For example, the importance of the political stability and the absence of violence to reduce the prevalence of undernourishment was more evident in the second (2014–2020) period (quantiles 0.50 and 0.60). The impacts from the temperature change were less visible in the second period; nonetheless, when these impacts are statistically significant, the coefficient is positive (quantiles 0.60 and 0.90), showing that in the most recent years, the impacts from climate change are more dramatic and related with increasing levels of undernourishment.

Table 9. Results for panel data quantile regressions with the variables in growth rates (encompassing climate change) worldwide over the periods of 2000–2013 and 2014–2020.

Prevalence of Undernourishment (Percent) (3-Year Average)	2000–2013			2014–2020		
	Coefficient	z	$p > \lvert z \rvert$	Coefficient	z	$p > \lvert z \rvert$
Quantile 0.10						
Gross domestic product per capita, PPP, dissemination (constant 2011 international $)	0.018	0.190	0.851	0.726 **	2.400	0.017
Average protein supply (g/cap/day) (3-year average)	−2.444 *	−42.520	0.000	−3.049 *	−2.660	0.008
Political stability and absence of violence/terrorism (index)	0.004 ***	1.710	0.087	0.007	0.440	0.658
Percentage of population using safely managed drinking water services (Percent)	0.336 *	2.990	0.003	1.596	1.110	0.267
Temperature change (°C)	0.000	0.210	0.836	−0.013	−0.500	0.615
Quantile 0.20						
Gross domestic product per capita, PPP, dissemination (constant 2011 international $)	0.007	0.160	0.877	0.430	1.640	0.100
Average protein supply (g/cap/day) (3-year average)	−2.527 *	−39.840	0.000	−3.248 *	−2.870	0.004
Political stability and absence of violence/terrorism (index)	0.003 *	4.690	0.000	0.010	1.270	0.205
Percentage of population using safely managed drinking water services (Percent)	−0.206 *	−5.320	0.000	3.117	1.430	0.153
Temperature change (°C)	0.000	−0.200	0.844	0.001	0.040	0.965
Quantile 0.30						
Gross domestic product per capita, PPP, dissemination (constant 2011 international $)	−0.147 *	−10.660	0.000	0.557 **	2.210	0.027
Average protein supply (g/cap/day) (3-year average)	−2.042 *	−28.890	0.000	−4.736 *	−4.110	0.000
Political stability and absence of violence/terrorism (index)	0.003 *	11.760	0.000	0.010	1.460	0.145
Percentage of population using safely managed drinking water services (Percent)	−0.012	−0.850	0.397	6.153 **	2.310	0.021
Temperature change (°C)	−0.002 *	−11.510	0.000	−0.011	−0.320	0.748
Quantile 0.40						
Gross domestic product per capita, PPP, dissemination (constant 2011 international $)	−0.049 **	−1.970	0.049	0.300	0.810	0.418
Average protein supply (g/cap/day) (3-year average)	−2.310 *	−28.600	0.000	−3.971 *	−3.500	0.000
Political stability and absence of violence/terrorism (index)	0.003 *	18.460	0.000	−0.009	−1.390	0.164
Percentage of population using safely managed drinking water services (Percent)	−0.101	−0.820	0.411	6.022 ***	1.720	0.085
Temperature change (°C)	−0.001 *	−2.580	0.010	−0.015	−0.480	0.631

Table 9. *Cont.*

Prevalence of Undernourishment (Percent) (3-Year Average)	2000–2013			2014–2020		
	Coefficient	z	$p > \lvert z \rvert$	Coefficient	z	$p > \lvert z \rvert$
Quantile 0.50						
Gross domestic product per capita, PPP, dissemination (constant 2011 international $)	−0.059 *	−3.180	0.001	−0.076	−0.800	0.422
Average protein supply (g/cap/day) (3-year average)	−2.387 *	−20.700	0.000	−1.758 *	−11.230	0.000
Political stability and absence of violence/terrorism (index)	0.003 *	6.670	0.000	−0.002 *	−4.270	0.000
Percentage of population using safely managed drinking water services (Percent)	−0.246 *	−3.630	0.000	0.482 *	3.530	0.000
Temperature change (°C)	−0.002 *	−31.530	0.000	0.005	0.450	0.654
Quantile 0.60						
Gross domestic product per capita, PPP, dissemination (constant 2011 international $)	−0.040	−1.500	0.132	−0.130	−1.410	0.158
Average protein supply (g/cap/day) (3-year average)	−2.526 *	−19.960	0.000	−2.682 *	−11.300	0.000
Political stability and absence of violence/terrorism (index)	0.004 *	8.610	0.000	−0.009 *	−3.040	0.002
Percentage of population using safely managed drinking water services (Percent)	0.006	0.040	0.971	0.776 *	4.590	0.000
Temperature change (°C)	−0.001 *	−2.980	0.003	0.012 *	5.760	0.000
Quantile 0.70						
Gross domestic product per capita, PPP, dissemination (constant 2011 international $)	−0.033 ***	−1.780	0.074	0.245	1.400	0.162
Average protein supply (g/cap/day) (3-year average)	−2.441 *	−17.780	0.000	−2.323 *	−3.620	0.000
Political stability and absence of violence/terrorism (index)	0.004 *	9.720	0.000	−0.008	−1.200	0.230
Percentage of population using safely managed drinking water services (Percent)	−0.046	−0.890	0.371	1.874 *	2.890	0.004
Temperature change (°C)	−0.001 *	−6.520	0.000	−0.003	−0.270	0.787
Quantile 0.80						
Gross domestic product per capita, PPP, dissemination (constant 2011 international $)	−0.041 **	−2.240	0.025	0.154	0.860	0.388
Average protein supply (g/cap/day) (3-year average)	−2.336 *	−34.910	0.000	−3.278 *	−3.660	0.000
Political stability and absence of violence/terrorism (index)	0.003 *	4.650	0.000	−0.002	−0.590	0.553
Percentage of population using safely managed drinking water services (Percent)	0.053	0.400	0.688	0.978	0.840	0.401
Temperature change (°C)	−0.002 *	−3.050	0.002	0.003	0.230	0.816
Quantile 0.90						
Gross domestic product per capita, PPP, dissemination (constant 2011 international $)	−0.044	−1.300	0.194	−0.065	−1.050	0.294
Average protein supply (g/cap/day) (3-year average)	−2.314 *	−22.000	0.000	−3.601 *	−12.630	0.000
Political stability and absence of violence/terrorism (index)	0.005 *	16.760	0.000	−0.001	−0.630	0.531
Percentage of population using safely managed drinking water services (Percent)	−0.212 **	−2.050	0.040	−1.012 *	−2.890	0.004
Temperature change (°C)	0.000	0.170	0.863	0.021 ***	1.740	0.083
Mean VIF = 1.030						

Note: *, statistically significant at 1%; **, statistically significant at 5%; ***, statistically significant at 10%.

6. Discussion and Conclusions

The research here carried out intended to bring more insights into the world food security context and its relationship with sustainability. In other words, it was aimed to characterise the current context to understand if the four pillars associated with the concept of food security are enough and well-designed to deal with the present challenges, in a framework of sustainable development. To achieve these objectives, data from the FAOSTAT were considered, for the period of 2000–2020, that were assessed through, namely quantile regressions.

The food security concept is defined by four dimensions (availability, accessibility, utilization, and stability) and is interrelated with the Sustainable Development Goals. In fact, food security is part of the SDGs (goal 2 for zero hunger). Nonetheless, the question is if the sustainability dimensions are adequately considered by the food security pillars. The scientific literature highlights some concerns about these interrelationships and proposes either a new pillar [9] or considering other variables [12], namely those associated with climate change [14]. In addition, the several relationships between food security and sustainability are complex and need further multidisciplinary assessments. Agricultural practices, alternative sources of food supply, and public policies are determinants for more sustainable food security.

The data used shows that, worldwide over the period taken into account, the average protein supply was 78.804 g/cap/day, the average supply of protein of animal origin was 35.464 g/cap/day, the share of dietary supply derived from cereals, roots, and tubers accounted for 46.807 kcal/cap/day, and the political stability and absence of violence/terrorism (index) was −0.032. These results reveal that there is still much more to account for to improve the nutritional conditions and the political stability and absence of violence/terrorism. If we consider that the Western Balkans and the European Union had, over the period of 2016–2020, values of 101.340 for average protein supply (g/cap/day), 57.300 average supply of protein of animal origin (g/cap/day), and 0.550 for political stability and absence of violence/terrorism (index), there is a long way to run worldwide [89]. These findings highlight that a part of the world deals with problems of undernutrition while the other part has difficulties in reducing the prevalence of obesity. In any case, political stability seems to be one of the most important variables to improve the contexts of undernutrition.

The results from the regressions, for the period 2000–2020, show that undernutrition is principally impacted by the growth rates for average protein supply and for the political stability and absence of violence/terrorism (index). However, the political stability and absence of violence/terrorism are not enough to guarantee reductions in undernutrition. On the other hand, the economic and social conditions present evidences of statistically significance, namely in the countries with extreme values for undernourishment growth rates, but do not reduce the undernutrition in the countries with lower growth rates for the prevalence of undernourishment (first two quantiles). Finally, the indicator for the climate changes (temperature changes), in growth rates, contributed to reducing undernutrition growth rates. Considering that we used dynamic models with panel data, these results highlight, namely, that the undernutrition mitigation has been achieved through approaches associated with increases in the temperature. In fact, in specific contexts, adopting healthy diets may increase the environmental impacts [110]. In addition, the temperature changes impacts depend on local conditions [111]. With shorter periods (2000–2013 and 2014–2020), to benchmark, the findings highlight the importance of the political stability and absence of violence/terrorism to reduce the undernutrition and the negative impacts from the temperature change in the most recent years. Spearman's rank correlation matrix [112] highlights the importance of the dependent variable considered for food insecurity in the regressions.

In terms of practical implications, it can be seen that the food security concept is part of the Sustainable Development Goals. However, sustainability needs to be more appropriately addressed inside the food security pillars, creating new dimensions or renaming the current pillars. This is important to give adjusted signs for several stakeholders. In terms of policy recommendations, it could be important to design instruments and measures that mitigate undernutrition through sustainable and more environmentally compatible practices. Regarding the limitations, the main constraints found to develop in this research were associated with the availability of statistical information. For future research, it is suggested to assess how the approaches used to mitigate undernutrition negatively impact the environment and contribute to climate change.

Author Contributions: Conceptualization, R.d.P.F.G., M.L.d.J.P., C.A.d.C., D.d.V.T.A.d.C., P.B.C.d.S. and V.J.P.D.M.; methodology, V.J.P.D.M.; software, V.J.P.D.M.; validation, R.d.P.F.G., M.L.d.J.P., C.A.d.C., D.d.V.T.A.d.C., P.B.C.d.S. and V.J.P.D.M.; formal analysis, R.d.P.F.G., M.L.d.J.P., C.A.d.C., D.d.V.T.A.d.C., P.B.C.d.S. and V.J.P.D.M.; investigation, R.d.P.F.G., M.L.d.J.P., C.A.d.C., D.d.V.T.A.d.C., P.B.C.d.S. and V.J.P.D.M.; writing—original draft preparation, V.J.P.D.M.; writing—review and editing, R.d.P.F.G. and C.A.d.C.; supervision, V.J.P.D.M.; funding acquisition, R.d.P.F.G. All authors have read and agreed to the published version of the manuscript.

Funding: This work is funded by National Funds through the FCT—Foundation for Science and Technology, I.P., within the scope of the project Refª UIDB/00681/2020.

Data Availability Statement: The data presented in this study are available on request from the corresponding author.

Acknowledgments: Furthermore we would like to thank the CERNAS Research Centre and the Polytechnic Institute of Viseu for their support.

Conflicts of Interest: The authors declare no conflict of interest.

Appendix A

Table A1. Pairwise correlation matrix between the variables related to the food security and sustainability worldwide over the period of 2000–2020.

	A	B	C	D	E	F	G	H	I	J	K	L	M	N	O	P	Q	R	S	T	U	V	W
A	1.000																						
B	0.4924 * (0.000)	1.000																					
C	0.6711 * (0.000)	0.7896 * (0.000)	1.000																				
D	0.7497 * (0.000)	0.6022 * (0.000)	0.8918 * (0.000)	1.000																			
E	0.1955 * (0.000)	0.3961 * (0.000)	0.3968 * (0.000)	0.4231 * (0.000)	1.000																		
F	0.6292 * (0.000)	0.9521 * (0.000)	0.8823 * (0.000)	0.7537 * (0.000)	0.4405 * (0.000)	1.000																	
G	−0.6645 * (0.000)	−0.4572 * (0.000)	−0.6577 * (0.000)	−0.8441 * (0.000)	−0.4024 * (0.000)	−0.6219 * (0.000)	1.000																
H	−0.3220 * (0.000)	−0.8478 * (0.000)	−0.7302 * (0.000)	−0.6002 * (0.000)	−0.4028 * (0.000)	−0.8871 * (0.000)	0.4659 * (0.000)	1.000															
I	−0.032 (0.105)	−0.1346 * (0.000)	−0.1154 * (0.000)	−0.1082 * (0.000)	−0.3980 * (0.000)	−0.1510 * (0.000)	0.0507 * (0.009)	0.027 (0.275)	1.000														
J	0.0909 * (0.000)	0.2301 * (0.000)	0.2487 * (0.000)	0.2585 * (0.000)	0.5546 * (0.000)	0.2768 * (0.000)	−0.2774 * (0.000)	−0.2669 * (0.000)	−0.4459 * (0.000)	1.000													
K	−0.0393 * (0.023)	−0.0845 * (0.000)	−0.0627 * (0.001)	−0.0491 * (0.009)	−0.0945 * (0.000)	−0.0698 * (0.000)	0.014 (0.446)	0.0939 * (0.000)	0.0650 * (0.001)	0.008 (0.677)	1.000												
L	0.1909 * (0.000)	0.0809 * (0.000)	0.1254 * (0.000)	0.0661 * (0.001)	0.018 (0.342)	0.1118 * (0.000)	−0.0397 * (0.042)	−0.3093 * (0.000)	0.2841 * (0.000)	−0.0629 * (0.001)	0.025 (0.202)	1.000											
M	0.5224 * (0.000)	0.3307 * (0.000)	0.5045 * (0.000)	0.6388 * (0.000)	0.2257 * (0.000)	0.4623 * (0.000)	−0.6006 * (0.000)	−0.3959 * (0.000)	0.039 (0.053)	0.1966 * (0.000)	−0.0729 * (0.000)	−0.0586 * (0.003)	1.000										
N	−0.2110 * (0.000)	−0.2366 * (0.000)	−0.2111 * (0.000)	−0.1593 * (0.000)	−0.1905 * (0.000)	−0.2559 * (0.000)	0.0803 * (0.000)	0.1231 * (0.000)	0.2130 * (0.000)	−0.0904 * (0.000)	0.0670 * (0.000)	−0.016 (0.409)	0.022 (0.219)	1.000									
O	0.1773 * (0.000)	0.2945 * (0.000)	0.4240 * (0.000)	0.3412 * (0.000)	0.2640 * (0.000)	0.3978 * (0.000)	−0.3242 * (0.000)	−0.3133 * (0.000)	−0.1089 * (0.000)	0.2206 * (0.000)	0.1737 * (0.000)	0.1224 * (0.000)	0.0605 * (0.002)	−0.0718 * (0.000)	1.000								
P	−0.5857 * (0.000)	−0.5895 * (0.000)	−0.6998 * (0.000)	−0.7703 * (0.000)	−0.3767 * (0.000)	−0.7459 * (0.000)	0.7829 * (0.000)	0.6194 * (0.000)	−0.005 (0.827)	−0.2767 * (0.000)	−0.035 (0.074)	−0.2787 * (0.000)	−0.4937 * (0.000)	0.0590 * (0.003)	−0.3245 * (0.000)	1.000							
Q	0.5292 * (0.000)	0.5721 * (0.000)	0.6739 * (0.000)	0.7133 * (0.000)	0.2926 * (0.000)	0.7096 * (0.000)	−0.7021 * (0.000)	−0.7030 * (0.000)	−0.026 (0.178)	0.2225 * (0.000)	−0.009 (0.607)	0.3283 * (0.000)	0.4861 * (0.000)	−0.1007 * (0.000)	0.3804 * (0.000)	−0.7775 * (0.000)	1.000						
R	0.5875 * (0.000)	0.5611 * (0.000)	0.7268 * (0.000)	0.7483 * (0.000)	0.3185 * (0.000)	0.7201 * (0.000)	−0.7517 * (0.000)	−0.6554 * (0.000)	−0.0585 * (0.003)	0.2391 * (0.000)	0.013 (0.466)	0.3646 * (0.000)	0.4581 * (0.000)	−0.1524 * (0.000)	0.4691 * (0.000)	−0.7935 * (0.000)	0.8803 * (0.000)	1.000					

Table A1. *Cont.*

	A	B	C	D	E	F	G	H	I	J	K	L	M	N	O	P	Q	R	S	T	U	V	W
S	0.7033 *	0.6639 *	0.7982 *	0.8070 *	0.3286 *	0.7996 *	−0.7625 *	−0.6652 *	−0.047	0.1855 *	−0.0468 *	0.1635 *	0.5583 *	−0.3489 *	0.3454 *	−0.8148 *	0.8565 *	0.8691 *	1.000				
	(0.000)	(0.000)	(0.000)	(0.000)	(0.000)	(0.000)	(0.000)	(0.000)	(0.057)	(0.000)	(0.033)	(0.000)	(0.000)	(0.000)	(0.000)	(0.000)	(0.000)	(0.000)					
T	0.6890 *	0.5886 *	0.7427 *	0.7695 *	0.3229 *	0.7129 *	−0.7042 *	−0.4581 *	−0.0704	0.1756 *	−0.0882	0.1051 *	0.5376 *	−0.2559 *	0.0788 *	−0.6304 *	0.6457 *	0.7157 *	0.7704 *	1.000			
	(0.000)	(0.000)	(0.000)	(0.000)	(0.000)	(0.000)	(0.000)	(0.000)	(0.004)	(0.000)	(0.000)	(0.000)	(0.000)	(0.000)	(0.001)	(0.000)	(0.000)	(0.000)	(0.000)				
U	−0.5772 *	−0.4827 *	−0.6490 *	−0.7266 *	−0.4296 *	−0.6293 *	0.7274 *	0.4283 *	0.1163 *	−0.2737 *	0.0595 *	−0.1550 *	−0.5036 *	0.1357 *	−0.3943 *	0.5935 *	−0.6826 *	−0.7485 *	−0.7316 *	−0.6544 *	1.000		
	(0.000)	(0.000)	(0.000)	(0.000)	(0.000)	(0.000)	(0.000)	(0.000)	(0.000)	(0.000)	(0.001)	(0.000)	(0.000)	(0.000)	(0.000)	(0.000)	(0.000)	(0.000)	(0.000)	(0.000)			
V	−0.4089 *	−0.4257 *	−0.6067 *	−0.6238 *	−0.3829 *	−0.5464 *	0.6363 *	0.3434 *	0.2153 *	−0.2452 *	−0.0754 *	−0.1101 *	−0.3686 *	0.2607 *	−0.4712 *	0.5874 *	−0.4861 *	−0.6692 *	−0.6149 *	−0.4964 *	0.6058 *	1.000	
	(0.000)	(0.000)	(0.000)	(0.000)	(0.000)	(0.000)	(0.000)	(0.000)	(0.000)	(0.000)	(0.001)	(0.000)	(0.000)	(0.000)	(0.000)	(0.000)	(0.000)	(0.000)	(0.000)	(0.000)	(0.000)		
W	0.2944 *	0.5396 *	0.5513 *	0.5516 *	0.2140 *	0.6017 *	−0.6189 *	−0.4926 *	−0.014	0.2042 *	0.0425 *	0.2172 *	0.3923 *	0.0501 *	0.3431 *	−0.5739 *	0.5552 *	0.5537 *	0.4631 *	0.3194 *	−0.4476 *	−0.4938 *	1.000
	(0.000)	(0.000)	(0.000)	(0.000)	(0.000)	(0.000)	(0.000)	(0.000)	(0.490)	(0.000)	(0.025)	(0.000)	(0.000)	(0.008)	(0.000)	(0.000)	(0.000)	(0.000)	(0.000)	(0.000)	(0.000)	(0.000)	(0.000)

Note: A, gross domestic product per capita, PPP, dissemination (constant 2011 international $); B, average dietary energy supply adequacy (percent) (3-year average); C, average protein supply (g/cap/day) (3-year average); D, average supply of protein of animal origin (g/cap/day) (3-year average); E, average value of food production (constant 2004–2006 I$/cap) (3-year average); F, dietary energy supply used in the estimation of prevalence of undernourishment (kcal/cap/day) (3-year average); G, share of dietary energy supply derived from cereals, roots and tubers (kcal/cap/day) (3-year average); H, prevalence of undernourishment (percent) (3-year average); I, cereal import dependency ratio (percent) (3-year average); J, per capita food production variability (constant 2004–2006 thousand int$ per capita); K, per capita food supply variability (kcal/cap/day); L, percent of arable land equipped for irrigation (percent) (3-year average); M, political stability and absence of violence/terrorism (index); N, value of food imports in total merchandise exports (percent) (3-year average); O, percentage of children under 5 years of age who are overweight (modelled estimates) (percent); P, percentage of children under 5 years of age who are stunted (modelled estimates) (percent); Q, percentage of population using at least basic drinking water services (percent); R, percentage of population using at least basic sanitation services (percent); S, percentage of population using safely managed drinking water services (Percent); T, percentage of population using safely managed sanitation services (Percent); U, prevalence of anaemia among women of reproductive age (15–49 years); V, prevalence of low birthweight (percent); W, prevalence of obesity in the adult population (18 years and older). *, statistically significant at 5%.

References

1. Pérez-Escamilla, R. Food security and the 2015–2030 sustainable development goals: From human to planetary health: Perspectives and opinions. *Curr. Dev. Nutr.* **2017**, *1*, e000513. [CrossRef]
2. Wischnath, G.; Buhaug, H. On climate variability and civil war in Asia. *Clim. Chang.* **2014**, *122*, 709–721. [CrossRef]
3. FAO The State of Food and Agriculture. 1996. Available online: http://www.fao.org/3/w1358e/w1358e00.htm (accessed on 29 August 2021).
4. Magrini, E.; Vigani, M. Technology adoption and the multiple dimensions of food security: The case of maize in Tanzania. *Food Secur.* **2016**, *8*, 707–726. [CrossRef]
5. Nosratabadi, S.; Khazami, N.; Abdallah, M.B.; Lackner, Z.; Band, S.S.; Mosavi, A.; Mako, C. Social capital contributions to food security: A comprehensive literature review. *Foods* **2020**, *9*, 1650. [CrossRef]
6. Oakley, A.R.; Nikolaus, C.J.; Ellison, B.; Nickols-Richardson, S.M. Food insecurity and food preparation equipment in US households: Exploratory results from a cross-sectional questionnaire. *J. Hum. Nutr. Diet.* **2019**, *32*, 143–151. [CrossRef]
7. Houessou, M.D.; Cassee, A.; Sonneveld, B.G.J.S. The effects of the COVID-19 pandemic on food security in rural and urban settlements in benin: Do allotment gardens soften the blow? *Sustainability* **2021**, *13*, 7313. [CrossRef]
8. FAO. *Building a Common Vision for Sustainable Food and Agriculture: Principles and Approaches*; FAO: Rome, Italy, 2014. ISBN 978-92-5-108471-7.
9. Berry, E.M.; Dernini, S.; Burlingame, B.; Meybeck, A.; Conforti, P. Food Security and Sustainability: Can One Exist without the Other? *Public Health Nutr.* **2015**, *18*, 2293–2302. [CrossRef] [PubMed]
10. Brooks, J. Food security and the Sustainable Development Goals. In *Debate the Issues: New Approaches to Economic Challenges*; Love, P., Ed.; OECD Publishing: Paris, France, 2016.
11. United Nations. The Sustainable Development Goals Report. 2021. Available online: https://unstats.un.org/sdgs/report/2021/The-Sustainable-Development-Goals-Report-2021.pdf (accessed on 6 September 2021).
12. Richardson, R.B. Ecosystem services and food security: Economic perspectives on environmental sustainability. *Sustainability* **2010**, *2*, 3520–3548. [CrossRef]
13. Allen, T.; Prosperi, P.; Cogill, B.; Padilla, M.; Peri, I. A delphi approach to develop sustainable food system metrics. *Soc. Indic. Res.* **2019**, *141*, 1307–1339. [CrossRef]
14. Singh, N.P.; Anand, B.; Singh, S.; Srivastava, S.K.; Rao, C.S.; Rao, K.V.; Bal, S.K. Synergies and trade-offs for climate-resilient agriculture in india: An agro-climatic zone assessment. *Clim. Chang.* **2021**, *164*, 11. [CrossRef]
15. Maxwell, S.; Smith, M. Household food security: A conceptual review. In *Household Food Security: Concepts, Indicators, Measurements: A Technical Review*; Maxwell, S., Frankenberger, T.R., Eds.; UNICEF: New York, NY, USA; IFAD: Rome, Italy, 1992.
16. Peng, W.; Berry, E.M. The Concept of food security. In *Encyclopedia of Food Security and Sustainability*; Ferranti, P., Berry, E.M., Anderson, J.R., Eds.; Elsevier: Oxford, UK, 2019; pp. 1–7. ISBN 978-0-12-812688-2.
17. Eakin, H.; Connors, J.P.; Wharton, C.; Bertmann, F.; Xiong, A.; Stoltzfus, J. Identifying attributes of food system sustainability: Emerging themes and consensus. *Agric. Hum. Values* **2017**, *34*, 757–773. [CrossRef]
18. Guma, I.P.; Rwashana, A.S.; Oyo, B. Food security indicators for subsistence farmers sustainability: A system dynamics approach. *Int. J. Syst. Dyn. Appl.* **2018**, *7*, 45–64. [CrossRef]
19. Van den Brande, K.; Happaerts, S.; Bruyninckx, H. Multi-level interactions in a sustainable development context: Different routes for flanders to decision-making in the UN commission on sustainable development. *Environ. Policy Gov.* **2011**, *21*, 70–82. [CrossRef]
20. Green, A.; Nemecek, T.; Chaudhary, A.; Mathys, A. Assessing nutritional, health, and environmental sustainability dimensions of agri-food production. *Glob. Food Secur.-Agric.Policy* **2020**, *26*, 100406. [CrossRef]
21. Omotayo, A.O.; Ijatuyi, E.J.; Ogunniyi, A.I.; Aremu, A.O. Exploring the resource value of transvaal red milk wood (*Mimusops zeyheri*) for food security and sustainability: An appraisal of existing evidence. *Plants* **2020**, *9*, 1486. [CrossRef]
22. Lin, J.; Frank, M.; Reid, D. No home without hormones: How plant hormones control legume nodule organogenesis. *Plant Commun.* **2020**, *1*, 100104. [CrossRef]
23. Liu, L.; Yao, S.; Zhang, H.; Muhammed, A.; Xu, J.; Li, R.; Zhang, D.; Zhang, S.; Yang, X. Soil nitrate nitrogen buffer capacity and environmentally safe nitrogen rate for winter wheat-summer maize cropping in Northern China. *Agric. Water Manag.* **2019**, *213*, 445–453. [CrossRef]
24. Pereira-Dias, L.; Gil-Villar, D.; Castell-Zeising, V.; Quinones, A.; Calatayud, A.; Rodriguez-Burruezo, A.; Fita, A. Main root adaptations in pepper germplasm (*Capsicum* spp.) to phosphorus low-input conditions. *Agronomy* **2020**, *10*, 637. [CrossRef]
25. Aharon, S.; Peleg, Z.; Argaman, E.; Ben-David, R.; Lati, R.N. Image-based high-throughput phenotyping of cereals early vigor and weed-competitiveness traits. *Remote Sens.* **2020**, *12*, 3877. [CrossRef]
26. Liu, Y.; Zhuo, L.; Yang, X.; Ji, X.; Yue, Z.; Zhao, D.; Wu, P. Crop production allocations for saving water and improving calorie supply in China. *Front. Sustain. Food Syst.* **2021**, *5*, 632199. [CrossRef]
27. Mangu, V.R.; Ratnasekera, D.; Yabes, J.C.; Wing, R.A.; Baisakh, N. Functional screening of genes from a halophyte wild rice relative porteresia coarctata in arabidopsis model identifies candidate genes involved in salt tolerance. *Curr. Plant Biol.* **2019**, *18*, 100107. [CrossRef]

28. Qi, M.; Zheng, W.; Zhao, X.; Hohenstein, J.D.; Kandel, Y.; O'Conner, S.; Wang, Y.; Du, C.; Nettleton, D.; MacIntosh, G.C.; et al. QQS orphan gene and its interactor NF-YC4 reduce susceptibility to pathogens and pests. *Plant Biotechnol. J.* **2019**, *17*, 252–263. [CrossRef] [PubMed]

29. Campi, M.; Duenas, M.; Fagiolo, G. Specialization in food production affects global food security and food systems sustainability. *World Dev.* **2021**, *141*, 105411. [CrossRef]

30. Scherer, L.A.; Verburg, P.H.; Schulp, C.J.E. Opportunities for sustainable intensification in European agriculture. *Glob. Environ. Chang.-Hum. Policy Dimens.* **2018**, *48*, 43–55. [CrossRef]

31. Cui, Z.; Zhang, H.; Chen, X.; Zhang, C.; Ma, W.; Huang, C.; Zhang, W.; Mi, G.; Miao, Y.; Li, X.; et al. Pursuing sustainable productivity with millions of smallholder farmers. *Nature* **2018**, *555*, 363–366. [CrossRef] [PubMed]

32. Bhat, M.A.; Kumar, V.; Bhat, M.A.; Wani, I.A.; Dar, F.L.; Farooq, I.; Bhatti, F.; Koser, R.; Rahman, S.; Jan, A.T. Mechanistic insights of the interaction of Plant Growth-Promoting Rhizobacteria (PGPR) with plant roots toward enhancing plant productivity by alleviating salinity stress. *Front. Microbiol.* **2020**, *11*, 1952. [CrossRef]

33. Borchert, F.; Emadodin, I.; Kluss, C.; Rotter, A.; Reinsch, T. Grass growth and N2O emissions from soil after application of jellyfish in coastal areas. *Front. Mar. Sci.* **2021**, *8*, 711601. [CrossRef]

34. Buller, H.; Blokhuis, H.; Jensen, P.; Keeling, L. Towards farm animal welfare and sustainability. *Animals* **2018**, *8*, 81. [CrossRef] [PubMed]

35. Ren, S.; Guo, B.; Wu, X.; Zhang, L.; Ji, M.; Wang, J. Winter wheat planted area monitoring and yield modeling using modis data in the Huang-Huai-Hai Plain, China. *Comput. Electron. Agric.* **2021**, *182*, 106049. [CrossRef]

36. Pott, L.P.; Carneiro Amado, T.J.; Schwalbert, R.A.; Corassa, G.M.; Ciampitti, I.A. Satellite-based data fusion crop type classification and mapping in Rio Grande Do Sul, Brazil. *ISPRS-J. Photogramm. Remote Sens.* **2021**, *176*, 196–210. [CrossRef]

37. Rasool, R.; Fayaz, A.; ul Shafiq, M.; Singh, H.; Ahmed, P. Land use land cover change in kashmir himalaya: Linking remote sensing with an indicator based dpsir approach. *Ecol. Indic.* **2021**, *125*, 107447. [CrossRef]

38. Boyer, D.; Sarkar, J.; Ramaswami, A. Diets, food miles, and environmental sustainability of urban food systems: Analysis of nine indian cities. *Earth Future* **2019**, *7*, 911–922. [CrossRef]

39. Kakar, K.; Xuan, T.D.; Haqani, M.I.; Rayee, R.; Wafa, I.K.; Abdiani, S.; Tran, H.-D. Current situation and sustainable development of rice cultivation and production in afghanistan. *Agriculture* **2019**, *9*, 49. [CrossRef]

40. Raheem, D.; Dayoub, M.; Birech, R.; Nakiyemba, A. The contribution of cereal grains to food security and sustainability in africa: Potential application of UAV in Ghana, Nigeria, Uganda, and Namibia. *Urban Sci.* **2021**, *5*, 8. [CrossRef]

41. Aiking, H.; de Boer, J. Protein and sustainability—The potential of insects. *J. Insects Food Feed* **2019**, *5*, 3–7. [CrossRef]

42. Guiné, R.P.F.; Correia, P.; Coelho, C.; Costa, C.A. The role of edible insects to mitigate challenges for sustainability. *Open Agric.* **2021**, *6*, 24–36. [CrossRef]

43. Aiking, H.; de Boer, J. The next Protein Transition. *Trends Food Sci. Technol.* **2020**, *105*, 515–522. [CrossRef]

44. Bessa, L.W.; Pieterse, E.; Sigge, G.; Hoffman, L.C. Insects as human food; from farm to fork. *J. Sci. Food Agric.* **2020**, *100*, 5017–5022. [CrossRef]

45. Armanda, D.T.; Guinee, J.B.; Tukker, A. The second green revolution: Innovative urban agriculture's contribution to food security and sustainability—A review. *Glob. Food Secur.-Agric.Policy* **2019**, *22*, 13–24. [CrossRef]

46. Ceppa, F.; Faccenda, F.; De Filippo, C.; Albanese, D.; Pindo, M.; Martelli, R.; Marconi, P.; Lunelli, F.; Fava, F.; Parisi, G. Influence of essential oils in diet and life-stage on gut microbiota and fillet quality of rainbow trout (*Oncorhynchus mykiss*). *Int. J. Food Sci. Nutr.* **2018**, *69*, 318–333. [CrossRef]

47. Csortan, G.; Ward, J.; Roetman, P. Productivity, resource efficiency and financial savings: An investigation of the current capabilities and potential of south australian home food gardens. *PLoS ONE* **2020**, *15*, e0230232. [CrossRef]

48. Domingo, A.; Charles, K.-A.; Jacobs, M.; Brooker, D.; Hanning, R.M. Indigenous community perspectives of food security, sustainable food systems and strategies to enhance access to local and traditional healthy food for partnering williams treaties first nations (Ontario, Canada). *Int. J. Environ. Res. Public Health* **2021**, *18*, 4404. [CrossRef]

49. Fowles, T.M.; Nansen, C. Artificial selection of insects to bioconvert pre-consumer organic wastes. A review. *Agron. Sustain. Dev.* **2019**, *39*, 31. [CrossRef]

50. Kamal, H.; Le, C.F.; Salter, A.M.; Ali, A. Extraction of protein from food waste: An overview of current status and opportunities. *Compr. Rev. Food. Sci. Food Saf.* **2021**, *20*, 2455–2475. [CrossRef]

51. Park, H.; Higgs, E. A criteria and indicators monitoring framework for food forestry embedded in the principles of ecological restoration. *Environ. Monit. Assess.* **2018**, *190*, 113. [CrossRef]

52. Saraiva, A.; Carrascosa, C.; Raheem, D.; Ramos, F.; Raposo, A. Natural sweeteners: The relevance of food naturalness for consumers, food security aspects, sustainability and health impacts. *Int. J. Environ. Res. Public Health* **2020**, *17*, 6285. [CrossRef] [PubMed]

53. Scott, C. Sustainably sourced junk food? Big food and the challenge of sustainable diets. *Glob. Environ. Polit.* **2018**, *18*, 93–113. [CrossRef]

54. Sisodia, G.S.; Alshamsi, R.; Sergi, B.S. Business valuation strategy for new hydroponic farm development—A proposal towards sustainable agriculture development in united arab emirates. *Br. Food J.* **2021**, *123*, 1560–1577. [CrossRef]

55. Smith, O.M.; Cohen, A.L.; Rieser, C.J.; Davis, A.G.; Taylor, J.M.; Adesanya, A.W.; Jones, M.S.; Meier, A.R.; Reganold, J.P.; Orpet, R.J.; et al. Organic farming provides reliable environmental benefits but increases variability in crop yields: A global meta-analysis. *Front. Sustain. Food Syst.* **2019**, *3*, 82. [CrossRef]
56. Song, Q.; Joshi, M.; Joshi, V. Transcriptomic analysis of short-term salt stress response in watermelon seedlings. *Int. J. Mol. Sci.* **2020**, *21*, 6036. [CrossRef]
57. Stark, P.B.; Miller, D.; Carlson, T.J.; de Vasquez, K.R. Open-source food: Nutrition, toxicology, and availability of wild edible greens in the east bay. *PLoS ONE* **2019**, *14*, e0202450. [CrossRef]
58. Xue, L.; Liu, X.; Lu, S.; Cheng, G.; Hu, Y.; Liu, J.; Dou, Z.; Cheng, S.; Liu, G. China's food loss and waste embodies increasing environmental impacts. *Nat. Food* **2021**, *2*, 519–528. [CrossRef]
59. Appleby, M.C.; Mitchell, L.A. Understanding human and other animal behaviour: Ethology, welfare and food policy. *Appl. Anim. Behav. Sci.* **2018**, *205*, 126–131. [CrossRef]
60. Eyduran, S.P.; Akin, M.; Eyduran, E.; Celik, S.; Erturk, Y.E.; Ercisli, S. Forecasting Banana harvest area and production in turkey using time series analysis. *Erwerbs-Obstbau* **2020**, *62*, 281–291. [CrossRef]
61. Galanakis, C.M.; Rizou, M.; Aldawoud, T.M.S.; Ucak, I.; Rowan, N.J. Innovations and technology disruptions in the food sector within the COVID-19 pandemic and post-lockdown era. *Trends Food Sci. Technol.* **2021**, *110*, 193–200. [CrossRef]
62. Knorr, D.; Augustin, M.A. Food processing needs, advantages and misconceptions. *Trends Food Sci. Technol.* **2021**, *108*, 103–110. [CrossRef]
63. Markandya, A.; Salcone, J.; Hussain, S.; Mueller, A.; Thambi, S. COVID, the Environment and food systems: Contain, cope and rebuild better. *Front. Environ. Sci.* **2021**, *9*, 674432. [CrossRef]
64. Jagtap, S.; Garcia-Garcia, G.; Duong, L.; Swainson, M.; Martindale, W. Codesign of food system and circular economy approaches for the development of livestock feeds from insect larvae. *Foods* **2021**, *10*, 1701. [CrossRef] [PubMed]
65. Aliabadi, M.M.F.; Kakhky, M.D.; Sabouni, M.S.; Dourandish, A.; Amadeh, H. Food production diversity and diet diversification in Rural and Urban area of Iran. *J. Agric. Environ. Int. Dev.* **2021**, *115*, 59–70. [CrossRef]
66. Arshad, M.; Amjath-Babu, T.S.; Aravindakshan, S.; Krupnik, T.J.; Toussaint, V.; Kaechele, H.; Mueller, K. Climatic variability and thermal stress in Pakistan's rice and wheat systems: A stochastic frontier and quantile regression analysis of economic efficiency. *Ecol. Indic.* **2018**, *89*, 496–506. [CrossRef]
67. Asfaw, S.; Pallante, G.; Palma, A. Distributional impacts of soil erosion on agricultural productivity and welfare in Malawi. *Ecol. Econ.* **2020**, *177*, 106764. [CrossRef]
68. Barnwal, P.; Kotani, K. Climatic impacts across agricultural crop yield distributions: An application of quantile regression on rice crops in Andhra Pradesh, India. *Ecol. Econ.* **2013**, *87*, 95–109. [CrossRef]
69. Bhattacharya, P.; Mitra, S.; Siddiqui, M.Z. Dynamics of foodgrain deficiency in India. *Margin* **2016**, *10*, 465–498. [CrossRef]
70. Bhuyan, B.; Sahoo, B.K.; Suar, D. Quantile regression analysis of predictors of calorie demand in india: An implication for sustainable development goals. *J. Quant. Econ.* **2020**, *18*, 825–859. [CrossRef]
71. Cupak, A.; Pokrivcak, J.; Rizov, M. Diversity of food consumption in Slovakia. *Polit. Ekon.* **2016**, *64*, 608–626. [CrossRef]
72. D'Souza, A.; Jolliffe, D. Food insecurity in vulnerable populations: Coping with food price shocks in afghanistan. *Am. J. Agr. Econ.* **2014**, *96*, 790–812. [CrossRef]
73. Djella, A.; Cembalo, L.; Furno, M.; Caraccaccacciolo, F. Is oil export a curse in developing economies? Evidence of paradox of plenty on food dependency. *New Medit* **2019**, *18*, 51–63. [CrossRef]
74. Hobbs, S.; King, C. The unequal impact of food insecurity on cognitive and behavioral outcomes among 5-year-old urban children. *J. Nutr. Educ. Behav.* **2018**, *50*, 687–694. [CrossRef]
75. Ikudayisi, A.; Okoruwa, V.; Omonona, B. From the lens of food accessibility and dietary quality: Gaining insights from urban food security in Nigeria. *Outlook Agric.* **2019**, *48*, 336–343. [CrossRef]
76. Irshad, C.; Dash, U.; Muraleedharan, V.R. Healthy ageing in india; a quantile regression approach. *J. Popul. Ageing* **2021**, 1–22. [CrossRef]
77. Ma, W.; Wang, X. Internet use, sustainable agricultural practices and rural incomes: Evidence from China*. *Aust. J. Agric. Resour. Econ.* **2020**, *64*, 1087–1112. [CrossRef]
78. Mahadevan, R.; Hoang, V. Is there a link between poverty and food security? *Soc. Indic. Res.* **2016**, *128*, 179–199. [CrossRef]
79. Mahadevan, R.; Suardi, S. Is there a role for caste and religion in food security policy? A look at Rural India. *Econ. Model.* **2013**, *31*, 58–69. [CrossRef]
80. Mahadevan, R.; Suardi, S. Regional differences pose challenges for food security policy: A case study of India. *Reg. Stud.* **2014**, *48*, 1319–1336. [CrossRef]
81. Mahajan, S.; Sousa-Poza, A.; Datta, K.K. Differential effects of rising food prices on indian households differing in Income. *Food Secur.* **2015**, *7*, 1043–1053. [CrossRef]
82. Mahmood, N.; Arshad, M.; Kaechele, H.; Ullah, A.; Mueller, K. Economic efficiency of rainfed wheat farmers under changing climate: Evidence from Pakistan. *Environ. Sci. Pollut. Res.* **2020**, *27*, 34453–34467. [CrossRef] [PubMed]
83. Tesfaye, W.; Tirivayi, N. Crop diversity, household welfare and consumption smoothing under risk: Evidence from Rural Uganda. *World Dev.* **2020**, *125*, 104686. [CrossRef]
84. Ugur, A.A.; Ozocakli, D. Some determinants of food insecurity (comparison of quantile regression method and fixed effect panel method). *Sosyoekonom* **2018**, *26*, 195–205. [CrossRef]

85. Van den Broeck, G.; Swinnen, J.; Maertens, M. Global value value chains, large-scale farming, and poverty: Long-term effects in Senegal. *Food Policy* **2017**, *66*, 97–107. [CrossRef]
86. Vilar-Compte, M.; Sandoval-Olascoaga, S.; Bernal-Stuart, A.; Shimoga, S.; Vargas-Bustamante, A. The impact of the 2008 financial crisis on food security and food expenditures in mexico: A disproportionate effect on the vulnerable. *Public Health Nutr.* **2015**, *18*, 2934–2942. [CrossRef]
87. FAOSTAT Several Statistics. Available online: http://www.fao.org/faostat/en/#data (accessed on 19 August 2021).
88. Mengel, M.; Treu, S.; Lange, S.; Frieler, K. ATTRICI v1.1-counterfactual climate for impact attribution. *Geosci. Model Dev.* **2021**, *14*, 5269–5284. [CrossRef]
89. Matkovski, B.; Đokić, D.; Zekić, S.; Jurjević, Ž. Determining food security in crisis conditions: A comparative analysis of the western balkans and the EU. *Sustainability* **2020**, *12*, 9924. [CrossRef]
90. Galton, F. Co-relations and their measurement, chiefly from anthropometric data. *Proc. R. Soc. Lond.* **1888**, *45*, 135–145.
91. Pearson, K. Mathematical contributions to the theory of evolution—III. Regression, heredity, and panmixia. *Philos. Trans. R. Soc. Lond. Ser. A* **1896**, *187*, 253–318. [CrossRef]
92. Pearson, K.; Filon, L.N.G. Mathematical contributions to the theory of evolution.— IV. On the probable errors of frequency constants and on the influence of random selection on variation and correlation. *Philos. Trans. R. Soc. London. Ser. A* **1898**, *191*, 229–311. [CrossRef]
93. StataCorp. *Stata 15 Base Reference Manual*; Stata Press: College Station, TX, USA, 2017.
94. StataCorp. *Stata Statistical Software: Release 15*; StataCorp LLC: College Station, TX, USA, 2017.
95. Stata Stata: Software for Statistics and Data Science. Available online: https://www.stata.com/ (accessed on 19 August 2021).
96. Bhargava, A. Nutrition, health, and economic development: Some policy priorities. *Food Nutr Bull* **2001**, *22*, 173–177. [CrossRef]
97. Vincent, J.L. *Factor Analysis in International Relations: Interpretation, Problem, Areas, and an Application*; Univ Pr of Florida: Gainesville, FL, USA, 1971; ISBN 978-0-8130-0315-3.
98. Kim, J.-O.; Mueller, C.W. *Factor Analysis: Statistical Methods and Practical Issues*; SAGE Publications, Inc: Newbury Park, CA, USA, 1978; ISBN 978-0-8039-1166-6.
99. Kim, J.-O.; Mueller, C.W. *Introduction to Factor Analysis: What It Is and How To Do It*, 1st ed.; SAGE Publications, Inc: Beverly Hills, CA, USA, 1978; ISBN 978-0-8039-1165-9.
100. Torres-Reyna, O. Getting Started in Factor Analysis (Using Stata) (Ver. 1.0 Beta/Draft). Available online: http://www.princeton.edu/~{}otorres/Stata/Factor (accessed on 19 August 2021).
101. Koenker, R. *Quantile Regression*; Cambridge University Press: Cambridge, UK, 2005.
102. Torres-Reyna, O. Panel Data Analysis Fixed and Random Effects Using Stata (v. 4.2). Available online: https://www.princeton.edu/~{}otorres/Panel101.pdf (accessed on 19 August 2021).
103. Alejo, J.; Galvao, A.; Montes-Rojas, G.; Sosa-Escudero, W. Tests for normality in linear panel-data models. *Stata J.* **2015**, *15*, 822–832. [CrossRef]
104. Verdoorn, P.J. Fattori che regolano lo sviluppo della produttivitá del lavoro. *L'Industria* **1949**, *1*, 3–10.
105. Kaldor, N. *Causes of the Slow Rate of Economic Growth of the United Kingdom: An Inaugural Lecture*; Cambridge University Press: Cambridge, UK, 1966.
106. Kaldor, N. *Strategic Factors in Economic Development*; New York State School of Industrial and Labor Relations, Cornell University: Ithaca, NY, USA, 1967; ISBN 978-0-87546-024-6.
107. Wells, H.; Thirlwall, A.P. Testing kaldor's growth laws across the countries of Africa. *Afr. Dev. Rev.* **2003**, *15*, 89–105. [CrossRef]
108. Martinho, V.J.P.D. Relationships between agricultural energy and farming indicators. *Renew. Sustain. Energy Rev.* **2020**, *132*, 110096. [CrossRef]
109. Rodriguez, R.N.; Yao, Y. Five things you should know about quantile regression. In Proceedings of the SAS Global Forum 2017 Conference, Orlando, FL, USA, 2–5 April 2017; SAS Institute Inc.: Cary, NC, USA.
110. Aleksandrowicz, L.; Green, R.; Joy, E.J.M.; Harris, F.; Hillier, J.; Vetter, S.H.; Smith, P.; Kulkarni, B.; Dangour, A.D.; Haines, A. Environmental impacts of dietary shifts in india: A modelling study using nationally-representative data. *Environ. Int.* **2019**, *126*, 207–215. [CrossRef] [PubMed]
111. Ebi, K.L.; Hasegawa, T.; Hayes, K.; Monaghan, A.; Paz, S.; Berry, P. Health risks of warming of 1.5 degrees C, 2 degrees C, and higher, above pre-industrial temperatures. *Environ. Res. Lett.* **2018**, *13*, 063007. [CrossRef]
112. Spearman, C. The Proof and Measurement of Association between Two Things. *Am. J. Psychol.* **1904**, *15*, 72–101. [CrossRef]

 *foods*

Article

Meta-Learning for Few-Shot Plant Disease Detection

Liangzhe Chen [1], Xiaohui Cui [2] and Wei Li [1,*]

[1] School of Artificial Intelligence and Computer Science & Jiangsu Key Laboratory of Media Design and Software Technology & Science Center for Future Foods, Jiangnan University, Wuxi 214122, China; lzchen@stu.jiangnan.edu.cn

[2] School of Cyber Science and Engineering, Wuhan University, Wuhan 430072, China; xcui@whu.edu.cn

* Correspondence: cs_weili@jiangnan.edu.cn

Abstract: Plant diseases can harm crop growth, and the crop production has a deep impact on food. Although the existing works adopt Convolutional Neural Networks (CNNs) to detect plant diseases such as Apple Scab and Squash Powdery mildew, those methods have limitations as they rely on a large amount of manually labeled data. Collecting enough labeled data is not often the case in practice because: plant pathogens are variable and farm environments make collecting data difficulty. Methods based on deep learning suffer from low accuracy and confidence when facing few-shot samples. In this paper, we propose local feature matching conditional neural adaptive processes (LFM-CNAPS) based on meta-learning that aims at detecting plant diseases of unseen categories with only a few annotated examples, and visualize input regions that are 'important' for predictions. To train our network, we contribute Miniplantdisease-Dataset that contains 26 plant species and 60 plant diseases. Comprehensive experiments demonstrate that our proposed LFM-CNAPS method outperforms the existing methods.

Keywords: food security; plant disease detection; convolutional neural networks; few-shot; meta-learning

Citation: Chen, L.; Cui, X.; Li, W. Meta-Learning for Few-Shot Plant Disease Detection. *Foods* **2021**, *10*, 2441. https://doi.org/10.3390/foods10102441

Academic Editors: António Raposo, Renata Puppin Zandonadi and Raquel Braz Assunção Botelho

Received: 10 September 2021
Accepted: 12 October 2021
Published: 14 October 2021

Publisher's Note: MDPI stays neutral with regard to jurisdictional claims in published maps and institutional affiliations.

1. Introduction

Food shortages may increase in many regions of the world. Coupled with pests and crop failures, food prices have soared. A lot of people may face severe hunger and death. In order to solve the food shortage, it is necessary to ensure the food security and sustainability. Due to pests, diseases [1,2], and lack of horticultural expertise [3–5], food yield loss is greater than 50% [6]. Food security is increasingly affected by crop production [7]. With the increase of agricultural intensification and the continuous strengthening of the agricultural industry chain, the risks related to viruses and pollution will increase. For the goal of global food security and sustainable development, by 2050, the current demand of crop disease detection needs to increase by 50% [8].

The traditional method of plant disease detection is manual inspection by farmers or experts. The method of plant disease diagnosis through optical observation of the symptoms on plant leaves incorporates a significantly high degree of complexity [2]. The method laboratory-based such as polymerase chain reaction (PCR), immunofluorescence (IF), and fluorescence in-situhybridization (FISH) require professional laboratory equipment and mass sampling work [9]. Due to this complexity and to the large number of cultivated plants and their existing phytopathological problems, manual plant disease detection can be time-consuming and expensive [10]. By contrast, images under analysis were obtained by employing cameras operating in the visible portion of the electromagnetic spectrum (400–700 nm). In this way, costly equipment or trained personnel are not required for obtaining the input data [11]. Therefore, future users of the developed protocol can acquire data through affordable/cost-effective, portable (thus in situ), and rapid means. With the development of computational systems in recent years, and in particular Graphical Processing Units (GPU) embedded processors, Convolutional Neural Networks (CNNs) [12] is often applied for image classification.

CNNs belong to a stackable feedforward neural network community [12]. The method of image classification through multi-layer CNNs is also called deep learning [13–15]. CNNs have good characterization learning ability, so they are mostly used for feature extraction, and the extracted features have the characteristics of translation invariance. The research on CNNs began in the 1980s and 1990s, and the time delay network and LeNet-5 were the earliest CNNs [12]. For a convolution operation, the essence is a traversal of the convolution kernel on the feature image. The convolution kernel will multiply and add the value at the corresponding position of the input feature image. In recent years, CNNs have been increasingly incorporated in plant phenotyping concepts. They have been very successful in modeling complicated systems, owing to their ability of distinguishing patterns and extracting regularities from data. Examples further extend to the variety identification in seeds [16] and in intact plants by using leaves [17]. Some research [18] collected and published the datasets of plant diseases that provided data sources for other methods. There are also some works [10] using image segmentation technology to separate the foreground and the background that can further improve the classification accuracy, and also solve the problem of poor performance on the online test.

Although the above deep learning methods have good performance related to plant disease detection, they still have the following problems. The first is the common problem of deep learning; model training requires a large amount of manually labeled datasets. The above-mentioned methods are currently based on the support of a large amount of data. Each category requires more than 1000 pictures. Data collection and marking require manpower and time. There is not enough data to support network training for plant pathogen variables in time, space, and genotype [19]. The second problem is that many of the more than 700 known plant viruses cause devastating diseases and often have wide host ranges. Barley yellow dwarf viruses (BYDV), for example, are distributed worldwide and infect over 150 species of the Poaceae, including most of the staple cereals—wheat, barley, oats, rye, rice, and maize [19]. It is unrealistic to identify all plant diseases at once through one task. However, the emergence of new tasks requires retraining the network. The above methods all limit the total number of categories for specific classification of several plant diseases. Every time a new task is encountered, based on traditional deep learning methods, it is necessary to rearrange the data and train the network to adapt to the task. For different sample numbers and image sizes, professional knowledge is needed to fine-tune the hyperparameters in the network structure. The last problem is the poor interpretability of the method. Compared with manual detection, experts can provide the basis for plant disease detection such as Oval-shaped irregular brown spots appearing on the leaves of plants with rust, and the leaf color on the leaves gradually becoming lighter, and, using fluorescence imaging, temporal and spatial variations of chlorophyll fluorescence were analyzed for precise detection of leaf rust and powdery mildew infections in wheat leaves at 470 nm [9]. Although the deep-learning-based methods show their effectiveness, it cannot explain their decisions and actions to human users. Therefore, the methods should give visual explanations to illustrate that our approach focuses on diseases' classification. This paper proposed a meta-learning [20] method to solve the challenge of plant disease detection.

Meta-learning is the science of systematically observing how different machine learning approaches perform on a wide range of learning tasks, and then learning from this experience, or meta-data, to learn new tasks much faster than otherwise possible [21]. There is a lot of research on meta-learning. Meta-learning is transfer learning in a broad sense [22], which chooses data from different sources to train the network so that the model has a good classification effect on all kinds of tasks. Few-shot learning [23], which is the problem of making predictions based on a limited number of samples, is an important application direction of meta-learning. The network is trained through other multi-source and sufficient datasets, so that it can deal with the task with few training samples. There is a lot of research on few-shot learning. For example, CNAPS [24] and Simple-CNAPS [25] use forward propagation instead of back propagation to solve the problem of overfitting, modular adaptation method [26], and Meta Fine-Tuning [27], which is also called Cross-Domain Few-Shot learning, can be trained to perform both tasks across domains. There is also some research on metric learning [28] such as MatchingNet [29] and ProtoNet [23] to solve the problem of insufficient samples and the poor performance of the classifier. Although there have been many works on few-shot learning, most of the works are more theoretically focused, and do not focus on specific applications. Based on the previous work, this paper applied meta-learning to plant diseases detection.

2. Materials and Methods

In this section, initially, the datasets chosen for training and testing are introduced. Afterwards, the meta-learning method proposed for plant detection called LFM-CNAPS is presented. Finally, visual explanations technology called TAM is introduced.

2.1. Datasets

The key to few-shot plant disease detection lies in the generalization ability of the pertinent model when presented with novel disease categories. Thus, high-diversity datasets are necessary for training the model that can detect unseen plant diseases. In this paper, Meta-Dataset [30] and Miniplantdisease-Dataset are chosen for model training.

A Meta-Dataset [30] is composed of 10 public datasets including ILSVRC-2012 (ImageNet) [31], Omniglot [32], FGVC-Aircraft (Air-craft) [33], CUB-200-2011 (Birds) [34], Describable Textures (DTD) [35], QuickDraw [36], FGVCx Fungi (Fungi)[37], VGG Flower (Flower) [38], Traffic Signs (Signs) [39], and MSCOCO [40]. Meta-Dataset is comprised of multiple existing datasets that contains more than 110,000 few-shot classification tasks. The tasks span a variety of visual concepts (natural and human-made) and vary in how fine-grained the class definition is [30]. Through Meta-Dataset training, the models' ability to leverage diverse training sources will be improved.

Miniplantdisease-Dataset proposed in this paper is composed of Apple foliar disease Dataset [41] and PlantVillage-Dataset [18]. The Apple foliar disease Dataset contains 3651 high-quality and real photos of various apple foliar diseases. The PlantVillage-Dataset has released more than 50,000 specialized images through the online platform Plantvillage [18]. The PlantVillage-Dataset contains various diseases per plant categories, while the Apple foliar disease Dataset only contains healthy and unhealthy two labels per plant categories. In keeping with a results report, the PlantVillage is divided into two parts with a ratio of 8:2 to compose Miniplantdisease-Dataset and test model, respectively. We report results using 48 plant diseases for training called in-domain Miniplantdisease-Dataset, reserving other 12 plant diseases for out-of-domain performance evaluation.

2.2. LFM-CNAPS

The method proposed in this paper to solve few-shot plant disease recognition is local feature matching conditional neural adaptive processes (LFM-CNAPS). As shown in Figure 1, it contains four main parts: input task, conditional adaptive feature extractor, and local feature matching classifier and parameters optimizer.

2.2.1. Task

Miniplantdisease-Dataset contains many different meta-tasks. The model learns the generalization ability from meta-tasks. When facing new categories, the classification can be completed without changing the existing model. The Miniplantdisease-Dataset contains 60 plant disease categories, with multiple samples in each category. For any meta-task in Miniplantdisease-Dataset, five plant disease categories will be randomly selected, with five samples for each category (a total of 25 samples). These samples with their labels will be constructed as support images and support labels of meta-task. In addition, then extract 50 samples for test from the remaining five categories samples as query images and query labels. That is, the model is required to learn how to distinguish these five categories from 25 samples. Such a task is called a 5-way 5-shot problem.

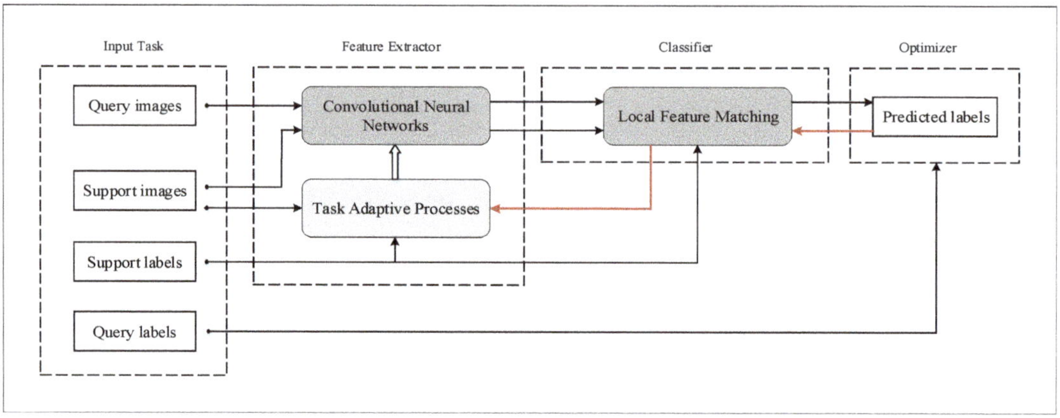

Figure 1. Flow chart of LFM-CNAPS. In the figure, the white solid squares are data variable, the black arrows are the data flow direction, the white hollow arrows are forward propagation, and the red arrows are backward propagation.

2.2.2. Conditional Adaptive Feature Extractor

The feature extractor chosen in this paper consists of two parts, a CNN framework named RESNET18 [42] and task adaptive processes [24]. Among them, RESNET18 is a stackable CNN layer with a batch normalization layer to prevent vanishing gradient and exploding gradient. The task adaptive process [24] is an effective method of impacting CNN intermediate variables to adapt the task [43]. The core of the task adaptive process is to choose forward propagation instead of back propagation to prevent overfitting due to few samples.

RESNET18 mainly contains CNNs and a Batch Normalization layer. The CNNs are essentially to do a dot product between the filter and the local area of the input data. The convolution kernel will multiply and add the value at the corresponding position of the input data, as shown in (1):

$$Con(I_{x,y}, K) = \sum_{c=1}^{C} \sum_{h=1}^{H} \sum_{w=1}^{W} I_{x+h,y+w,c} \times K_{h,w} \tag{1}$$

Among them, I represents the input data, K represents the convolution kernel, and x and y represent the position of the convolution kernel on the feature map I. H, W, and C respectively represent the length and width of the convolution kernel and the number of channels. The CNNs will extract specific local features according to the convolution kernel parameters. For CNNs, there are many hyperparameters such as the size of the convolution kernel, sliding step size, and the number of CNN layers. Different hyperparameter settings will have a great impact on the accuracy of the model. RESNET18 gives the hyperparameters suitable for most image feature extractions [42]. The core of RESNET18 [42] is stackable CNN layers, and the hyperparameters of the CNNs are fixed. On this basis, Batch Normalization layer is applied. In the Batch Normalization layer, such an operation is shown in (2):

$$H(X) = F(X) + X \tag{2}$$

Among them, $H(X)$ represents the Batch Normalization operation, $F(X)$ represents the corresponding CNNs, and X represents the input. It can be seen intuitively that the process of Batch Normalization layer is to add the input X and the result of the CNNs. Batch Normalization effectively solves the problem of gradient vanishing and degradation caused by the network being too deep. Batch Normalization layers make deep network training possible.

The task adaptive process contains the task encoder and FILM layer [44]. The task encoder is composed of CNN layers and fully connected layers that take the support set as input and FILM layer parameters as output. The task encoder provides FILM layer parameters to make CNNs better adapt tasks. For traditional deep learning methods, back propagation is an important method of updating parameters. However, most deep learning methods require a large number of labeled samples. For a few-shot task, there are only a few labeled samples for training. Too few samples to update the parameters through back propagation will cause overfitting. That is, the accuracy of the training set is very high, and the result of the test set is very poor. To avoid this, the FILM layer [44] is proposed to perform affine transformation on the intermediate features of the CNNs, as shown in (3):

$$\sum_{c=1}^{C} \sum_{x=1}^{X} \sum_{y=1}^{Y} I_{x,y,c} \times \gamma_c + \beta_c \tag{3}$$

Among them, I is the middle feature map, and X, Y, and C represent the length and width of the feature and the number of channels. γ and β are the parameters of the FILM layer and they are generated by the task encoder. The parameters updated by back propagation are proportional to the volume of the convolution kernel, and the forward propagation only needs to update the parameters that are proportional to the number of channels. Therefore, the depth of the overall network has not changed, but the number of updated parameters are reduced, avoiding the overfitting caused by few samples.

2.2.3. Local Feature Matching Classifier

For image classification methods, classifiers are indispensable [45]. For the traditional method, after handcrafted features are extracted, a separate classifier such as SVM is needed. For deep learning, a fully connected layer and activation function are generally chosen as a classifier. For SVM, the parameters need to be trained separately [46]. For deep learning classifier that can be trained end-to-end, the fully connected layer contains hundreds of parameters that need to be optimized. When samples are not enough, parameter optimization can be difficult. Therefore, this paper chooses metric learning as the classifier for few-shot plant disease detection.

The obvious advantage of the metric learning classifier is that there are no parameters be optimized. For metric learning, the distance between the feature value and the prototype [23] is calculated to determine which category the query sample belongs to. The concept of prototype comes from the prototype network [23], and the most common definition of prototype is the average of each category's features. To output the labels of query set, methods usually calculate the metric distance between query set and each prototype.

This paper chooses the local feature matching classifier [47]. This method has two advantages. First, for other metric learning methods, the extracted features need to be pooled that destroy the original spatial information of the features. The local feature matching classifier directly takes the extracted high-dimensional features for classification. Secondly, the local feature matching method can reduce the impact of occlusion or noise on classification to a certain extent that improves the robustness of the algorithm. The calculation process of the local feature matching classifier is as follows, as shown in (4):

$$\sum_{x1=1}^{W} \sum_{y1=1}^{H} Max_K \left(\left\{ \frac{F^q_{x1,y1} \cdot F^c_{x2,y2}}{\left| F^q_{x1,y1} \right| \times \left| F^c_{x2,y2} \right|} \mid 1 \leqslant x2 \leqslant W, 1 \leqslant y2 \leqslant H \right\} \right) \tag{4}$$

Among them, F^q and F^c respectively represent the feature of query set and the prototype, and H and W are the length and width of the feature map. $Max_K()$ represents the function that is to select top K maximums. The classifier regards each pixel of the feature map as a local feature of the image. The calculation process of (4) is to traverse all local features on F^q. Calculate the cosine distance between the local features on F^q and all the local features on F^c. The maximum K values summation is selected as the matching value, and the final summation of all matching values is the metric distance between F^q and F^c. The larger the metric distance value represents, the closer F^q and F^c is. When all the category prototypes are traversed, the category with the largest metric value is selected as the category of the query image F^q.

2.2.4. Parameters Optimizer

The meta-learning method chosen in this paper contains the following parameters: the CNN parameters in RESNET18, the task adaptive encoder parameters, and the FILM layer parameters. Among them, the parameters of RESNET18 are pre-trained and do not participate in the update, and the parameters of the FILM layer are generated by the task encoder. Therefore, for the LFM-CNAPS, it is the parameters in the task adaptive encoder that need to be trained and updated. The parameter update is reflected in two parts. First, during the meta-training process, the parameters in the task adaptive encoder are updated through back propagation. Secondly, during the meta-test process, the parameters in the FILM layer are updated through forward propagation. The parameters optimizer is proposed for back propagation.

For the optimizer, the most important thing is the loss function and optimization method. The loss function and optimization algorithm chosen in this paper are cross entropy loss [48] and Adam algorithms [49]. The cross entropy loss is calculated as (5):

$$\sum y_c log(p_c) \tag{5}$$

where y represents the category label, c represents the category name, and p represents the predicted probability. If the query image this time is of category c, then the value of y_c is 1; otherwise, it is 0. For the prediction result of the algorithm, various probabilities p_c are obtained through the sigmoid activation function. In summary, the cross-entropy loss obtains a loss value from the label predicted by the model and the actual label.

2.3. Task Activation Mapping

For deep learning, most algorithms are black box. They reduce the loss through back propagation and improve the test accuracy through a large number of samples. However, deep models are not easy to visualize and could not give the basis of classification results. For CNNs, there have been many studies on visual explanations [50].

The TAM algorithm proposed in this paper is modified on the basis of the Grad-CAM [50]. The Grad-CAM process is as follows: first, a test image is needed as input, and the classification probability is obtained through the trained network. Grad-CAM will select the channel where the back propagation is located through the label. When the back propagation reaches the last layer of the CNNs, Grad-CAM would record the parameter gradient of the last layer. The gradient tensor will be averaged in the channel direction, and a one-dimensional variable whose length is the number of channels will be obtained. Grad-CAM would multiply the one-dimensional variable with the intermediate variable of the last layer to obtain the activated intermediate variable. The intermediate variables will be averaged in the direction of the feature map to obtain a two-dimensional activation layer. Grad-CAM will convert the two-dimensional activation layer mapping from 0 to 255 into a heat map. The heat map will be mapped to the input image to get a visual CNN heat map. Grad-CAM obtains the influence of various features by the degree of the convolution gradient. The brighter the red in the figure, the greater this part of the feature effect on the result.

However, for the method in this paper, the use of Grad-CAM has been restricted. Since the classifier does not contain parameters, the back propagation starts directly from the last layer of the CNNs and the parameters updated by back propagation are part of the task encoder. TAM is proposed based on the Grad-CAM. It can be known from the Grad-CAM that the pooled gradient one-dimensional variable needs to be obtained, and, from (3),

γ generated by the encoder is such a variable. For the task encoder, its function is to generate parameters through task features, and interfere with the intermediate variables of RESNET18. Therefore, the γ of the last layer of CNNs is chosen to average the intermediate variables in the direction of the feature map to obtain the two-dimensional activation layer. Then, TAM will perform its mapping to get the CNN heat map. Compared with Grad-CAM, TAM does not choose categories for gradient transformation but task features. Secondly, Grad-CAM is done through back propagation gradients and TAM is through forward propagation. Visual explanations of tomato disease output by TAM are shown in Figure 2.

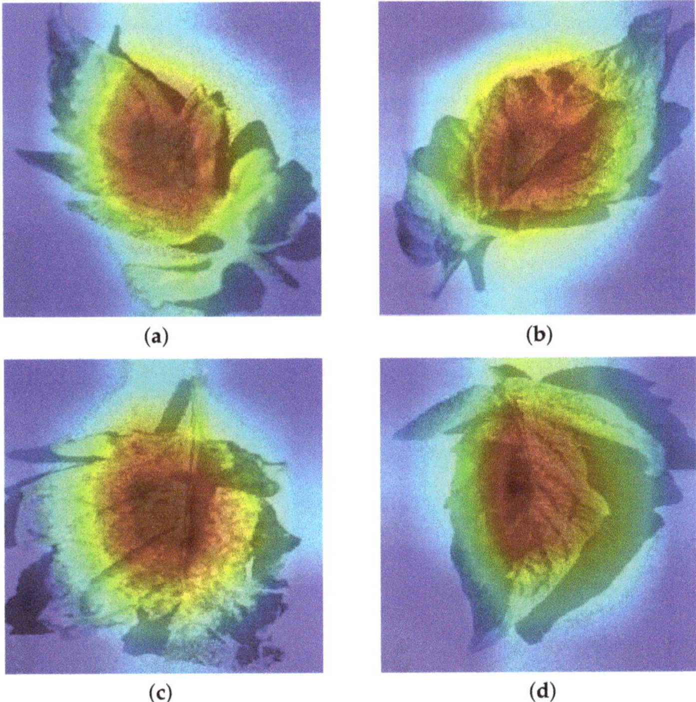

Figure 2. Visual explanations of tomato disease. (**a**) Leaf mold; (**b**) Late blight; (**c**) Septoria leaf spot; (**d**) two spotted spider mite.

3. Results

We evaluate LFM-CNAPS on the Miniplantdisease-Dataset family of datasets, demonstrating improvements by ablation experiment. Two prediction visual explanations are also given.

3.1. Performance of Plant Disease Detection

We train LFM-CNAPS on the Meta-Dataset and Miniplantdisease-Dataset, evaluate it on the Miniplantdisease-Dataset and PlantVillage-Dataset. To investigate the performance of LFM-CNAPS proposed, six comparison algorithms are adopted. First, the deep learning method [2] composed of CNNs and a fully connected layer is chosen (RESNET18 + FC). To control variables, CNNs use RESNET18 that is the same as LFM-CNAPS. Next, the model composed of RESNET18 and local feature matching classifier is chosen (RESNET18 + LFM) to show the effect of task adaptive processes of LFM-CNAPS. Finally, four few-shot learning methods are adopted including: MatchingNet [29], ProtoNet [23], Simple-CNAPS [25], and Meta Fine-Tuning [27]. The information of machine specifications and time cost is shown in Table 1.

Table 1. Machine specifications and time cost.

Name	Value
Video Memory	11G
Graphics	NVIDIA GeForce GTX 1080 Ti
Processor	Intel(R) Xeon(R) CPU E5-2640
Operating system	Windows 10 Home 64
Training time	17,548.73 s
Test time	57.36 s

The training results of the Meta-Dataset are shown in Table 2. Table 2 is the result of the method performance on Meta-Dataset which was also trained on Meta-Dataset. Among them, cifar10 and cifar100 are not included in the training dataset and are only used for testing. The results in Table 2 can reflect the performance of the approach on general classification tasks. These include classification of animal species, classification of objects and tools, and classification of handwritten fonts. The task format of the Meta-Dataset is not fixed. Through 110,000 times of training, the algorithm has a better classification effect.

Table 2. LFM-CNAPS test accuracy on the Meta-Dataset.

Dataset Name	Accuracy (%)
ilsvrc 2012	55.0+/−1.0
omniglot	92.0+/−0.6
aircraft	82.4+/−0.6
cu birds	74.3+/−0.8
dtd	65.3+/−0.7
quickdraw	75.5+/−0.8
fungi	48.0+/−1.1
Vgg flower	89.4+/−0.5
Traffic sign	68.2+/−0.7
mscoco	51.1+/−1.0
mnist	93.3+/−0.4
cifar10	71.1+/−0.7
cifar100	57.3+/−1.0

The training results of the in-domain Miniplantdisease-Dataset are shown in Table 3. A total of 20,000 tasks were randomly generated, including 60 types of plant diseases. The names of various types of plants, the number of their diseases, and the number of corresponding samples are declared in the table. The data sources are distinguished in the table. It can be seen that the two datasets do not contain the same plant categories. Each plant of Apple foliar disease has only two categories: healthy and diseased. The number of plant diseases in PlantVillage is relatively random, as many as 10 and as few as one. Through cross-domain dataset training, the model can be more robust. Secondly, from the perspective of sample size, the sample size of Apple foliar disease is much smaller than that of PlantVillage. The unbalanced sample distribution is more practical for application because there is no absolutely balanced sample in reality, and most plant classification samples are random. From the results, after 20,000 trainings, the average accuracy of the algorithm reached 97.5%.

Table 3. LFM-CNAPS train results on in-domain Miniplantdisease-Dataset.

Species	Number of Plant Diseases	Number of Samples
Apple foliar disease		
Alstonia Scholaris	2	433
Arjun	2	452
Bael	2	266
Chinar	2	223
Gauva	2	419
Jamun	2	624
Jatropha	2	257
Lemon	2	236
PlantVillage		
Apple	4	7169
Blueberry	1	1816
Cherry	2	3509
Corn	4	7316
Grape	4	7222
Orange	1	2010
Peach	2	3566
Pepper	2	3901
Potato	2	3763
Tomato	10	18,345
Number of training steps		**Training accuracy (%)**
10,000		97.0
20,000		97.5

Table 4 shows the test results of out-of-domain Miniplantdisease-Dataset. Out-of-domain and in-domain datasets do not contain the same plant diseases. Testing the algorithm through untrained plant diseases can better reflect the robustness of the algorithm. The accuracy of algorithms for newly emerged plant diseases is also more practical. The out-of-domain datasets include 12 plant disease categories and 600 random tasks. LFM-CNAPS has an average accuracy rate of 93.3% on out-of-domain dataset.

Table 5 shows ablation studies of LFM-CNAPS on an Out-of-Domain Miniplantdisease-Dataset. Our model mainly includes conditional adaptive feature extractor and local feature matching classifier components. A conditional adaptive feature extractor extracts meaningful features via forward propagation, which helps the model learn those features even in the few-shot dataset. A local feature matching classifier replaces the fully connected layers with metric learning to avoid overfitting which deeply hurts the performance of the neural network. The result of ablation experiments is shown in Table 5. LFM-CNAPS with only conditional adaptive feature extractor holds 86.1% accuracy and LFM-CNAPS with only a local feature matching classifier reaches 85.2% accuracy. However, LFM-CNAPS with the two components has 93.9% accuracy.

The test results of PlantVillage are shown in Table 6, which contains a total of 38 plant diseases. Although some plant disease categories in PlantVillage are used for meta-training, the test pictures are different from the training pictures. The number of categories and the number of samples are more than the first test. Therefore, the test on PlantVillage is more challenging than the first test, and the average accuracy of LFM-CNAPS is 89%.

Table 4. LFM-CNAPS test results on an Out-of-Domain Miniplantdisease-Dataset.

Plant State	Number of Samples
Mango diseased	265
Mango healthy	170
Pomegranate diseased	272
Pomegranate healthy	287
Pongamia Pinnata diseased	276
Pongamia Pinnata healthy	322
Potato Late blight	1939
Raspberry healthy	1781
Soybean healthy	2022
Squash Powdery mildew	1736
Strawberry healthy	1824
Strawberry Leaf scorch	1774
Method	**Accuracy (%)**
RESNET18 + FC	20.0+/−0.5
MatchingNet	19.5+/−0.5
ProtoNet	20.5+/−0.6
RESNET18 + LFM	85.2+/−0.7
Simple-CNAPS	92.5+/−0.4
Meta Fine-Tuning	91.14+/−0.5
LFM-CNAPS	93.9+/−0.4

Table 5. Ablation studies of LFM-CNAPS on an Out-of-Domain Miniplantdisease-Dataset.

Feature Extractor	Classifier	Accuracy (%)
		20.0+/−0.5
✓		86.1+/−0.6
	✓	85.2+/−0.7
✓	✓	93.9+/−0.4

Table 6. LFM-CNAPS test results on a PlantVillage-Dataset.

Plant Category	Number of Plant Diseases	Number of Samples
Apple	4	1943
Blueberry	1	454
Cherry	2	877
Corn	4	1829
Grape	4	1805
Orange	1	503
Peach	2	891
Pepper	2	975
Potato	3	1426
Raspberry	1	445
Soybean	1	505
Squash	1	434
Strawberry	2	900
Tomato	10	4585
Method		**Test accuracy (%)**
RESNET18 + FC		19.8+/−0.5
RESNET18 + LFM		81.7+/−0.7
LFM-CNAPS		89.0+/−0.5

3.2. Visual Explanations

Abnormal phenotype can be caused by either abiotic or biotic stress. The former is caused, for instance, by lack or excess of nutrients or water [51]. The latter can be caused by fungi, bacteria, and viruses. The typical symptomatology of (abiotic or biotic) stress includes discoloration, necrosis, decay, wilting, and atypical forms. Most of the existing deep learning methods for plant disease classification pay more attention to the test accuracy, and do not pay much attention to the classification basis. However, for practical application, a reasonable classification basis is more convincing and more acceptable. While our method gives the classification results, it also can save the classification's heat map by TAM technology. As shown in Figure 3, (a) is a sample map of Alstonia Scholari affected by pests, and (b) is a sample map of potato with late blight. The red part is the part that the algorithm pays more attention to, and it is also the core part that affects the classification result. For (a), the red part mainly appears near the wormhole, and, for (b), the red part is also where the leaves turn yellow and wither. Although it is impossible to give a detailed description of the causes of plant diseases, the focus of a heat map can be used to visualize the parameters of the CNNs, and the interpretability of the black box network can be improved to a certain extent.

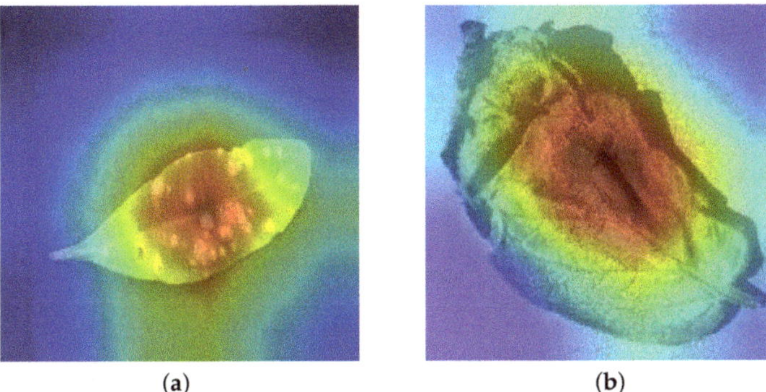

(a) (b)

Figure 3. LFM-CNAPS classification basis is visualized by TAM. (**a**) the sample of diseased Alstonia Scholari; (**b**) the sample of potato with late blight.

4. Discussion

Pests and diseases seriously threaten crop yields, leading to food shortages, e.g., more than 800 million people do not have adequate food; 1.3 billion live on less than $1 a day and at least 10% of global food production is lost to plant disease [19]. In order to combat the losses, the emerging plant disease needs to be detected before it has a large-scale impact on crop growth. This paper intends to propose a method that can detect plant diseases with few samples. The results showed that LFM-CNAPS proposed has an average accuracy of 93.9% on detecting unseen plant disease with only 25 annotated examples. The method RESNET18 + FC based on deep learning [10] only has an accuracy of 20.0%. More specifically, when classifying the five diseases of tomato: bacterial spot, early blight, healthy, late blight, and leaf mold, the performance of LFM-CNAPS is much better than the other two methods. LFM-CNAPS can give visual explanations through TAM, similar to optical observation of the symptoms on plant leaves. As show in Figure 3, Alstonia Scholari's wormhole is presented. However, due to complexity, even experienced agronomists often fail to successfully diagnose specific diseases [2]. For example, it is difficult for people to distinguish the corn with northern leaf blight from the corn with gray leaf spot. LFM-CNAPS adaptively adjusts and extracts the potential differences between the two categories by task adaptive processes. This feature will be more abstract rather than simple geometric features. On a commercial scale, evidently, a capital investment is initially required for adopting the employed approach [52]. Nevertheless, the wide-ranging large-scale commercial applications can provide high returns through considerable improvements in process enhancement and cost reduction.

Limitations of the study are due to a single task format. All tests are based on using 25 samples for plant disease detection. Recommendations for further research are to deal with a different scale of samples for better classification results.

5. Conclusions

This paper proposed LFM-CNAPS to solve few-shot plant disease detection and made the following contributions: first, a Miniplantdisease-Dataset suitable for meta-learning is provided, including two public datasets, 60 plant disease categories. Secondly, the LFM-CNAPS proposed is evaluated on the Miniplantdisease-Dataset, with an accuracy rate of 93.9% . Finally, TAM was proposed for CNN visualization. Without affecting the classification results and time cost, the classification heat map is saved to realize visual explanations.

Author Contributions: W.L. contributed to the conceptualization design and the design of the methodology. L.C. contributed to the algorithm achievement and paper writing. X.C. provides the data and revises this paper. All authors have read and agreed to the published version of the manuscript.

Funding: This work was supported by the Fundamental Research Funds for the Central Universities (No. JUSRP121073, JUSRP521004).

Institutional Review Board Statement: Not applicable.

Informed Consent Statement: Not applicable.

Data Availability Statement: The dataset can be obtained from https://www.kaggle.com/vipoooool/new-plant-diseases-dataset (accessed on 14 July 2021) and https://www.kaggle.com/c/plant-pathology-2020-fgvc7 (accessed on 14 July 2021) .

Conflicts of Interest: The authors declare no conflict of interest.

References

1. Muimba-Kankolongo, A. *Food Crop Production by Smallholder Farmers in Southern Africa | | Climates and Agroecologies*; Academic Press: Cambridge, MA, USA, 2018; pp. 5–13.
2. Ferentinos, K.P. Deep learning models for plant disease detection and diagnosis. *Comput. Electron. Agric.* **2018**, *145*, 311–318. [CrossRef]
3. Mohameth, F.; Bingcai, C.; Sada, K.A. Plant disease detection with deep learning and feature extraction using plant village. *J. Comput. Commun.* **2020**, *8*, 10–22. [CrossRef]
4. Klauser, D. Challenges in monitoring and managing plant diseases in developing countries. *J. Plant Dis. Prot.* **2018**, *125*, 235–237. [CrossRef]
5. Kader, A.A.; Kasmire, R.F.; Reid, M.S.; Sommer, N.F.; Thompson, J.F. *Postharvest Technology of Horticultural Crops*; University of California Agriculture and Natural Resources: Davis, CA, USA, 2002.
6. Teng, P.S.; James, W.C. Disease and yield loss assessment. In *Plant Pathologists Pocketbook*; CABI: Wallingford, UK, 2002.
7. Maxwell, S. Food security: A post-modern perspective. *Food Policy* **1996**, *21*, 155–170. [CrossRef]
8. Hunter, M.C.; Smith, R.G.; Schipanski, M.E.; Atwood, L.W.; Mortensen, D.A. Agriculture in 2050: Recalibrating targets for sustainable intensification. *Bioscience* **2017**, *67*, 386–391. [CrossRef]
9. Fang, Y.; Ramasamy, R.P. Current and prospective methods for plant disease detection. *Biosensors* **2015**, *5*, 537–561. [CrossRef] [PubMed]
10. Sharma, P.; Berwal, Y.P.S.; Ghai, W. Performance analysis of deep learning CNN models for disease detection in plants using image segmentation. *Inf. Process. Agric.* **2020**, *7*, 566–574. [CrossRef]
11. Taheri-Garavand, A.; Nejad, A.R.; Fanourakis, D.; Fatahi, S.; Majd, M.A. Employment of artificial neural networks for non-invasive estimation of leaf water status using color features: A case study in Spathiphyllum wallisii. *Acta Physiol. Plant.* **2021**, *43*, 1–11. [CrossRef]
12. LeCun, Y.; Bengio, Y. Convolutional networks for images, speech, and time series. *Handb. Brain Theory Neural Netw.* **1995**, *3361*, 1995.
13. LeCun, Y.; Bengio, Y.; Hinton, G. Deep learning. *Nature* **2015**, *521*, 436–444. [CrossRef]
14. Li, W.; Xu, L.; Liang, Z.; Wang, S.; Cao, J.; Lam, T.C.; Cui, X. JDGAN: Enhancing generator on extremely limited data via joint distribution. *Neurocomputing* **2021**, *431*, 148–162. [CrossRef]

15. Li, W.; Fan, L.; Wang, Z.; Ma, C.; Cui, X. Tackling mode collapse in multi-generator GANs with orthogonal vectors. *Pattern Recognit.* **2021**, *110*, 107646. [CrossRef]
16. Taheri-Garavand, A.; Nasiri, A.; Fanourakis, D.; Fatahi, S.; Omid, M.; Nikoloudakis, N. Automated In Situ Seed Variety Identification via Deep Learning: A Case Study in Chickpea. *Plants* **2021**, *10*, 1406. [CrossRef]
17. Nasiri, A.; Taheri-Garavand, A.; Fanourakis, D.; Zhang, Y.D.; Nikoloudakis, N. Automated Grapevine Cultivar Identification via Leaf Imaging and Deep Convolutional Neural Networks: A Proof-of-Concept Study Employing Primary Iranian Varieties. *Plants* **2021**, *10*, 1628. [CrossRef] [PubMed]
18. Hughes, D.; Salathé, M. An open access repository of images on plant health to enable the development of mobile disease diagnostics. *arXiv* **2015**, arXiv:1511.08060.
19. Strange, R.N.; Scott, P.R. Plant disease: a threat to global food security. *Annu. Rev. Phytopathol.* **2005**, *43*, 83–116. [CrossRef] [PubMed]
20. Vilalta, R.; Drissi, Y. A perspective view and survey of meta-learning. *Artif. Intell. Rev.* **2002**, *18*, 77–95. [CrossRef]
21. Vanschoren, J. Meta-learning: A survey. *arXiv* **2018**, arXiv:1810.03548.
22. Pan, S.J.; Qiang, Y. A Survey on Transfer Learning. *IEEE Trans. Knowl. Data Eng.* **2010**, *22*, 1345–1359. [CrossRef]
23. Snell, J.; Swersky, K.; Zemel, R.S. Prototypical networks for few-shot learning. *arXiv* **2017**, arXiv:1703.05175.
24. Requeima, J.; Gordon, J.; Bronskill, J.; Nowozin, S.; Turner, R.E. Fast and flexible multi-task classification using conditional neural adaptive processes. *Adv. Neural Inf. Process. Syst.* **2019**, *32*, 7959–7970.
25. Bateni, P.; Goyal, R.; Masrani, V.; Wood, F.; Sigal, L. Improved few-shot visual classification. In Proceedings of the IEEE/CVF Conference on Computer Vision and Pattern Recognition, Seattle, WA, USA, 13–19 June 2020; pp. 14493–14502.
26. Lin, X.; Ye, M.; Gong, Y.; Buracas, G.; Basiou, N.; Divakaran, A.; Yao, Y. Modular Adaptation for Cross-Domain Few-Shot Learning. *arXiv* **2021**, arXiv:2104.00619.
27. Cai, J.; Shen, S.M. Cross-domain few-shot learning with meta fine-tuning. *arXiv* **2020**, arXiv:2005.10544.
28. Davis, J.V.; Kulis, B.; Jain, P.; Sra, S.; Dhillon, I.S. Information-theoretic metric learning. In Proceedings of the 24th International Conference on Machine Learning, Corvallis, OR, USA, 20–24 June 2007; pp. 209–216.
29. Vinyals, O.; Blundell, C.; Lillicrap, T.; Wierstra, D. Matching networks for one shot learning. *Adv. Neural Inf. Process. Syst.* **2016**, *29*, 3630–3638.
30. Triantafillou, E.; Zhu, T.; Dumoulin, V.; Lamblin, P.; Evci, U.; Xu, K.; Goroshin, R.; Gelada, C.; Swersky, K.; Manzagol, P.A.; et al. Meta-dataset: A dataset of datasets for learning to learn from few examples. *arXiv* **2019**, arXiv:1903.03096.
31. Russakovsky, O.; Deng, J.; Su, H.; Krause, J.; Satheesh, S.; Ma, S.; Huang, Z.; Karpathy, A.; Khosla, A.; Bernstein, M.; et al. Imagenet large scale visual recognition challenge. *Int. J. Comput. Vis.* **2015**, *115*, 211–252. [CrossRef]
32. Lake, B.M.; Salakhutdinov, R.; Tenenbaum, J.B. Human-level concept learning through probabilistic program induction. *Science* **2015**, *350*, 1332–1338. [CrossRef]
33. Maji, S.; Rahtu, E.; Kannala, J.; Blaschko, M.; Vedaldi, A. Fine-grained visual classification of aircraft. *arXiv* **2013**, arXiv:1306.5151.
34. Wah, C.; Branson, S.; Welinder, P.; Perona, P.; Belongie, S. *The Caltech-Ucsd Birds-200-2011 Dataset*; California Institute of Technology: Pasadena, CA, USA, 2011.
35. Cimpoi, M.; Maji, S.; Kokkinos, I.; Mohamed, S.; Vedaldi, A. Describing textures in the wild. In Proceedings of the IEEE Conference on Computer Vision and Pattern Recognition, Columbus, OH, USA, 23–28 June 2014, pp. 3606–3613.
36. Wheatley, G. *Quick Draw*; Mathematics Learning: Bethany Beach, DE, USA, 2007.
37. Schroeder, B.; Cui, Y. Fgvcx Fungi Classification Challenge 2018. Available online: github.com/visipedia/fgvcx_fungi_comp (accessed on 14 July 2021).
38. Nilsback, M.E.; Zisserman, A. Automated flower classification over a large number of classes. In Proceedings of the 2008 Sixth Indian Conference on Computer Vision, Graphics & Image Processing, Bhubaneswar, India, 16–19 December 2008; pp. 722–729.
39. Houben, S.; Stallkamp, J.; Salmen, J.; Schlipsing, M.; Igel, C. Detection of traffic signs in real-world images: The German Traffic Sign Detection Benchmark. In Proceedings of the 2013 International Joint Conference on Neural Networks (IJCNN), Dallas, TX, USA, 4–9 August 2013; pp. 1–8.
40. Lin, T.Y.; Maire, M.; Belongie, S.; Hays, J.; Perona, P.; Ramanan, D.; Dollár, P.; Zitnick, C.L. Microsoft coco: Common objects in context. In Proceedings of the European Conference on Computer Visio, Zurich, Switzerland, 6–12 September 2014; pp. 740–755.
41. Thapa, R.; Zhang, K.; Snavely, N.; Belongie, S.; Khan, A. The Plant Pathology Challenge 2020 data set to classify foliar disease of apples. *Appl. Plant Sci.* **2020**, *8*, e11390. [CrossRef]
42. He, K.; Zhang, X.; Ren, S.; Sun, J. Deep residual learning for image recognition. In Proceedings of the IEEE Conference on Computer Vision and Pattern Recognition, Las Vegas, NV, USA, 27–30 June 2016; pp. 770–778.
43. Finn, C.; Abbeel, P.; Levine, S. Model-agnostic meta-learning for fast adaptation of deep networks. In Proceedings of the International Conference on Machine Learning, PMLR, Sydney, Australia, 6–11 August 2017; pp. 1126–1135.
44. Perez, E.; Strub, F.; De Vries, H.; Dumoulin, V.; Courville, A. Film: Visual reasoning with a general conditioning layer. In Proceedings of the AAAI Conference on Artificial Intelligence, New Orleans, LA, USA, 2–7 February 2018; Volume 32.
45. Li, W.; Liu, X.; Liu, J.; Chen, P.; Wan, S.; Cui, X. On improving the accuracy with auto-encoder on conjunctivitis. *Appl. Soft Comput.* **2019**, *81*, 105489. [CrossRef]
46. Joachims, T. Making Large-Scale SVM Learning Practical. Technical Report. 1998. Available online: https://www.econstor.eu/handle/10419/77178 (accessed on 9 March 2021).

47. Li, W.; Wang, L.; Xu, J.; Huo, J.; Gao, Y.; Luo, J. Revisiting local descriptor based image-to-class measure for few-shot learning. In Proceedings of the IEEE/CVF Conference on Computer Vision and Pattern Recognition, Long Beach, CA, USA, 15–20 June 2019; pp. 7260–7268.
48. Zhang, Z.; Sabuncu, M.R. Generalized cross entropy loss for training deep neural networks with noisy labels. In Proceedings of the 32nd Conference on Neural Information Processing Systems (NeurIPS), Montreal, QC, Canada, 3–8 December 2018.
49. Kingma, D.P.; Ba, J. Adam: A method for stochastic optimization. *arXiv* **2014**, arXiv:1412.6980.
50. Selvaraju, R.R.; Cogswell, M.; Das, A.; Vedantam, R.; Parikh, D.; Batra, D. Grad-cam: Visual explanations from deep networks via gradient-based localization. In Proceedings of the IEEE International Conference on Computer Vision, Venice, Italy, 22–29 October 2017; pp. 618–626.
51. Chatzistathis, T.; Fanourakis, D.; Aliniaeifard, S.; Kotsiras, A.; Delis, C.; Tsaniklidis, G. Leaf Age-Dependent Effects of Boron Toxicity in Two Cucumis melo Varieties. *Agronomy* **2021**, *11*, 759. [CrossRef]
52. Taheri-Garavand, A.; Mumivand, H.; Fanourakis, D.; Fatahi, S.; Taghipour, S. An artificial neural network approach for non-invasive estimation of essential oil content and composition through considering drying processing factors: A case study in Mentha aquatica. *Ind. Crop. Prod.* **2021**, *171*, 113985. [CrossRef]

Article

The Effects of Using Pineapple Stem Starch as an Alternative Starch Source and Ageing Period on Meat Quality, Texture Profile, Ribonucleotide Content, and Fatty Acid Composition of Longissimus Thoracis of Fattening Dairy Steers

Chanporn Chaosap [1,*], Katatikarn Sahatsanon [2], Ronachai Sitthigripong [2], Suriya Sawanon [3] and Jutarat Setakul [2]

1. Department of Agricultural Education, Faculty of Industrial Education and Technology, King Mongkut's Institute of Technology Ladkrabang, Bangkok 10520, Thailand
2. Department of Animal Technology and Fishery, Faculty of Agricultural Technology, King Mongkut's Institute of Technology Ladkrabang, Bangkok 10520, Thailand; katatikarnnamwa3@gmail.com (K.S.); ronachai.sit@kmitl.ac.th (R.S.); ksejutar@gmail.com (J.S.)
3. Department of Animal Science, Faculty of Agriculture at Kamphaeng Saen, Kasetsart University, Nakhon Pathom 73140, Thailand; agrsusa@ku.ac.th
* Correspondence: chanporn.ch@kmitl.ac.th; Tel.: +66-838829217

Citation: Chaosap, C.; Sahatsanon, K.; Sitthigripong, R.; Sawanon, S.; Setakul, J. The Effects of Using Pineapple Stem Starch as an Alternative Starch Source and Ageing Period on Meat Quality, Texture Profile, Ribonucleotide Content, and Fatty Acid Composition of Longissimus Thoracis of Fattening Dairy Steers. *Foods* **2021**, *10*, 2319. https://doi.org/10.3390/foods10102319

Academic Editors: António Raposo, Renata Puppin Zandonadi and Raquel Braz Assunção Botelho

Received: 7 September 2021
Accepted: 27 September 2021
Published: 29 September 2021

Publisher's Note: MDPI stays neutral with regard to jurisdictional claims in published maps and institutional affiliations.

Abstract: The effects of different starch sources (ground corn (CO), ground cassava (CA) and pineapple stem starch (PI)) and ageing period (14 and 21 days) on meat characteristics of Holstein steers were investigated. Starch sources had no effect on meat characteristics, while meat aged for 14 days had less thawing loss than that aged for 21 days. Meat from steers fed PI had higher levels of inosine monophosphate (IMP) than the others ($p < 0.05$). With increasing duration of ageing, the content of IMP and guanosine monophosphate in the meat decreased, while the content of hypoxanthine increased ($p < 0.05$). Meat from steers fed CO had the highest oleic acid but the lowest erucic acid ($p < 0.05$) in contrast to meat from steers fed PI, which had the lowest oleic acid but the highest erucic acid. Steers fed CO appeared to produce healthier meat as this was positively associated with monounsaturated fatty acid content. Meat from steers fed PI had higher levels of IMP, which may be associated with good taste.

Keywords: starch source; Holstein steers; meat characteristics; meat flavour; fatty acid profile

1. Introduction

Pineapple is one of the most important economic crops in the world, and Thailand is one of the countries with a large pineapple production. In 2020, 1.39 million tons of pineapple fruit were produced, ranking sixth in the world for pineapple production [1]. Pineapples are usually consumed fresh or as pineapple juice, pulp or canned. By-products from pineapple production and canneries can account for 70–75% (w/w) of the crop, including peel, crown, core, stem and leaf [2]. For economic reasons, the large quantities of by-products from pineapple production should be properly utilized. Otherwise, improper waste management would lead to long-term environmental deterioration, especially through large-scale land clearing, which usually involves biomass burning and soil and water pollution.

In intensive production systems, strategic nutritional supplementation may be undertaken to compensate for deficiencies in the quantity and quality of pasture or forage fed to cattle prior to finishing [3]. High-concentrate commercial cattle fattening systems can provide a sustainable and adequate supply of live cattle to meet domestic consumption requirements with excellent meat quality [3–5]. High-concentrate fattening is easy to operate, and the results are predictable [5,6]. The main problem with high-concentrate finishing is the high cost and it is only profitable if the starting value of the animal is low, concentrate costs are low or there is excellent carcass and meat quality [3,5,7]. Male dairy calves that are not desirable for dairy production can therefore be used for beef production, with lower initial cost of animals compared to beef breeds such as Charolais crosses, which are popularly fattened for premium beef in Thailand [8]. However, dairy breeds require 20% more energy and 25% more feed than beef breeds [9]. To achieve cost advantages, agro-industrial feed by-products can be used as alternative feed sources instead of using a main feed source of both roughage and concentrate. Several studies suggest that pineapple by-products such as pineapple cannery by-products [10], pineapple stem [11], pineapple stem by-product silage [12] and ensiled pineapple waste [2] can be used as a promising roughage source to improve rumen function and production performance of animals without compromising production quality and also reduce feed costs. In the past, the pineapple stem was left as waste after replanting before being dried in the sun and burned. Recently, the pineapple stem has been used to extract the enzyme bromelain [13] which can be used as a potential phytomedicinal agent, so the rest of the by-product can be used as a potential starch source for cattle. Pineapple stem starch has similar starch content, but higher amylose content and smaller starch particle size compared to corn and cassava starch: 98, 101 and 100% (w/w) for starch content; 34.4, 16.2 and 15.3% (w/w) of whole sample for amylose content; and 9.7, 15.8 and 15.8 μm for starch particle size of pineapple stem, corn and cassava starch, respectively [14]. Khongpradit et al., 2020 [15] found that pineapple stem starch can be a useful starch source with cost advantages for beef cattle to improve growth performance and increase rumen fermentation when formulated as a 40% concentrate compared to other starch sources, ground corn and ground cassava. In addition, Khongpradit et al. (in press) [7] found that the use of pineapple stem starch as a starch source in the concentrate for 206-day fattening dairy steers had a lower feed cost per gain compared to ground cassava and ground corn (2.90, 3.62, and 4.02 USD/kg, respectively).

The most important principles for the palatability of meat are tenderness, juiciness, and flavour, all of which are related to consumer satisfaction. Tenderness was the most important factor influencing beef palatability, but previous research has shown that when tenderness is adequate, flavour is the most important aspect of consumer satisfaction [16]. In addition, the nutritional value of meat, such as fatty acid composition, was an important preference factor for health-conscious customers [17]. Monounsaturated fatty acids (MUFA) are considered better for health, while saturated fatty acids (SFA) can cause health problems [18]. Concentrate-fed beef had more tender meat, higher intramuscular fat with higher MUFA, less connective tissue, and higher palatability scores compared to grass-fed beef [4]. Previous studies have shown that high intramuscular fat can improve flavour, juiciness, and tenderness [4]. Meat flavour is also related to the content of ribonucleotides, such as inosine monophosphate (IMP) and guanosine monophosphate (GMP), as they have an umami flavour [19,20]. However, hypoxanthine, the degradation product of IMP during post-mortem ageing, can cause a bitter taste. Apart from the feeding regime, post-mortem ageing may also serve to improve the tenderness of the meat due to post-mortem proteolysis by specific endogenous enzymes, especially calpains [21], which yield specific protein degradation products such as the 30 kDa degradation product of troponin T [22].

Due to the limited information on the effect of pineapple stem starch in cattle concentrate diet on meat characteristics, the objective of this study was to investigate the effect of pineapple stem starch and also ageing duration on meat quality, texture profile and ribonucleotide content of dairy steers. In addition, the effect of pineapple stem starch on fatty acid composition was investigated.

2. Materials and Methods

2.1. Animal Ethics

Muscle samples used in this study were from 36 Holstein steers that were part of a study funded by the Agricultural Research and Development Agency of Thailand (ARDA) (project code PRP6205031930), Khongpradit et al., 2020 and Khongpradit et al., (in press).

The pineapple stem starch used in this study was obtained from Hong Mao Biochemicals Co, Ltd. in Rayong, Thailand. The Animal Use and Ethics Committee of Kasetsart University, Thailand approved the experimental procedure (ACKU62-AGK-007).

2.2. Experimental Cattle and Muscle Collection

Thirty-six Holstein steers with an average weight of 453 ± 35.3 kg and age of 22 months were divided into three treatment groups of 12 steers each and each steer was fed individually. The treatments were three different sources of starch in the concentrate: ground corn (CO), ground cassava (CA), and pineapple stem starch (PI), all at a concentration of 40% (Table S1). From day 1 to day 74, all experimental animals were fed the concentrate ad libitum with Napier grass silage (3 kg/head/day in DM) in a ratio of 75:25. From day 75 to 206, steers were fed the same concentrate with roughage of Napier grass silage (2 kg/head/day in DM) and rice straw (0.9 kg/head/day in DM) in a ratio of concentrate to roughage of 76:24. Cattle were fed twice daily for up to 206 days. The steers were ready for slaughter at an average age of 33 months. Before slaughter, all animals were housed in an enclosure with access to water for 12 h. The animals were weighed, stunned, bled, skinned, eviscerated and washed. Carcasses were split lengthwise and stored at 1 °C for 14 days. Samples of the longissimus thoracis (LT) were taken 14 days post-mortem from the left side of each carcass. All visible fat was removed from the samples, which were then divided into two 3-cm-thick subsamples and two 1.5-cm-thick subsamples. The first 3-cm-thick subsample was used for colour measurement on post-mortem day 14 and then weighed before being vacuum packed and stored at -20 °C to further measure thawing loss, cooking loss, shear force and texture profile. The second 3-cm-thick subsample was vacuum packed, stored at 1 °C for up to 7 days, then unwrapped, meat colour measured 21 days post-mortem, weighed, vacuum packed again, and stored at -20 °C to measure thawing loss, cooking loss, shear force, and texture profile 21 days post-mortem. The first 1.5-cm-thick subsample was vacuum packed and stored at -80 °C to further analyse ribonucleotide content and fatty acid composition on post-mortem day 14. The second 1.5-cm-thick subsample was vacuum packed, stored at 1 °C for up to 7 days, and then stored at -80 °C to further analyse ribonucleotide content on post-mortem day 21.

2.3. Meat Characteristics

2.3.1. Colour Measurement

The CIE L*, a*, and b* values of three measurement positions of the cut surface of 3-cm-thick LT steak were measured 14 and 21 days post-mortem after 30 min of blooming at 25 ± 2 °C using a handheld colorimeter with illuminant D65 and 8-mm aperture (MiniScan®EZ 45/0 LAV, Hunter Associates Laboratory Inc., Reston, VA, USA).

2.3.2. Thawing Loss, Cooking Loss and Shear Force Analysis

Two 3-cm-thick subsamples, aged 14 and 21 days, were stored at -20 °C for about 2 weeks, thawed at 4 °C for 24 h and then weighed. Weight changes after thawing were expressed as a percentage of thawing loss. Thawed muscle samples were placed in a high-density polyethylene bag before sealing and then placed in a water bath set at 80 °C and boiled for approximately 30 min until they reached an internal temperature of 70 °C. The internal temperature was measured using a thermometer (TM-19475D, Lutron Electronics, Taipei City, Taiwan). The cooked samples were allowed to cool under running tap water for 30 min before being weighed. After cooking, the percentage weight loss was calculated. The cooked samples were sliced parallel to the fibre orientation to obtain eight slices (width $\times$ length $\times$ height, $1.25 \times 1.25 \times 3$ cm rectangle) before measuring shear force using a texture analyser (Model EZ-SX, Shimadzu, Kyoto, Japan) equipped with a 50 kg load cell and a crosshead speed of 50 mm/min. The shear force value of each sample was averaged from 8 slices.

2.3.3. Texture Profile Analysis

Texture profile analysis (TPA) samples were analysed in the same manner as the shear force samples. After cooling, the cooked surface was removed by cutting with a knife to avoid a hard, dry surface, and each sample was cut into 15-mm cubes. The fibre axis of each cube was perpendicular to the direction of the probe. A texture analyser was used to determine the TPA (model EZ-SX, Shimadzu, Kyoto, Japan). Using a 36-mm-diameter cylindrical probe, each sample was placed between special stainless-steel plates and compressed perpendicular to the muscle fibre orientation in two consecutive cycles of 50% compression (based on sample width, 127 mm/min crosshead speed) with a 1 s pause between cycles. The probe moved downward at a constant speed of 127 mm/min. The force–time data of each test were collected, and the mean values for the TPA parameters of each sample—hardness, springiness, gumminess, chewiness and cohesiveness—were calculated from at least four tests.

2.4. Ribonucleotide Analysis

Frozen 1.5-cm-thick subsamples from 14 and 21 days post-mortem were pulverized through cryogenic grinding using a micro-Waring Blender (50–250 mL). One gram of the pulverized muscle sample was homogenized in 6 mL of cold 0.6 M perchloric acid at 23,000 g for 10 s (T25 Ultra-Turrax®, Ika, Staufen, Germany) according to [15]. After cooling for 15 min, the homogenate was neutralized with 5.4 mL of 0.8 M KOH and 0.25 mL of KH_2PO_4 buffer. The pH of the combined sample was raised to 7 with 0.8 M KOH and the volume was increased to 15 mL with HPLC water. An amount of 1 mL of the supernatant was aspirated into a small tube and frozen at -80 °C after centrifugation at $10,000 \times g$ for 10 min at 4 °C (Scanspeed 1580R, Labogene, Lillerod, Denmark). The supernatants were analysed by HPLC (Chromaster, Hitachi, Tokyo, Japan) with a UV detector for IMP, inosine, hypoxanthine and GMP after the frozen sample was thawed and centrifuged at $10,000 \times g$ for 5 min at 4 °C (Scanspeed 1580R, Labogene, Denmark). The stationary phase was a TSK Gel Amide -80 column (Tosoh, Tokyo, Japan) and the elution phase was a buffer of acetonitrile: KH_2PO_4, 70:30. External standards were used to calculate ribonucleotide content from a standard curve (57510 inosine-5-monophosphate disodium salt hydrate, 14,125 inosine, H9377 hypoxanthine, and G8377 guanosine-5-monophosphate disodium salt hydrate, Sigma-Aldrich, St. Louis, MO, USA).

2.5. Fatty Acid Analysis

The lipid extraction with chloroform was performed as described by [23]. Lipid was extracted from pulverized muscle samples at 14 days post-mortem with chloroform. Methyl Nona decanoate (SFA-013N, Accu Standard, New Haven, CT, USA) was added as an internal standard during the extraction procedure. A fused silica capillary column (100 m × 0.25 mm × 0.2 μm film thickness, model SPTM-2560, Supelco, Bellfonte, PA, USA) was used to evaluate the fatty acid methyl esters (FAM) by gas chromatography (model 7890B, Agilent, Santa Clara, CA, USA). The conditions for gas chromatography were as follows: temperature program: starting temperature 60 °C, followed by an increase of 20 °C/min to 170 °C, 5 °C/min to 220 °C, and 2 °C/min to 240 °C; carrier gas, He; split ratio, 10:1. Fatty acid methyl ester peaks were detected by comparing retention times with authentic standards (F.A.M.E. Mix, C4-C24, Supelco) and measured with an internal standard, nonadecanoic acid (C19:0).

2.6. Statistical Analysis

The data were considered as a 3 × 2 factorial arrangement in a completely randomised design, with 3 starch sources (CO, CA, and PI) and 2 ageing periods (14 and 21 days post-mortem). The general linear model procedure (SAS Institute Inc., Cary, NC, USA) was used for analysis of variance to analyse meat quality, texture profile, and ribonucleotide content, including the effects of starch source, ageing period, and their interaction. For fatty acid composition, only starch source was defined as a treatment. The PDIFF option was used to separate least square means. P values less than 0.05 were considered statistically significant. Principal component analysis (PCA) was used to evaluate the relationship between ribonucleotide content and fatty acid composition in relation to the different starch sources and ageing period using XLSTAT software (Addinsoft, Long Island City, NY, USA).

3. Results and Discussion

3.1. Meat Quality and Texture Profile

In this study, there was no interaction between starch sources and ageing time on meat quality traits ($p > 0.05$), as shown in Table 1. Neither concentrate starch sources nor ageing time had any effect on meat colour ($p > 0.05$) as there were no significant differences in L*, a*, and b* when compared among starch sources or ageing time. There was no significant effect of starch sources on thawing loss and cooking loss ($p > 0.05$). However, ageing time had an effect on thawing loss because thawing loss was higher for a longer ageing time of 21 days than for an ageing time of 14 days, 4.06% and 3.08%, respectively. Shear force was not affected by starch sources and ageing time ($p > 0.05$). Texture profile analysis was performed to investigate whether the starch sources of the concentrate or the ageing time affected the meat texture of LT muscle of dairy steers. No significant interactive effect of starch sources and ageing time on meat texture was found ($p > 0.05$). Neither starch sources nor ageing time had any effect on the meat texture profile ($p > 0.05$). The average hardness of LT muscle in this study was 44.05 N/cm^2, springiness 0.99 cm, gumminess 24.87 N/cm^2, chewiness 24.46 N/cm and cohesiveness 0.57.

Table 1. Effect of starch source of concentrate and ageing period on meat quality and texture profile of fattening dairy steers.

Trait	Treatment (T) [1]			Ageing (A)		RMSE [2]	p-Value		
	CO	CA	PI	14 Days	21 Days		T	A	T × A
Meat Quality									
Colour									
L*	35.32	36.51	36.41	35.67	36.49	2.32	0.347	0.268	0.933
a*	14.46	15.29	14.78	14.96	14.72	1.79	0.509	0.676	0.339
b*	41.36	42.13	42.24	41.44	42.36	3.21	0.739	0.358	0.641
Thawing loss (%)	3.31	3.48	3.93	3.08 [b]	4.06 [a]	1.18	0.430	0.019	0.777
Cooking loss (%)	19.23	21.01	21.73	19.60	21.72	4.22	0.287	0.121	0.712
Shear force (kg)	5.72	4.97	5.03	5.41	5.07	1.08	0.151	0.328	0.949
Texture Profile									
Hardness (N/cm^2)	40.57	42.74	48.85	40.92	47.19	22.80	0.615	0.392	0.890
Springiness (cm)	0.99	0.99	0.99	0.99	0.99	0.00	0.473	0.694	0.636
Gumminess (N/cm^2)	24.29	23.95	26.38	22.98	26.76	12.60	0.863	0.351	0.931
Chewiness (N/cm)	23.48	23.94	25.95	22.27	26.64	12.46	0.858	0.275	0.945
Cohesiveness(ratio)	0.57	0.59	0.55	0.57	0.56	0.06	0.206	0.880	0.980

[1] CO, ground corn; CA, ground cassava; PI, pineapple stem starch. [2] root mean square error; [a,b] Lsmeans having different superscripts within the same main effect are significantly different ($p < 0.05$).

The fact that no significant effect of starch sources on meat colour, thawing loss, cooking loss, and shear force was observed in the current study could be due to the similar pH of muscle from steers fed different starch sources as mentioned in the study by [7], which used the same sample sources as the current study. According to Hughes et al. 2019 [24], pH was found to be negatively correlated with water holding capacity and lightness. According to Moller et al. (2010) [25], muscle pH was negatively correlated with shear force, implying that shear force tends to decrease as pH increases. Another possible explanation for the non-significant shear force in beef from steers fed different starch sources could be the similar intramuscular fat content, as indicated by the similar amount of total fatty acids in this study. There was a positive correlation between intramuscular fat and tenderness of meat [26].

Proteolysis of muscle fibres can be impaired by ageing, leading to degradation of cytoskeletal proteins and then impairing the ability of muscle fibres to bind water [27]. As muscle structures loosen due to the degradation of myofibrillar and cytoskeletal proteins, the ability to bind water decreases, resulting in a gradual release and drainage of intracellular fluid with prolonged ageing [27]. According to Ledward et al. 1992 [28], the colour of muscle tissue depends on the reflectivity and oxygenation of myoglobin. With ageing, the ability of muscle fibres to retain water decreases, which may improve reflectivity properties and thus increase brightness. Prolonged ageing may also improve the redness of steaks when exposed to oxygen. Ageing may affect mitochondrial function, resulting in decreased competition for oxygen between mitochondria and myoglobin, allowing oxygen to reach tissues [29]. In this study, no effect of ageing on meat colour was observed. This could be due to the fact that the effects of ageing usually occur in the early phase of ageing and myoglobin may not be able to bind oxygen after the early phase. As Colle et al., 2015 [30] found, L* increased post-mortem with longer ageing from 2 to 14 days, but there was no difference in L* during ageing from 14 to 63 days. In the current study, cooking loss was not affected by duration of ageing, but thawing loss was, because a longer ageing of 21 days resulted in higher thawing loss than ageing of 14 days. The higher thawing loss at the longer ageing duration in the present study might be related to the lower ability of the degraded muscle proteins to bind water [27].

Shear force is an objective measurement of tenderness, measured physically by the force required to cut muscle fibres. Shear force has been shown to be inversely related to tenderness. A previous study has shown that tenderness of meat improves when shear force decreases with increasing ageing time [30]. Proteolytic degradation of certain myofibrillar proteins by calpain proteases is responsible for the improved tenderness during ageing [21,31,32]. The presence of the 30 kDa polypeptide and the degradation of troponin-T indicate not only post-mortem proteolysis but also post-mortem decay of the muscle Z-disc [22,31,32]. Duration of ageing had no significant effect on shear force in the present study, which may be due to the fact that little or no post-mortem proteolysis occurred after 14 days of ageing. This is in agreement with the findings of [30] who found that post-mortem ageing of Longissimus lumborum steaks from 2 to 14 days improved tenderness, but no further improvement occurred after 14 days. Ageing also did not affect meat texture in this study as there were no significant differences in hardness, springiness, gumminess, chewiness and cohesiveness of 14-day and 21-day aged beef. Palka, 2003 [33] found that the hardness and chewiness of raw meat was two times lower compared between 5 and 12 days of ageing, but no significant difference was found in cooked meat.

3.2. Ribonucleotides

The interaction between starch sources and ageing time on ribonucleotide content was not significant in the present study ($p > 0.05$), as shown in Table 2. There was no significant effect of starch sources on the content of hypoxanthine, inosine and GMP, but there was a significant effect of starch sources on IMP content ($p < 0.001$) as meat from cattle fed with PI had a higher content of IMP than those fed with CO and CA, 107.21, 71.82 and 55.42 mg/100g, respectively. Ageing time affected the content of hypoxanthine, IMP and GMP ($p < 0.05$) but not the content of inosine ($p > 0.05$). At ageing of 21 days, higher hypoxanthine content than ageing for 14 days was found, 34.80 and 27.13 mg/100 g, respectively. At 14 days of ageing, the content of IMP and GMP was higher than at 21 days ageing, 107.44 and 48.86 mg/100 g for IMP and 3.47 and 2.33 mg/100 g for GMP, respectively.

Muscle is known to be turned to meat as food during post-mortem aging. Post-mortem aging improves the flavour and texture of meat. The increase in free amino acids and peptides caused by endogenous proteolytic enzymes in meat during post-mortem aging is associated with improved meat flavour [34]. It is believed that the increase in free amino acids helps to enhance brothy flavour, especially the umami taste, while the increase in peptides is responsible for the mildness of the meat [34]. During post-mortem meat aging, nucleotide triphosphates such as adenosine triphosphate (ATP) and guanosine triphosphate (GTP) are degraded, resulting in umami taste-related compounds such as IMP and guanosine monophosphate (GMP) [19,20]. Adenosine triphosphate is degraded to adenosine diphosphate (ADP) and adenosine monophosphate (AMP), which are subsequently degraded to IMP. Inosine and hypoxanthine are produced once IMP is degraded [19,20]. Inosine is a tasteless substance, but hypoxanthine has a bitter taste [20,35]. Regarding the aging effect, IMP and GMP decreased with increasing hypoxanthine during post-mortem ageing in this study, which is consistent with previous studies [19]. These authors also found a significant decrease in inosine levels during ageing. This differs slightly from this study, which found lower levels of inosine in beef aged 21 days than in beef aged 14 days, with no statistical difference. The lower levels of IMP and GMP and the higher levels of hypoxanthine during ageing in the current study suggest that prolonged ageing has negative effects on beef flavour. However, there are many other factors that affect meat flavour, such as fatty acid composition. Melton et al. (1982) [36] reported that fatty acids such as myristoleic acid, palmitoleic acid, stearic acid, oleic acid, linoleic acid and alpha-linolenic acid can cause good meat flavour.

Table 2. Effect of starch source of concentrate and ageing period on ribonucleotide of fattening dairy steers.

Trait [1]	Treatment (T) [2]			Ageing (A)		RMSE [3]	*p*-Value		
	CO	CA	PI	14 Days	21 Days		T	A	T x A
Hypo [4]	31.76	33.52	27.61	27.13 [d]	34.80 [c]	9.80	0.294	0.019	0.507
Inosine	39.18	36.43	38.06	40.31	35.47	12.12	0.848	0.216	0.900
IMP [5]	71.82 [b]	55.42 [b]	107.2 [a]	107.44 [a]	48.86 [b]	35.20	0.002	<0.0001	0.629
GMP [6]	2.49	3.09	3.10	3.47 [c]	2.33 [d]	1.67	0.552	0.039	0.079

[a,b] Lsmeans having different superscripts within the same main effect are significantly different (*p* < 0.01).; [c,d] Lsmeans having different superscripts within the same main effect are significantly different (*p* < 0.05).; [1] mg/100 g.; [2] CA = ground cassava; CO = ground corn; PI = pineapple stem starch.; [3] root mean square error.; [4] hypoxanthine.; [5] inosine monophosphate.; [6] guanosine monophosphate.

Propionate production in the rumen and the uptake of glucose from the rumen bypass concentrate in the small intestine are both major sources of glucose in concentrates, while roughage is an important source of acetate produced by the fermentation process in the rumen [5]. Theurer, 1986 [37] reported that dietary ingested starch can reach the small intestine between 4% and 60% in cattle depending on the grain source and processing. Optimizing starch fermentation in the rumen to produce propionate while increasing starch digestion in the small intestine bypassed by the rumen helps improve glucose supply. A very interesting finding of this study is the significantly higher content of the umami substance IMP in LT muscle of dairy steers fed pineapple stem starch in concentrate compared to ground corn and ground cassava as starch sources. This result suggests that cattle fed pineapple stem starch may be more palatable than the other two starch sources. The higher content of IMP might be related to the higher content of its precursor, glucose, in PI fed steers. Khongpradit et al., 2020 [15] found a significantly higher content of Ruminococcus bromii C1 and total short chain fatty acids (SCFA) in the rumen of cattle fed PI than CO and CA, suggesting that PI is more fermented in the rumen, which could be related to the smaller particle size of starch, lower neutral detergent fibre (NDF) and lower crude protein and lipid content associated with starch, making this starch source more degradable and fermentable than others. However, they found that the percentage of propionic acid in the rumen fermentation profile, the main source of glucose production, did not differ between starch sources. Therefore, the possible explanation for the higher IMP content in PI could be related to the higher glucose supply through small intestinal digestion. In a previous study, significantly higher amylose content was found in pineapple stem starch than corn and cassava [14]. Goats fed high amylose corn in the total mixed ration had more starch that was not degradable in the rumen and therefore more starch entered the small intestine for digestion, resulting in higher blood glucose levels [38]. Further studies are needed on the effects of the different starch sources in this study that could affect blood glucose levels.

3.3. Fatty Acid Composition

The fatty acid composition of the different starch sources in the concentrate is shown in Table 3. There was no effect of starch source on fatty acid composition except that cattle fed CO had the highest oleic acid but the lowest erucic acid ($p < 0.05$), in contrast to cattle fed PI, which had the lowest oleic acid but the highest erucic acid. The content of oleic acid was highest in the meat of cattle fed CO and lowest in cattle fed PI, while cattle fed CA were not significantly different from the others, 43.70, 42.36 and 40.65% of total fatty acids, respectively. The content of erucic acid was higher in PI than the others, 0.45, 0.47 and 0.74% of total fatty acids for cattle fed CO, CA and PI, respectively. In addition, cattle fed PI tended to have more palmitic acid than the others (P = 0.092), 27.8, 27.90 and 29.59% of total fatty acids for meat from cattle fed CO, CA and PI, respectively. The most abundant fatty acid in the meat of this study was monounsaturated oleic acid, followed by saturated palmitic acid and stearic acid, which ranged from 40.64 to 43.70, 27.81 to 29.59, and 10.68 to 11.39% of total fatty acids, respectively. Starch sources had no effect on MUFA, PUFA (polyunsaturated fatty acid) and SFA (saturated fatty acid) ($p > 0.05$). However, meat from cattle fed PI tended to have lower levels of desirable fatty acids (DFA), which include MUFA, PUFA and stearic acid, than others.

Table 3. Effect of starch source of concentrate on fatty acid composition of fattening dairy steers.

Trait		Starch Source [1]			RMSE	*p*-Value
		CO	CA	PI		
Fatty Acid Composition (% of Total Fatty Acids)						
Lauric acid	C12:0	0.18	0.19	0.17	0.05	0.776
Myristic acid	C14:0	4.37	4.84	4.83	0.46	0.133
Myristoleic acid	C14:1	1.55	1.81	1.54	0.45	0.501
Pentadecylic acid	C15:0	0.25	0.25	0.24	0.08	0.961
Palmitic acid	C16:0	27.81	27.90	29.59	1.57	0.092
Palmitoleic acid	C16:1	6.28	6.23	6.10	0.81	0.917
Margaric acid	C17:0	0.57	0.55	0.60	0.08	0.477
Heptadecenoic acid	C17:1	0.55	0.51	0.56	0.11	0.657
Stearic acid	C18:0	10.93	11.39	10.68	1.38	0.650
Oleic acid	C18:1n9c	43.70 [a]	42.36 [ab]	40.65 [b]	1.91	0.027
Linoleic acid	C18:2n6c	1.15	1.10	1.18	0.24	0.821
α-Linolenic acid	C18:3n3	0.23	0.23	0.18	0.07	0.383
Heneicosylic acid	C21:0	0.24	0.23	0.19	0.10	0.580
Erucic acid	C22:1n9	0.45 [b]	0.47 [b]	0.74 [a]	0.10	0.021
Arachidonic acid	C20:4n6	0.15	0.18	0.19	0.07	0.659
Lignoceric acid	C24:0	0.11	0.13	0.17	0.08	0.413
Nervonic acid	C24:1	1.49	1.64	2.40	0.82	0.116
Total fatty acid [2]		8.60	9.37	9.07	1.82	0.748
SFA		44.45	45.49	46.45	1.99	0.200
MUFA		54.02	53.01	51.99	2.02	0.202
PUFA		1.53	1.51	1.55	0.27	0.956
UFA		55.55	54.51	53.55	1.99	0.200
DFA		66.48	65.91	64.22	1.76	0.072

[a,b] Lsmeans having different superscripts within the same main effect are significantly different ($p < 0.05$).; [1] CA = ground cassava; CO = ground corn; PI = pineapple stem starch.; [2] g/100g muscle.; SFA (saturated fatty acids): C12:0+C14:0+C15:0+C16:0+C17:0+ C18:0+C21:0+C24:0; MUFA (monounsaturated fatty acids): C14:1+C16:1 + C17:1 + C18:1n9c+C22:1+C24:1; PUFA (polyunsaturated fatty acids): C18:2n6c + C18:3n3 + C20:4n6; UFA (unsaturated fatty acids): MUFA + PUFA; DFA (desirable fatty acids): MUFA + PUFA + C18:0.

Feeding cereal grains is one way to improve net energy supply because energy-rich grains can be used to form volatile fatty acids and glucose in the rumen and small intestine. When IMF preferentially uses glucose as a substrate for fatty acid synthesis while subcutaneous fat prefers acetate, the deposition of IMF is greater in diets with higher concentrate content as explained by [39]. According to Park et al. (2018) [5], the uptake of excess net energy is a key component of intramuscular fat (IMF) deposition. Konpradit et al.

(2020) [15] reported the higher degradability of PI starch in the rumen resulting in higher weight gain, average daily gain and feed conversion ratio in PI-fed steers compared to CO- and CA-fed steers. However, the different sources of starch in the concentrate ration did not affect the intramuscular fat content, as indicated by the similar amount of total fatty acids in the present study. This is in agreement with Kongpradit et al., 2021 (inpress) [7] who reported non-significant intramuscular fat content in the meat of cattle fed different starch sources of CO, CA and PI. Regarding the fatty acid composition in the intramuscular fat of cattle, the concentration of MUFA in adipose tissue is catalysed by the enzyme Δ9-desaturase (stearoyl-CoA desaturase) [40]. Palmitic acid and stearic acid are the preferred substrates that are converted to palmitoleic acid and oleic acid, respectively [41]. The higher concentration of palmitic acid and the lower concentration of oleic acid in the intramuscular fat of meat from steers fed PI in the present study might be due to the lower activity of Δ9-desaturase. Therefore, further studies to determine Δ9-desaturase activity in the meat of cattle fed PI are needed to understand how PI affects lipogenesis. The other possible explanation for the higher oleic acid in the meat of cattle fed CO, and the lower content in the meat of cattle fed PI could be due to the energy balance between the different starch sources of the concentrate, which is why a higher proportion of rice bran was used in the CA and PI concentrate diets than in the CO concentrate diet, as detailed in Supplementary Table S1 [15]. Since corn contains a higher percentage of linoleic acid than rice bran (56.5% and 34.8%, respectively), as mentioned by [35], this fatty acid would be converted to oleic acid by biohydrogenation in the rumen and then accumulate in the muscles of cattle [36].

Grain-fed beef is one of the most plentiful sources of MUFAs in the form of oleic acid and may be a significant source of MUFAs in the human diet (18:1, n-9) [37]. The importance of MUFAs in cardiovascular health has been extensively established. Higher oleic acid content in beef is beneficial as it can increase blood HDL cholesterol levels and reduce cardiovascular disease risk factors [12]. In the present study, it was found that the significantly lower oleic acid content, a trend toward lower DFA content and a trend toward higher palmitic acid content in the meat of cattle fed PI could be a risk factor for human health. However, previous studies suggest that the oleic acid content of grain-fed diets is higher than that of grass-fed beef [13,37], including the concentrate base in this study. The oleic acid content in the meat of cattle fed PI is still higher compared to natural grass-fed beef and is therefore healthier.

3.4. Ribonucleotide Content and Fatty Acid Composition in Relation to Different Starch Sources in Concentrate and Ageing Period

PCA was performed to evaluate the relationships between ribonucleotide content and fatty acid composition of LT muscle from dairy steers fed different starch sources and aged for different periods of time, 14 days or 21 days (Figure 1). There were two principal components (PC1 and PC2) that explained 100% of the total variance (68.39% and 31.61%, respectively). The first component, PC1, was strongly positively loaded by inosine at 21 days of ageing, IMP at 14 days and 21 days of ageing, PUFA, SFA and C22:1n9, but strongly negatively loaded by hypoxanthine content at 14 days and 21 days of ageing and MUFA, UFA, DFA and C18:1n9c content. The second component, PC2, was strongly positively loaded by GMP content at 21 days ageing and inosine content at 14 days ageing but was strongly negatively loaded by GMP content at 14 days ageing.

The PCA bi-plot showing ribonucleotide content and fatty acid composition at different ageing times varied among the different starch sources. The different starch sources exhibited different characteristics. It was found that CO was strongly associated with MUFA, UFA, DFA and C18:1n9c content. Conversely, PI was strongly associated with fatty acid composition, PUFA, SFA, C22:1n9 and ribonucleotide content; inosine at 21 days of ageing and IMP at 14 and 21 days of ageing. CA showed significant differences in fatty acid composition and ribonucleotide content from others, especially hypoxanthine at 14 and 21 days of ageing. From the PCA bi-plot, the use of CO as a starch source in concentrate may provide healthier beef as it is positively strongly associated with the content of MUFA,

UFA and DFA, while the use of PI as a starch source may provide tastier beef, as it is positively strongly associated with the content of IMP.

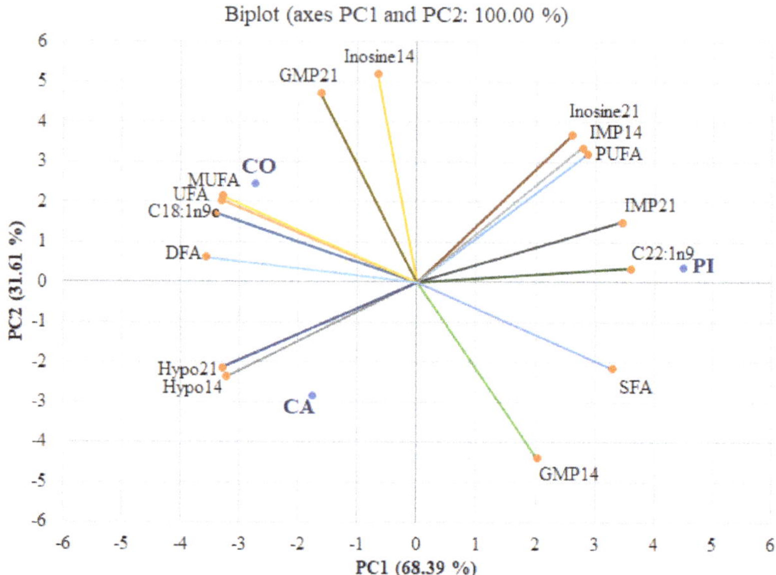

Figure 1. The results of principal component analysis (PCA) of ribonucleotide content and fatty acid composition at different ageing times varied among the different starch sources: CA = ground cassava, CO = ground corn, PI = pineapple stem starch, SFA (saturated fatty acids), MUFA (monounsaturated fatty acids), PUFA (polyunsaturated fatty acids), UFA (unsaturated fatty acids), DFA (desirable fatty acids), Hypo14 or Hypo21 (hypoxanthine content measured at day 14 or 21 post-mortem), inosine14 or inosine21 (inosine content measured on day 14 or 21 post-mortem), IMP14 or IMP21 (inosine monophosphate content measured on day 14 or 21 post-mortem), GMP14 or GMP21 (guanosine monophosphate measured on day 14 or 21 post-mortem).

4. Conclusions

Pineapple stem starch can be used as an alternative starch source in cattle concentrates without affecting meat quality, and it can also reduce feed costs. PI may also improve meat flavour, as meat from cattle fed PI may have a higher content of IMP than meat from cattle fed CO and CA. However, the meat from cattle fed CO appears to be healthier as it contains more oleic acid than others. The optimum ageing time to improve meat quality in dairy steers could not be more than 14 days because the improvement in meat quality did not occur after 14 days of ageing, and a longer ageing time could have a negative effect on meat flavour because the IMP and GMP content is lower after 14 days of ageing with higher hypoxanthine content.

Supplementary Materials: The following are available online at https://www.mdpi.com/article/10.3390/foods10102319/s1, Table S1: Ingredients and nutrient composition of concentrates and roughage (% DM).

Author Contributions: Conceptualization, Methodology, Resources, Formal analysis, Visualization, Writing—original draft, Writing—review and editing, C.C.; Project administration, Visualization, Supervision, R.S., J.S. and S.S.; Investigation, Methodology, Resources, K.S. All authors have read and agreed to the published version of the manuscript.

Funding: This research did not receive any specific grants from funding agencies in the public, commercial, or not-for-profit sectors.

Institutional Review Board Statement: Animal Use and Ethics Committee of Kasetsart University, Thailand approved the experimental procedure (ACKU62-AGK-007).

Informed Consent Statement: Not applicable.

Data Availability Statement: The data presented in this study are available on request from the corresponding author.

Conflicts of Interest: The authors declare no conflict of interest.

References

1. Department of Trade Negotiations, Ministry of Commerce, Thailand. Product Profile Pineapple. Available online: https://api.dtn.go.th/files/v3/60fa587cef41404be4216b7f/download (accessed on 1 September 2021).
2. Suksathit, S.; Wachirapakorn, C.; Opatpatanakit, Y. Effects of levels of ensiled pineapple waste and pangola hay fed as roughage sources on feed intake, nutrient digestibility and ruminal fermentation of Southern Thai native cattle. *Songklanakarin J. Sci. Technol.* **2011**, *33*, 281–289.
3. Greenwood, P.L. Review: An overview of beef production from pasture and feedlot globally, as demand for beef and the need for sustainable practices increase. *Animal* **2021**, 100295. [CrossRef] [PubMed]
4. Hwang, Y.-H.; Joo, S.-T. Fatty Acid Profiles, Meat Quality, and Sensory Palatability of Grain-fed and Grass-fed Beef from Hanwoo, American, and Australian Crossbred Cattle. *Food Sci. Anim. Resour.* **2017**, *37*, 153–161. [CrossRef] [PubMed]
5. Park, S.J.; Beak, S.-H.; Jung, D.J.S.; Kim, S.Y.; Jeong, I.H.; Piao, M.Y.; Kang, H.J.; Fassah, D.M.; Na, S.W.; Yoo, S.P.; et al. Genetic, management, and nutritional factors affecting intramuscular fat deposition in beef cattle—A review. *Asian Australas. J. Anim. Sci.* **2018**, *31*, 1043–1061. [CrossRef] [PubMed]
6. Duckett, S.K.; Neel, J.P.S.; Lewis, R.M.; Fontenot, J.P.; Clapham, W.M. Effects of forage species or concentrate finishing on animal performance, carcass and meat quality1,2. *J. Anim. Sci.* **2013**, *91*, 1454–1467. [CrossRef]
7. Khongpradit, A.; Boonsaen, P.; Homwong, N.; Matsuba, K.; Kobayashi, Y.; Sawanon, S. Effect of Starch Source in Concentrate Diets and of Days on Feed on Growth Performance, Carcass Characteristics, Meat Quality, and Economic Return of Feedlot Steers. *Tropic. Anim. Health Product.* **2021**, in press.
8. Sawanon, S.; Boonsaen, P.; Khongpradit, A. *A Guide for High. Quality Beef Production from Fattening Male Dairy Cattle*; Department of Animal Science, Faculty of Agriculture, Kamphaeng Saen Campus; Kasetsart University: Nakhon Pathom, Thailand, 2017.
9. Duff, G.C.; McMurphy, C.P. Feeding Holstein steers from start to finish. *Vet. Clin. North Am. Food Anim. Pract.* **2007**, *23*, 281–297. [CrossRef]
10. Choi, Y.; Lee, S.; Na, Y. Effects of a pineapple (*Ananas comosus* L.) cannery by-product on growth performance and carcass characteristics in finishing Hanwoo steers. *Anim. Biosci.* **2021**, *34*, 233–242. [CrossRef]
11. Pintadis, S.; Boonsaen, P.; Hattakum, C.; Homwong, N.; Sawanon, S. Effects of concentrate levels and pineapple stem on growth performance, carcass and meat quality of dairy steers. *Trop. Anim. Health Prod.* **2020**, *52*, 1911–1917. [CrossRef]
12. Hattakum, C.; Kanjanapruthipong, J.; Nakthong, S.; Wongchawalit, J.; Piamya, P.; Sawanon, S. Pineapple stem by-product as a feed source for growth performance, ruminal fermentation, carcass and meat quality of Holstein steers. *S. Afr. J. Anim. Sci.* **2019**, *49*, 147. [CrossRef]
13. Ketnawa, S.; Chaiwut, P.; Rawdkuen, S. Pineapple wastes: A potential source for bromelain extraction. *Food Bioprod. Process.* **2012**, *90*, 385–391. [CrossRef]
14. Nakthong, N.; Wongsagonsup, R.; Amornsakchai, T. Characteristics and potential utilizations of starch from pineapple stem waste. *Ind. Crop. Prod.* **2017**, *105*, 74–82. [CrossRef]
15. Khongpradit, A.; Boonsaen, P.; Homwong, N.; Suzuki, Y.; Koike, S.; Sawanon, S.; Kobayashi, Y. Effect of pineapple stem starch feeding on rumen microbial fermentation, blood lipid profile, and growth performance of fattening cattle. *Anim. Sci. J.* **2020**, *91*, e13459. [CrossRef] [PubMed]
16. Felderhoff, C.; Lyford, C.; Malaga, J.; Polkinghorne, R.; Brooks, C.; Garmyn, A.; Miller, M. Beef Quality Preferences: Factors Driving Consumer Satisfaction. *Foods* **2020**, *9*, 289. [CrossRef] [PubMed]
17. Flowers, S.; McFadden, B.R.; Carr, C.C.; Mateescu, R.G. Consumer preferences for beef with improved nutrient profile1. *J. Anim. Sci.* **2019**, *97*, 4699–4709. [CrossRef]
18. Gilmore, L.A.; Walzem, R.L.; Crouse, S.F.; Smith, D.R.; Adams, T.H.; Vaidyanathan, V.; Cao, X.; Smith, S.B. Consumption of High-Oleic Acid Ground Beef Increases HDL-Cholesterol Concentration but Both High- and Low-Oleic Acid Ground Beef Decrease HDL Particle Diameter in Normocholesterolemic Men. *J. Nutr.* **2011**, *141*, 1188–1194. [CrossRef]
19. Muroya, S.; Oe, M.; Ojima, K.; Watanabe, A. Metabolomic approach to key metabolites characterizing postmortem aged loin muscle of Japanese Black (Wagyu) cattle. *Asian Australas. J. Anim. Sci.* **2019**, *32*, 1172–1185. [CrossRef]
20. Tikk, M.; Tikk, K.; Tørngren, M.A.; Meinert, L.; Aaslyng, M.D.; Karlsson, A.H.; Andersen, H.J. Development of Inosine Monophosphate and Its Degradation Products during Aging of Pork of Different Qualities in Relation to Basic Taste and Retronasal Flavor Perception of the Meat. *J. Agric. Food Chem.* **2006**, *54*, 7769–7777. [CrossRef]
21. Koohmaraie, M.; Geesink, G. Contribution of postmortem muscle biochemistry to the delivery of consistent meat quality with particular focus on the calpain system. *Meat Sci.* **2006**, *74*, 34–43. [CrossRef]

22. Ho, C.; Stromer, M.; Robson, R. Identification of the 30 kDa polypeptide in post mortem skeletal muscle as a degradation product of troponin-T. *Biochimie* **1994**, *76*, 369–375. [CrossRef]
23. Folch, J.; Lees, M.; Stanley, G.S. A simple method for the isolation and purification of total lipides from animal tissues. *J. Biol. Chem.* **1957**, *226*, 497–509. [CrossRef]
24. Hughes, J.M.; Clarke, F.M.; Purslow, P.P.; Warner, R.D. Meat color is determined not only by chromatic heme pigments but also by the physical structure and achromatic light scattering properties of the muscle. *Compr. Rev. Food Sci. Food Saf.* **2019**, *19*, 44–63. [CrossRef] [PubMed]
25. Moeller, S.; Miller, R.; Edwards, K.; Zerby, H.; Logan, K.; Aldredge, T.; Stahl, C.; Boggess, M.; Box-Steffensmeier, J. Consumer perceptions of pork eating quality as affected by pork quality attributes and end-point cooked temperature. *Meat Sci.* **2010**, *84*, 14–22. [CrossRef] [PubMed]
26. Frank, D.; Joo, S.-T.; Warner, R. Consumer Acceptability of Intramuscular Fat. *Food Sci. Anim. Resour.* **2016**, *36*, 699–708. [CrossRef] [PubMed]
27. Huff-Lonergan, E.; Lonergan, S.M. Mechanisms of water-holding capacity of meat: The role of postmortem biochemical and structural changes. *Meat Sci.* **2005**, *71*, 194–204. [CrossRef] [PubMed]
28. Ledward, D.A.; Johnston, D.E.; Knight, M.K. *The Chemistry of Muscle-Based Foods*; Royal Society of Chemistry: Cambridge, UK, 1992.
29. Mancini, R.; Ramanathan, R. Effects of postmortem storage time on color and mitochondria in beef. *Meat Sci.* **2014**, *98*, 65–70. [CrossRef] [PubMed]
30. Colle, M.; Richard, R.; Killinger, K.; Bohlscheid, J.; Gray, A.; Loucks, W.; Day, R.; Cochran, A.; Nasados, J.; Doumit, M. Influence of extended aging on beef quality characteristics and sensory perception of steaks from the gluteus medius and longissimus lumborum. *Meat Sci.* **2015**, *110*, 32–39. [CrossRef]
31. Chaosap, C.; Sivapirunthep, P.; Sitthigripong, R.; Tavitchasri, P.; Maduae, S.; Kusee, T.; Setakul, J.; Adeyemi, K. Meat quality, post-mortem proteolytic enzymes, and myosin heavy chain isoforms of different Thai native cattle muscles. *Anim. Biosci.* **2021**. online ahead of print. [CrossRef] [PubMed]
32. Chaosap, C.; Sitthigripong, R.; Sivapirunthep, P.; Pungsuk, A.; Adeyemi, K.; Sazili, A.Q. Myosin heavy chain isoforms expression, calpain system and quality characteristics of different muscles in goats. *Food Chem.* **2020**, *321*, 126677. [CrossRef]
33. Palka, K. The influence of post-mortem ageing and roasting on the microstructure, texture and collagen solubility of bovine semitendinosus muscle. *Meat Sci.* **2002**, *64*, 191–198. [CrossRef]
34. Nishimura, T. Mechanism Involved in the Improvement of Meat Taste during Postmortem Aging. *Food Sci. Technol. Int.* **1998**, *4*, 241–249. [CrossRef]
35. Jones, N.R. Meat and fish flavors; significance of ribomononucleotides and their metabolites. *J. Agric. Food Chem.* **1969**, *17*, 712–716. [CrossRef]
36. Melton, S.L.; Black, J.M.; Davis, G.W.; Backus, W.R. Flavor and Selected Chemical Components of Ground Beef from Steers Backgrounded on Pasture and Fed Corn up to 140 Days. *J. Food Sci.* **1982**, *47*, 699–704. [CrossRef]
37. Theurer, C.B. Grain Processing Effects on Starch Utilization by Ruminants. *J. Anim. Sci.* **1986**, *63*, 1649–1662. [CrossRef] [PubMed]
38. Wang, S.; Wang, W.; Tan, Z. Effects of dietary starch types on rumen fermentation and blood profile in goats. *Czech. J. Anim. Sci.* **2016**, *61*, 32–41. [CrossRef]
39. Smith, S.B.; Crouse, J.D. Relative Contributions of Acetate, Lactate and Glucose to Lipogenesis in Bovine Intramuscular and Subcutaneous Adipose Tissue. *J. Nutr.* **1984**, *114*, 792–800. [CrossRef] [PubMed]
40. Griinari, J.M.; Corl, B.A.; Lacy, S.H.; Chouinard, P.Y.; Nurmela, K.V.V.; Bauman, D.E. Conjugated Linoleic Acid Is Synthesized Endogenously in Lactating Dairy Cows by Δ9-Desaturase. *J. Nutr.* **2000**, *130*, 2285–2291. [CrossRef]
41. Nakamura, M.T.; Nara, T.Y. Structure, function, and dietary regulation of δ6, δ5, and δ9 desaturases. *Annu. Rev. Nutr.* **2004**, *24*, 345–376. [CrossRef]

 *foods*

MDPI

Article

Blockchain Technology for Trustworthy Operations in the Management of Strategic Grain Reserves

Adnan Iftekhar, Xiaohui Cui * and Yiping Yang

Key Laboratory of Aerospace Information Security and Trusted Computing, Ministry of Education,
School of Cyber Science and Engineering, Wuhan University, Wuhan 430072, China; adnan@whu.edu.cn (A.I.);
yangyiping@whu.edu.cn (Y.Y.)
* Correspondence: xcui@whu.edu.cn

Abstract: Food is a daily requirement for everyone, while production patterns are seasonal. Producing sufficient nutrients is becoming more difficult because water and soil resources are already stressed and are becoming increasingly strained by climate change. Improving food security requires expertise in various areas, including sophisticated climate models, genetics research, market and household behaviour modelling, political shock modelling, and comprehensive environmental research. Additionally, governments stockpile grains to enhance national food security. These reserves should engage in markets only according to clear and transparent regulations and within defined price ranges to facilitate market functioning. It increases the demand for better technology in public administration to boost the management and distribution capacity while concentrating on improved controls and transparent governance systems. Blockchain technology emerges as a promising technology to enhance visibility, transparency, and data integrity with an immutable distributed ledger to increase trust in the parties' business transactions. This paper discusses blockchain technology and its potential role in strategic grain reserve management.

Keywords: food security; strategic grain reserve; blockchain technology; hyperledger fabric

Citation: Iftekhar, A.; Cui, X.; Yang, Y. Blockchain Technology for Trustworthy Operations in the Management of Strategic Grain Reserves. *Foods* **2021**, *10*, 2323. https://doi.org/10.3390/foods10102323

Academic Editors: António Raposo, Renata Puppin Zandonadi and Raquel Braz Assunção Botelho

Received: 29 August 2021
Accepted: 27 September 2021
Published: 29 September 2021

Publisher's Note: MDPI stays neutral with regard to jurisdictional claims in published maps and institutional affiliations.

1. Introduction

The strategic grain reserves are stockpiles of food in the form of grains. In developing countries, strategic grain reserves are an integral part of agricultural policies to stabilize food prices [1]. These grain reserves sources serve food-insecure households at subsidized prices during some shortage in the market and supply food to the areas struck by natural or human-generated disasters. The rapid increased in the prices of many commodities during 2007–2008 shocked the governments and consumers in many parts of the world [2]. Figure 1 shows a significant surge in food prices during 2007–2008 and 2011–2012 based on FAO data. According to the Food and Agriculture Organization of the United Nations (FAO), 35 countries with strategic grain reserves, including Cambodia, Cameroon, China, Ethiopia, India, Kenya, Nigeria, Pakistan, and Senegal, used their stocks to provide food at subsidized prices to their people. These countries responded more quickly and economically than those with limited or no reserves [3]. The 2009 G-8 and G20 summits and the FAO World Food Security Summit acknowledged the strategic grain reserves' significance [4]. Regional organizations such as the Association of Southeast Asian Nations (ASEAN) and the Economic Community of West African States (ECOWAS) in West Africa also stated their intention to establish regional strategic grain reserves. It will help these governments manage food surplus and food shortages within the same region in a short period.

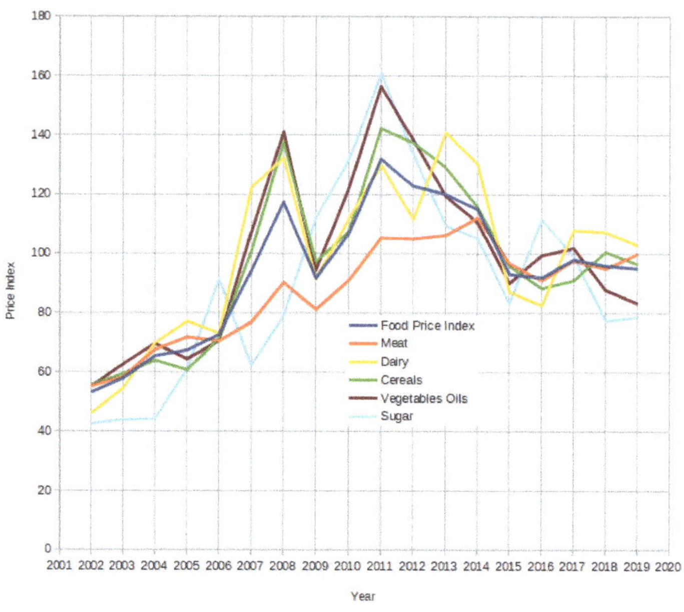

Figure 1. The price surge of food commodities during 2007–2008 and 2011–2012.

Many governments across the globe have either abandoned or significantly reduced their grain reserve initiatives during the last three decades. The food crisis exposed the weakness of depending only on the market to handle growing uncertainty and fluctuating pricing in agricultural markets. Food security is contingent upon a healthy balance of commerce and self-sufficiency. In developing countries, reserve creates a guaranteed market that encourages investment in agricultural products and distribution systems. Governments procure significant crops such as wheat, corn, and rice to encourage farmers to increase agricultural production. At the same time, the export of excessive output is a significant source of foreign exchange for countries [5]. However, the larger companies engaged in the agricultural and food businesses have superior access to resources and market knowledge than the governments. These companies impact food policy in their favour by using their lobby in the governments. The grain reserves need to be transparent and well-governed.

Grain reserves cost money. The management and release of public stocks are often coupled with subsidised sales of food. There is a lack of effective coordination, auditing, and sharing of data across departments and offices in developing countries. Moreover, most of the procedure is manual and paper-based. No department takes any responsibility for missing or corruption in records or data in the case of any incident or inquiry. The most significant challenge for the good governance of grain reserves is to ensure linked data throughout the supply chain. The existing IT solutions reduced some of these challenges. However, there is still a lot of undetected frauds and transparency levels. Blockchain technology is a novel digital method to guaranteeing data integrity and preventing tampering. It is a peer-to-peer distributed ledger system that prevents single point failure by providing fault-tolerance, immutability, trust, transparency, and full traceability of the stored transactions. A potential solution to alleviate the above issues and concerns is the use of blockchain technology in the management of the strategic grain reserves. The system attempts to ensure data sharing and integrity by storing it in a single shared ledger, thus meeting the transparency criteria necessary to avoid corruption and mismanagement.

The rest of this paper is arranged as follows. In Section 3, we describe the Blockchain Technology without involving technical details to make it understandable for government officials and policy makers. In Section 5, we discuss the generic architecture of the proposed solution. In Section 6, we present the potential qualitative analysis of the proposed solution followed by the conclusion of the article in Section 7.

2. Challenges in Emerging Economies

In developing countries, government purchases of important crops are a kind of compensation for farmers. Fraud and mismanagement in the purchase and distribution system also directly impact farmers and their ability to earn a living and producing food. Farmers are forced to sell their produce at a lower price than the government-set rates because of the malpractices of intermediaries and private sector business organizations [6]. The intermediaries, for the most part, charge a large commission from the farmers to facilitate sales of theirs crops. Farmers in emerging countries suffer directly due to the government's inability to regulate the grain markets. Farmers are also compelled to sell their production at cheap rates to intermediaries due to the poor management of the government's food and agriculture departments. These intermediaries create fake shortages to drive up prices and then resell the product at inflated rates back to the market. It has been reported that thousands of individuals who work in the agricultural industry committed suicide in India due to these issues [7]. The Indian farmers' protest is an ongoing demonstration against the 2020 Indian agricultural reforms, as they describe these reforms as putting the farmers at the mercy of multinational corporations [8]. Some of the most frequent issues are listed below.

- **Corruption & Bribe**
  Bribery is a common form of procurement fraud that occurs on a large scale. After receiving the money for crops, the farmer pays this bribe to the administrative personnel on the other side. It may range anywhere from 5 percent to 10 percent. The majority engaged in this operation are intermediaries and brokers, who also get a portion of the proceeds.
- **Influence Usage**
  Over time, lower quality goods are bought at exorbitant costs. The vendor subsequently made a payment to the administrative personnel in the amount of the percentage. This strategy was used by the influenced brokers and intermediates, including the landlords engaged in politics and government.
- **Ghost Billing**
  The billing will be in the form of overweight. It is also found mud and bricks into the crop's bags instead of the crop for making overweight billing and later removed before storage. The excess payment for the overweight later goes to the administrative staff.
- **Conflict of Interest**
  Many of the individuals who hold positions of power in the government are industrialists. They set the pricing of crops according to their interests and encourage corruption to purchase products from farmers at lower prices than those set by the government.

An abstract model, loosely based on the traditional procurement system in the subcontinent, is illustrated in Figure 2. The governments have land resources or a revenue department that holds barren or fertile agricultural land, commercial, and residential lands. The agricultural department conducts the survey twice a year to record the farmer, the crop, and its expected yield. The government makes a procurement and export policy based on these reports. This policy then executes through the food departments. The food department creates a consortium with state banks and other regional organizations to purchase directly from the farmers. The farmer brings their production to the designated purchase centers, confirming the farmer's purchase receipt's quality assurance and issues. The purchase center sends a payment scroll to the bank to make payments to the farmer through the state-owned banks. The food department is also responsible for releasing these stocks to the flour mills and public through designated centers.

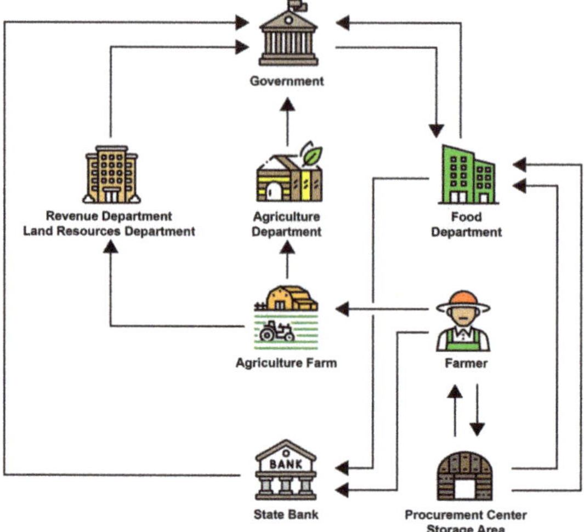

Figure 2. A traditional system of reserve grain management in the subcontinent region.

There is no proper coordination, auditing, and sharing of the data between the departments and the offices throughout the regions and country. It also causes the following problems to design an effective big data analytics tool.

- The lack of data accuracy
- The outdated data and low latency of data transferring
- Security and Privacy of the data

To eliminate these issues, we proposed adopting blockchain technology due to several advantages, such as all the nodes in blockchain are connected to a single logical channel. Its records and shared all the time-stamped and cryptographically secured transactions with all the channel members in almost real-time without the requirement of any additional auditing and involvement of third party department (Figure 3).

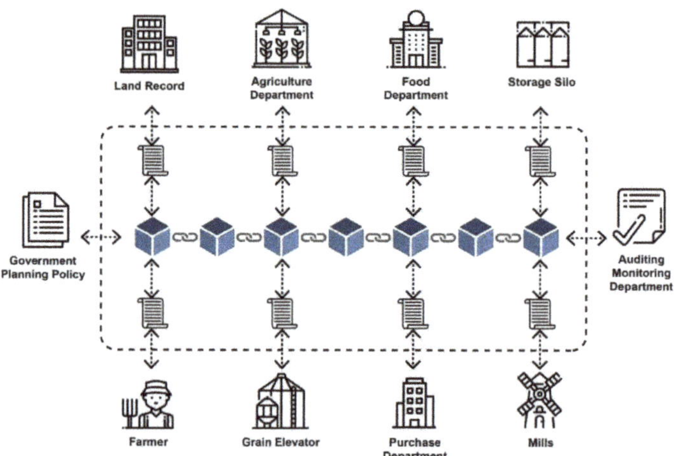

Figure 3. Logical overview of the proposed blockchain-based system.

We already published several articles on blockchain applications and deployed blockchain technology in the food safety area with an alliance of food producers, and distributors in China's Hubei province [9–14]. The purpose of our work is more about the horizontal applications of blockchain technology in social computing in developing countries. It is more about shaping a food system for tomorrow's digital world rather than making a blockchain technological stack but where it is needed. We developed a design-based research approach with the concept of mindful use of information technology [9]. The mindful use of technology emphasizes employing the most efficient and cost-effective technological elements to solve problems. The design-oriented approach focuses on analyzing real-world practical issues via researchers and professionals to create a solution that uses current design principles, and technology advances [15]. These solutions are subsequently developed and improved with the research and development needed to address a specified issue in the production environment. Figure 4 shows the whole process of our proposed technique based on a design-based research approach.

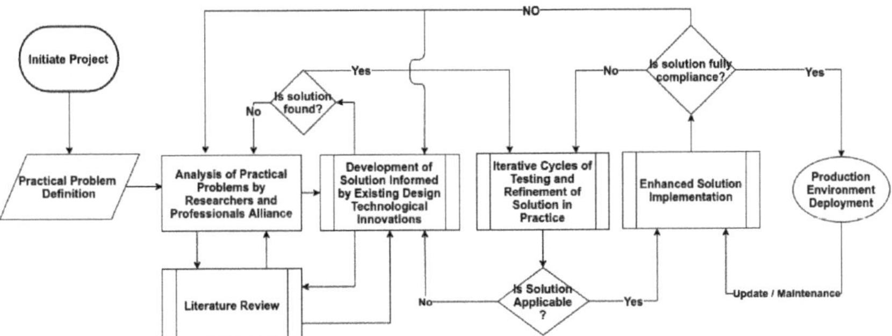

Figure 4. Research Methodology.

3. Blockchain Technology

In 2013, the German government decided the start of the Fourth Industrial Revolution (Industry 4.0) [16]. Traditional manufacturing and industrial processes are being continuously automated as part of the Fourth Industrial Revolution, which is being driven by contemporary smart technologies. In today's world, sensors are the building blocks of technologies such as intelligent homes and cities, smart grids, intelligent health systems, and wearable technology [17]. The Internet of Things (IoT) achieved considerable momentum in the automation industry. As the number of IoT devices grows, the volume of data generated by them will also grow. Managing these rapidly expanding IoT devices and enormous data efficiently to be available to all authorized users without compromising its integrity will become essential in the near future. On the other side, many information security incidents have been recorded, increasing the requirement for countermeasures. While safeguards against hostile third parties have been commonplace until now, operators and parties have seen an increase in demand for data falsification detection and blocking. Blockchain technology is well-known for its privacy, immutability, and decentralized nature.

Since the introduction of cryptocurrency Bitcoin [18] in response to the 2007–2008 global financial crises, Blockchain technology has become a point of interest among researchers and developers. A blockchain is a decentralized, trustless verification system based on cryptography, peer-to-peer distributed ledger, forged by consensus, and combined with a system for smart contracts and other assistive technologies. Each party participating in this network has an exact copy of the ledger throughout the network. It also allows creating a single instantaneous source of truth. These transactions are secure and verifiable in a transparent way. The main advantage of blockchain over the existing technologies is that it enables the two parties to make transactions over the Internet securely without any intermediary party's interference [19].

A blockchain consist of blocks of information. The block consists of data, the hash of the previous block, and the current block's hash. The data depends on the type of the blockchain. In the bitcoin blockchain, the data consists of the transactions and have senders address, receiver address, and coins. A block also has a hash. The hash of the block is always unique. Once a block is created, its hash is being calculated. Changing something inside the block will cause the hash to change. The third element in the block is the hash of the previous block. It creates a chain of blocks that makes the blockchain secured from tampering [20].

In Figure 5, we have a chain of three blocks as we can see that each block has a hash and hash of the previous block. So the block number 21 contains the hash of the block number 20 and block number 22 contains the hash of the block number 21. If we tampered with any block, the hash of the block would be changed. It will become invalid, and the following blocks will also become invalid as they no longer hold the previous block's hash.

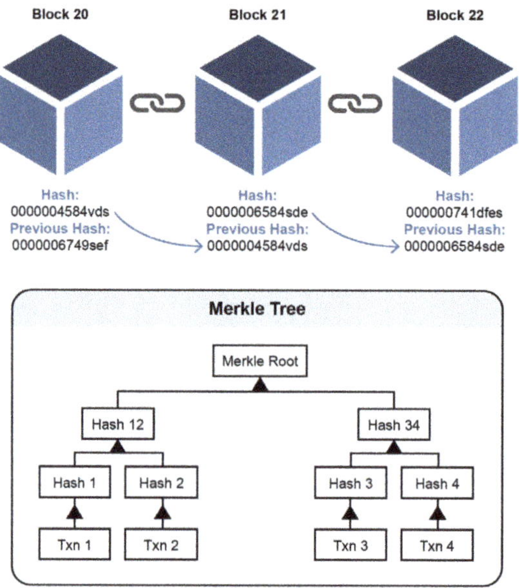

Figure 5. Blocks forming blockchain using hash signature.

Blockchain is not going to replace existing business logic. It is more of a log-based system, recording what has happened, as it is happening, rather than being the source of what to do with money from a customer, etc. It is a linked list type log data that is read and writeable by many participants. Blockchain technology passed through three eras. The blockchain 1.0 era passed through Bitcoin, the blockchain 2.0 era was marked by smart contracts, and the blockchain 3.0 era for applications in the social field [21]. Blockchains are classified into three categories. They are public, private, and permissioned. Bitcoin is a classic example of a public blockchain, in which any participant can join, read, and write data without requiring permission from a central authority. The private blockchain is restricted to members of an organization who are recognized and trustworthy. Permissioned blockchains are an example of a collection of businesses or consortiums where members are required to sign a legal contract in order to obtain access to, read, and write the blockchain. A summary of this classification is summarized in Table 1. Mohammad Dabbagh et al. published a barometric study on the evolution of blockchain. It describes the distribution of blockchain publications, most investigated research areas, the most influential papers, and the prominent funding agencies for research on blockchain [22]. Jameela AL-Jaroodi and Nader Mohamed surveyed the acceptance of blockchain in the industry. This survey shows that a wide range of industrial domains is starting to adopt or consider adopting blockchain to facilitate their operations to streamlining processes, enhancing security and data sharing, increasing efficiency, and ultimately reducing costs to gain a competitive advantage [23].

Table 1. Classification of Blockchains.

	Public Blockchain	**Private Blockchain**	**Permission-Ed Blockchain**
Read Access	Not Required	Within the Organization	Controlled
Write Access	Not Required	Within the Organization	Controlled
Consensus Process	Open Access	Within Organization	Within Consortium
Scalability	High	Low	Medium

3.1. Basics of Blockchain Technology

This section will introduce the basic concepts that need to understand blockchain technology.

3.1.1. Cryptography

Cryptography is the process of using Encryption and Decryption techniques in communications. Encryption is the process of converting our message or information into some coded message that cannot be understood by unauthorized parties. It is the core of the blockchain technology.

3.1.2. Cryptographic Hash Function

A cryptographic hash function is a mathematical algorithm that maps data of arbitrary size to a bit array of a fixed size. SHA256 is an algorithm developed by the National Security Agency of the USA. SHA256 and MD5 algorithms takes the data input and processes it through a sequence of complex mathematical calculations and other transformations [24]. For computers, the data are just the binaries 1 and 0. This process outputs a fixed string of characters called a hash value or digest of the data. The hashing process is designed so that even the tiniest change in the document will result in a completely different hash. It is not encrypted because the SHA256 algorithm does not encode the information, and the hash cannot be reversed into the original data. The primary purpose of hash is a comparison, not the encryption. Figure 6 is showing the hash value of different strings generated by some famous hashing algorithms.

```
String      : Hello World
SHA256 Hash: A591A6D40BF420404A011733CFB7B190D62C65BF0BCDA32B57B277D9AD9F146E
SHA512 Hash: 2C74FD17EDAFD80E8447B0D46741EE243B7EB74DD2149A0AB1B9246FB30382F2
             7E853D8585719E0E67CBDA0DAA8F51671064615D645AE27ACB15BFB1447F459B
MD5    Hash: b10a8db164e0754105b7a99be72e3fe5

string      : hello world
SHA256 Hash: B94D27B9934D3E08A52E52D7DA7DABFAC484EFE37A5380EE9088F7ACE2EFCDE9
SHA512 Hash: 309ECC489C12D6EB4CC40F50C902F2B4D0ED77EE511A7C7A9BCD3CA86D4CD86F
             989DD35BC5FF499670DA34255B45B0CFD830E81F605DCF7DC5542E93AE9CD76F
MD5    Hash: 5eb63bbbe01eeed093cb22bb8f5acdc3

string      : Hello world
SHA256 Hash: 64EC88CA00B268E5BA1A35678A1B5316D212F4F366B2477232534A8AECA37F3C
SHA512 Hash: B7F783BAED8297F0DB917462184FF4F08E69C2D5E5F79A942600F9725F58CE1F
             29C18139BF80B06C0FFF2BDD34738452ECF40C488C22A7E3D80CDF6F9C1C0D47
MD5    Hash: 3e25960a79dbc69b674cd4ec67a72c62
```

Figure 6. Hash values generated by different Hashing Algorithms.

3.1.3. Symmetric abd Asymmetric Encryption

In Symmetric Encryption, a single passphrase (key) is used to encrypt and decrypt the message. For example, Alice wants to send a classified document to Bob. She uses a key to encrypt the document and send it to Bob. Bob cannot read this document as he does not have the passphrase or key to decrypt the document. Alice needs to send this passphrase to Bob, but she cannot send it securely over the email, etc. as the attacker can attack it. Therefore, she has to find a secure way to handover this key to Bob.

In Asymmetric Encryption, both Alice and Bob generate a keypair (public key + private key) on their machines [25]. The RSA Algorithm generates a public and private key pair that is mathematically linked to each other. The public key is used to encrypt the data, and only the matching private key can decrypt that data as illustrated in Figure 7. The private key can not be derived from the public key. Bob and Alice can now exchange the public keys without any fear of each other, over the Internet or via any public means. Alice now encrypts the classified document with Bob's public key. Alice herself now cannot decrypt that document. That document can only be decrypted with Bob's private key.

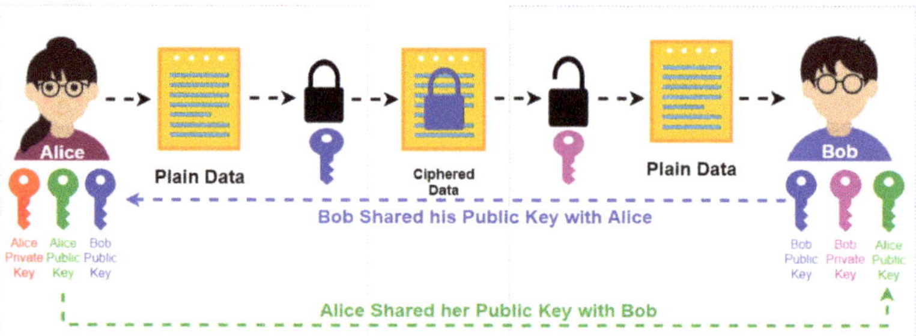

Figure 7. The Asymmetric Encryption Process.

3.1.4. Digital Signature and Digital Certificate

Bob wants to send a document to Alice by electronic means. There is nothing secret about the document. Alice wants to make sure that Bob sends it, and nobody else made any changes to it on the way. Before sending the document to Alice, Bob will digitally sign this document [26]. Bob's machine program will create SHA256 of the document and then encrypt that hash with Bob's private key. The resulted output is called a digital signature [27]. The algorithm will embed this digital signature into the document before sending it to Alice. On receiving the document, Alice's machine will extract the digital signature from the document and decrypt it with Bob's public key. If the decryption is successful, that means Bob signs the document. Now the machine will calculate the hash of the document, excluding the digital signature. If the calculated hash matches with the decrypted hash, the document is original and intact. Remember that Bob and Alice do not care if someone else reads the document.

A hacker may be pretending to be Bob. He could generate a pair of public/private keys and a fake document, hash it with SHA256 and digitally sign it. That is where a digital certificate comes in handy. Bob can apply for a digital certificate from a well-known, and well-trusted organization called a certification authority. Bob will send their public key and other information to the certification authority as part of this process. The certification authority carefully checks that Bob is the one he pretends to be in the data. They then send them a particular type of file called a digital certificate [28]. It contains details about Bob alongside the information about the certification authority. This certificate also has an expiry date and Bob's public key. Now Bob can send the digital certificate with the document to Alice. Alice can inspect the certificate first, and then after she trusts that the certificate is about Bob, she can use the public key extracted from the certificate to check the digital signature of the document. The certification authority is watching Bob carefully. Anything the certification authority sends to Bob is signed with their digital certificate provided by the higher certification authority as illustrated in Figure 8. In 2000, a law was passed in the UK called the electronic communication act. This law makes the digital signature legally binding, and this allowed businesses to thrive on the web [29,30].

Figure 8. Digital Certification Authority Issues Digital Certificate After Verification of the User.

3.1.5. Proof of Work

The proof of work concept was initially introduced in 1993 to prevent denial of service attacks and spam on the network by requiring some work from the service user [31]. It requires some minimum processing time by the computer. Bitcoin introduced an innovative way of proof of work called consensus algorithm. It needs to solve a complex mathematical puzzle. For example, the hash value of a block must start with 13 zero, which is also called its difficulty. A particular set of nodes called miners put a string of numbers at the end of the block data and calculate its hash until they meet the requirements. The number which fulfills this requirement is called the nonce. The probability of such a number is about 1 in a billion digits. Therefore, the only way to find such a number is guess and check. The current time of creating a bitcoin block is about 10 min.

3.1.6. Peer-to-Peer Network

A network is a group of interconnected communication devices connected by the wire or wireless media by switching or over the Internet. The one model is a client-server model. The server is held by a single company that performs all the tasks requested over the network. A central database is an example where a central authority stores all the data. To transfer one device's data to other devices, it first goes to the central server and then is distributed to the other devices. In a peer-to-peer network, the devices are directly connected with each other [32]. The connected devices act as server and client simultaneously. In this kind of network, all computer devices are working together to share data with each other. The blockchain is a peer-to-peer network of devices. There is no central storage in the blockchain network, and all the connected devices have an identical copy of the data passes to each other, including network information. The devices on the blockchain network are called nodes or peers.

3.1.7. Smart Contracts

Smart contracts are prevalent at present. Nick Szabo first used the term smart contract in 1997 before bitcoin was created. Smart contracts are similar to contracts in the real world. The only difference is that they are entirely digital. In fact, a smart contract is a small computer program that is stored inside the blockchain. The smart contracts code cannot be tampered with and distributed; everyone validates the smart contract's output [33]. A single person can not force the smart contract code change as the output from the tampered smart contract will be identified as invalid on the blockchain.

4. Hyperledger Fabric

Hyperledger Fabric is a programmable blockchain network that encapsulates the business logic implementation, and application of the business network by way of smart contracts or chaincode. Some more essential components and services provided by Hyperledger Fabric, an open-source blockchain platform for enterprises, which made it a strong candidate of choice for this work, are described below.

- **Permissioned Network**

  Individuals downloaded the program and instantly started transacting anonymously on a public blockchain network. This is an acceptable way to operate in corporate networks. Anonymity is not permitted on business networks. Members of a business network are continuously recognized by their unique identifiers and given responsibilities. The Hyperledger Fabric is a permission-based network that routes transactions according to established identities and responsibilities. Authentication is required for all users and components on the Hyperledger Fabric network. The Hyperledger Fabric assigns these entities their network identities through Membership Service Providers (MSP) and Certification Authorities (CA), which use Public Key Infrastructure (PKI) to approve and verify users and components.

- **Confidential Transactions**
  Confidentiality with unrelated parties is critical in a wide variety of business settings. Corporate networks often seek to keep their transactions very confidential from unrelated parties and reveal them only to the counter party. Hyperledger Fabric has a channel feature that allows transactions between defined parties to remain private. Each channel has its own ledger, and a consortium member connecting to the same network may have several channels.

- **Consensus and Policy Support**
  Members of the consortium create many policies, decisions, rules, and regulations that regulate the consortium's functioning. Typically, consortium makes decentralized choices. Numerous administrators from member organizations vote by majority to make network modifications that affect the members of the consortium or business network. This kind of decentralized decision-making system necessitates the establishment of governance and decision-making structures. Hyperledger Fabric technology supports decentralized administration by way of policies.

- **Identity Management**
  PKI (Public Key Infrastructure) is used by Hyperledger Fabric to manage identities. It includes two tools for identity management: Active Directory integration and Fabric-CA Server. Figure 8 illustrates a common method for creating IDs. The identity owner provides the registration authority with evidence of identification. The registration authority verified and sent the user's details to the certification authority. The certification authority generates an x509 certificate and returns it to the owner and validation authority for validation purposes. The identity is also required for the other components, such as peers and orderer nodes, to participate in the network.

- **Efficient Processing**
  To allow the network to operate in a concurrent and parallelism-friendly manner, transaction execution is kept independent from transaction ordering and commitment. Nodes in the network are assigned a role that corresponds to one of these activities, which is determined by the kind of node in the network. As a result of this concurrent execution, the processing efficiency of each peer is increased, resulting in faster transaction delivery.

- **Fabric Channel**
  Essentially, a Hyperledger Fabric channel is a private subnet of communication that is established between two or more particular network users for the aim of facilitating private and confidential transactions. Members (organizations), the shared ledger, chaincode apps, and the ordering service node all contribute to the definition of a channel (s). In the network, every transaction is carried out via a channel in which all parties are required to be verified and allowed to transact on that channel. Each peer that enters a channel has its own identity, which is provided by a Membership Service Provider (MSP), which authenticates each peer in relation to the other peers and services in the channel.

- **Endorsement**
  Every chaincode is connected with an endorsement policy, which applies to all of the smarts contracts that are specified inside that chaincode. This policy specifies which organizations in a blockchain network are required to sign a transaction produced by a certain smart contract in order for that transaction to be deemed legitimate by the network.

- **Valid transaction**

  During the validation process of a transaction that is disseminated to all peer nodes on the network, there are two stages. First and foremost, the transaction is verified to verify that it has been signed by the organizations that have been stated in accordance with the endorsement policy. Second, it is verified that the current value of the world state corresponds to the read set of the transaction that was signed by the endorsing peer nodes at the time of the transaction signing. If a transaction satisfies both of these criteria, it is considered to be legitimate. Transactions are added to the blockchain history in both legitimate and invalid ways, but only valid transactions result in a change to the global state.

- **Ordering Service**

  When it comes to Hyperledger Fabric networks, there is another kind of node known as an orderer, which is in charge of overseeing "transaction ordering" and constructing the final block of transactions before they are transmitted to be committed in each peer's ledger. The Ordering Service is made up of a collection of orderer nodes, which are connected together. It is the Ordering Service nodes' responsibility to accept transactions from numerous applications after the endorsement flow has happened, and to organize batches of these transactions into a well-defined order and bundle them into the final block of transactions. In addition to their ordering responsibilities, the orderers are in responsibility of maintaining the "consortium" list, which is comprised of a list of organizations that have the ability to establish channels and are in charge of enforcing the most basic access control for the channels.

- **Network Authentication**

  Peers, orderers, client apps, and administrators are some of the various entities that may be found in a blockchain network. X.509 digital certificates are required for each of these organizations in order for them to be identified digitally. These identities are a critical component of the authentication process since they define the precise rights over resources and access to information that entities have in the network, making them a very essential component.

- **Application Development and Integration**

  Integrating the Hyperledger Fabric blockchain with an existing business system is very important to encourage the organizations adopt blockchain technology. Each company in the network may customize the interaction system to meet its own requirements. Fabric front end applications can be developed independently using RESTful APIs as middle-ware, or the custom middleware can also be designed using one of the SDKs provided by Hyperledger Fabric (Figure 9).

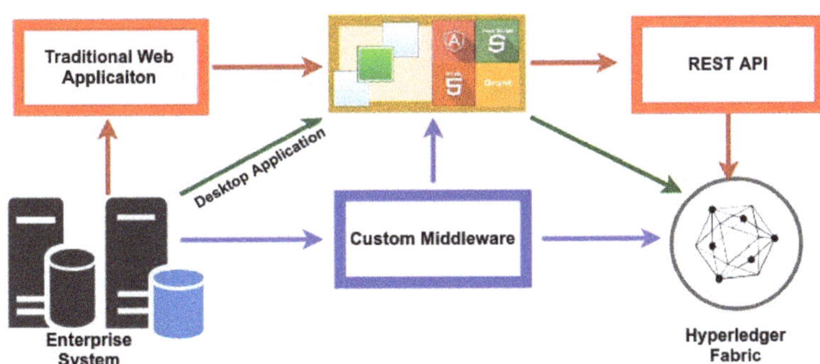

Figure 9. Hyperledger Fabric Application Development.

4.1. Transactions Workflow in Hyperledger Fabric

The transaction process in a Hyperledger Fabric network may be split into three major stages, which are described below.

1. **Proposal**

   When a client application requests approval from the necessary group of peers, this is referred to as the first step of the transaction approval process. On receiving a transaction proposal, each of these peers (who are acting in the role of endorsing peers) executes a chaincode to produce a transaction proposal response, after verifying and signing the incoming transaction proposal. The first step is completed after the application has obtained a sufficient number of signed proposals (depending on the requirement stated in the endorsement policy), and the second phase is completed once the application has received a sufficient number of signed proposals.

2. **Ordering Transactions**

   Transactions comprising approved transaction proposal replies are sent to an ordering service node by application clients at this step of the process. In subsequent blocks, the ordering service nodes collaborate to collectively establish a well-defined sequence, which is then packaged into the next block that is added to the blockchain. This step concludes with the blocks being stored to the orderers' personal ledgers and then broadcasting the block to all of the peers who have connected to the orderers' ledger.

3. **Validation and Commit**

   It is at this phase that the orderers distribute the blocks to all of the peers who are linked to them. While each peer will process this block separately, they will do it in precisely the same manner as every other peer on the channel, ensuring that the ledger remains consistent. After that, the peers will verify that the transaction has been endorsed and that it conforms with the endorsement policy in place (such as the application has done in Phase 1). After verifying the endorsement of the transaction and that the current state of the ledger is compliant with the state when the proposed change was produced, it commits and updates its copies of the ledger before committing and updating its copies of the ledger. Finally, the block events produced by the peers bring this phase to a close, events to which the application may listen in order to be notified when a transaction is committed to the ledger.

4.2. Rationale of Choice

An organization's choice on which technology to use is a long-term investment. One of the most significant advantages of using an open source business application is the great degree of flexibility that is allowed by open source code, modular components, and standard compliance, among other things. This allows a business to quickly and readily change technology in order to achieve real user-friendliness. For example, if a company currently employs Kafka clusters as a message protocol in their system, we may use it as a consensus mechanism in the Hyperleger Fabric blockchain network in place of the RAFT orderering service, saving both time and money. Furthermore, we may use other existing consensus algorithms or build our own light weight consensus algorithm for IoT-platforms in order to overcome the processing power limitations of the platform. No one vendor needs to be used by enterprise and corporate networks; instead, they may choose from among the most innovative and active communities, taking advantage of the fast pace of innovation in the blockchain technology industry.

5. Blockchain-Based Approach for Strategic Grain Reserves Management

The grain reserves management is a multi-link sector. It includes exploration, procurement, processing, storage, and reserves release. A large number of reconciliations between different departments and track of the work and transactions require. Moreover, it is incredibly flexible and cost-efficient as it does not require some particular infrastructure and costly servers. We have developed cost-effective solutions for developing countries where finance is a big hurdle in accepting modern technology. The following subsections briefly describe working of our proposed architecture.

5.1. Business Consortium

Hyperledger Fabric is a consortium-oriented blockchain platform. Figure 10 shows the consortium configuration of our proposed system. The consortium consists of an agriculture department, food department, and finance department. The agriculture department has the record of the agricultural land and the farmer currently farming on that piece of land. They also maintain a record of the current crop on a particular land and expected yield too. The department is also responsible for purchasing, storing, and release of the grains to the mills and food product producers on a quota basis. The finance department is responsible for providing finance and an overview of the payments. The grain elevator operator determines the grain's grade and quality and purchases the grain from the farmer. Some critical factors to be considered while storing the grain are temperature, moisture, and storage duration. The consortium makes policies to conduct operations and updates its planning and procedure to carry on all the operations. To ensure the secure tracking of all the operations using Hyperledger Fabric blockchain and chaincode the agriculture department update all the record of the agriculture land, current crop, and farmer on the system. The agriculture department also needs to update the crops' growth and expected yield as it conducts the survey. The food department purchases the crop from the farmer by financing from the finance department. The crop samples are brought to the site, where they are passed or rejected by the quality control lab. The crop is then brought to the center where it passes through the weighing bridge cleaning and drying and then stored in the storage silo. IoT sensors and GPS systems can monitor the conditions of the silo in real time. The department is also responsible for releasing the crop to the mills or other events.

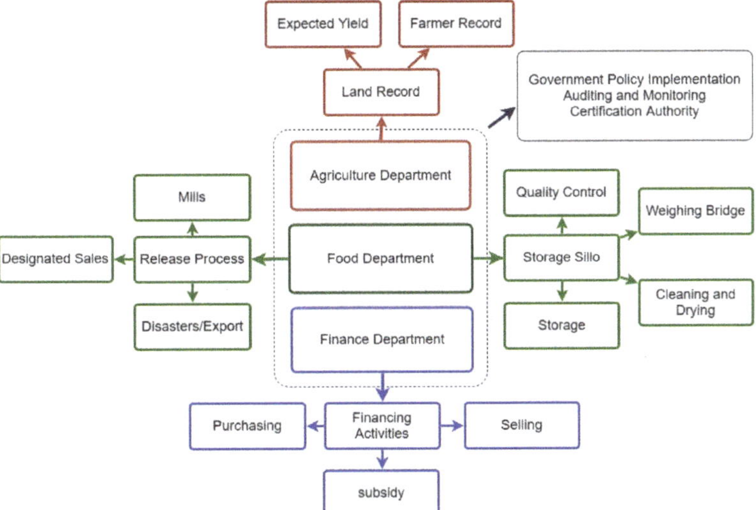

Figure 10. An overview of the proposed consortium.

The auditing and monitoring department is receiving the live update and situation at all the stages, transaction by transaction. The audit can be conducted at any time while we have an entire tamper-proof transactions record of all the facilities. The Hyperledger Fabric blockchain uses Smart Contracts called Chaincode. The chaincode guarantees that regulations are transparent to all parties, and appropriate criteria should prompt interventions in food reserves. Chain codes are the functions in a blockchain network that accept input in the form of transactions and deliver alerts to the network's members to monitor and detect any rule violations. The Hyperledger Fabric is an entirely permissioned network. Each component of the system, including users and the operator, must have its own identity. So no attempt of violation can be hidden in any way.

5.2. Application Server

Figure 11 is demonstrating the general application server components. The application server is based on NodeJS, which hosts Fabric SDK, Express.js, MQTT.js, and SQL/Mongoose. Express.js provides RESTful APIs and various functions to send, receive, and access data from ERP infrastructure at the organizations. This application server provides a generic interface that can generate a new user interface or integrate it into an already existing application interface. Every current organization or future organization joining this venture can integrate this solution into their existing system according to their needs with very minimum effort due to its modular nature and simplicity.

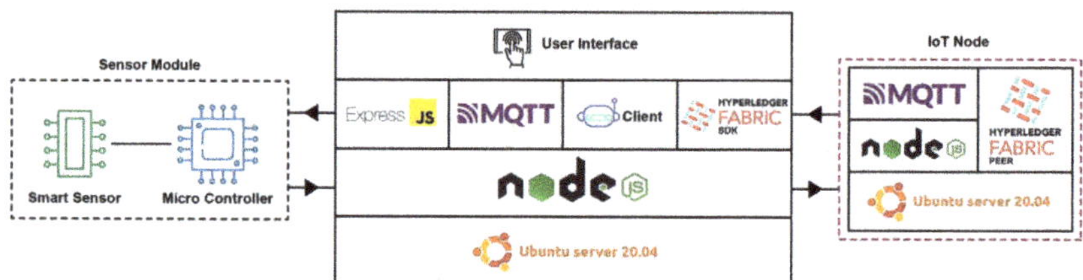

Figure 11. Application Node Architecture.

5.3. Blockchain Network Architecture

Figure 12 is an illustration of a blockchain network diagram. The peer is a bridge between the fabric network and the real world. Organizations are connected with their peers to obtain access to the network. It is also a place where the ledger is stored. Remember that it is distributed to all of the peers connected to a particular channel. Every channel has one single ledger, and every participant has an exact copy of the ledger. The ledger is logically stored on the channel, but it is stored on the connected peer nodes. We still need smart contracts to interact with the ledger.

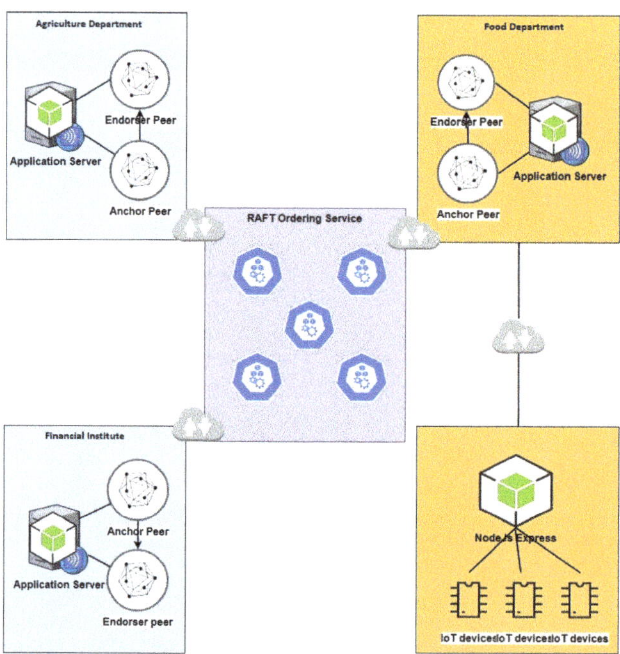

Figure 12. Hyperledger Fabric Blockchain Network.

The Hyperledger Fabric blockchain uses Smart Contracts called Chain Code. The chain code can ensure the transparency of the rules for everyone, and relevant criteria should trigger food reserves interventions. In a blockchain network, the chain codes are the functions that take input in transactions and sent notifications to the network's participants to monitor and notice any violation of the rules. The Hyperledger Fabric is a fully permission-ed network. Each component of the system requires carrying its own identity, including users and operators. So no attempt of violation can be hidden in any way. The chaincode is deployed on all the peer nodes. The chaincode is not connected to the channel, but it is hosted on the peer nodes. Now our network is ready where all of the three organizations have an equal right on the network.

5.4. Internet of Things Integration

The Internet of Things is a broad term that refers to a network of all these sensors capable of collecting data from their surroundings and feeding it to algorithms running on processing nodes located anywhere in the world through the Internet or a local area network. Data has surpassed oil as the most valuable resource. Industry 4.0 and agriculture 4.0 are built on the Internet of Things. The growing usage of intelligent technologies such as machine learning, cyber-physical systems, and the Internet of Things is changing human existence into entirely reliant on data generated by surrounding objects and information production. The world's technical advances are now directly proportionate to global automation, facilitated by the Internet of Things and specific other physical systems. While the Internet of Things manages connectivity between items and machines, cyber-physical systems are machines that control or monitor a mechanism by computer-based algorithms. These linked objects will communicate with one another and make their own choices. The algorithms are fed data from a variety of sources, including smart sensor devices.

A large number of food grains are wasted due to improper temperature and humidity at storage facilities. The grains become infested with mold and insects due to the lack of environmental monitoring techniques. When the grains go through the drying process, temperature and moisture present in the grain can cause stress, cracks on the grain kernel. The IIoT [34] focuses on machines, sensors, and devices to provide various services such as tracking objects [35], environmental monitoring [36], healthcare industry [37], traffic management [38], smart cities [39], and many more. International Society of Automation (ISA) provides various standards for implementing automation in the industry and so on [40]. Figure 13 is self describing the IoT system implementation.

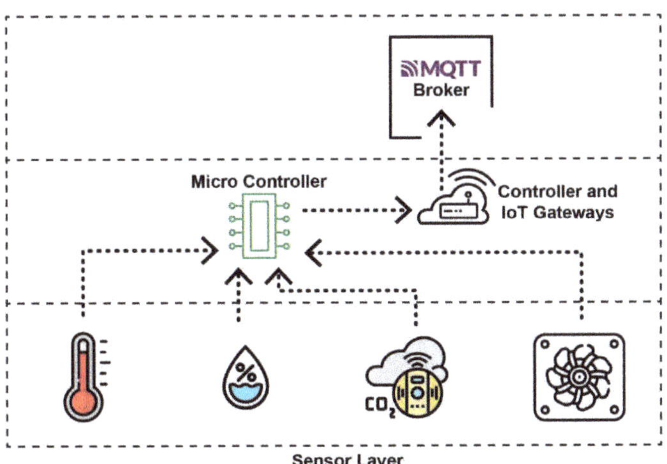

Figure 13. IoT Layer Configuration.

6. Qualitative Analysis of the Proposed Blockchain Solution

Many existing technological options are available for traceability and control. However, blockchain technology provides another level of trust by a tamper-proof shared ledger. The transparency it gives and its tamper-evident nature creates trust in low-cost IT solutions. It stores every transaction in the network, and all the parties in the system have the same data. It maintains an undeniable record of what is added to the data, by whom, in the chronological order. The data cannot be removed in the blockchain. By enabling each party to see the same data, in near real-time, blockchain can help eliminate complicated and costly data reconciliation required by most systems in the world today.

6.1. Decentralized Database

Blockchain is a decentralized platform for conducting transactions. In traditional databases, we have a centralized authority to store all the data. The conventional client-server model is a perfect example of this. The conventional model is an easy task for hackers or criminals as all the data is stored in one place, and the worst scenario is if the centralized entity became corrupt. The main idea behind blockchain is if we want to make a transaction with anyone, and we can do it directly without any third party, and both of us have a tamperproof record of the transaction. All the nodes maintain the data on the blockchain system, and no node has full access to it.

6.2. Persistency

The data on the blockchain is persistent as these data are distributed along with the network. Each node has control of its data. All the transactions are transparent and tamperproof, which makes the complete blockchain persistence.

6.3. Validity

The transactions put on blockchain may be validated by some of the other nodes in the network. Any mistakes as malicious transactions can be detected very quickly. The system consists of three primary roles. (1) propose a transaction (2) validates which transactions to accept (3) Lerner who accept the chase value.

6.4. Anonymity and Identity

The public blockchains such as bitcoin and Ethereum provide complete anonymity. A single user can obtain multiple identities. The public blockchain has not any central entity to maintain the identities. The private blockchains, on the other hand, required the identity to access the blockchain network. In permission-ed blockchains, the identities are provided by the consortium which is operating the blockchain under consortium policies.

6.5. Audibility

All the blockchain transactions are time-stamped and tightly linked with the previous transaction in the chain protected by the cryptographic hash functions. Therefore, it is easy for a node to trace the history of the transaction. The blockchain's audibility also depends on the type of blockchain, such as the public blockchain, which is fully accessible, and any node can fully trace any transaction. On the private and permission-ed blockchain, the parties' agreements may not allow it to become fully audible for everyone.

6.6. Closeness and Openness

The open blockchain consists of public nodes for consensus and maintenance of transactions. Anyone can join these open blockchains and start transactions or can become part of the consensus process. The private and permission-ed blockchains are closed or semi-opened. There are preselected nodes for the consense on the data. These blockchains rely on a single entity called a consortium for the permissions.

6.7. Traceability and Audit

In our proposed blockchain-based system, all the information regarding expected yield, finance, available stock and conditions at silo are available to all stakeholders in real time. Beginning with the land record and expected yield the policy makers can make their policy with real time data availability in minimum time. The crop sold by a farmer cannot be greater than a set percentage of expected yield and have a complete record of the land and farmer. In addition, the grains of different qualities cannot be mixed and total volume of purchase and sale is known by all the stakeholders in real time. It is possible that some stakeholder entered the wrong and fraudulent information into the system. If that data is caught at some stage, all the stakeholders know with 100% confidence who originated that data. Continuous monitoring of the silo location and quality control with traceable identifiers per lot and the realtime transaction traceability and availability to all the stakeholders makes it more corruption secure.

7. Conclusions

We proposed a blockchain-based solution to manage the strategic grain reserves of a country in this article. We presented possible general motivations to apply blockchain technology in food reserves management. We described the very generic detail of the blockchain systems overview and how our system can be applied applied to provide tamper-proof data availability. Moreover, many governments are also showing an interest in using blockchain technology in governance. Future needs include the development of a system where the different blockchains can transfer the data to each other or can be merged. Future work includes the technological demonstration of the project with the Artificial Intelligence for data analytic and data visualization.

Author Contributions: Conceptualization, A.I.; investigation, A.I. and Y.Y.; writing—original draft, A.I.; writing—review and editing, Y.Y. and X.C.; Funding, X.C. All authors have read and agreed to the published version of the manuscript.

Funding: This work was supported by the National Key R&D Program of China (No. 2018YFC1604000) and the National Natural Science Foundation of China (No. 61572374 & No. U163620068).

Institutional Review Board Statement: Not Applicable.

Informed Consent Statement: Not Applicable.

Data Availability Statement: Data sharing is not applicable for this article.

Conflicts of Interest: The authors have no affiliation with any organization with a direct or indirect financial interest in the subject matter discussed in the manuscript.

References

1. Wiggins, S.; Keats, S. *Grain Stocks and Price Spikes*; Overseas Development Institute (ODI): London, UK, 2010. Available online: https://odi.org/en/publications/grain-stocks-and-price-spikes/ (accessed on 10 September 2021).
2. Bohstedt, J. Food riots and the politics of provisions from early modern Europe and China to the food crisis of 2008. *J. Peasant. Stud.* **2016**, *43*, 1035–1067. [CrossRef]
3. Demeke, M.; Pangrazio, G.; Maetz, M. Initiative on soaring food prices. Country responses to the food security crisis: Nature and preliminary implications of the policies pursued. *EASYPol Modul.* **2009**. Available online: http://www.globalbioenergy.org/uploads/media/0812_FAO_-_Country_responses_to_the_crisis.pdf (accessed on 26 September 2021).
4. Declaration, M. Action plan on food price volatility and agriculture. In Proceedings of the Meeting of G20 Agriculture Ministers, Paris, France, 23 June 2011; Volume 22.
5. Timmer, C.P. Managing Public Grain Reserves. In *Issues and Challenges of Inclusive Development: Essays in Honor of Prof. R. Radhakrishna*; Saleth, R.M., Galab, S., Revathi, E., Eds.; Springer: Singapore, 2020; pp. 103–120. [CrossRef]
6. Vorley, B. *Corporate Concentration from Farm to Consumer UK Food Group*; International Institute for Environment and Development: London, UK, 2003.
7. Parvathamma, G. Farmers suicide and response of the Government in India—An Analysis. *IOSR J. Econ. Financ.* **2016**, *7*, 1–6.
8. Narayanan, S. Understanding Farmer Protests in India. *Acad. Stand Poverty* **2020**, *1*, 137–144.
9. Iftekhar, A.; Cui, X.; Hassan, M.; Afzal, W. Application of blockchain and Internet of Things to ensure tamper-proof data availability for food safety. *J. Food Qual.* **2020**, doi: 10.1155/2020/5385207. [CrossRef]

10. Tao, Q.; Chen, Q.; Ding, H.; Adnan, I.; Huang, X.; Cui, X. Cross-department Secure Data Sharing in Food Industry Via Blockchain-Cloud Fusion Scheme. *Secur. Commun. Netw.* **2021**, doi: 10.1155/2021/6668339. [CrossRef]
11. Tao, Q.; Cui, X.; Huang, X.; Leigh, A.M.; Gu, H. Food safety supervision system based on hierarchical multi-domain blockchain network. *IEEE Access* **2019**, *7*, 51817–51826. [CrossRef]
12. Cui, X. Wuhan University Food Safety and Food Quality Blockchain Alliance by Cui Xiaohui. Available online: foodblockchain.com.cn (accessed on 26 September 2021).
13. Iftekhar, A.; Cui, X. Blockchain-Based Traceability System That Ensures Food Safety Measures to Protect Consumer Safety and COVID-19 Free Supply Chains. *Foods* **2021**, *10*, 1289. [CrossRef] [PubMed]
14. Iftekhar, A.; Cui, X.; Tao, Q.; Zheng, C. Hyperledger fabric access control system for internet of things layer in blockchain-based applications. *Entropy* **2021**, *23*, 1054. [CrossRef] [PubMed]
15. Ashford-Rowe, K. Applying a design-based research approach to the determination and application of the critical elements of an authentic assessment. In Proceedings of the Emerging Technologies Conference, Wollongong, Australia, 18–20 June 2008.
16. Wang, D.; Wang, H.; Fu, Y. Blockchain-based IoT device identification and management in 5G smart grid. *EURASIP J. Wirel. Commun. Netw.* **2021**, *2021*, 1–19. [CrossRef]
17. Nižetić, S.; Šolić, P.; González-de, D.L.D.I.; Patrono, L. Internet of Things (IoT): Opportunities, issues and challenges towards a smart and sustainable future. *J. Clean. Prod.* **2020**, *274*, 122877. [CrossRef]
18. Nakamoto, S. Bitcoin: A peer-to-peer electronic cash system. *Decentralized Bus. Rev.* **2008**, 21260. Available online: https://www.ussc.gov/sites/default/files/pdf/training/annual-national-training-seminar/2018/Emerging_Tech_Bitcoin_Crypto.pdf (accessed on 26 September 2021).
19. Yli-Huumo, J.; Ko, D.; Choi, S.; Park, S.; Smolander, K. Where Is Current Research on Blockchain Technology?—A Systematic Review. *PLoS ONE* **2016**, *11*, e0163477. [CrossRef] [PubMed]
20. Swan, M. *Blockchain: Blueprint for a New Economy*; O'Reilly Media, Inc.: Newton, MA, USA, 2015.
21. Mohsin, A.; Zaidan, A. Blockchain authentication of network applications: Taxonomy, classification, capabilities, open challenges, motivations, recommendations and future directions. *Comput. Stand. Interfaces* **2019**, *64*, 41–60. [CrossRef]
22. Dabbagh, M.; Sookhak, M.; Safa, N.S. The Evolution of Blockchain: A Bibliometric Study. *IEEE Access* **2019**, *7*, 19212–19221. [CrossRef]
23. Al-Jaroodi, J.; Mohamed, N. Blockchain in Industries: A Survey. *IEEE Access* **2019**, *7*, 36500–36515. [CrossRef]
24. Rachmawati, D.; Tarigan, J.; Ginting, A. A comparative study of Message Digest 5 (MD5) and SHA256 algorithm. *J. Phys. Conf. Ser.* **2018**, *978*, 012116. [CrossRef]
25. Li, N. Asymmetric Encryption. Encyclopedia of Database Systems. 2018.
26. Subramanya, S.; Yi, B.K. Digital signatures. *IEEE Potentials* **2006**, *25*, 5–8. [CrossRef]
27. Pointcheval, D.; Stern, J. Security Arguments for Digital Signatures and Blind Signatures. *J. Cryptol.* **2015**, *13*, 361–396. [CrossRef]
28. Feghhi, J.; Feghhi, J.; Williams, P. *Digital Certificates: Applied Internet Security*; Addison Wesley Longman, Inc.: Boston, MA, USA, 1998.
29. Parliament of U.K. Electronic Communications Act 2000.
30. Sari, P.K.; Prasetio, A. Customer Awareness towards Digital Certificate on E-Commerce: Does It Affect Purchase Decision? In Proceedings of the 2018 Third International Conference on Informatics and Computing (ICIC), Palembang, Indonesia, 6–8 September 2018; pp. 1–4.
31. Gervais, A.; Karame, G.; Wüst, K.; Glykantzis, V.; Ritzdorf, H.; Capkun, S. On the Security and Performance of Proof of Work Blockchains. In Proceedings of the 2016 ACM SIGSAC Conference on Computer and Communications Security, Vienna, Austria, 24–28 October 2016.
32. Leuf, B. *Peer to Peer: Collaboration and Sharing over the Internet*; Addison-Wesley Longman Publishing Co., Inc.: Boston, MA, USA, 2002.
33. Zheng, Z.; Xie, S.; Dai, H.; Chen, W.; Chen, X.; Weng, J.; Imran, M. An Overview on Smart Contracts: Challenges, Advances and Platforms. *Future Gener. Comput. Syst.* **2020**, *105*, 475–491. [CrossRef]
34. Sisinni, E.; Saifullah, A.; Han, S.; Jennehag, U.; Gidlund, M. Industrial internet of things: Challenges, opportunities, and directions. *IEEE Trans. Ind. Inform.* **2018**, *14*, 4724–4734. [CrossRef]
35. Velandia, D.M.S.; Kaur, N.; Whittow, W.G.; Conway, P.P.; West, A.A. Towards industrial internet of things: Crankshaft monitoring, traceability and tracking using RFID. *Robot. Comput. Integr. Manuf.* **2016**, *41*, 66–77. [CrossRef]
36. Li, W.; Kara, S. Methodology for monitoring manufacturing environment by using wireless sensor networks (WSN) and the internet of things (IoT). *Procedia CIRP* **2017**, *61*, 323–328. [CrossRef]
37. Yuehong, Y.; Zeng, Y.; Chen, X.; Fan, Y. The internet of things in healthcare: An overview. *J. Ind. Inf. Integr.* **2016**, *1*, 3–13.
38. Elkin, D.; Vyatkin, V. IoT in Traffic Management: Review of Existing Methods of Road Traffic Regulation. In Proceedings of the Computer Science Online Conference, Zlin, Czech Republic, 15 July 2020; Springer: Berlin/Heidelberg, Germany, 2020; pp. 536–551.
39. Zhang, C. Design and application of fog computing and Internet of Things service platform for smart city. *Future Gener. Comput. Syst.* **2020**, *112*, 630–640. [CrossRef]
40. Find ISA Standards: In Numerical Order—ISA. Available online: https://www.isa.org/standards-and-publications/isa-standards/find-isa-standards-in-numerical-order (accessed on 3 February 2021).

Article

Suggestions for a Systematic Regulatory Approach to Ocean Plastics

Margherita Paola Poto [1],*, Edel Oddny Elvevoll [2], Monica Alterskjær Sundset [3], Karl-Erik Eilertsen [2], Mathilde Morel [1] and Ida-Johanne Jensen [2,4]

[1] Faculty of Law, UiT—The Arctic University of Norway, N-9037 Tromsø, Norway; mathilde.d.morel@uit.no
[2] The Norwegian College of Fishery Science, Faculty of Biosciences, Fisheries and Economics, UiT—The Arctic University of Norway, N-9037 Tromsø, Norway; edel.elvevoll@uit.no (E.O.E.); karl-erik.eilertsen@uit.no (K.-E.E.); idaj.jensen@ntnu.no (I.-J.J.)
[3] Department of Arctic and Marine Biology, UiT—The Arctic University of Norway, N-9037 Tromsø, Norway; monica.a.sundset@uit.no
[4] Department of Biotechnology and Food Science, Norwegian University of Science and Technology, NTNU, N-7491 Trondheim, Norway
* Correspondence: margherita.p.poto@uit.no

Abstract: The research investigates the problems and maps the solutions to the serious threat that plastics pose to the oceans, food safety, and human health, with more than eight million tons of plastic debris dumped in the sea every year. The aim of this study is to explore how to better improve the regulatory process of ocean plastics by integrating scientific results, regulatory strategies and action plans so as to limit the impact of plastics at sea. Adopting a problem-solving approach and identifying four areas of intervention enable the establishment of a regulatory framework from a multi-actor, multi-issue, and multi-level perspective. The research methodology consists of a two-pronged approach: 1. An analysis of the state-of-the-art definition of plastics, micro-, and nanoplastics (respectively, MPs and NPs), and 2. The identification and discussion of loopholes in the current regulation, suggesting key actions to be taken at a global, regional and national level. In particular, the study proposes a systemic integration of scientific and regulatory advancements towards the construction of an interconnected multi-tiered (MT) plastic governance framework. The milestones reached by the project SECURE at UiT - The Arctic University of Norway provide evidence of the strength of the theory of integration and rights-based approaches. The suggested model holds substantial significance for the fields of environmental protection, food security, food safety, and human health. This proposed MT plastic governance framework allows for the holistic and effective organization of complex information and scenarios concerning plastics regulation. Containing a clear definition of plastics, grounded on the precautionary principle, the MT plastic framework should provide detailed mitigation measures, with a clear indication of rights and duties, and in coordination with an effective reparatory justice system.

Keywords: Plasticene; seas; science; regulation; sustainability; multi-tiered; system; food; health

Citation: Poto, M.P.; Elvevoll, E.O.; Sundset, M.A.; Eilertsen, K.-E.; Morel, M.; Jensen, I.-J. Suggestions for a Systematic Regulatory Approach to Ocean Plastics. *Foods* **2021**, *10*, 2197. https://doi.org/10.3390/foods10092197

Academic Editors: António Raposo, Renata Puppin Zandonadi, Raquel Braz Assunção Botelho and Theodoros Varzakas

Received: 8 July 2021
Accepted: 14 September 2021
Published: 16 September 2021

Publisher's Note: MDPI stays neutral with regard to jurisdictional claims in published maps and institutional affiliations.

1. Introduction

This study takes place within the broad context of the 'Plastics Age', also known as the 'Plasticene', characterized by an exponential increase of plastic deposits on the planet since the mid-1940s [1,2]. Scientists have been investigating the effects of the escalation of plastic pollution on the planet [2–4], and the impacts of plastic on the marine environment. Currently, there is at least eight million tons of plastic dumped in the ocean every year (IUCN 2021 [5]). There is consensus among the preeminent studies on the topic for the need for further research to explore the development of an effective and coordinated regulatory strategy to efficiently reduce plastic pollution on seas and lands [6–11].

A prerequisite and *condicio sine qua non* of the regulation process, in the absence of scientific certainty of harmful impacts of plastics at sea, is the reference to the precautionary principle and the need to take preventive urgent action. The precautionary principle is enshrined in Article 15 of the Rio Declaration on Environment and Development [12] and Article 191(2) of the Treaty on the Functioning of the EU [13]. The need for urgent action is especially evident in the case of the unknown impacts of the smallest particles of plastics: microplastics (MPs, less than 5 mm in size) and nanoplastics (NPs with a size defined as either less than 100 or 1000 nm) [14]. Due to the dearth of coordination between the research on the harmful impacts, one may argue that there is not yet sufficient evidence to substantiate a regulatory effort that tackles all the risks that plastics pose to food security, food safety, and ultimately to human health [14]. Consequently, the precautionary principle and the need to take preventive measures can legitimize the policymakers' anticipatory action even under scientific uncertainty.

Based on this premise, the plastic regulatory effort needs to be methodologically sound and constructed under the multilevel environmental governance (MEG) system [15,16]. MEG combines top-down and bottom-up approaches to achieve an effective reduction of some forms of plastics at the global, regional, and local levels [15]. In this framework, we argue that plastics governance can be developed in at least in three dimensions: through *multi-level*, *multi-actor*, and *multi-issues* perspectives. For clarity, hereinafter these three dimensions or perspectives are grouped under the broad category of 'multi-tiered (MT) plastics governance'. An MT enforcement system provides criminal, civil, and administrative sanctions in case of non-compliance on the one side (hence the reference to a *multi-issues* dimension). On the other side, it brings together a diverse group of responsible and impacted actors (here the *multi-actor* and *multi-level* dimensions), in the environmental justice movement [17]. The reasoning behind the latter approach stems from the conclusions reached by a UNEP Report that demonstrates how micro and nanoplastic pollution is now in the public domain as an issue of global concern [18]. Consequently, the global fight against plastic pollution, much like the environmental justice movement, is characterized by a multi-actor-driven process [18]. A pluri-subjective regulatory system entails the need for coordinated actions, for provisions on civil society suit authority, collective actions, and promotion of youth campaigns against global inaction on plastics waste [19].

In addition, the provisions of such a complex framework are to be coordinated with the pre-existing legislation on marine environmental protection, food safety, and, in general, conceptualized through the lens of the Agenda 2030 and its sustainable development goals (SDGs): in particular, but not exclusively, SDG 2 (Zero Hunger), 3 (Health), 12 (Sustainable Consumption and Production) and 14 (Life Below Water) [20].

For all the above-mentioned reasons, the present study intends to address the needs highlighted by the preeminent literature. It does so by prospecting the need to develop a regulatory model for the reduction and the sustainable management of marine plastic pollution. The model integrates the "multitude of multi-tiered approaches" (da Costa et al. [11]) thus preventing some of the impending long-term impacts that pollution of the marine environment can have on food security, nutrition, and health.

2. Materials and Methods

2.1. Problem-Solving Approach

It is possible to identify at least four issues and four corresponding approaches in the problem-solving process related to plastic regulation. Two issues are related to the previously mentioned difficulty of integrated approaches to plastics from hard sciences and law: first, scientific research on the impacts of plastics on seafood security, food safety, and human health is still in its infancy (1). Second, this circumstance hinders regulatory efforts causing a substantive divide between scientific research and regulatory interventions (2). There are two additional issues resulting from endogenous legal factors: plastic regulation has not yet been fully framed and consistently constructed at the global, regional, and local level, which results in fragmented legislation and consequently uncoordinated enforcement (3). Moreover, plastic regulation is generally characterized by a weak level of effectiveness both in terms of time and enforceability of the intervention measures (4) [14,15].

Such highlighted issues can be solved through a regulatory initiative that is grounded on the precautionary principle and that develops along the lines of a systematic and rights-based approach. In particular, the precautionary principle and preventive measures can help overcome issues (1) and (2) and support a framework that overcomes the divide between science and law. A way out of fragmentation and scarce enforceability (respectively, issues (3) and (4)) can be achieved in the systems thinking approach in conjunction with the rights-based approach.

The research design involves the following steps: a. Drawing insights from the definition of plastics and the impacts on the marine environment, nutrition, and health (Section 2.1); b. Understanding the state-of-the-art of plastic regulation (Section 2.2); and c. Combining the two understandings with the proposal of an integrated framework based on systematic approaches (Section 2.3). In particular this third step builds upon several aspects of the preliminary results from the project SECURE at UiT—The Arctic University of Norway on Novel Marine Resources for Food Security and Food Safety [21].

2.2. Drawing Insights from the Definition of Plastics and Its Impacts on the (Marine) Environment, Food Security, Food Safety, and Human Health

2.2.1. Definition

Steps towards the creation of a working definition of plastics that could be enshrined in legislation have been facilitated by the efforts of Nanna Hartmann et al. [10]. The research group has established seven criteria for the identification of plastics that threaten the marine environment at large, criteria that must be considered in the assessment of plastic materials: 1. Chemical composition; 2. Solid-state; 3. Solubility; 4. Size; 5. Shape and structure; 6. Color; and 7. Origin [10]. In the view of the researchers, the combination of these criteria and their specifications should facilitate the definition of what plastic is (an inert synthetic polymer) and of its potential harms to the environment [10]. Further research has been conducted by Mitrano et al. (2020) towards the construction of an epistemological basis for regulation, specifically focusing on the impacts of plastic on the marine food web [9]. According to the study, microplastics (MPs) are debris with sizes below 5 mm, resulting from the release of small particles and the fragmentation of larger plastics (respectively, primary and secondary MPs [9]). As for the definition of nanoplastics (NPs), experts refer to the European Food Safety Authority (EFSA)'s opinion of 2016 [22] in which the term is associated with any plastic debris with size between 1 and 100 nm, produced by the degradation of MPs or released directly from domestic and industrial sources [23]. The research highlights the importance of adopting a regulatory approach on the onset of MPs and NPs definitions and throughout the development of a grounded theory for a coordinated framework of the impacts of these on the environment, food security, food safety and human health.

2.2.2. Impacts

Petroleum-based polymer materials, such as polypropylene, polyester, polyethylene and polystyrene have been used as plastics packaging for many years, replacing other packaging materials and increasing by 20-fold since the mid-sixties [24,25]. The scientific community has not reached a consensus on the full extent of threats that plastic—and especially MPs and NPs—pose to the (marine) environment, the food web, and ultimately human health [26,27]. Hence, very little is known about how acute and chronic exposure to MPs and NPs (through, e.g., seafood or plastic bottle drinks) will affect global health [26]. Humans are estimated to ingest 52,000 MPs per year and inhale 69,000 MPs per year through plastic-contaminated air [28]. Most of the ingested MPs and NPs are expected to pass through the digestive tract, and only smaller NPs may potentially enter the blood stream circulation. The impact of a continual presence of MPs and NPs in the human gut is unknown, but to date the research indicates that very high intakes can cause inflammation of the gut lining [29]. Furthermore, organic pollutants can also be absorbed onto MPs and NPs and may enhance the effective uptake and toxicity of these pollutants. Similarly, metallic toxins (cadmium and mercury) and toxic trace elements may also interact with plastic particles serving as vectors for toxic uptake by living organisms [29].

Bio-based plastic packaging materials (natural biopolymers extracted from plants, animals, or microorganisms, including cellulose, starch, milk proteins, polysaccharide gums), can reduce climate impact. However, they may also have less favorable effects when it comes to other environmental impacts such as eutrophication (the gradual increasing of plant nutrients in an aging aquatic ecosystem), use of water, pesticides, and effects on biodiversity [25].

The described set of impacts is heterogeneous and fragmented, resulting in a lack of coordination between scientific research and regulatory intervention [3–7]. At the root of the inconsistencies between measured data and consequent actions is the lack of a common terminology that supports scientific evidence on such threats [7,8]. Moreover, reviews from the EFSA [22], the Food and Agriculture Organization (FAO) [30], the Science Advice for Policy by European Academies (SAPEA) [31], and the Norwegian Scientific Committee for Food and Environment (VKM) [32], provide little knowledge about the presence of NPs in the marine water and seafood, since the methods for either their detection or quantification have not been established yet [22,33]. NPs are expected to be absorbed in the gastrointestinal tract and may end up damaging different organs in marine organisms [30]. Thus, NPs are considered the most dangerous plastics possibly posing the highest risk of being deleterious to human health. MPs, on the other hand, are typically detected only in the gastrointestinal tract of marine organisms, commonly not consumed. Small low trophic organisms, like mussels, shellfish, and mesopelagic species, however, are often eaten whole and could thus contribute to ingested MPs [30,34,35]. Despite these preliminary studies, the VKM concluded that the available information regarding MPs and NPs does not provide a sufficient basis to characterize potential toxicity in humans [32].

Acknowledging the lack of scientific certainty and the fragmentation of the preliminary results, the cited research reaches a consensus on the urgency of intervening at the regulatory level. It does so by applying the precautionary principle in the areas where there is still insufficient scientific evidence, especially regarding the impacts of NPs on food safety and human health, "given the unavoidable increase in the coming decades of micro and nanoplastics in the marine environment" [36].

A legal reform that seeks to effectively tackle the plastic emergency should take into consideration the highlighted advances in research, and by doing so contribute to the improvement of the science-policy dialogue in the subject matter.

2.3. Understanding the State-Of-The-Art of Plastics Regulation

Plastic governance—when compared to the governance of other environmental problems —has at least three shortcomings: 1. It is not yet a fully, integrated and globally steered process [10]; 2. It is mainly enacted through soft-law mechanisms; and 3. It spans a time range of 10 years (up until 2030), a circumstance that hinders prompt and decisive action. Despite the efforts to tackle marine litter at the international level [37], the initiatives undertaken so far appear to be characterized by very weak enforceability (they belong to the category of 'soft law instruments') while covering a very wide margin of time intervention. Similar drawbacks are also present at the EU level, where the EU Plastics Strategy has been launched under the umbrella of the Green Deal and the Circular Economy Plan [38]. Beyond the statement of general principles, only one among the announced actions has been implemented to date. It is the Directive EU 2019/904 on the reduction of the impact of certain plastic products on the environment [39]. Hence the concern that the EU's efforts—albeit commendable—are not sufficiently timely nor pervasively binding [40]. Accordingly, this lack of effective regulatory mechanisms, at both international and regional levels, illustrates that plastic governance provides, in fact, just another area where law is (too) far behind.

Despite the general reference to the principles of circular economy, silo, uni-sectoral and individualistic approaches still prevail in the choice of regulatory tools. This preference for individualistic approaches, in turn, creates several small bubbles of circularity with no communication between them: the circular system of plastic, of food, of human health, and even of the Agenda 2030. The challenge is to connect them, contributing to the 'porosity of the governance soil', and create synergies between sectors. The limitations of a singular and reductionist regulatory perspective can be clearly seen as one of the contributory causes of poor legal enforcement. This understanding reinforces the argumentation on the importance to unwaveringly push towards systematic and integrated approaches when facing the challenges posed by the complexity of sustainability.

2.4. Combining the Two Understandings with the Proposal of an Integrated Framework

The preliminary results of the project SECURE establish a blueprint which rectifies the fragmentation, and reductionist and weakly enforceable approaches to the MT plastic system [21]. Under this interdisciplinary project, research is conducted on the legal framework for environmentally conscious harvesting practices of new low-trophic marine species, the composition of nutrients and contaminants in these, and their effects on the gastrointestinal microbiome and cardiometabolic health [21]. The research spans complex sustainability challenges, such as environmental protection, food safety and security, human health, and innovation [21]. Thus, the SECURE team has been developing pioneering research approaches across different systematic disciplines under conditions of scientific uncertainty [41], offering insights to complex problem solving including those related to ocean plastics.

The first leg of the project focuses on the need to develop a matrix that conceptually encompasses the objectives of the Agenda 2030 and its 17 SDGs from a regulatory perspective on the one hand, and allows the upscaling of climate-smart practices applied to the ocean—as it is the case of the SECURE, on the other hand (Figure 1) [41]. The observed lack of a systematic mapping of the SDGs' interactions and the dearth of climate-smart practices applied to the ocean spurred reflections on the need to create a critical conceptual framework as a reading key for the SDGs' interactions and for upscaling successful climate-smart practices [41]. The application of a matrix based on systems thinking within a legal fabric opened the door to reflections about integrated regulatory approaches applied to the ocean.

Figure 1. Illustration of Step 1 of the Project SECURE [41], representing the need to tackle novel marine resources through a systematic approach of the Agenda 2030 (with a main focus on SDG 14, on Life Below Water). Illustration for SECURE by Camilla Neema Haule 2021.

Based on this, the second step of the legal research in SECURE has argued that the adoption of the specific systematic lens of integral ecology (IE) overcomes regulatory fragmentation caused by 'glocal' environmental challenges (where the adjective 'glocal' is meant to comprise both worldwide ranging and local issues) [36]. In particular, the integral ecology is the approach that observes the interconnections of the living systems, be they living organisms, social systems, or ecosystems with the aim to regulate them (Figure 2) [42]. Under IE, for example, there is no distinction between the natural world and human beings (both have agency, rights to life and regenerate), neither is there a distinction between health of the environment (water, plants, animals, soil) and health of humans. Through the lens of IE, a risk posed to the natural environment has consequences for all living beings [43]. In essence, IE advances the development and application of comprehensive and integrated approaches to sustainability issues, providing comprehensive, far-sighted, and flexible solutions for the environment—solutions that can restore the relationship, at multiple scales, with the earth [44].

Figure 2. A case study relevant to the EU food on the integral ecology approach to food novelty applied to seaweed (frame of the illustration) through the lens of the systematic approach to the Agenda 2030 [42]. Illustration for SECURE by Camilla Neema Haule 2021.

The integral approach suggested by IE offers further insights into the complex problems and the uncertainties posed by plastics at sea, laying the foundation for systematic and integrated regulation. Construing the issue of plastic pollution through the lens of IE redefines the health of the ecosystem as the health of all beings, unlocking solutions in all those cases where the risks posed by plastics to food security and health are not yet proved beyond doubt. Grounded on the precautionary principle, a systematic and integrated regulation of plastic pollution at sea is justified by the need to prevent potential harms to the human rights to a healthy environment [43], healthy food, and to the rights of the Earth (and its living beings) to life, be respected and regenerate (Figure 3) [44].

Figure 3. The systematic approach of the Agenda 2030 applied to ocean plastics: overcoming loopholes in regulation through a multilevel governance system applied to the SDG 14, on Life Below Water [41,42]. Illustration by Camilla Neema Haule for SECURE 2021.

3. Results

The suggestion for a systematic regulatory approach to ocean plastics stems from the research on the systemic approaches to environmental and societal challenges. The current narrative of understanding complex challenges is developed through a systemic approach based on resilient relationships rather than focusing on individual components of systems [45]. The issues related to sustainable development are often referred to by the scholars as 'wicked problems' (Figure 4) [46,47], defined as "trivial or lasting situations that cannot be overcome immediately due to their inner complexity or exogenous/endogenous relations" [46]. Plastic pollution, with its multiple potential impacts on the environment, food security, and food safety, as well as human health, is certainly one example of a wicked problem that needs to be tackled through a process-based, multi-tiered and systemic approach. The research, therefore, takes a crucial step forward towards the regulation of the wicked problem of plastics, concluding with recommendations for policymakers. Such recommendations are to be approved in the form of an MT governance framework.

A MT common regulatory framework tackling plastic pollution should:

a. Contain a definition of plastics developed from the most advanced research in the field;
b. Be grounded on the precautionary principle and the legitimacy of preventive actions, according to the international and European legal provisions [13];
c. Be based on the premise that plastics pollution violates both human and nature rights;
d. Provide the most detailed range of measures, with a clear indication of rights and duties (rights of lands, coasts, rivers, and marine ecosystem to be protected and duties of the human communities to act as nature's guardians);
e. Contain a wide range of bans, restrictions, and limits (e.g., broadening the current EU restrictions on certain single-use plastics to all single-use, and expanding such bans outside the EU), sanctions in case of infringement in coordination with the national criminal, civil, and administrative law systems;
f. Promote, coordinate with, and implement systems that emphasize circular waste management practices and the use of alternative (renewable) resources as supplements or replacements for plastics; be coordinated with an effective reparatory justice

system, allowing judicial and administrative reviews through actions initiated by the plurality of actors involved in the plastic justice movement; and

g. Be constructed following the systems thinking approach and therefore develop multiple interconnections to the SDGs in the Agenda 2030 [47,48].

Figure 4. Plastic as a wicked problem, affecting the marine ecosystem as a whole. Illustration by Rudi Caeyers for SECURE, UiT Grafiske tjenester, 2021.

This research study demonstrates how a methodological approach applying systematic thinking to science and regulation helps to coordinate the most advanced results in science and the need to regulate the impacts of plastic pollution at sea. Multi-tiered and systemic methods for monitoring and ultimately restricting and banning plastics are key requirements to better understand, manage, and regulate such impacts, as well as to develop regulatory strategies and action plans at the international, regional, and national levels.

Studies on the complex challenges posed by sustainability highlight the movement from disciplinary regulatory baselines that tackle one objective at a time towards integrated approaches that acknowledge how the different problems in a system are interlinked [49]. Further research on empirical applications of the systemic approach is needed to assess the effectiveness of holistic approaches to tackle the complex problems of sustainability.

4. Discussion

4.1. Towards the Adoption of a Cross-Disciplinary Systems Thinking Approach

This study signifies a step forward in joint research concerning plastic pollution. Its reach extends to the exploration of the potential positive impacts of a systematic regulatory approach to plastics for the environment, human health, and in general the wellness of all living organisms. In this regard, it also marks a step forward in the project SECURE, at the crossroads of environment, food, and health. It does so by specifying in which terms systemic thinking applied to law and marine science can help tackle plastic pollution at sea.

Consequently, actors and disciplines can find a field of cross-cooperation and continue to develop ways to explore synergies. Two major reflections spur from this systemic approach: 1. How to effectively put it into practice; and 2. How to develop control mechanisms for its proper functioning.

The research community plays a key role in the solutions of the two questions. To test the robustness of systemic thinking and its applications to cross-boundaries issues, it is necessary to invest efforts towards systemic research. The steps of systemic research include developing standards to conduct systemic research, the establishment of a common language (to the extent possible) between the different research disciplines, as well as the implementation of review panels that overview, assist, and review the developments of the research. Examples of the latter can be found within research in healthcare [50].

Such review panels are to be formed by researchers, as well as key users and stakeholders, with the skills and content knowledge to produce and assess systematic research. The tasks in the systemic research fields should therefore be conducted and reviewed by multiple individuals with a wide range of expertise, from scoping studies to developing systemic methods that synthesize the research findings. To reinforce the role of systems thinking in solving complex problems, it would be desirable to introduce such an approach also in education, across the physical, natural, and social sciences [51,52].

While additional research is needed to build empirical support for effective approaches to systems thinking, this article has set the groundwork to facilitate the adoption of such an approach in research, education, and ultimately in decision-making.

Improving the use of systemic research and reviews in decision-making has the potential to provide multi-faceted interventions to limit the impacts of plastic pollution at sea. Collaboration between hard scientists and legal researchers is essential for developing consensus and clarity on regulatory strategies and action plans, thus maximizing the coordination and effectiveness of the intervention measures [9].

Finally, the advancements in systemic approaches are expected to further not only the plastic pollution research field, but also all the related and expectedly impacted ambits, such as food security, food safety, and human health.

4.2. From Legislation to Implementation: A Comparative Overview of the First Implementing Efforts of Directive EU 2019/904

France and Germany offer two virtuous examples of implementation of the Directive EU 2019/904 [39]. In particular, France has been a pioneer in the introduction of obligations on plastic waste, with the approval of the Law on the Circular economy (Law No. 2020-105 of 10 February 2020 — 'Circular Economy Law') [53]. Among the measures introduced by the law to curb plastics waste, it is worth listing the progressive ban from January 2021 on single-use plastic products; the ban on the import and manufacture of single-use plastic bags intended for sale or for giving-out for free; general limits on the use of plastic [53].

In the wake of implementing the European law, Germany has adopted the Ordinance on Single-Use Plastics (Einwegkunststoffverbotsverordnung, EWKVerbotsV), entered into force on 3 July 2021 [54]. The EWKVerbotsV implements Article 6 (1), (2) and (4) of Directive EU 2019/904, requiring Member States to ensure that the single-use plastic products listed in part C of the Annex to that Directive, whose closures and lids are made of plastic, are placed on the market only if the closures and lids remain attached to the containers during the period of use. The Ordinance also implements Article 7 (1) and (3) of Directive (EU) 2019/904, requiring the EU member states to ensure that the containers listed in Part D of the Annex are only placed on the market if they bear a marking on the packaging or on the product itself.

Both examples bode well in terms of improvements of a MT common regulatory framework, involving multi-actors and multi-action strategies.

5. Conclusions

The known and unknown challenges posed by plastics pollution are putting pressure on hard scientists and legislators. Bridging scientific advancements and regulatory needs is the challenge that researchers are asked to address in a subject that is still, defined "at its infancy" [10]. The complex challenges that plastics pollution poses require the adoption of systemic approaches that interpret and analyze the most advanced research from marine biology, food sciences, and law. Within such a scenario, methodologies based on systems thinking and integral approaches need to be adopted as the key to intervening, marking a significant and effective regulatory change in a way that can expectantly neutralize or at least limit the devastation of plastic pollution in the ocean. The model suggested in this article has the potential to be applied in the different fields of environmental protection, food security, food safety, and human health to organize information about complex scenarios holistically and effectively. This method transcends the plastic pollution scene: systemic approaches can be implemented by other scholars and researchers in different sectors, thereby facilitating cross-disciplinary discussions on systemic environmental and societal challenges. Systemic research dealing with cross-cutting issues, such as environmental, food, and health challenges requires coordinated efforts among a multitude of actors and institutions across sectors, levels, and jurisdictions. Achieving environmental sustainability, health, food security, and safety for all beings requires a whole-of-system approach and coordinating efforts from all the actors involved.

While additional empirical studies are needed to assess whether the implementation of the described model can effectively address the challenges posed by marine pollution, this study has set the baseline for further exploration of the systems thinking practices to solve wicked interconnected problems.

Author Contributions: Conceptualization, M.P.P., I.-J.J., E.O.E.; methodology, M.P.P.; investigation, M.P.P., E.O.E., M.A.S., I.-J.J.; writing—original draft preparation, M.P.P.; writing—review and editing, M.P.P., E.O.E., M.A.S., M.M.; supervision, E.O.E., M.A.S., K.-E.E., M.M., I.-J.J.; project administration, E.O.E., M.A.S., K.-E.E.; funding acquisition, E.O.E., I.-J.J., M.A.S., K.-E.E., M.P.P. All authors have read and agreed to the published version of the manuscript.

Funding: This research was funded by UiT The Arctic University of Norway, grant ID—2061344.

Institutional Review Board Statement: Not applicable.

Informed Consent Statement: Not applicable.

Data Availability Statement: Not applicable.

Acknowledgments: M.P.P. thanks Linda Finska, juris at the UiT The Arctic University of Norway for her encouragement to take the plastic regulation path; Nicola Nurra, University of Turin, Italy, and his research team for the conversation on the impacts of the Plasticene for the marine ecosystems. The authors thank Juliana Hayden for her work of proof reading. Their gratitude goes also to the illustrators Camilla Neema Haule (Figures 1–3), Rudi Caeyers (Figure 4) and Valentina Bongiovanni, creator of the seaweed frame.

Conflicts of Interest: The authors declare no conflict of interest.

References

1. Haram, L.E.; Carlton, J.T.; Ruiz, G.M.; Maximenko, N.A. A Plasticene Lexicon. *Mar. Pollut. Bull.* **2020**, *150*, 110714. [CrossRef]
2. Nurra, N.; Battuello, M.; Mussat Sartor, R.; Favaro, L. Plasticene: Tra Plancton Blu e Micromeduse le Tracce di Una Nuova Epoca, Frida Forum Della Ricerca di Ateneo, Unito. 2020. Available online: https://frida.unito.it/wn_pages/contenuti.php/748_ecosistemi-biodiversit-e-comportamento-animale/540_plasticene-tra-plancton-blu-e-micro-meduse-le-tracce-di-una-nuova-epoca/ (accessed on 15 May 2021).
3. Bucci, K.; Tulio, M.; Rochman, C.M. What is known and unknown about the effects of plastic pollution: A meta-analysis and systematic review. *Ecol. Appl.* **2020**, *30*, e02044. [CrossRef]
4. Thomas, P.J.; Perono, G.; Tommasi, F.; Pagano, G.; Oral, R.; Burić, P.; Lyons, D.M. Resolving the effects of environmental micro-and nanoplastics exposure in biota: A knowledge gap analysis. *Sci. Total Environ.* **2021**, *1*.
5. IUCN brief. Available online: https://www.iucn.org/resources/issues-briefs/marine-plastics (accessed on 29 August 2021).
6. Sorensen, R.M.; Jovanović, B. From nanoplastic to microplastic: A bibliometric analysis on the presence of plastic particles in the environment. *Mar. Pollut. Bull.* **2021**, *163*, 111926. [CrossRef]
7. Haward, M. Plastic pollution of the world's seas and oceans as a contemporary challenge in ocean governance. *Nat. Commun.* **2018**, *9*, 1–3. [CrossRef]

8.   Ostle, C.; Thompson, R.C.; Broughton, D.; Gregory, L.; Wootton, M.; Johns, D.G. The rise in ocean plastics evidenced from a 60-year time series. *Nat. Commun.* **2019**, *10*, 1622. [CrossRef]
9.   Mitrano, D.M.; Wohlleben, W. Microplastic regulation should be more precise to incentivize both innovation and environmental safety. *Nat. Commun.* **2020**, *11*, 5324. [CrossRef]
10. Hartmann, N.B.; Hüffer, T.; Thompson, R.C.; Hassellöv, M.; Verschoor, A.; Daugaard, A.E.; Rist, S.; Karlsson, T.; Brennholt, N.; Cole, M.; et al. Are We Speaking the Same Language? Recommendations for a Definition and Categorization Framework for Plastic Debris. *Environ. Sci. Technol.* **2019**, *53*, 1039–1047. [CrossRef]
11. Da Costa, J.P.; Mouneyrac, C.; Costa, M.; Duarte, A.C.; Rocha-Santos, T. The Role of Legislation, Regulatory Initiatives and Guidelines on the Control of Plastic Pollution. *Front. Environ. Sci.* **2020**, *8*. [CrossRef]
12. UNGA. Available online: https://www.un.org/en/development/desa/population/migration/generalassembly/docs/globalcompact/A_CONF.151_26_Vol.I_Declaration.pdf (accessed on 14 June 2021).
13. Consolidated Versions of the Treaty on European Union and the Treaty on the Functioning of the European Union (TFEU). *Off. J.* 2016; C202/1.
14. Toussaint, B.; Raffael, B.; Angers-Loustau, A.; Gilliland, D.; Kestens, V.; Petrillo, M.; Rio-Echevarria, I.M.; van den Eede, G. Review of micro- and nanoplastic contamination in the food chain. *Food Addit. Contam.* **2019**, *36*, 639–673. [CrossRef]
15. Di Gregorio, M.; Fatorelli, L.; Paavola, J.; Locatelli, B.; Pramova, E.; Nurrochmat, D.R.; May, P.H.; Brockhaus, M.; Sari, I.M.; Kusumadewi, S.D. Multi-level governance and power in climate change policy networks. *Glob. Environ. Change* **2019**, *94*, 64–77. [CrossRef]
16. Jordan, A. The Politics of Multilevel Environmental Governance: Subsidiarity and Environmental Policy in the European Union. *Environ. Plan. A Econ. Space* **2000**, *32*, 1307–1324.
17. Purdy, J. The long environmental justice movement. *Ecol. LQ* **2017**, *44*, 809.
18. UNEP Press Release: Plastic Pollution Is an Environmental Injustice to Vulnerable Communities—New Report. March 2021. Available online: https://www.unep.org/news-and-stories/press-release/plastic-pollution-environmental-injustice-vulnerable-communities-new (accessed on 14 June 2021).
19. Poto, M.P. Salvare la Nostra Casa Comune È l'Affaire Du Siècle. *Resp. Civ. Prev.* **2021**, *3*, 1–24.
20. Transforming Our World: The 2030 Agenda for Sustainable Development. Available online: https://sdgs.un.org/2030agenda (accessed on 6 June 2021).
21. SECURE Novel Marine Resources for Food Security and Food Safety. Available online: https://uit.no/research/seafood/project?pid=667623 (accessed on 6 June 2021).
22. EFSA. Presence of microplastics and nanoplastics in food, with particular focus on seafood. *EFSA J.* **2016**, *14*. [CrossRef]
23. UNEP Guide on Tackling Plastic Pollution: Tackling Plastic Pollution: Legislative Guide for the Regulation of Single-Use Plastic Products. 2020. Available online: https://wedocs.unep.org/bitstream/handle/20.500.11822/34570/PlastPoll.pdf.pdf?sequence=3&isAllowed=y (accessed on 20 June 2021).
24. Salwa, H.N.; Sapuan, S.M.; Mastura, M.T.; Zuhri, M.Y.M. Green bio composites for food packaging. *Int. J. Recent Technol. Eng.* **2019**, *8*, 450–459. [CrossRef]
25. Mendes, A.C.; Pedersen, G.A. Perspectives on sustainable food packaging: Is bio-based plastics the solution? *Trends Food Sci. Technol.* **2021**, *112*, 839–846. [CrossRef]
26. Yong, C.Q.Y.; Valiyaveetill, S.; Tang, B.L. Toxicity of microplastics and nanoplastics in mammalian systems. *Int. J. Environ. Res. Public Health* **2020**, *17*, 1509. [CrossRef] [PubMed]
27. Piccardo, M.; Renzi, M.; Terlizzi, A. Nanoplastics in the oceans: Theory, experimental evidence and real world. *Mar. Pollut. Bull.* **2020**, *157*, 111317. [CrossRef]
28. Cox, K.D.; Covernton, G.A.; Davies, H.L.; Dower, J.F.; Juanes, F.; Dudas, S.E. Human consumption of microplastics. *Environ. Sci. Technol.* **2019**, *53*, 7068–7074. [CrossRef]
29. Wright, S.L.; Kelly, F.J. Plastic and human health: A micro issue? *Environ. Sci. Technol.* **2017**, *51*, 6634–6647. Available online: https://pubs.acs.org/doi/pdf/10.1021/acs.est.7b00423 (accessed on 26 June 2021). [CrossRef] [PubMed]
30. Lusher, A.; Hollman, P.; Mendoza-Hill, J. *Microplastics in Fisheries and Aquaculture: Status of Knowledge on Their Occurrence and Implications for Aquatic Organisms and Food Safety*; FAO: Rome, Italy, 2007.
31. SAPEA. A Scientific Perspective on Microplastics in Nature and Society. 2019. Available online: https://www.sapea.info/wp-content/uploads/report.pdf (accessed on 26 June 2021).
32. Microplastics, V.K.M. Occurrence, Levels and Implications for Environment and Human Health Related to Food (2535–4019). 2019. Available online: https://vkm.no/download/18.345f76de16df2bc85a513b4e/1571823698421/20191023%20Microplastics;%20occurrence,%20levels%20and%20implications%20for%20environment%20and%20human%20health%20related%20to%20food.pdf (accessed on 26 June 2021).
33. Koelmans, A.; Besseling, E.; Shim, W. Nanoplastics in the aquatic environment. Critical review. In *Marine Anthropogenic Litter*; Bergmann, M., Gutow, L., Klages, M., Eds.; Springer Open: Cham, Switzerland, 2015; pp. 325–340. [CrossRef]
34. Danopoulos, E.; Jenner, L.C.; Twiddy, M.; Rotchell, J.M. Microplastic Contamination of Seafood Intended for Human Consumption: A Systematic Review and Meta-Analysis. *Environ. Health Perspect.* **2020**, *128*, 126002. [CrossRef]
35. Gamarro, E.G.; Ryder, J.; Elvevoll, E.O.; Olsen, R.L. Microplastics in Fish and Shellfish—A Threat to Seafood Safety? *J. Aquat. Food Prod. Technol.* **2020**, *29*, 417–425. [CrossRef]

36. Gallo, F.; Fossi, C.; Weber, R.; Santillo, D.; Sousa, J.; Ingram, I.; Nadal, A.; Romano, D. Marine litter plastics and microplastics and their toxic chemicals components: The need for urgent preventive measures. *Environ. Sci. Eur.* **2018**, *30*, 1–14. [CrossRef]
37. IMO. Available online: https://wwwcdn.imo.org/localresources/en/MediaCentre/HotTopics/Documents/IMO%20marine%20litter%20action%20plan%20MEPC%2073-19-Add-1.pdf (accessed on 11 June 2021).
38. The Commission to The European Parliament, The Council, The European Economic and Social Committee and the Committee of the Regions a European Strategy for Plastics in a Circular Economy, COM/2018/028 Final. Available online: https://eur-lex.europa.eu/legal-content/EN/TXT/?qid=1516265440535&uri=COM:2018:28:FIN (accessed on 31 May 2021).
39. Directive (EU) 2019/904 of the European Parliament and of the Council of 5 June 2019 on the Reduction of the Impact of Certain Plastic Products on the Environment (Text with EEA Relevance). *Off. J. Eur. Union* **2019**, 1–19.
40. EU Monitor. Available online: https://www.eumonitor.eu/9353000/1/j9vvik7m1c3gyxp/vh75mdhkg4s0 (accessed on 11 June 2021).
41. Poto, M.P. A Conceptual Framework for Complex Systems at the Crossroads of Food, Environment, Health, and Innovation. *Sustainability* **2020**, *12*, 9692. [CrossRef]
42. Poto, M.P.; Morel, M.D. Suggesting an Extensive Interpretation of the Concept of Novelty That Looks at the Bio-Cultural Dimension of Food. *Sustainability* **2021**, *13*, 5065. [CrossRef]
43. Esbjörn-Hargens, S. Integral ecology: The what, who, and how of environmental phenomena. *World Futures* **2005**, *61*, 5–49. [CrossRef]
44. Borràs, S. New transitions from human rights to the environment to the rights of nature. *TEL* **2016**, *5*, 113. [CrossRef]
45. Bender, M. The Earth Law Framework for Marine Protected Areas: Adopting a Holistic, Systems, and Rights-Based Approach to Ocean Governance. *Earth Law Cent.* **2019**. Available online: https://static1.squarespace.com/static/55914fd1e4b01fb0b851a814/t/5adca14b352f538288f4ea67/1524408668126/Final+Draft+3.pdf (accessed on 15 May 2021).
46. Battistoni, C.; Giraldo Nohra, C.; Barbero, S. A systemic design method to approach future complex scenarios and research towards sustainability: A holistic diagnosis tool. *Sustainability* **2019**, *11*, 4458. [CrossRef]
47. De Sousa, F.D.B. The role of plastic concerning the sustainable development goals: The literature point of view. *Clean. Responsible Consum.* **2021**, *1*.
48. Walker, T.R. (Micro) plastics and the UN sustainable development goals. *Curr. Opin. Green Sustain. Chem.* **2021**, *1*.
49. Ramos, G.; Hynes, W.; Müller, J.M.; Lees, M. *Systemic Thinking for Policy Making: The Potential of Systems Analysis for Addressing Global Policy Challenges in the 21st Century;* OECD Publishing: Paris, France, 2019; Available online: https://www.oecd.org/naec/averting-systemic-collapse/SG-NAEC(2019)4_IIASA-OECD_Systems_Thinking_Report.pdf (accessed on 8 September 2021).
50. Eden, J.; Levit, L.; Berg, A.; Morton, S. *Standards for Initiating a Systematic Review. Finding What Works in Health Care: Standards for Systematic Reviews;* The National Academies Press: Washington, DC, USA, 2011.
51. Pierson-Brown, T. (Systems) Thinking Like A Lawyer. *Clin. L. Rev.* **2019**, *26*, 515.
52. Flood, R.L. The relationship of 'systems thinking' to action research. *Syst. Pract. Action Res.* **2010**, *23*, 269–284. [CrossRef]
53. Loi n° 2020-105 du 10 Février 2020 Relative à la Lutte Contre le Gaspillage et à L'économie circulaire, NOR: TREP1902395L, Available at JORF n°0035 du 11 Février. 2020. Available online: https://www.legifrance.gouv.fr/loda/id/JORFTEXT000041553759/ (accessed on 12 September 2021).
54. Verordnung über das Verbot des Inverkehrbringens von Bestimmten Einwegkunststoffprodukten und von Produkten aus Oxo-abbaubarem Kunststoff* (Einwegkunststoffverbotsverordnung-EWKVerbotsV) EWKVerbotsV Ausfertigungsdatum: 20.01.2021. Available online: https://www.gesetze-im-internet.de/ewkverbotsv/EWKVerbotsV.pdf (accessed on 12 September 2021).

Article

Eating Competence among Brazilian Adults: A Comparison between before and during the COVID-19 Pandemic

Fabiana Lopes Nalon de Queiroz [1,*], Eduardo Yoshio Nakano [2], Raquel Braz Assunção Botelho [1], Verônica Cortez Ginani [1], António Raposo [3,*] and Renata Puppin Zandonadi [1]

1   Department of Nutrition, Faculty of Health Sciences, Campus Universitário Darcy Ribeiro, University of Brasília, Brasilia 70910-900, DF, Brazil; raquelbotelho@unb.br (R.B.A.B.); vcginani@gmail.com (V.C.G.); renatapz@unb.br (R.P.Z.)
2   Department of Statistics, University of Brasilia, Brasilia 70910-900, DF, Brazil; eynakano@gmail.com
3   CBIOS (Research Center for Biosciences and Health Technologies), Universidade Lusófona de Humanidades e Tecnologias, Campo Grande 376, 1749-024 Lisboa, Portugal
*   Correspondence: fabinalon@hotmail.com (F.L.N.d.Q.); antonio.raposo@ulusofona.pt (A.R.)

Abstract: The coronavirus pandemic started a worldwide emergency, and tight preventive actions were necessary to protect the population, changing individuals' daily habits. Dwelling and working at home can change dietary habits, affect food choice and access, as well as the practice of physical activity. In this regard, this study's goal was to compare eating competence (EC) among Brazilian adults before and during the coronavirus pandemic, using the Brazilian version of the eating competence Satter inventory (ecSI2.0™BR) with the "retrospective post-then-pre" design. This cross-sectional study was performed from 30 April to 31 May 2021 among a convenience sample of the Brazilian adult population using an online platform (Google® Forms). In the studied sample (n = 302 in which 76.82% were females), EC total score lowered during the pandemic (31.69 ± 8.26 vs. 29.99 ± 9.72; $p < 0.005$), and the decrease was worst after the beginning of the pandemic among those who reported weight gain, decreased the consumption of fruit and vegetables, and increased the consumption of sugary beverages. The contextual skill component seems relevant in this scenario, where our life and routines were changed entirely, demonstrating that the ability to manage the food context is essential, especially when sanitary and economic situations represent a new challenge.

Keywords: eating competence; eating behavior; COVID-19; diet; food choice

Citation: Queiroz, F.L.N.d.; Nakano, E.Y.; Botelho, R.B.A.; Ginani, V.C.; Raposo, A.; Zandonadi, R.P. Eating Competence among Brazilian Adults: A Comparison between before and during the COVID-19 Pandemic. *Foods* **2021**, *10*, 2001. https://doi.org/10.3390/foods10092001

Academic Editor: Jean-Xavier Guinard

Received: 21 July 2021
Accepted: 24 August 2021
Published: 26 August 2021

Publisher's Note: MDPI stays neutral with regard to jurisdictional claims in published maps and institutional affiliations.

1. Introduction

The pandemic of coronavirus started a worldwide state of emergency. Until 31 May, Brazil had registered 16,545,554 confirmed cases of COVID-19 and 462,791 deaths [1]. The measures to control and prevent the disease differed among Brazilian regions. However, the most common was social distancing to control individuals movability, like the schools' and universities' closure, as well as non-essential commerce, public areas, and the use facial masks [2]. While this strict preventive measure is essential to protect people, it can radically change individuals' quotidian habits. Staying and working at home can affect diet, food choice, and access to food and limit the practice of physical activity [3].

Containment measures like self-isolation and social distancing substantially impact the population's everyday life and a wide range of psychosocial impacts on individuals [4]. Stress factors like uncertainty about the duration of the quarantine, fear of possible infections, boredom, and fear of financial losses impact lifestyle and dietary habits [5]. For the general population, the lack of physical activity and poor mental health are the most critical risk factors for morbidity by chronic diseases. However, it is more worrisome for elderly and chronically ill patients at a higher risk of COVID-19-induced mortality [6].

Eating is complex and composed of learned behavior, social expectations, acquired tastes, attitudes, and feelings about eating and/or food items [7]. Eating competence

(EC) is an attitudinal and behavioral view established by the Satter eating competence model (ecSatter). This concept is not centered on nutrients, portion size, or food groups, but on appreciating food and eating, with attention to the variability in the diet, paying attention to signals of hunger and satiety, and preparing nourishing meals with attention to the environment in which the food is consumed [7,8]. Four components compose the EC: eating attitudes; food acceptance; internal regulation; and contextual skills [7,8].

Considering eating behavior, home confinement leads to an altered food cue exposure, which can enhance impulsive eating behavior and emotional eating [9]. In a French study, the lockdown was associated to changes on food choice motives, remarkably increased mood as a food choice motive for 48% of the individuals [10]. Moreover, eliminating social eating practices could reduce mindful eating, which might negatively influence dietary choices and promote overeating [11]. Combining isolation and socioeconomic hardship with the deterioration of psychosocial health and a sedentary lifestyle might negatively affect eating behavior. Furthermore, the negative impact on metabolic health might worsen obesity and its metabolic comorbidities [9]. Regarding physical health, even a short period of overeating with abruptly reduced physical activity can have several consequences for health-related to a proinflammatory state [12]. It is associated with the development of metabolic syndrome leading to an increase in the risk of multiple chronic diseases and the development of the severe form of COVID-19 infection [9,12]. Nutrition becomes a priority during pandemic time, considering the situational stress-eating due to confinement [13].

Some studies evaluated the effect of lockdowns and restrictions connected to the COVID-19 pandemic on dietary–lifestyle behaviors shifts [3,6] and revealed substantial changes in dietary habits at the beginning of the pandemic. For example, in Poland, a significant percentage of individuals started eating and snacking more, leading to weight gain in previous overweight and obese subjects, aggravating this condition [14]. In Italy, a survey evaluating the connection between eating habits, mental and emotional mood showed that individuals tended to consume more comfort food and increase food intake to feel better [5]. On the other hand, Rodríguez-Pérez et al. found that Spanish adults have adopted healthier dietary behaviors closer to the Mediterranean diet during home confinement [15]. In Brazil, an increase in the ingestion of ultra-processed foods at the beginning of the pandemic was reported [16,17].

Given that the prolonged or intermittent social restriction may be needed until 2022 in some countries [18], it is essential to have the best possible understanding of its effects on people's everyday life. Due to the restrictions related to the measurement confinement, it is hypothesized that, during the pandemic, there were changes in EC and food consumption compared to the period prior to the pandemic. In this sense, our study focused on comparing eating competence (EC) among a sample of Brazilian adults before and during the coronavirus pandemic, using the Brazilian translation of the eating competence Satter inventory (ecSI2.0™BR). It is expected that the findings of changes in EC and its four components may support the development of appropriate recommendations for lifestyle modifications during this time.

2. Materials and Methods

This cross-sectional study was performed from 30 April to 31 May 2021 among the Brazilian adult population using an online platform (Google® Forms) accessible by any device with an Internet connection. The participants (convenience sample) were recruited using a link disseminated through social networks (WhatsApp, Instagram, Facebook, email list, and by researchers' personal contact). The inclusion criteria were: being Brazilian adults (20–59 years old, according to the Brazilian Health Ministry [19]); agreed to participate after they were told about the purpose of the survey; and completely filled out the questionnaire.

We used a self-administered survey using online application due to sanitary restrictions to perform a personal interview. It could also reach a more diverse population, despite geographical limitations. Participants were asked to answer the questionnaire regarding current and before the pandemic, using "retrospective post-then-pre" design, where information "during" and "before" pandemic were collected at the same time respondents were asked to answer at first the present information and then, the past information [20,21]. The post/pre-design was selected to offer a useful methodology for documenting the impact of the pandemic in everyday life and assess changes in EC over time, since a comparison of the pre-test scores with their posttest scores can result in inaccurate pre-test scores due to "response shift bias" [22].

The survey was composed of an initial page describing the aims and information on the ethics and the consent form. Participants were volunteers and did not receive any reward for their participation following the National Research Ethics Commission guidelines [23]. There was no time limit for completing the questionnaire. Individuals' responses were registered after clicking on the "submit" button. Participants could stop answering the questionnaire at any step before the submission; in this case, the responses would not be registered. Incomplete questionnaires were removed from the final sample.

Researchers used the Brazilian version of the ecSI2.0™BR questionnaire, previously validated with the Brazilian adult population to assess EC [24] which was approved by an Ethics Committee (CAAE 24415819.2.0000.8101), according to Declaration of Helsinki guidelines. No names or other information that could identify the participant was requested.

The structured questionnaire included forty questions, divided into four different sections. Participants were asked to answer the questions regarding the period during and before the pandemic using the "retrospective post-then-pre" design. In this model, information regarding before and after are collected simultaneously, after the event or intervention has occurred [20]. At first, participants are asked to report their behavior or understanding as a result of the event (post) and, second, participants are asked to report their behavior before the event (pre) [21].

The first part of the questionnaire had sociodemographic data: age, gender, region of residence, level of education, income, employment situation, family income changes, and the number of individuals living at home, using questions from the Brazilian National Institute of Geography and Statistics (IBGE) as a reference [25].

The second part had anthropometrics and health information as weight (current and before pandemic), height, and whether the respondent or a family member tested positive for COVID-19.

In the third part, EC was accessed using the Brazilian Satter eating competence inventory (ecSI2.0™BR), composed of 16 questions and validated for online use among the Brazilian adult population [24]. The ecSI2.0™ is available at https://www.needscenter.org/ upon approval of an application [26]. The NEEDs Center previously approved the use of ecSI2.0™BR for this study. Respondents were asked to answer "current" and "before" confinement. A minimum total score of 32 is the cut-point to consider the individual as a competent eater, without established score cutoffs for each component [27,28].

The fourth part of the questionnaire investigated food consumption and eating habits changes during the pandemic period: frequency of consumption of fruits, vegetables, and artificial drinks (sugary beverages such as industrialized juice, tea, and soft drinks). For this purpose, it was selected some questions used in the Brazilian Food Questionnaire (VIGITEL) [29] that is already validated for online use [30]. Respondents were asked to answer "current" and "before" confinement. We included a question asking if the respondent noticed any change in the amount and the frequency of food eaten. In addition, we asked who is responsible for food preparation at home (if meals are ready-to-eat; if the respondent prepares meals; and if meals are prepared at home by another person).

The scores of ecSI2.0™BR and its four components were presented by means and standard deviation. The comparison of the scores before and during the pandemic was performed by paired Student's t-test. The raw frequencies and percentages of respondents who showed eating competence (total EC score $\geq$ 32) were presented. The MCNemar test was used to compare these frequencies before and during the pandemic. Statistical analyses were performed using IBM SPSS Statistics, version 22. The tests took into consideration two-tailed hypotheses and a significance level of 5%.

3. Results

The study obtained 340 responses, but only responses from individuals aligned to the inclusion criteria and responded all the questions were considered, reducing the number of valid observations to 302 (88.8% of the total questionnaires). The sample consisted predominantly of women (n = 232; 76.82%), and mostly $\geq$ 40 years (Table 1). Most participants had a higher education level (54.6% graduate and 40.7% undergraduate), and most lived with other family members. During the pandemic, 66.5% (n = 201) of the respondents reported they did not reduce their income, and 42.5% did not change their job, but they were working from home.

Table 1. Sub-scores of the ecSI2.0™BR scale categorized by variables before and during the COVID-19 pandemic (n = 302, Brazil).

	Eating Attitude	Food Acceptance	Internal Regulation	Contextual Skills	Total	ecSI2.0™B R $\geq$ 32
	Mean (SD)	Mean (SD)	Mean (SD)	Mean (SD)	Mean (SD)	Freq. (%)
General (n = 302)						
Before pandemic	12.68 (3.74)	5.18 (2.47)	4.10 (1.43)	9.73 (3.31)	31.69 (8.26)	160 (53.0%)
During pandemic	11.77 (4.28)	4.95 (2.53)	3.74 (1.68)	9.53 (3.61)	29.99 (9.72)	143 (47.4%)
p	0.000 *	0.001 *	0.000 *	0.293 *	0.000 *	0.019 **
Gender *						
Female (n = 232)						
Before pandemic	12.62 (3.70)	5.40 (2.41)	3.97 (1.38)	10.06 (2.99)	32.04 (7.84)	128 (55.2%)
During pandemic	11.65 (4.00)	5.10 (2.47)	3.56 (1.66)	9.62 (3.56)	29.92 (9.30)	106 (45.7%)
p	0.000 *	0.000 *	0.000 *	0.032 *	0.000 *	0.000 **
Male (n = 70)						
Before pandemic	12.89 (3.92)	4.46 (2.56)	4.53 (1.52)	8.64 (4.01)	30.51 (9.48)	32 (45.7%)
During pandemic	12.17 (5.13)	4.47 (2.67)	4.34 (1.61)	9.24 (3.80)	30.23 (11.08)	37 (52.9%)
p	0.060 *	0.908 *	0.102 *	0.184 *	0.740 *	0.277 **
Age						
Up to 40 years (n = 145)						
Before pandemic	12.39 (4.03)	5.31 (2.61)	3.99 (1.47)	9.23 (3.49)	30.92 (9.00)	74 (51.0%)
During pandemic	11.32 (4.28)	5.09 (2.66)	3.55 (1.71)	8.93 (3.77)	28.89 (9.96)	61 (42.1%)
p	0.000 *	0.052 *	0.000 *	0.320 *	0.002 *	0.019 **
More than 40 years (n = 157)						
Before pandemic	12.95 (3.45)	5.06 (2.35)	4.2 (1.39)	10.19 (3.07)	32.39 (7.47)	86 (54.8%)
During pandemic	12.18 (4.26)	4.83 (2.4)	3.91 (1.63)	10.09 (3.38)	31.01 (9.42)	82 (52.2%)
p	0.000 *	0.006 *	0.000 *	0.659 *	0.003 *	0.503 **
Schooling level *						
High School (n = 14)						
Before pandemic	13.29 (4.18)	5.14 (2.68)	4.57 (1.45)	10.50 (3.74)	33.50 (9.34)	9 (64.3%)
During pandemic	12.43 (4.70)	5.21 (2.22)	4.07 (1.69)	9.43 (4.38)	31.14 (10.17)	8 (57.1%)
p	0.201 *	0.818 *	0.187 *	0.059 *	0.164 *	1.000 **
Undergraduate (n = 123)						
Before pandemic	13.02 (3.49)	5.08 (2.46)	4.32 (1.37)	9.27 (3.55)	31.68 (8.04)	61 (49.6%)
During pandemic	12.13 (4.23)	4.89 (2.52)	3.93 (1.74)	9.25 (3.75)	30.20 (9.88)	60 (48.8%)
p	0.001 *	0.122 *	0.000 *	0.964 *	0.038 *	1.000 **
Graduate (n = 165)						
Before pandemic	12.38 (3.88)	5.25 (2.48)	3.90 (1.45)	10.01 (3.05)	31.54 (8.36)	90 (54.5%)
During pandemic	11.44 (4.29)	4.98 (2.57)	3.56 (1.62)	9.75 (3.45)	29.74 (9.62)	75 (45.5%)
p	0.000 *	0.001 *	0.000 *	0.231 *	0.000 *	0.001 **
Family Income Changes						
Did not reduce (n = 201)						
Before pandemic	12.75 (3.68)	5.12 (2.43)	4.24 (1.40)	9.72 (3.18)	31.84 (7.77)	103 (51.2%)
During pandemic	12.00 (4.22)	4.91 (2.50)	3.93 (1.68)	9.77 (3.52)	30.60 (9.38)	95 (47.3%)
p	0.000 *	0.006 *	0.000 *	0.849 *	0.008 *	0.229 **
Reduced (n = 101)						
Before pandemic	12.54 (3.89)	5.29 (2.57)	3.81 (1.45)	9.75 (3.55)	31.4 (9.20)	57 (56.4%)
During pandemic	11.32 (4.39)	5.04 (2.60)	3.36 (1.61)	9.07 (3.76)	28.78 (10.30)	48 (47.5%)
p	0.000 *	0.076 *	0.000 *	0.032 *	0.000 *	0.022 **

Table 1. *Cont.*

	Eating Attitude	Food Acceptance	Internal Regulation	Contextual Skills	Total	ecSI2.0™B R ≥ 32
	Mean (SD)	Mean (SD)	Mean (SD)	Mean (SD)	Mean (SD)	Freq. (%)
COVID-19 infection						
No (*n* = 280)						
Before pandemic	12.54 (3.76)	5.13 (2.43)	4.09 (1.44)	9.74 (3.33)	31.50 (8.34)	148 (52.9%)
During pandemic	11.74 (4.22)	4.94 (2.50)	3.76 (1.67)	9.62 (3.56)	30.06 (9.65)	133 (47.5%)
p	0.000 *	0.006 *	0.000 *	0.534 *	0.000 *	0.032 **
Yes (*n* = 22)						
Before pandemic	14.45 (3.07)	5.77 (2.93)	4.18 (1.33)	9.68 (3.01)	34.09 (6.90)	12 (54.5%)
During pandemic	12.09 (5.11)	5.09 (2.96)	3.50 (1.77)	8.45 (4.19)	29.14 (10.85)	10 (45.5%)
p	0.016 *	0.065 *	0.032 *	0.210 *	0.041 *	0.625 **
COVID-19 infection in a family member						
No (*n* = 158)						
Before pandemic	12.39 (3.68)	5.15 (2.35)	3.94 (1.43)	9.54 (3.51)	31.03 (8.36)	81 (51.3%)
During pandemic	11.83 (4.16)	5.03 (2.46)	3.79 (1.60)	9.71 (3.75)	30.36 (9.77)	78 (49.4%)
p	0.002 *	0.200 *	0.059 *	0.508 *	0.173 *	0.701 **
Yes (does not live with me *n* = 113)						
Before pandemic	12.65 (3.85)	4.91 (2.47)	4.31 (1.39)	9.85 (3.19)	31.72 (8.25)	58 (51.3%)
During pandemic	11.49 (4.49)	4.58 (2.40)	3.69 (1.77)	9.12 (3.53)	28.88 (9.80)	48 (42.5%)
p	0.000 *	0.011 *	0.000 *	0.032 *	0.000 *	0.013 **
Yes (live with me *n* = 31)						
Before pandemic	14.26 (3.42)	6.32 (2.84)	4.13 (1.52)	10.26 (2.56)	34.97 (7.14)	21 (67.7%)
During pandemic	12.48 (4.14)	5.90 (3.08)	3.65 (1.78)	10.13 (3.14)	32.16 (8.96)	17 (54.8%)
p	0.007 *	0.025 *	0.001 *	0.805 *	0.026 *	0.219 **
Weight changes						
Decreased (*n* = 35)						
Before pandemic	11.77 (4.66)	5.91 (2.23)	3.86 (1.73)	9.46 (4.10)	31.00 (10.48)	21 (60.0%)
During pandemic	11.77 (5.12)	5.63 (2.53)	4.11 (1.43)	10.03 (3.69)	31.54 (9.75)	18 (51.4%)
p	1.000 *	0.185 *	0.263 *	0.442 *	0.702 *	0.453 **
Maintained (*n* = 201)						
Before pandemic	13.81 (3.40)	5.17 (2.43)	4.10 (1.40)	9.84 (3.12)	32.42 (7.56)	110 (54.7%)
During pandemic	12.62 (3.93)	5.14 (2.45)	3.93 (1.52)	10.11 (3.25)	31.81 (8.63)	107 (53.2%)
p	0.000 *	0.597 *	0.000 *	0.137 *	0.069 *	0.690 **
Increased (*n* = 66)						
Before pandemic	11.24 (3.77)	4.80 (2.68)	4.21 (1.34)	9.56 (3.43)	29.82 (8.78)	29 (43.9%)
During pandemic	9.18 (4.22)	4.02 (2.55)	2.95 (2.00)	7.50 (3.95)	23.65 (10.31)	18 (27.3%)
p	0.000 *	0.001 *	0.000 *	0.000 *	0.000 *	0.007 **
Fruit consumption						
Decreased (*n* = 62)						
Before pandemic	12.19 (3.43)	4.76 (2.17)	4.23 (1.38)	9.92 (3.36)	31.10 (7.02)	31 (50.0%)
During pandemic	10.58 (4.53)	3.81 (2.13)	3.10 (1.91)	8.35 (4.15)	25.84 (10.84)	20 (32.3%)
p	0.000 *	0.000 *	0.000 *	0.024 *	0.000 *	0.052 **
Maintained (*n* = 195)						
Before pandemic	12.95 (3.79)	5.30 (2.53)	4.11 (1.45)	10.02 (3.26)	32.37 (8.56)	109 (55.9%)
During pandemic	12.10 (4.31)	5.21 (2.49)	3.91 (1.60)	9.89 (3.45)	31.11 (9.33)	98 (50.3%)
p	0.000 *	0.129 *	0.001 *	0.447 *	0.000 *	0.007 **
Increased (*n* = 45)						
Before pandemic	12.16 (3.92)	5.24 (2.60)	3.89 (1.40)	8.24 (3.07)	29.53 (8.26)	20 (44.4%)
During pandemic	11.98 (3.58)	5.42 (2.80)	3.87 (1.46)	9.60 (3.26)	30.87 (8.34)	25 (55.6%)
p	0.646 *	0.263 *	0.888 *	0.001 *	0.113 *	0.063 **
Vegetable consumption						
Decreased (*n* = 63)						
Before pandemic	11.54 (3.75)	4.92 (2.50)	3.83 (1.28)	9.78 (2.79)	30.06 (8.07)	26 (41.3%)
During pandemic	9.22 (4.01)	3.68 (2.37)	2.60 (1.62)	6.63 (3.40)	22.14 (9.22)	9 (14.3%)
p	0.000 *	0.000 *	0.000 *	0.000 *	0.000 *	0.000 **
Maintained (*n* = 202)						
Before pandemic	13.18 (3.67)	5.42 (2.48)	4.17 (1.45)	10.31 (3.19)	33.08 (8.02)	125 (61.9%)
During pandemic	12.40 (4.02)	5.35 (2.47)	4.03 (1.55)	10.41 (3.17)	32.18 (8.55)	113 (55.9%)
p	0.000 *	0.131 *	0.019 *	0.523 *	0.003 *	0.008 **
Increased (*n* = 37)						
Before pandemic	11.86 (3.65)	4.32 (2.24)	4.16 (1.54)	6.49 (2.94)	26.84 (7.71)	9 (24.3%)
During pandemic	12.65 (4.60)	4.97 (2.46)	4.08 (1.66)	9.70 (3.76)	31.41 (9.94)	21 (56.8%)
p	0.051 *	0.001 *	0.446 *	0.000 *	0.000 *	0.000 **

Table 1. *Cont.*

	Eating Attitude	Food Acceptance	Internal Regulation	Contextual Skills	Total	ecSI2.0™BR $\geq$ 32
	Mean (SD)	Mean (SD)	Mean (SD)	Mean (SD)	Mean (SD)	Freq. (%)
Sugary beverages						
Decreased ($n = 39$)						
Before pandemic	11.18 (3.93)	4.18 (2.16)	4.08 (1.51)	6.56 (3.28)	26.00 (7.90)	9 (23.1%)
During pandemic	10.62 (5.42)	4.56 (2.35)	3.49 (1.75)	9.10 (3.32)	27.77 (10.60)	16 (41.0%)
p	0.404 *	0.168 *	0.016 *	0.003 *	0.309 *	0.092 **
Maintained ($n = 210$)						
Before pandemic	13.24 (3.48)	5.44 (2.42)	4.20 (1.43)	10.46 (2.97)	33.34 (7.68)	128 (61.0%)
During pandemic	12.57 (3.65)	5.22 (2.48)	3.97 (1.62)	10.27 (3.33)	32.03 (8.73)	117 (55.7%)
p	0.000 *	0.000 *	0.001 *	0.236 *	0.000 *	0.027 **
Increased ($n = 53$)						
Before pandemic	11.55 (4.10)	4.89 (2.72)	3.72 (1.34)	9.19 (3.16)	29.34 (8.45)	23 (43.4%)
During pandemic	9.43 (4.70)	4.19 (2.70)	3.00 (1.65)	6.92 (3.70)	23.55 (9.80)	10 (18.9%)
p	0.000 *	0.006 *	0.001 *	0.000 *	0.000 *	0.000 **
Amount of food						
Reduced ($n = 56$)						
Before pandemic	13.02 (3.96)	5.36 (2.45)	4.34 (1.62)	9.23 (4.15)	31.95 (8.98)	30 (53.6%)
During pandemic	12.86 (4.37)	5.34 (2.34)	4.14 (1.47)	9.80 (3.83)	32.14 (9.46)	33 (58.9%)
p	0.747 *	0.914 *	0.213 *	0.362 *	0.875 *	0.648 **
Maintained ($n = 99$)						
Before pandemic	14.02 (3.14)	5.38 (2.38)	4.33 (1.34)	10.67 (2.91)	34.40 (7.45)	63 (63.6%)
During pandemic	13.76 (3.20)	5.36 (2.40)	4.27 (1.39)	10.85 (3.01)	34.24 (7.79)	63 (63.6%)
p	0.015 *	0.754 *	0.158 *	0.268 *	0.539 *	1.000 **
Increased ($n = 147$)						
Before pandemic	11.65 (3.75)	4.97 (2.55)	3.85 (1.38)	9.29 (3.07)	29.76 (8.02)	67 (45.6%)
During pandemic	10.01 (4.19)	4.53 (2.63)	3.22 (1.78)	8.54 (3.62)	26.31 (9.63)	47 (32.0%)
p	0.000 *	0.000 *	0.000 *	0.009 *	0.009 *	0.000 **
Meals during pandemic						
Bought ready-to-eat ($n = 24$)						
Before pandemic	11.54 (3.99)	3.88 (2.36)	3.67 (1.74)	8.29 (3.41)	27.38 (9.04)	10 (41.7%)
During pandemic	9.21 (4.64)	3.17 (2.28)	2.71 (1.76)	6.75 (3.31)	21.83 (9.76)	4 (16.7%)
p	0.001 *	0.023 *	0.005 *	0.015 *	0.001 *	0.031 **
Prepared by myself ($n = 135$)						
Before pandemic	12.93 (3.81)	5.83 (2.39)	4.18 (1.43)	10.51 (3.2)	33.44 (8.39)	85 (63.0%)
During pandemic	12.01 (4.22)	5.47 (2.48)	3.76 (1.74)	10.10 (3.76)	31.33 (10.00)	71 (52.6%)
p	0.000 *	0.003 *	0.000 *	0.157 *	0.001 *	0.003 **
Prepared at home by others ($n = 143$)						
Before pandemic	12.64 (3.63)	4.78 (2.42)	4.10 (1.37)	9.24 (3.23)	30.76 (7.62)	65 (45.5%)
During pandemic	11.97 (4.17)	4.77 (2.47)	3.90 (1.55)	9.47 (3.31)	30.10 (8.81)	68 (47.6%)
p	0.001 *	0.852 *	0.002 *	0.388 *	0.191 *	0.664 **

* Paired Student's *t*-test, comparing before and during pandemic. ** McNemar test.

In general, the EC differs before and during the pandemic considering its four components (eating attitude, food acceptance, internal regulation and contextual skills) and the total score for EC ($p < 0.05$). Considering gender, EC was lower for all components among females during the pandemic ($p < 0.05$) but did not differ among males. The EC (total) was lower during the pandemic compared to before it, regardless of age. However, contextual skills did not differ among the groups regardless of age and food acceptance among individuals up to 40 y/o.

Individuals with high school education did not differ EC (total and each component) during the pandemic. EC was lower among undergraduate and graduate individuals during the pandemic than before ($p < 0.05$), except for food acceptance and contextual skills among undergraduate and contextual skills among graduate individuals that did not differ.

Considering individuals who reduced the income during the pandemic, the EC lowered for all components and the total ($p < 0.05$), except for food acceptance that did not differ. However, the EC lowered during the pandemic regardless of income reduction (except for contextual skills among those with no income reduction, which did not differ) (Table 1).

EC lowered among individuals who had or had not COVID-19 ($p < 0.05$). However, it did not differ on contextual skills among those who did not have COVID-19 and food acceptance and contextual skills among those tainted by Sars-CoV-2. Individuals who had relatives tainted by Sars-CoV-2 presented lower EC during the pandemic than before ($p < 0.05$). Among those whose relatives were not infected, the EC did not differ.

Individuals who increased weight during the pandemic (21.8% of the sample) had a reduction in EC (total and components) ($p < 0.05$). The group that maintained weight during the pandemic (66.5% of the sample) did not change total EC, but showed a reduction in eating attitude and internal regulation ($p < 0.05$). EC did not differ among those who lost weight during the pandemic (11.5% of the sample).

Among those who reported a decrease in fruit and vegetable consumption, there was a reduction in total EC and all four components ($p < 0.05$). Among those who maintained fruit consumption, there was a reduction in total EC and internal regulation ($p < 0.05$). The group that maintained vegetable consumption showed a reduction in total EC and eating attitude and internal regulation ($p < 0.05$). The group that reported an increase in fruit consumption showed no changes in total EC and components, except for contextual skills, which increased. Those who increased vegetable consumption showed an increase in total EC and components ($p < 0.05$), except for internal regulation.

The group that reported a decrease in the consumption of industrialized beverages had no changes in total EC, eating attitude, and food acceptance but showed a reduction in internal regulation and increased contextual skills. The group that maintained the consumption of industrialized beverages showed a loss of total EC and components ($p < 0.05$), except for contextual skills, which remained the same. Among those who increased the consumption of industrialized beverages, there was a reduction in EC (total and all components) ($p < 0.05$).

EC (total and all components) lowered among individuals who increased food consumption during the pandemic period ($p < 0.05$). Among individuals that maintained or reduced food consumption, EC did not differ.

Individuals who buy ready-to-eat meals showed a reduction in total EC and all components ($p < 0.05$). There was no reduction in the contextual skills component among those who usually prepare their food. However, there was a reduction in the total EC and the other components ($p < 0.05$). Individuals with meals prepared at home by another person did not report differences in total EC, food acceptance, and contextual skills. However, they showed a decrease in eating attitude and internal regulation ($p < 0.05$).

4. Discussion

The impact of social isolation during the pandemic can vary by income, education, age, and gender [2]. However, the reduction in EC observed across the sample is alarming regardless of the variables evaluated. Considering the sample profile (mostly female, aged ≥ 40 years old, with a higher education level, and living with a family member), the outcomes reveal an isolation negative impact on EC that may reflect throughout the family and that should be the focus of public health strategies in the years that follow the pandemic.

Some aspects highlight the need to develop public policies capable of improving the situation. It is well known that higher educational level has been usually related with higher socioeconomic status, which has been associated with better diet quality [15]. In addition, a previous study with the Brazilian population showed that competent eaters (EC total score ≥ 32) are primarily individuals >40 years; graduated, and had a high income [31]. Despite the majority of this study's sample having a profile similar to that of the aforementioned studies, the present study showed that, in general, EC (total score) decreased (31.60 ± 8.26 before pandemic vs. 29.00 ± 9.72 during pandemic p 0.000), revealing that restrictive measures during the pandemic play a negative impact on EC. The same has occurred analyzing the four components of EC, which have also decreased.

The data obtained appears to be a trend as a COVID-19 impact. Resembling results were published by Górnicka et al. [3], who investigated dietary changes among Polish adults during the pandemic and their associations with sociodemographic data and lifestyle changes. Although related to a sample of partially distinct characteristics, authors found that those >40 years old, having children, not working, and not consuming homemade meals might be more exposed to unhealthy behaviors [3].

The effect of the COVID-19 on the worldwide economy is significant and financial loss during quarantine is a difficult socioeconomic harm in Brazil [2]. The pandemic crisis might throw millions of workers into unemployment [32]. Since the pandemic generated an economic crisis, unemployment rose extensively, worsening health and social insecurity [33]. Our results showed that among individuals who reduced income during the pandemic (33.4%), all components and total EC lowered ($p < 0.05$), except for food acceptance. This finding is aligned with prior studies showing that low-income individuals have less favorable dietary patterns and show less EC [34,35].

Food insecurity is inversely associated with diet quality [36]. A recent study showed that females, with lower education and incomes, have higher overweight and obesity levels in Brazil. There is an association between unhealthy food intake and poverty [37]. The ability to be a competent eater is affected by food insecurity and/or income restrictions, as having worries about money for food is associated with a lack of EC [38,39]. Bezerra et al. (2020) investigated the perception of social isolation during the COVID-19 pandemic with a sample of 16.440 Brazilian respondents (69% females, mean age 41 y/o). Authors found that social interactivity was the most affected field among individuals with higher education and income, and financial problems generated a more significant impact among those with low income and education [2]. Considering that part of society has experienced a reduction of their income, it is expected to reflect their consumption patterns. On the other hand, those who have not changed their income may have changed to convenience foods when purchasing or cooking their meals [32].

Regarding health data, individuals who had relatives infected by Sars-CoV-2 had a decrease in EC during the pandemic, probably due to the stress about the outcome of the infection since stress can be associated with negative changes in eating behavior [40,41]. Adding to the pandemic scenario as a whole, these changes can be leveraged. Therefore, the decrease in EC can be a consequence of the experience of individuals with relatives affected by the virus.

The group that maintained weight during the pandemic (66.5% of the sample) did not change total EC, but showed a reduction in eating attitude and internal regulation ($p < 0.05$). EC did not differ among those who lost weight during the pandemic (11.5% of the sample). The association between weight and EC is well documented in studies that found an inverse association of EC and weight [8,38,42–45]. It must be highlighted that obesity was one of the worse global health problems even before the COVID-19 pandemic and it is a risk factor for other metabolic diseases [13]. Moreover, it is known that individuals with obesity and metabolic disorders have a higher risk of developing critical COVID-19 outcome [12]. Thus, the adequate control of weight and metabolic complications is vital to reduce the risk of severe forms of COVID-19.

Energy balance primarily defines weight gain or loss [46]. However, some foods are directly associated with energy supply, which can affect body weight. The consumption of fruits and vegetables has a positive effect in this regard [47], while sugary beverages can contribute to weight gain by increasing caloric intake [48]. As most of the sample did not report weight change, the majority's consumption of these foods was also not changed.

In addition, fruit and vegetable intake (FV) is essential to ensure micronutrients that can boost immune function [13]. FV consumption is used as an indicator of health, and it is adequate when consumed at least five times per week [47]. In our study, EC (total and all four components) decreased among those who reported decreased FV consumption during the pandemic. On the contrary, EC did not change among those who reported an increase in FV consumption, showing an increase in the contextual skills component. Those who increased vegetable consumption showed an increase in total EC and components ($p < 0.05$), except for internal regulation which did not differ. Similar results were reported by Queiroz et al. (2020), who found a positive association between EC and FV consumption among Brazilian adults [31]. These results are consistent with prior studies in which EC was related to diet quality [34,35,42,49]. Thus, considering that the pandemic negatively affects EC, a worsening in food quality is expected. In Brazil, Malta et al. (2020) noticed a decrease in the frequency of consumption of healthy foods during the pandemic, and the most significant decrease was observed in the regular consumption of vegetables [16]. The decrease in FV consumption in the pandemic leads to a deficiency in micronutrients, which can harm the immune function, increasing susceptibility to infection [50].

It is also important to consider problems related to food distribution and the changes in global supply chains, which have adverse effects like the lack of some fresh food supplies [32]. The dominance of the food industry and supermarket chains and the reduction in the activities of the local markets put the local food producer at a disadvantage compared to the supply of processed foods [16,17,32]. Furthermore, it can explain the decrease in the consumption of fresh food like FV. In Brazil, Ribeiro-Silva et al. (2020) analyzed the effects of pandemic on food and nutrition security. They highlighted losses in the supply of fresh food from family farming, especially FV, and reduced consumption of fresh foods, increasing weight gain or eating disorders associated with physical inactivity and social distancing [51].

The regular intake of soft drinks is a mark of unhealthy eating habits. Soft drinks consumption is connected with higher energy intake and glycemic index, lower intake of fruits and fiber, and higher intake of fast foods [48]. In our study, those who increased their consumption of industrialized beverages showed a reduction in total EC and all components. It is essential to highlight that sugar intake improves serotonin production, positively affecting mood momentarily and transiently, encouraging people to consume foods with sugar frequently [13]. This unhealthy nutritional habit could increase the risk of developing obesity. The group that reported a decrease in the consumption of industrialized beverages had no changes in total EC, eating attitude, and food acceptance but showed a reduction in internal regulation and increased contextual skills. Those who maintained the consumption of industrialized beverages showed a loss of total EC and components, except for contextual skills, which remained the same. Similar results were reported by Queiroz et al. (2020), who found an inverse association between EC and artificial juice or soda consumption among Brazilian adults [31].

Restrictive measures conduct to irregular eating patterns and frequent snacking, associated with higher caloric ingestion and risk of excess weight [12,41]. Since quarantine is associated with the interruption of the routine, this could result in boredom and stress, associated with a greater energy intake [13,36,41]. In our sample, individuals who increased food consumption showed decreased EC (total and all components). On the contrary, individuals that maintained or reduced food consumption showed no difference in EC. Previous studies investigating dietary practices during pandemics showed similar results. Ammar et al. [6] conducted an online study in April 2020, in seven languages, to investigate the behavioral and lifestyle consequences of COVID-19 restrictions. Authors found an increase in unhealthy diet during COVID-19, with food consumption and meal patterns (the type of food, eating out of control, snacks between meals, number of main meals) more unhealthy in the period of the pandemic [6]. Górnicka et al. (2020) also found similar results among Polish adults, who showed increased food consumption and snacking during restrictive measures, especially among those with higher BMI [3]. Similar results were noticed by Cosgrove and Wharton (2021), who found that participants most often reported purchasing more food in various food categories [36]. In France, the nutritional quality of the diet was also higher before the lockdown than during the pandemic [10].

On the contrary, Rodríguez-Pérez et al. observed an improvement in Spanish adults' dietary behaviors during the COVID-19 restrictions, like a higher intake of FV compared to their habits [15]. Lohse et al. (2010) showed that, among elderly Spanish, EC was positively associated with greater adherence to a Mediterranean diet, which is considered a healthy eating pattern [52]. Cosgrove and Wharton (2021) also noticed positive changes among US adults since participants reported improvements in dietary healthfulness during the pandemic [36]. There are some possible explanations for these findings: the greater possibility of preparing a meal at home, more free time to take care of health during home confinement, or food choice motives. For example, a French study with 938 adults revealed that improved diet quality was associated with an increase in the importance of weight control. Reduced diet quality was associated with the increased importance of mood, signing that food choice motives are essential factors related to changes in diet quality [10].

Directly staying at home affects lifestyle, including dietary habits and eating patterns [4,5]. Cosgrove and Wharton (2021), in an online survey with 958 US respondents, found changes concerning food shopping from grocery stores, farmer's markets, sit-down restaurants, and fast-food restaurants. All have decreased and increases were observed in online grocery shopping and ordering takeout [36]. Usually, a turbulent lifestyle and less time in everyday life are barriers to incorporating a healthy diet in daily life. It was expected that, while staying at home, individuals should have had more time to cook and organize their meals [15]. However, individuals who are not used to managing the food context may have found it more difficult when faced with this new reality. In fact, in our study, individuals who buy ready-to-eat meals showed a reduction in total EC and all components. There was no reduction in the contextual skills component among those who usually prepare their food, which is expected to be higher among individuals that usually cook [7]. Individuals with meals prepared at home by another person did not report differences in total EC, food acceptance, and contextual skills. However, they showed a decrease in eating attitude and internal regulation, which are not related to managing the food context. This is probably because worries about food, whether for aesthetic reasons or health reasons, favor negative eating attitudes [53].

Although home cooking might have increased, it is not clear whether quarantine and social isolation have improved the nutrition quality of dishes. Individuals of different ages and lifestyle situations need skills in food purchasing and preparation to maintain or achieve a healthy lifestyle [54]. Therefore, the contextual skill component seems to be an important aspect that would help to deal with eating in this pandemic scenario since it is linked to the ability to plan meals, develop food purchasing strategies, and have culinary skills that enable food autonomy, managing the time dedicated to preparing and consuming meals [7,38].

A previous study showed that the contextual skills component has a positive association among the Brazilian adult population with age, education level, income, BMI, and quality of food consumption [31]. There is some evidence that improving Contextual Skills in weight management interventions helps weight loss (Lohse et al., 2018). The pandemic shows the necessity for nutrition and food preparation education programs to deal with feeding in everyday life and during an emergency [3].

Cosgrove (2021) noted that most studies assessing COVID-19-related changes in lifestyle behaviors have focused on physical activity and sedentary behavior, neglecting the influence of diet during this health crisis [36]. This study is the first to investigate changes in EC during the COVID-19 pandemic, making the comparison with other results more challenging. Although the questionnaire was directed to all Brazilian regions and different segments of the adult population, the main limitation of this study is the convenience sample. Our sample had a higher proportion of female respondents, limiting the generalizability of the results to the Brazilian population. The socioeconomic level of the sample is also a potential bias. The pandemic does not uniformly affect all people, and the highest number of deaths is usually observed among the most impoverished populations because they are more likely to have chronic conditions that put them at a higher risk of mortality [33].

Individuals with low access to information are more predisposed to ignore government health warnings since they are more likely to be affected by misinformation and miscommunication [33]. The present research was carried out one year after the beginning of the containment measures, during the worst period of the second wave of the pandemic in Brazil. There was limited opportunity to conduct studies involving respondents. Individuals' adherence to isolation may have been reduced over time [18], and, at present, the measures to counter the pandemic have already led to changes in purchasing and consumption habits [32]. A great discrepancy was observed among individuals with higher income and education (the majority) and individuals with lower income and education (the minority). Thus, this work does not represent the EC of the whole Brazilian population during the pandemic but only of the sample studied.

The COVID-19 pandemic threatens the health of individuals and affects social and economic well-being around the world, revealing the fragility of the global supply chains and the new challenges that local administrations have to face. In addition to the obligation to combat the increase in obesity and malnutrition, there is a tendency for new problems, which will require greater attention [32]. The COVID-19 pandemic imposed new challenges to support a healthy diet [50]. In this context, the challenge of public policies is to promote and ensure food and nutrition security after the social and health crisis created by the pandemic, which involves an effort in planning the development of distribution of food based on international health standards, but focusing on the preservation of food production and consumption's local culture, knowledge, and traditions [2,32,50].

The context of the pandemic may allow to introduce new questions on the feeding habits and patterns [32]. It is time to think about the relationship between food and eating. EC is an indicator of positive eating-related behavior, is associated with health-promoting food consumption and health outcomes [7,38,55]. Thus, more than ever, the time seems to be favorable to invest in improvements in EC since it involves eating attitudes and behaviors, internal regulation of intake, food acceptance, and skills related to selecting, preparing, and planning meals.

The present study's strength is that we had responses from all Brazilian regions. In addition, this is an unprecedented study that draws attention to the aspect of EC changes caused by the pandemic. However, our study presents some limitations as the non-representativeness of the sample and the results cannot be generalized to the entire adult population since we used a convenience sample mainly composed of females. Finally, the duration of the research could influence the results due to the very dynamic feature of the COVID-19. Another potential study bias is the internet survey dissemination. However, during the pandemic period, the internet is the primary way to achieve participants.

5. Conclusions

Eating competence, is an indicator of positive eating-related behavior, associated with health-promoting food consumption. In the studied sample, our hypothesis was confirmed in which total EC score lowered during the pandemic, and the decrease was worst among those who reported weight gain, decreased the consumption of fruit and vegetables, and increased the consumption of sugary beverages, after the beginning of the pandemic. The contextual skill component seems relevant in this scenario, where our life and routines were changed entirely, demonstrating that the ability to manage the food context is essential, primarily when sanitary and economic situations represent a new challenge. Further studies are necessary to evaluate EC after the pandemic, aiming to focus on recommendations for lifestyle modifications after this challenging time.

Author Contributions: Conceptualization, F.L.N.d.Q. and R.P.Z.; methodology, F.L.N.d.Q., E.Y.N., R.B.A.B., V.C.G. and R.P.Z.; validation, F.L.N.d.Q., E.Y.N., R.B.A.B., V.C.G. and R.P.Z.; formal analysis, F.L.N.d.Q., E.Y.N., R.B.A.B. and R.P.Z.; investigation, F.L.N.d.Q., E.Y.N., R.B.A.B. and R.P.Z.; resources, F.L.N.d.Q., E.Y.N. and R.P.Z.; data curation, F.L.N.d.Q. and R.P.Z.; writing—original draft preparation, F.L.N.d.Q., E.Y.N. and R.P.Z.; writing—review and editing, F.L.N.d.Q., E.Y.N., R.B.A.B., V.C.G., A.R. and R.P.Z.; visualization, F.L.N.d.Q., E.Y.N., R.B.A.B., A.R. and R.P.Z.; supervision, R.P.Z.; project administration, F.L.N.d.Q. All authors have read and agreed to the published version of the manuscript.

Funding: This research received no external funding.

Institutional Review Board Statement: The study was conducted according to the guidelines of the Declaration of Helsinki, and approved by the Ethics Committee of Ethics Committee (CAAE 24415819.2.0000.8101).

Informed Consent Statement: Informed consent was obtained from all subjects involved in the study.

Data Availability Statement: The study did not report any data.

Acknowledgments: The authors acknowledge the National Council for Scientific and Technological Development (CNPq) and the National Council for the Improvement of Higher Education (CAPES), and the NEEDs Center for granting permission to use ecSI2.0™BR in this research.

Conflicts of Interest: The authors declare no conflict of interest.

References

1. BRASIL. Ministério da Saúde Coronavírus Brasil. Available online: https://covid.saude.gov.br/ (accessed on 17 June 2021).
2. Bezerra, A.C.V.; da Silva, C.E.M.; Soares, F.R.G.; da Silva, J.A.M. Factors associated with people's behavior in social isolation during the covid-19 pandemic. *Cienc. Saude Coletiva* **2020**, *25*, 2411–2421. [CrossRef]
3. Górnicka, M.; Drywien, M.E.; Zielinska, M.A.; Hamulka, J. Dietary and Lifestyle Changes During COVID-19 and the Subsequent Lockdowns among Polish Adults: PLifeCOVID-19 Study. *Nutriets* **2020**, *12*, 2324. [CrossRef] [PubMed]
4. Hossain, M.M.; Sultana, A.; Purohit, N. Mental health outcomes of quarantine and isolation for infection prevention: A systematic umbrella review of the global evidence. *Epidemiol. Health* **2020**, *42*, 1–11. [CrossRef]
5. di Renzo, L.; Gualtieri, P.; Cinelli, G.; Bigioni, G.; Soldati, L.; Attinà, A.; Bianco, F.F.; Caparello, G.; Camodeca, V.; Carrano, E.; et al. Psychological aspects and eating habits during covid-19 home confinement: Results of ehlc-covid-19 italian online survey. *Nutrients* **2020**, *12*, 2152. [CrossRef] [PubMed]
6. Ammar, A.; Brach, M.; Trabelsi, K.; Chtourou, H.; Boukhris, O.; Masmoudi, L.; Bouaziz, B.; Bentlage, E.; How, D.; Ahmed, M.; et al. Effects of COVID-19 Home Confinement on Eating Behaviour and Physical Activity: Results of the ECLB-COVID19 International Online Survey. *Nutrients* **2020**, *12*, 1583. [CrossRef]
7. Satter, E. Eating Competence: Definition and Evidence for the Satter Eating Competence Model. *J. Nutr. Educ. Behav.* **2007**, *39*, S142–S153. [CrossRef] [PubMed]
8. Satter, E. Eating Competence: Nutrition Education with the Satter Eating Competence Model. *J. Nutr. Educ. Behav.* **2007**, *39*, S189–S194. [CrossRef]
9. Clemmensen, C.; Petersen, M.B.; Sørensen, T.I.A. Will the COVID-19 pandemic worsen the obesity epidemic? *Nat. Rev. Endocrinol.* **2020**, *16*, 469–470. [CrossRef]
10. Marty, L.; de Lauzon-Guillain, B.; Labesse, M.; Nicklaus, S. Food choice motives and the nutritional quality of diet during the COVID-19 lockdown in France. *Appetite* **2021**, *157*, 105005. [CrossRef]
11. Higgs, S.; Thomas, J. Social influences on eating. *Curr. Opin. Behav. Sci.* **2016**, *9*, 1–6. [CrossRef]
12. Martinez-Ferran, M.; De La Guía-Galipienso, F.; Sanchis-Gomar, F.; Pareja-Galeano, H. Metabolic Impacts of Confinement during the COVID-19 Pandemic Due to Modified Diet and Physical Activity Habits. *Nutrients* **2020**, *12*, 1549. [CrossRef] [PubMed]
13. Muscogiuri, G.; Barrea, L.; Savastano, S.; Colao, A. Nutritional recommendations for CoVID-19 quarantine. *Eur. J. Clin. Nutr.* **2020**, *74*, 850–851. [CrossRef] [PubMed]
14. Sidor, A.; Rzymski, P. Dietary choices and habits during COVID-19 lockdown: Experience from Poland. *Nutrients* **2020**, *12*, 1657. [CrossRef] [PubMed]
15. Rodríguez-Pérez, C.; Molina-Montes, E.; Verardo, V.; Artacho, R.; García-Villanova, B.; Jesús Guerra-Hernández, E.; Dolores Ruíz-López, M. Changes in Dietary Behaviours during the COVID-19 Outbreak Confinement in the Spanish COVIDiet Study. *Nutrients* **2020**, *12*, 1730. [CrossRef] [PubMed]
16. Malta, D.C.; Szwarcwald, C.L.; Barros, M.B.d.a.; Gomes, C.S.; Machado, Í.E.; de Souza Júnior, P.R.B.; Romero, D.E.; Lima, M.G.; Damacena, G.N.; Pina, M.d.F.; et al. A pandemia da COVID-19 e as mudanças no estilo de vida dos brasileiros adultos: Um estudo transversal, 2020. *Epidemiol. Serv. Saude Rev. Sist. Unico Saude Bras.* **2020**, *29*, e2020407.
17. Steele, E.M.; Rauber, F.; dos Santos Costa, C.; Leite, M.A.; Gabe, K.T.; da Costa Louzada, M.L.; Levy, R.B.; Monteiro, C.A. Dietary changes in the NutriNet Brasil cohort during the covid-19 pandemic. *Rev. Saude Publica* **2020**, *54*, 1–8. [CrossRef]

18. Kissler, S.M.; Tedijanto, C.; Goldstein, E.; Grad, Y.H.; Lipsitch, M. Projecting the transmission dynamics of SARS-CoV-2 through the postpandemic period. *Science* **2020**, *368*, 860–868. [CrossRef]
19. Brasil. *Ministério da Saude. Secretaria de Atenção à Saúde. Departamento de Atenção Básica. Orientações Para a Coleta e Análise de Dados Antropométricos Em Serviços de Saúde: Norma Técnica do Sistema de Vigilância Alimentar e Nutricional-SISVAN/Ministério da Saúde, Secretaria de Atenção à Saúde, Departamento de Atenção Básica*, 1st ed.; Ministerio da Saúde: Brasília, Brazil, 2011; ISBN 9788533418134.
20. Kaushal, K. Response Shift Bias in Pre- and Post-test Studies. *Indian J. Dermatol.* **2016**, *61*, 91.
21. Rohs, F.R.; Langone, C.A.; Coleman, R.K. Response shift bias: A problem in evaluating nutrition training using self-report measures. *J. Nutr. Educ. Behav.* **2001**, *33*, 165–170. [CrossRef]
22. Drennan, J.; Hyde, A. Controlling response shift bias: The use of the retrospective pre-test design in the evaluation of a master's programme. *Assess. Eval. High. Educ.* **2008**, *33*, 699–709. [CrossRef]
23. Brasil, M.d.S. Cartilha dos Direitos dos Participantes de Pesquisa. Available online: http://gg.gg/l70q5 (accessed on 16 August 2021).
24. de Queiroz, F.L.N.; Nakano, E.Y.; Ginani, V.C.; Botelho, R.B.A.; Araújo, W.M.C.; Zandonadi, R.P. Eating competence among a select sample of Brazilian adults: Translation and reproducibility analyses of the satter eating competence inventory. *Nutrients* **2020**, *12*, 2145. [CrossRef]
25. IBGE. IBGE-Instituto Brasileiro de Geografia: Pesquisa Nacional por Amostra de Domicílio Contínua (PNAD Contínua). Available online: http://www.ibge.gov.br/estatisticas-novoportal/sociais/educacao/1727-pnad-continua.html (accessed on 22 April 2020).
26. NEEDs Center Protocol for the Use of the ecSatter Inventory 2.0. Available online: https://www.needscenter.org/wp-content/uploads/2019/09/ecSI-2.0-Usage-Protocol-2-1.pdf (accessed on 9 March 2020).
27. NEEDs Center Satter Eating Competence Model (ecSatter). Available online: https://www.needscenter.org/satter-eating-competence-model-ecsatter/ (accessed on 9 March 2020).
28. Godleski, S.; Lohse, B.; Krall, J.S. Satter Eating Competence Inventory Subscale Restructure After Confirmatory Factor Analysis. *J. Nutr. Educ. Behav.* **2019**, *51*, 1003–1010. [CrossRef]
29. Brasil. *Ministério da Saúde. VIGITEL 2019: Vigilância de Fatores de Risco e Proteção para Doenças Crônicas por Inquérito Telefônico*, 1st ed.; Ministerio da Saúde: Brasília, Brazil, 2020; ISBN 9788533427655.
30. Hargreaves, S.M.; Araújo, W.M.C.; Nakano, E.Y.; Zandonadi, R.P. Brazilian vegetarians diet quality markers and comparison with the general population: A nationwide cross-sectional study. *PLoS ONE* **2020**, *15*, e0232954.
31. de Queiroz, F.L.N.; Nakano, E.Y.; Botelho, R.B.A.; Ginani, V.C.; Cançado, A.L.F.; Zandonadi, R.P. Eating Competence Associated with Food Consumption and Health Outcomes among Brazilian Adult Population. *Nutrients* **2020**, *12*, 3218. [CrossRef]
32. Belik, W. Sustainability and food security after COVID-19: Relocalizing food systems? *Agric. Food Econ.* **2020**, *8*, 4. [CrossRef]
33. Ahmed, F.; Ahmed, N.; Pissarides, C.; Stiglitz, J. Why inequality could spread COVID-19. *Lancet Public Health*. **2020**, *5*, e240. [CrossRef]
34. Krall, J.S.; Lohse, B. Cognitive testing with female nutrition and education assistance program participants informs validity of the satter eating competence inventory. *J. Nutr. Educ. Behav.* **2010**, *42*, 277–283. [CrossRef] [PubMed]
35. Krall, J.S.; Lohse, B. Interviews with Low-Income Pennsylvanians Verify a Need to Enhance Eating Competence. *J. Am. Diet. Assoc.* **2009**, *109*, 468–473. [CrossRef] [PubMed]
36. Cosgrove, K.; Wharton, C. Predictors of COVID-19-Related Perceived Improvements in Dietary Health: Results from a US Cross-Sectional Study. *Nutrients* **2021**, *13*, 2097. [CrossRef]
37. Alves, C.E.; Dal'Magro, G.P.; Viacava, K.R.; Dewes, H. Food Acquisition in the Geography of Brazilian Obesity. *Front. Public Health*. **2020**, *8*, 37. [CrossRef]
38. Lohse, B.; Satter, E.; Horacek, T.; Gebreselassie, T.; Oakland, M.J. Measuring Eating Competence: Psychometric Properties and Validity of the ecSatter Inventory. *J. Nutr. Educ. Behav.* **2007**, *39*, S154–S166. [CrossRef]
39. Lohse, B.A.; Arnold, K.N. Measuring Eating Competence: Congruence between Two Satter Inventories Supports Supplanting the Original Version with the Low-Income Adaptation. *J. Acad. Nutr. Diet.* **2012**, *112*, A63. [CrossRef]
40. Yau, Y.H.C.; Potenza, M.N. Stress and eating behaviors. *Minerva Endocrinol.* **2013**, *38*, 255–267. [PubMed]
41. Cummings, J.R.; Ackerman, J.M.; Wolfson, J.A.; Gearhardt, A.N. COVID-19 stress and eating and drinking behaviors in the United States during the early stages of the pandemic. *Appetite* **2021**, *162*, 105163. [CrossRef] [PubMed]
42. Krall, J.S.; Lohse, B. Validation of a measure of the Satter eating competence model with low-income females. *Int. J. Behav. Nutr. Phys. Act.* **2011**, *8*, 1–10. [CrossRef]
43. Clifford, D.; Keeler, L.A.; Gray, K.; Steingrube, A.; Morris, M.N. Weight Attitudes Predict Eating Competence among College Students. *Fam. Consum. Sci. Res. J.* **2010**, *39*, 184–193. [CrossRef]
44. Quick, V.; Byrd-Bredbenner, C.; Shoff, S.; White, A.A.; Lohse, B.; Horacek, T.; Colby, S.; Brown, O.; Kidd, T.; Greene, G. Relationships of Sleep Duration With Weight-Related Behaviors of U.S. College Students. *Behav. Sleep Med.* **2016**, *14*, 565–580. [CrossRef]
45. Lohse, B.; Krall, J.S.; Psota, T.; Kris-Etherton, P. Impact of a Weight Management Intervention on Eating Competence: Importance of Measurement Interval in Protocol Design. *Am. J. Health. Promot.* **2018**, *32*, 718–728. [CrossRef]
46. Hall, K.D.; Heymsfield, S.B.; Kemnitz, J.W.; Klein, S.; Schoeller, D.A.; Speakman, J.R. Energy balance and its components: Implications for body weight regulation. *Am. J. Clin. Nutr.* **2012**, *95*, 989–994. [CrossRef]
47. WHO. *Diet, Nutrition and the Prevention of Chronic Diseases*; WHO Technical Report Series; WHO: Geneva, Switzerland, 2003.

48. Vartanian, L.R.; Schwartz, M.B.; Brownell, K.D. Effects of Soft Drink Consumption on Nutrition and Health: A Systematic Review and Meta-Analysis. *Am. J. Public Health* **2007**, *97*, 667–675. [CrossRef]
49. Tilles-Tirkkonen, T.; Outi, N.; Sakari, S.; Jarmo, L.; Kaisa, P.; Leila, K. Preliminary Finnish measures of Eating Competence suggest association with health-promoting eating patterns and related Psychobehavioral factors in 10–17 year old adolescents. *Nutrients* **2015**, *7*, 3828–3846.
50. Naja, F.; Hamadeh, R. Nutrition amid the COVID-19 pandemic: A multi-level framework for action. *Eur. J. Clin. Nutr.* **2020**, *74*, 1117–1121. [CrossRef] [PubMed]
51. Ribeiro-Silva, R.d.C.; Pereira, M.; Campello, T.; Aragão, É.; Guimarães, J.M.d.M.; Ferreira, A.J.F.; Barreto, M.L.; dos Santos, S.M.C. Covid-19 pandemic implications for food and nutrition security in Brazil. *Cienc. Saude Coletiva* **2020**, *25*, 3421–3430. [CrossRef] [PubMed]
52. Lohse, B.; Psota, T.; Zazpe, I.; Sorli, V.; Salas-salvado, J.; Ros, E. Eating Competence of Elderly Spanish Adults Is Associated with a Healthy Diet and a Favorable Cardiovascular Disease Risk Profile 1–3. *J. Nutr.* **2010**, 1322–1327. [CrossRef]
53. American Dietetic Association Position of the American Dietetic Association: Nutrition intervention in the treatment of anorexia nervosa, bulimia nervosa, and other eating disorders. *J. Am. Diet. Assoc.* **2006**, *106*, 2073–2082. [CrossRef]
54. Burton, E.T.; Smith, W.A. Mindful Eating and Active Living: Development and Implementation of a Multidisciplinary Pediatric Weight Management Intervention. *Nutrients* **2020**, *12*, 1425. [CrossRef] [PubMed]
55. Lohse, B.; Bailey, R.L.; Krall, J.S.; Wall, D.E.; Mitchell, D.C. Diet quality is related to eating competence in cross-sectional sample of low-income females surveyed in Pennsylvania. *Appetite* **2012**, *58*, 645–650. [CrossRef] [PubMed]

foods

MDPI

Article

Could *Japonica* Rice Be an Alternative Variety for Increased Global Food Security and Climate Change Mitigation?

Daniel Dooyum Uyeh [1,2,3], Senorpe Asem-Hiablie [4], Tusan Park [1,3,*], Kyungmin Kim [5], Alexey Mikhaylov [6], Seungmin Woo [1,2,3] and Yushin Ha [1,2,3,*]

1   Department of Bio-Industrial Machinery Engineering, Kyungpook National University, Daegu 41566, Korea; uyehdooyum@gmail.com (D.D.U.); woosm7571@gmail.com (S.W.)
2   Upland-Field Machinery Research Centre, Kyungpook National University, Daegu 41566, Korea
3   Smart Agriculture Innovation Center, Kyungpook National University, Daegu 41566, Korea
4   Institutes of Energy and the Environment, The Pennsylvania State University, University Park, PA 16802, USA; senorpe.ah@gmail.com
5   Division of Plant Biosciences, School of Applied Biosciences, College of Agriculture & Life Science, Kyungpook National University, Daegu 41566, Korea; kkm@knu.ac.kr
6   Financial Research Institute of Ministry of Finance of the Russian Federation, 127006 Moscow, Russia; Alexeyfa@ya.ru
*   Correspondence: tusan.park@knu.ac.kr (T.P.); yushin72@knu.ac.kr (Y.H.)

Abstract: The growing importance of rice globally over the past three decades is evident in its strategic place in many countries' food security planning policies. Still, its cultivation emits substantial greenhouse gases (GHGs). The *Indica* and *Japonica* sub-species of *Oryza sativa* L. are mainly grown, with *Indica* holding the largest market share. The awareness, economics, and acceptability of *Japonica* rice in a food-insecure *Indica* rice-consuming population were surveyed. The impact of parboiling on *Japonica* rice was studied and the factors which most impacted stickiness were investigated through sensory and statistical analyses. A comparison of the growing climate and greenhouse gas emissions of *Japonica* and *Indica* rice was carried out by reviewing previous studies. Survey results indicated that non-adhesiveness and pleasant aroma were the most preferred properties. Parboiling treatment altered *Japonica* rice's physical and chemical properties, introducing gelatinization of starch and reducing adhesiveness while retaining micronutrient concentrations. Regions with high food insecurity and high consumption of *Indica* rice were found to have suitable climatic conditions for growing *Japonica* rice. Adopting the higher-yielding, nutritious *Japonica* rice whose cultivation emits less GHG in these regions could help strengthen food security while reducing GHGs in global rice cultivation.

Keywords: rice cultivation; rice carbon emissions; hidden hunger; parboiling; rice quality; rice preference

Citation: Uyeh, D.D.; Asem-Hiablie, S.; Park, T.; Kim, K.; Mikhaylov, A.; Woo, S.; Ha, Y. Could *Japonica* Rice Be an Alternative Variety for Increased Global Food Security and Climate Change Mitigation? *Foods* **2021**, *10*, 1869. https://doi.org/10.3390/foods10081869

Academic Editors: António Raposo, Renata Puppin Zandonadi and Raquel Braz Assunção Botelho

Received: 25 June 2021
Accepted: 10 August 2021
Published: 12 August 2021

1. Introduction

Recent World Bank reports show that half of the world's poorest people are in just five countries on two continents [1]. These are India and Bangladesh in South Asia, and Nigeria, the Democratic Republic of Congo, and Ethiopia in Sub-Saharan Africa. Rice is a highly consumed staple in three of these countries with reported values as follows: Bangladesh (260 kg/capita and 36.1 Megatonnes); India (103 kg/capita and 104 Megatonnes), Nigeria (46.3 kg/capita 6.60 Megatonnes), Democratic Republic of Congo (29.7 kg/capita and 0.47 Megatonnes), and Ethiopia (4.69 kg/capita and 0.66 Megatonnes) [1,2]. Two of these countries are also among the ten top global importers of rice, with Nigeria at 1.8 Megatonnes spending USD 8.5 billion and Bangladesh at 0.9 Megatonnes spending USD 4.03 billion [3]. In these countries, there are high rates of malnourishment and food insecurity [4].

Food security has been defined in various ways. As reported by the Food and Agriculture Organization (FAO) of the United Nations, the initial focus was on the volume and stability of food supplies. This definition has evolved into "when all people always have physical, social and economic access to enough, safe and nutritious food which meets their dietary needs and food preferences for an active and healthy life" (FAO, 2019) [5]. According to this latest definition, 820 million food-insecure people [6] live in poor countries. However, "hidden hunger" becomes pertinent due to challenges such as inadequate intake of calories to support an individual's daily activities, insufficient protein to support growth, and deficiencies in mineral and vitamins intake [7]. Serious health outcomes of hidden hunger have been recorded in most developing countries [8,9] and have been predicted to grow exponentially in Africa at a rate of 18% annually [10].

Rice, a monocot cereal customarily grown as an annual plant, is an excellent source of complex carbohydrates. As a staple for more than half of the world's population, worldwide production has been growing steadily [11] in response to increasing demands over the past three decades. The growing importance of rice globally is evident in the strategic food security planning policies of many countries. While, currently, nine out of every ten people in the world who consume rice are Asian, rice is ranked as the fastest growing staple in Africa. In Latin America and the Caribbean, it is gaining increased acceptance [12].

By 2039, the total land area of rice harvest is projected to increase by 11% to meet the world's needs [13]. However, rice cultivation is also a major source of greenhouse gas (GHG) emissions, with annual levels in 2017 recorded at 924 Megatonnes of carbon dioxide equivalents (CO_2e), ahead of hard coal mining (790 Megatonnes CO_2e) [14]. The total land area used to cultivate cereals globally is nearly 7.31×10^8 ha [15]. Rice harvests account for about 23.5% of the total harvested cereal area and 27% of global cereal production and consumption [16].

Rice is cultivated in more than 100 countries [17]. Still, except for a few countries that have attained self-sufficiency, demand currently exceeds production in many, and large quantities of rice are imported to meet needs. As many as 21 of the 39 rice-producing countries in Africa import between 50 and 99% of their rice needs at substantial economic costs [18,19], making it unaffordable for the populace living in poverty. Furthermore, supply chain disruptions caused by occurrences such as conflicts and pandemics drive calls for securing local stores of domestic production of food staples. For example, the World Food Program [20] has warned of risks of widespread famines due to the COVID-19 global disease outbreak caused by the severe acute respiratory syndrome coronavirus 2 (SARS-CoV-2).

Nigeria, Africa's second most populous country, was reported to be the second largest rice importer in the world [21] and in terms of calories consumed, rice ranked fourth. Although Nigeria was reported as the largest rice producer in West Africa, productivity is relatively low. The major constraint to growing rice in Africa is reliable water availability, as unsteady and low rainfall patterns results in crop vulnerability to drought.

Only two rice species (the African *Oryza glaberrima* and the Asian *Oryza sativa*) are reportedly cultivated in the world [22]. These are morphologically similar but differ ecologically. *O. glaberrima* has the important characteristics of being more disease and pest resistant compared to the Asian *O. sativa* [22]. Additionally, *O. glaberrima* is less susceptible to weeds given its broader leaves that shade weeds out. Furthermore, the African variety has a better tolerance for fluctuations in water depth, severe climates, iron toxicity, and infertile soils. A faster maturity rate gives *O. glaberrima* high potential as an important emergency food [23]. The higher yields of the Asian *O. sativa*, however, present an advantage over *O. glaberrima* whose grains are brittle and more challenging to mill. Hybrids of the two main species have been produced to preserve the desirable traits in both. Cultivation of *O. glaberrima* is predominantly in Sub-Saharan Africa but it is being replaced by *O. sativa*.

The Asian rice species, *Oryza sativa*, is the primary type cultivated worldwide and its subspecies, *Indica* and *Japonica*, are the most grown. *Indica* rice, also known as long grain, grows well near the equator with a kernel four to five times longer than its width and it holds over 80% of the global rice market share [24].

Rice quality is primarily assessed based on physical properties such as head rice yield, chalkiness, grain size and shape, grain color, adhesiveness, protein and amylose content, and aroma [25–27]. These properties of rice affect its physicochemical and cooking properties and, consequently, desirability among consumers. Amylose content and alkali digestion are widely recognized as important determinants of consumption quality because they reflect the functionality of the rice grain's starch [28]. Protein content also plays an essential role in determining quality, as does the grain texture and the surface hardness of cooked rice [29].

The parboiling process of harvested paddy improves quality by producing mostly desirable biochemical, physical, and sensory changes according to reports by Oli et al. [30]; Mir and Bosco [31]; Bello et al. [32] and da Fonseca et al. [33]. The gelatinization of starch in the rice occurs during parboiling, resulting in loss of adhesiveness and producing a desirable aroma. Parboiling involves partially boiling the paddy in its husk and consists of three basic steps: soaking, steaming, and drying, as Miah et al. [34] described. During the parboiling process, the first stage of soaking rice in warm water allows the water-soluble nutrients to penetrate the endosperm from the husk. The husk contains 0.05–0.07 mg of riboflavin, 0.09–0.21 mg thiamin, and 1.6–4.2 mg niacin [35,36]. Riboflavin is slightly soluble in water, thiamin is highly soluble, and niacin is poorly soluble in water.

During the cooking process, starch granules in the grain expand, and amylose leaches into the boiling water to form a viscous solution. As the cooked rice cools, the leached amylose chains line up, lock together, and form a gel, resulting in adhesiveness. In theory, the higher the amylose content of rice, the firmer the cooked grain of rice would be. Amylose is also responsible for the way that rice hardens upon cooling. When rice cools to room temperature or lower, retrogradation occurs as the amylose chains and linear parts of amylopectin crystallize by forming hydrogen bridges.

The *Indica* paddy is usually treated to a parboiling process, resulting in the milled grains being fluffy with separate kernels when cooked. The rice seeds are also processed into other products, including pasta, rice cakes, flour, and animal feed. *Japonica* rice, commonly referred to as medium and short grain, has kernels two to three times longer than its width and grows mostly in temperate regions. It is commonly found in the Republic of Korea, Japan, the Democratic People's Republic of Korea, and parts of China, Australia, and the United States of America. The *Japonica* rice paddy is usually not treated with any process, and the milled grains are moist, adhesive, and bright white in color. The glutinous (adhesiveness) nature of *Japonica* makes it most ideal for Mediterranean and Asian dishes that require adhesiveness, such as paella, risotto, and sushi.

The objectives of this study were to: (1) determine the awareness, acceptability, and economics of *Japonica* rice using structured questionnaires in a major parboiled *Indica* rice-consuming market; (2) analyze the differences in nutritional and physical properties of untreated *Japonica*, treated *Japonica*, and treated *Indica* rice; (3) compare the growing climate and greenhouse gas emissions of *Japonica* and *Indica* rice using previous studies; and discuss the implications for food security in food-insecure *Indica* rice-consuming markets.

2. Materials and Methods

A survey of consumer rice preference was performed, and analyses of the impact of parboiling on the quality of *Japonica* rice considering consumer preference were carried out. Following Adhikari et al. [37], moisture, temperature, adhesion, and quantity of low molecular sugars were measured to assess adhesiveness. Furthermore, color and nutrient levels in the rice samples were measured using spectrophotometric and wet chemistry methods.

2.1. Survey of Consumer Rice Preferences Using a Structured Questionnaire

A survey was administered to 1057 literate and non-literate randomly selected participants in Nigeria's six main geographical regions (North-east, North-west, North-central, South-east, South-west, and South-south). A total of 506 literate respondents participated via an online questionnaire and in-person interviews were administered to the 551 non-literate respondents. To determine the participants' rice preferences when presented with choices, the information gathered through the survey (Appendix B) included the following:

- knowledge of the rice sub-species and preference of sub-species,
- length of grain, adhesiveness, aroma, and taste after cooking as well as the cooking time,
- willingness to purchase healthier rice at a relatively higher cost (USD 1 = NGN 410).

The survey results were summarized using descriptive statistics.

2.2. Analyses of the Impact of Parboiling on the Properties of Japonica Rice

2.2.1. Rice Parboiling Procedure

Samples for this study were collected from a *Japonica* paddy harvested in Dalseong gun (Republic of Korea). The rice samples were parboiled using hydrothermal and mechanical treatments as described below.

Nine *Japonica* rice samples weighing 200 g each were soaked in a water bath shaker as described in [38]. Fabricated parboiling equipment [38] was used to treat the samples under the different conditions, given in Table 1. The equipment consisted of a steaming chamber with a perforated aluminum sample holder, a digital control system, a relay switch, pressure gauge, pressure valve, safety valve, and water inlet and outlet points. Heat loss was reduced by insulating the parboiling device with a 10 mm layer of polyurethane foam. The digital control system regulated steam temperature. The pressure was controlled with an installed pressure gauge and safety valves.

Table 1. Conditions for parboiling *Japonica* rice.

Exp. No.	Moisture Content after Soaking (% Wet Basis)	Steaming Temperature (°C)	Steaming Duration (min)	Drying Temperature (°C)
1	24	50	20	35
2	24	70	40	45
3	24	90	60	55
4	27	70	20	55
5	27	90	40	35
6	27	50	60	45
7	30	90	20	45
8	30	50	40	55
9	30	70	60	35

Following the hydrothermal treatments, 100 g each of the nine rice samples were dried, and their moisture contents determined. Drying was carried out using a hot air oven (VS-120203N, Vision Scientific, Daejeon, Korea) [39] which was also used in moisture content determination [40], using Equations (1)–(3).

$$MC_{wb} = \frac{MC_i - MC_f}{MC_i} \times 100 \tag{1}$$

$$MC_{db} = \frac{MC_i - MC_f}{MC_f} \times 100 \tag{2}$$

Weight loss during drying:

$$MC_f = MC_i \times \frac{100 \times MC_{wb}}{100 - MC_{wb}} \tag{3}$$

where MC_{wb} is moisture content on a wet basis, MC_{db} is moisture content on a dry basis, MC_i is initial weight, and MC_f are final weight.

The dried *Japonica* rice samples were milled following the method reported in [39]. Milling equipment included a huller (Model SY-88-TH, Ssang Yong, Yongin-si, Korea), abrasive type rice mill (Model TM05, Satake, Hiroshima-ken, Japan), friction type mill (Model SY-92-TR, Ssang Yong, Yongin-si, Korea), and large and small broke separator (Model SY2000-HRG, Ssang Yong, Yongin-si, Korea). The rice yield after milling was calculated as:

$$Y_{milling} = \frac{W_{m.rice}}{W_{p.rice}} \times 100 \tag{4}$$

where $Y_{milling}$ is milled rice yield, $W_{m.rice}$ is milled rice weight, and $W_{p.rice}$ is the initial weight of *Japonica* paddy.

Sorting out of foreign materials and other quality control and quality assurance measures followed the mechanical and hydrothermal treatments. The pre-treatment processes undergone by the rice samples are shown in Figure 1.

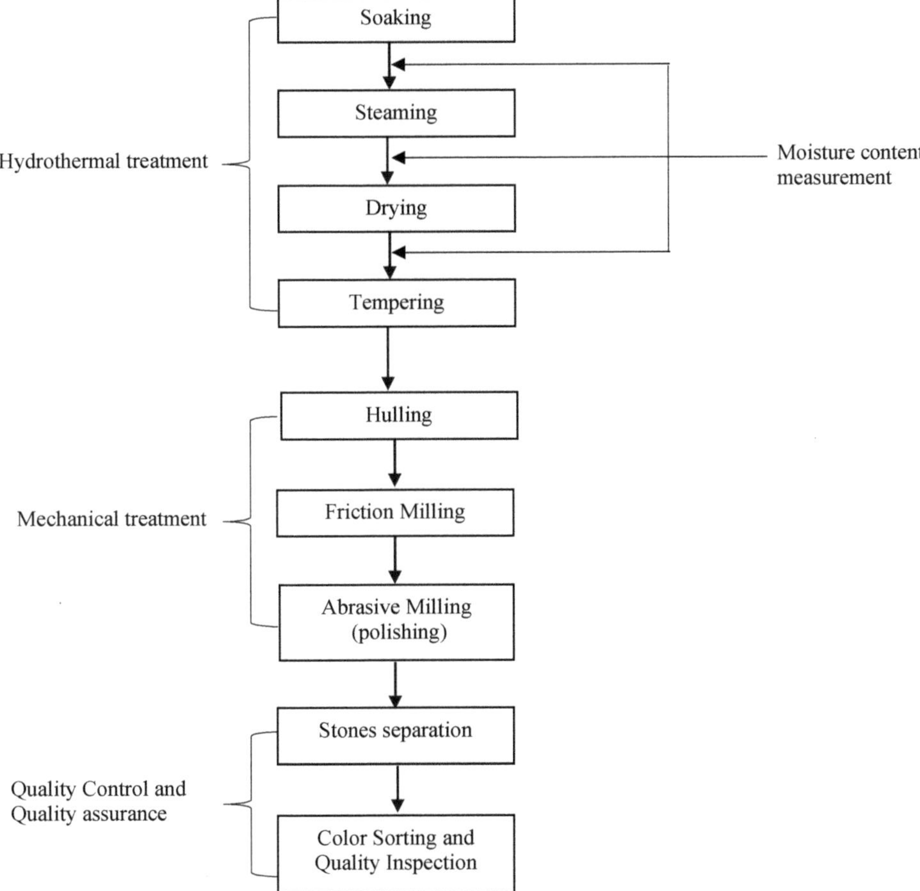

Figure 1. Process flow for parboiling and milling *Japonica* rice.

During the parboiling treatment, the milled rice yield of *Japonica* rice was tested following an adaptation of the Taguchi method as previously described by Uyeh et al. [38], using three different settings each for moisture, steaming temperature, drying temperature, and duration of treatment (Table 1).

2.2.2. Analysis of the Differences in Nutritional and Physical Properties of Milled Rice Samples

The differences in nutritional and physical properties of untreated *Japonica*, treated *Japonica*, and treated *Indica rice* were measured and analyzed. The properties measured were adhesiveness, amylose content, alkali digestion, nutrients, and color.

a.   Analysis of adhesiveness in milled parboiled and non-parboiled rice samples

To measure the adhesiveness of *Japonica* and *Indica* rice, 25 g of milled triplicate samples of parboiled and non-parboiled *Japonica* and parboiled *Indica* were transferred into 100 mL beakers and placed in a home type rice cooker (DWX-220K, Morningcom, Seoul, Korea). A 40 mL volume of water was immediately transferred into each beaker of the milled rice samples, and the mixtures were stirred to a uniform consistency and evenly spread. The three beakers of samples were then placed into a rice cooker, and 400 g of water was transferred into the cooker's pot. A thermometer was inserted into the pot to measure the cooking temperature. Twenty minutes after the temperature reached 100 °C or water vapor formed, the rice cooker was turned off and the set-up was allowed to stand for another 10 min. At this point, the rice was considered "cooked" for human consumption. The beakers were then left to stand upside-down (Figure 2) on a flat, straight, and stable surface for one hour [41]. The rice samples that stuck to the beaker were considered to be sticky.

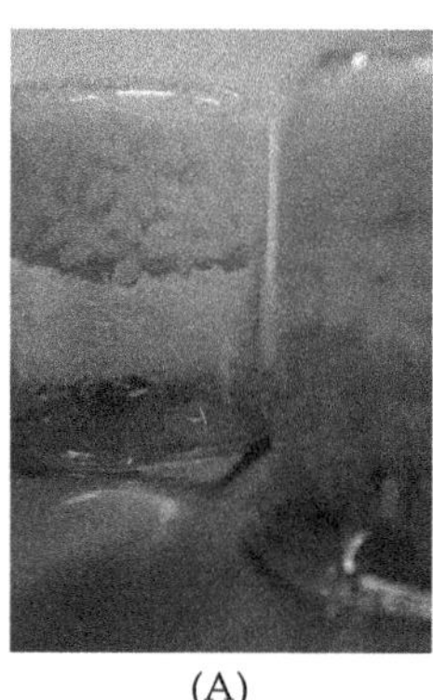
(A)

(B)

(C)

Figure 2. Beakers of cooked rice samples placed in inverted positions to investigate adhesion; (**A**) non-parboiled *Japonica* rice; (**B**) parboiled *Japonica* rice; and (**C**) *Indica* rice.

b.   Analysis of amylose content in milled parboiled and non-parboiled rice samples

A component analyzer (AN-820, KETT, Tokyo, Japan) was used to determine the amylose content of the milled parboiled *Japonica* and *Indica* rice and raw *Japonica* rice (control) samples. The analyzer was set to the "short-milled rice" option pre-calibrated by the manufacturers before measurement.

c.   Analysis of alkali digestion in milled parboiled and non-parboiled rice samples

Alkali digestion values of milled rice samples were measured. Twenty-five-gram triplicates of each type of milled rice (*Japonica* and *Indica* rice) were placed in Petri dishes (90 × 15 mm diameter) containing 20 mL of 1.4% KOH solution. The grains were allowed to stand at 30 °C for 18 h following Choe and Heu [42]. Alkali digestion was graded from 1 to 7 referencing the seven-point scale of the International Rice Research Institute [43], as shown in Table 2 below.

Table 2. Alkali digestion value seven-point scale.

Degree of Digestion	Alkali Digestion Classification	Alkali Digestion Value
Grain not affected	Low	1
Grain swollen	Low	2
Grain swollen, collar incomplete and narrow	Low	3
Grain swollen, collar complete and wide	Intermediate	4
Grain split or segmented, collar complete and wide	Intermediate	5
Grain dispersed, merging with collar	High	6
Grain completely dispersed and intermingled	High	7

d.   Nutritional analyses of milled parboiled and non-parboiled rice samples

Macronutrient (crude protein, crude fat, crude ash, and carbohydrates) and micronutrient (iron, magnesium, and zinc) contents of the parboiled *Japonica* rice and *Indica* rice and untreated *Japonica* rice were measured on a wet basis using standard methodology as described by the AOAC [44].

e.   Color measurement of milled parboiled and non-parboiled rice samples

The CIE color scales L*a*b* of cooked and uncooked for *Japonica* (parboiled and non-parboiled) and rice samples were measured with a reflectance colorimeter (CS-288, Konica Minolta, Tokyo, Japan) and a spectrum resolution of 10 nm. On the CIE color scale, L* indicates the degree of lightness or darkness from perfect black (0) to almost perfect white (100); a* indicates the degree of redness (+) and greenness ($-$); and b* indicates the degree of yellowness (+) and blueness ($-$). Calibration of the spectrometer was carried out with a standard white plate of L*a* b*.

2.3. Growing of Japonica Rice and the Environmental Impact

The climatic factors important for *Japonica* rice cultivation reported in previous studies [45–49] were compared among two countries with the highest global cultivation of this variety (Republic of Korea and Japan) and the three countries with the highest levels of malnutrition globally (India, Bangladesh, and Nigeria). The compared factors were temperature, sunshine, and precipitation. Additionally, data on the impact of cultivating the two major varieties of rice (*Indica* and *Japonica*) on the environment were retrieved from published studies, analyzed, and discussed.

2.4. Statistical Analyses

The impacts of moisture content after soaking, steaming temperature, steaming time, and drying temperature on the adhesiveness of *Japonica* rice were analyzed using the Taguchi orthogonal array in Minitab® Version 14. Using this method, the factors affecting adhesiveness were ranked from 1 to 4, with 1 having the most significant effect and 4, the least, and represented by graphs (Figure A1). Further categorizations were carried out using the signal to noise (S/N) ratio in Equations (5)–(7). The smaller the S/N ratio in Equation (5), the more significant the impact, while the converse was true for Equation (7). For Equation (6), a nominal value was taken as the best indication of a strong impact on adhesiveness.

$$S/N = -10 \times \log(\Sigma(Y^2)/n) \tag{5}$$

where Y is responses for the given factor level combination, and n, the number of responses in the factor level combination.

$$S/N = -10 \times \log(s^2) \tag{6}$$

where s is the standard deviation of the responses for all noise factors for the given factor level combination.

$$S/N = -10 \times \log(\Sigma(1/Y^2)/n) \tag{7}$$

where responses for the given factor level combination and number of responses in the factor level combination were Y and n, respectively.

The significant differences ($p \leq 0.05$) in measured nutrients among non-parboiled and parboiled *Japonica* and parboiled *Indica* rice were assessed using single-factor analysis of variance. Where significant differences were found, mean comparisons were carried out post hoc with Tukey's honestly significant difference test. The nutrients were protein, fat, carbohydrates, iron (Fe), magnesium (Mg), and zinc (Zn). Ash content was also compared among the rice samples.

3. Results

3.1. Rice Attributes Preferred by Consumers

Survey respondents ($n = 1057$) lived across the six geopolitical regions of Nigeria and ranged in age from 16 to 61 years, with 44.3% being female and 55.7% male. Their occupations included farming, manufacturing, and service provision. Sixty-eight percent of those surveyed were aware of the different rice varieties, 28% were not, and the remaining 4% were unsure. Respondents showed a varied preference for the rice properties tested; 32.9% were indifferent, 44.9% showed a preference for aroma, and 23.5% strongly preferred rice with aroma (Figure 3A). Generally, a total of 74.4% indicated that the non-adhesive rice was an important preference, with 32.2% indicating that it strongly was (Figure 3B). About 80% of respondents were also willing to pay a moderately higher price for non-adhesive rice (Figure 3C).

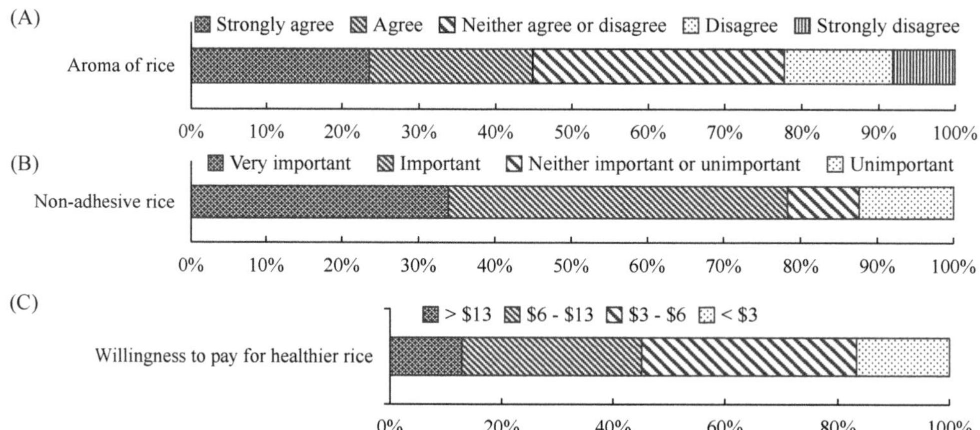

Figure 3. Survey respondents' preference for rice based on (**A**) aroma, (**B**) adhesiveness, and (**C**) costs in Nigeria (West Africa) per 50 kg of rice (USD 1 = NGN 410).

3.2. Effects of Parboiling on the Physiochemical Properties of Rice Samples

a.   Adhesiveness and amylose content of milled parboiled and non-parboiled rice samples

After one hour of the beakers of cooked rice samples standing upside-down on a flat surface, the contents of the beakers with the parboiled *Japonica* and *Indica* rice samples dropped onto the surface (Figure 2). In contrast, the non-parboiled rice sample remained at the bottom of the beaker. Parboiling reduced the adhesiveness of *Japonica* rice and increased the amylose content at all levels of treatment, as shown in Table 3. The Taguchi orthogonal array results showed that steaming temperature and moisture content after soaking were the factors with the highest impact on the adhesiveness of the parboiled *Japonica* rice (Figure A1).

Table 3. Amylose content of parboiled and milled *Japonica* rice.

Treatment	1	2	3	4	5	6	7	8	9
Amylose (%)	17.93	18.33	19.03	18.53	19.43	18.20	19.37	18.37	18.87
S/N Value	25.07	25.26	25.59	25.36	25.77	25.20	25.74	25.28	25.51

b.   *Alkali digestion in milled parboiled and non-parboiled rice samples*

The sample medians were 2, 5, and 5 for *Indica* (n = 12), parboiled *Japonica* (n = 12), and non-parboiled *Japonica* (n = 12), respectively (Figure 4). Both *p*-values (adjusted and non-adjusted for ties) observed were less than the significance level of 0.05, indicating that the median quality index differed in at least one of the *Indica* samples, and post hoc tests confirmed this.

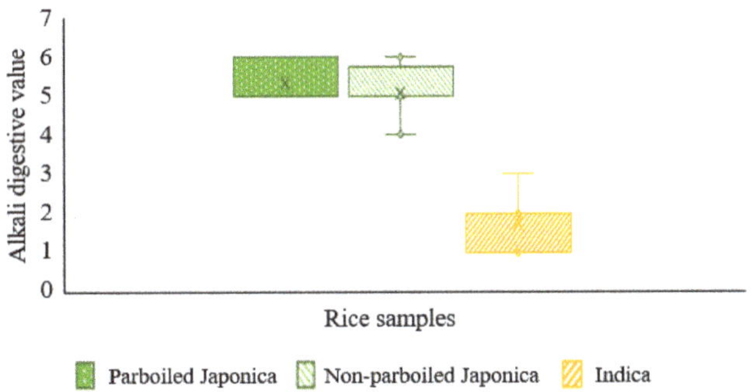

Figure 4. Kruskal–Wallis test for alkali digestion value of parboiled and non-parboiled *Japonica* rice and *Indica* rice.

c.   *Nutrient levels observed in milled parboiled and non-parboiled rice samples*

The carbohydrate, fat, and ash contents measured in rice samples were comparable in non-parboiled *Japonica*, parboiled *Japonica*, and parboiled *Indica* (Table 4). The concentration of the micronutrient Fe was also similar in the measured rice samples (Table 4). Protein was highest in the parboiled *Indica* samples and lowest in the non-parboiled *Japonica* ($p \leq 0.05$; Figure 5) at concentrations of 9.1 ± 0.03 g/100 g and 7.5 ± 0.03 g/100 g, respectively. However, magnesium and Zn levels were lowest in the parboiled *Indica* samples while similar in both *Japonica* rice samples ($p \leq 0.05$; Figure 5). Heinemann et al. [50] reported protein concentrations of 6.85 ± 0.34% and 6.76 ± 0.20% in non-parboiled and parboiled brown rice, respectively, with a moisture content of 12.6 ± 0.54% for non-parboiled and 12.07 ± 0.7% parboiled brown rice.

Table 4. Mean and standard deviation (SD) of chemical constituents measured on a dry weight basis in *Japonica* and *Indica* rice samples.

Constituent	Units	Non-Parboiled *Japonica* (n = 3)	Parboiled *Japonica* (n = 3)	Parboiled *Indica* (n = 3)	Reported Values in Literature (Range)
		Mean (SD)	Mean (SD)	Mean (SD)	
Crude Protein	g/100 g	7.5 (0.23)	8.1 (0.11)	9.1 (0.03)	5.71–6.71 [a]; 7.21–8.53 [b]
Crude Fat	g/100 g	0.4 (0.23)	0.4 (0.23)	0.5 (0.28)	0.31–0.47 [a]
Ash	g/100 g	0.5 (0.24)	0.5 (0.26)	0.5 (0.19)	0.49–0.60 [a]

Table 4. Cont.

Constituent	Units	Non-Parboiled *Japonica* (*n* = 3)	Parboiled *Japonica* (*n* = 3)	Parboiled *Indica* (*n* = 3)	Reported Values in Literature (Range)
		Mean (SD)	Mean (SD)	Mean (SD)	
Carbohydrates	g/100 g	91.6 (0.28)	90.9 (0.54)	89.9 (0.44)	
Fe	mg/100 g	0.8 (0.18)	0.9 (0.23)	1.0 (0.13)	0.12–0.78 [a]
Mg	mg/100 g	1.6 (0.04)	1.5 (0.09)	0.5 (0.02)	
Zn	mg/100 g	2.4 (0.11)	2.3 (0.08)	1.6 (0.20)	0.84–1.46 [a]

([a]) Kwarteng et al. [51] and ([b]) Heinemann et al. [50].

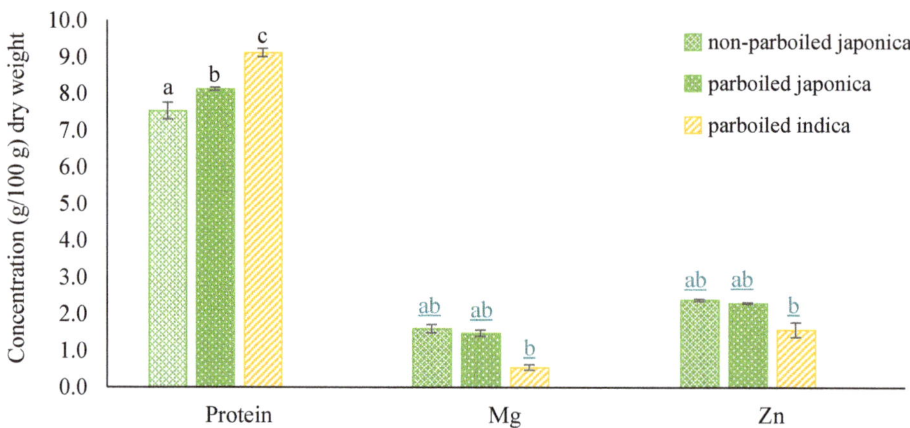

Figure 5. Bar graphs showing mean protein and micronutrient (Mg and Zn, mg/100 g) concentrations in non-parboiled *Japonica* (*n* = 3), parboiled *Japonica* (*n* = 3), and *Indica* (*n* = 3) rice samples. Error bars show standard deviations. Mean concentrations with different letters are significantly different ($p \leq 0.05$).

d.   Color of milled parboiled and non-parboiled rice samples

The degrees of whiteness (L) and yellowness (b) were similar among all three uncooked rice sample types. Whiteness values ranged between 18.8 in parboiled *Japonica* and 24.2 in *Indica* rice (Figure 6A). The degree of redness was significantly different ($p \leq 0.05$) in all three uncooked rice sample types with average values of 1.2 ± 0.35, 2.7 ± 0.82, and 3.22 ± 0.42 for parboiled *Indica*, non-parboiled *Japonica*, and parboiled *Japonica*, respectively. The whiteness increased when samples were cooked with *Indica*, still scoring the highest at 45.5 ± 0.54, although the differences among all cooked rice types were not statistically significant, $p \leq 0.05$ (Figure 6B). Redness changed into greenness in all three samples when cooked, but the differences remained significant ($p \leq 0.05$). Redness values recorded were −5.10 ± 0.39, 5.26 ± 0.29, and −3.92 ± 0.34 for parboiled *Indica*, non-parboiled *Japonica*, and parboiled *Japonica*, respectively.

3.3. Growing Conditions of Japonica Rice in Different Rice-Consuming Regions

The climatic factors important for *Japonica* rice cultivation (temperature, sunshine, and precipitation) which have been reported in previous studies [45,46]. for five countries are presented in Figure 7. The Republic of Korea and Japan are the two countries with the highest global cultivation of *Japonica* rice. India, Bangladesh, and Nigeria are the three countries with the highest levels of malnutrition globally. Published data on the greenhouse gas impacts of cultivating the two major varieties of rice (*Indica* and *Japonica*) were also reviewed and the implications for food-insecure regions are discussed below.

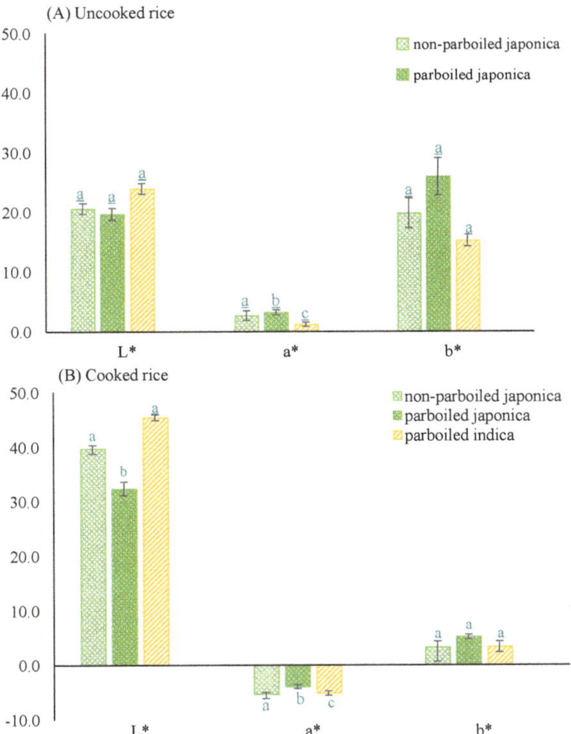

Figure 6. Bar graphs showing mean CIE L*a* b* in (**A**) cooked and (**B**) uncooked non-parboiled *Japonica* (*n* = 5), parboiled *Japonica* (*n* = 5), and *Indica* (*n* = 5) rice samples. Error bars show standard deviations. Means with different letters are significantly different ($p \leq 0.05$).

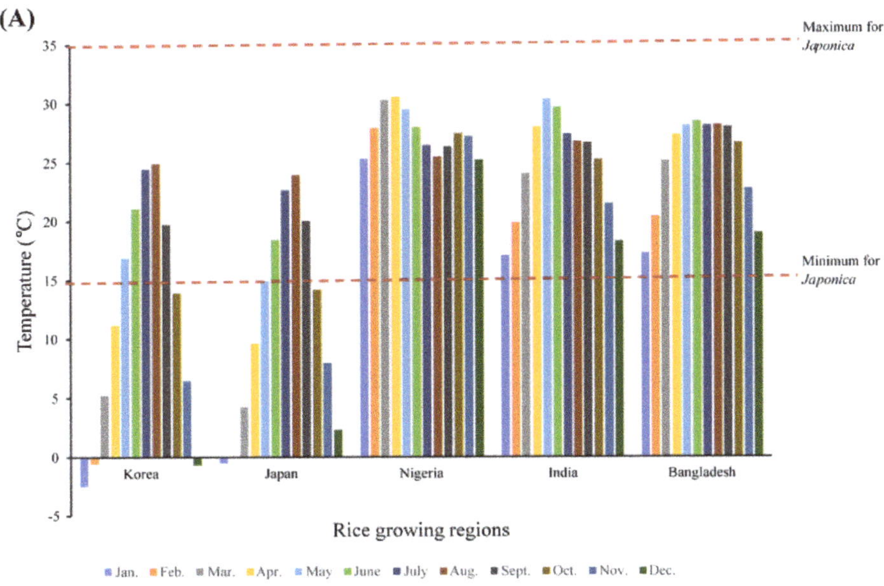

Figure 7. *Cont.*

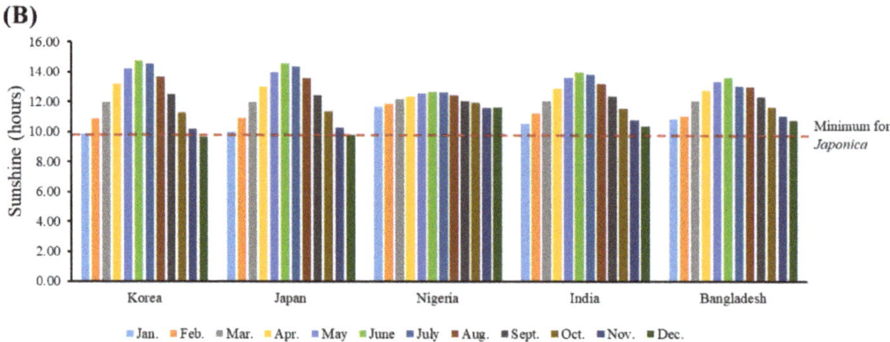

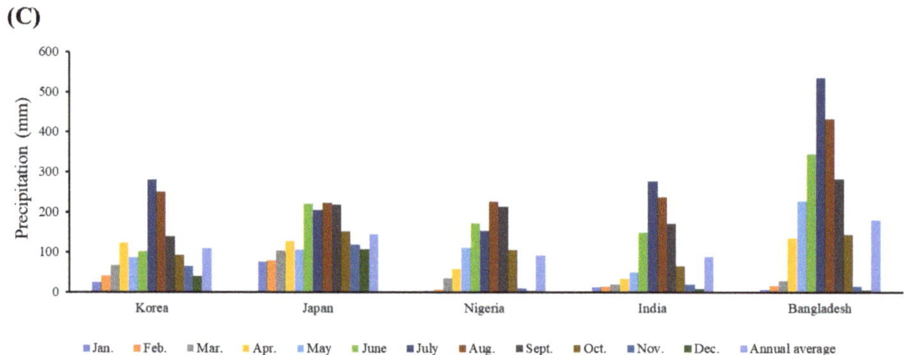

Figure 7. Major factors for growing *Japonica* rice: (**A**) mean temperature, (**B**) sunshine, and (**C**) precipitation [45,46].

4. Discussion

4.1. Rice Quality Attributes Preferred by Surveyed Consumers and the Effects of Parboiling on Physicochemical Properties

A larger proportion of the *Indica* rice-consuming population surveyed preferred non-stickiness (non-adhesiveness) and the pleasant aroma of parboiled rice. The survey also showed that the reduced level of adhesiveness and aroma that comes with parboiled rice [50] formed important criteria in determining eating quality. The majority (about 80%) of respondents were willing to pay more for rice that was seen as non-adhesive. The main reasons given by survey respondents for their choices included ease in cooking, taste satisfaction, customs and traditions, and appearance after cooking.

As the parboiling treatment has been demonstrated to alter the physicochemical properties of *Japonica* rice [50] and optimize rice quality parameters [51,52], the impacts of the parboiling treatment on preferred characteristics were further investigated in *Japonica* rice.

Amylose content in rice correlates with its adhesiveness and is seen as a very important rice consumption parameter. The *Japonica* rice variety usually contains low amylose, which is associated with increased adhesiveness. Parboiling increased the amylose content at all level of treatment in this study. Chinnaswamy and Hanna [53] recorded amylose content of 16.1% for non-parboiled *Japonica* rice (control) and 23.1% for *Indica* rice. The same study saw elevated rice amylose levels with increasing processing temperature.

Adequate steaming and soaking of rice before parboiling has been shown to eliminate the breakage of rice kernels caused by internal cracks acquired during maturity that are invisible to the naked eye. Soaking enables the void spaces in the hull to be filled with water [54], causing water absorption and subsequent swelling of the starch granules. Furthermore, rapid and uniform water absorption is enabled by soaking [55] and can be controlled with temperature and duration of soaking to achieve the desired moisture content. Similar to soaking, the results of steaming are also dependent on the temperature of the steam and its duration but, in addition, pressure plays a role in the final quality of the steamed paddy [56]. Steaming has also been recorded to greatly impact the final color of the milled rice and the head rice (whole grain) yield because it helps in hardening the rice kernel after soaking [57]. The rice quality from steaming is dependent on the prior soaking process [58]. Improper hydration of the paddy prior to steaming results in a steep moisture gradient from the outer to the inner core layers of the paddy, leading to cracks rather than hardening of the grain [59]. These cracks will result in a substantial loss of head rice. This implies that the two important factors affecting the adhesiveness of the rice also have substantial impacts on the head rice recovery rate.

The average ranks in a Kruskal–Wallis test showed that *Indica* rice differed the most from the average rank for all observations of alkali digestion and was lower than the overall median. Alkali digestion was inferred to be an influencer of the gelatinization temperature of rice [60].

With the exception of Mg, the chemical components measured in this study were comparable to values reported for other rice varieties [50,51]. In addition to pretreatment, protein and micronutrient concentrations in rice may be attributed to differences in genetic makeup, soil-water nutrient availability, and agronomic conditions during cultivation and harvesting. Conflicting results have been reported by studies of linkages between protein and micronutrient concentrations in rice. In a study of the micronutrient content of 274 milled rice genotypes, Jiang et al. [61] reported significant linkages between rice protein content and the micronutrients Mg and Zn, whereas Yang et al. [62] found no associations in a study of 285 polished rice grain genotypes. In the current study, the Mg and Zn contents were both higher in the *Japonica* rice samples (Figure 5). This indicated that *Japonica* rice could be an important source of micronutrients in helping solve hidden hunger.

In terms of visual appeal, rice color is an important measure, as reported in this study's survey and supported by [63]. White rice was preferred among consumers. The degree of whiteness increased when samples were cooked. Although *Indica* scored the highest, the differences among all cooked rice types were not statistically significant.

4.2. Growing Conditions Conducive for Japonica Rice in Different Rice-Consuming Regions

Since the domestication of Asian rice, it has been cultivated under diverse agroecological systems to meet different human demands [64,65]. This has resulted in a wide genetic diversity in rice around the world, as shown by molecular tools such as the analysis of restriction fragment length polymorphisms [66], simple sequence repeats [67], and genome-wide single nucleotide polymorphisms [68]. Consequently, many rice varieties with different characteristics have arisen under natural and human selection [69]. Although *Japonica* and *Indica* sub-species have been reported to have the same origin [70,71], they have been domesticated under widely different environmental conditions. Rice has evolved into different types, including upland and lowland paddy rice [72], as well as glutinous and non-glutinous rice [73]. Among these, *Indica* and *Japonica* rice represent the most important differentiated sub-species. For example, *Japonica* cultivars are mainly cultivated in temperate environments at high latitudes or altitudes with cool climatic conditions. In contrast, *Indica* cultivars are usually grown in tropical and sub-tropical regions at low latitudes and altitudes [74].

Indica and *Japonica* represent two partially isolated gene pools that serve as extremely valuable genetic resources in the improvement of rice varieties [75]. Being adapted to different environments, *Indica* and *Japonica* varieties have developed diverse morphological, agronomical, physiological, and molecular characteristics [76] that provide valuable genetic resources for breeding high-yielding rice [76]. Efficient utilization of such resources relies on understanding the genetic differentiation and geographical distribution patterns.

In [77], the relationships between average temperatures (range between 14 °C and 33 °C) at 17 rice cultivating sites with available precipitation and sunlight data and their latitudes (between 2° S to 44.5° N) were analyzed with a general linear model. This was carried out during the rice planting season (April–October) over 30 years (1961–1990). Furthermore, to estimate the relationships between the distribution of *Indica*, *Japonica*, and intermediate rice varieties and temperature, the average temperature correlation from nine latitude ranges against the percentage of alleles for every type of rice was analyzed. Results from [76] observed that as the latitude increased, the average temperature during the rice-growing season decreased. The study went on to examine the linkages between temperature and the spatial distribution of the different types of rice (i.e., *Indica*, *Japonica*, and intermediate) and found that, up to a point, with increasing temperature (and decreasing latitude), the proportion of *Indica* rice (in terms of total area cultivated) increased. This increase peaked at around 26 °C (corresponding to 30° N) and thereafter decreased with a further increase in latitudinal temperature, indicating that the distribution of *Indica* rice was not consistently correlated with the change in temperature. In contrast, *Japonica* rice varieties from different latitude ranges and their corresponding temperatures showed a negative correlation. This result indicated that *Japonica* rice, usually grown in temperate regions, is highly sensitive to temperature changes in terms of its geographical distribution across different latitudes.

Japonica rice is prevalently grown under sub-tropical and temperate climatic conditions, but in recent years, new strains have been adapted to tropical conditions and produced in countries such as Thailand and Vietnam. This could imply that the market segmentation into *Japonica* and *Indica* rice may become less marked over the longer term, with a larger number of countries able to shift from *Indica* to *Japonica* rice varieties or vice versa, as prices change.

Temperature is a major factor for rice cultivation, with its different growth stages varying in requirements. These include temperature for germination as follows: six days at 25 °C and two days at 27 °C to 37 °C. Lower temperatures take longer and the seedlings will not grow when it exceeds 40 °C [77]. At 30–35 °C, the plant height increases, the rate of grain growth is faster, and a shorter grain-filling period occurs [78,79]. In [80], low temperatures were reported to have a significant negative effect on all growth stages of rice of different varieties. In [80], three temperature classes (maximum, minimum, and optimum) for growing *Japonica* rice varieties were determined using leaf elongation rate. The base temperature (minimum) reported was between 13 °C and 14 °C, the maximum was between 38 °C and 42 °C, and the optimum was between 28 °C and 31 °C in most varieties.

Japonica rice is usually cultivated between May and September in the Republic of Korea and Japan. For five years (2012 to 2016), the temperatures required for rice plant development at different growth stages in the major *Japonica* rice-cultivating countries were recorded [80]. The Republic of Korea had a mean maximum temperature of 24.9 °C in August and the lowest of 16.9 °C in May (Figure 7A). As May is usually the first month of cultivation, this indicated that the farmers started the planting season in mild temperatures, which is suitable for germination but not necessarily the optimal, as stated above. A similar trend was recorded in Japan, with minimum and maximum mean temperatures ranging from 15.0 °C to 24.0 °C in May and August, respectively (Figure 7A). To achieve better performance, rice cultivation in Korea and Japan generally begins in nurseries with controlled temperatures and seedlings transplanted into fields as the weather warms during the planting season. The high energy costs and associated greenhouse gas contributions would be expected to be higher than in the tropical regions studied where controlled environments are not needed.

The amount of daily sunshine hours in the Republic of South Korea was a minimum of 13.7 h in August and 14.8 h in June, with a similar trend in Japan (Figure 7B). The Republic of South Korea and Japan had average annual precipitation of 109.2 mm and 144.8, respectively, as shown in Figure 7C.

In examining the temperatures in Africa (Figure 7A), the highest mean temperature in Nigeria for a period of five years (2012–2016) was 30.6 °C, and the lowest was 25.5 °C. This fell within the optimal temperature range required for *Japonica* rice [80]. The lowest and highest mean temperatures between 2012 and 2016 for Bangladesh and India were

17.3 °C and 28.5 °C, and 17.1 °C and 29.7 °C, respectively (Figure 7A). These data show a favorable climate for growing *Japonica* rice year-round in these countries with the world's largest populations of resource-poor and malnourished people.

Furthermore, *Japonica* rice could be cultivated at least twice a year in a four-month cultivation cycle in tropical countries. In [81], it was reported that more sunlight would improve the productivity of the rice. Rice is recorded to need 9–12 h of light daily [81], but no specific amounts have been given for *Japonica* rice to the authors' knowledge. Bangladesh, India, and Nigeria had the required amount of daily sunlight for rice growth all year round, as shown in Figure 7B. The only constraint in cultivating *Japonica* rice year-round in the high rice-consuming countries with vulnerable people might be the availability of water (Figure 7C). This problem could be mitigated by constructing dams and storing water during the peak raining periods for irrigation.

4.3. The Potential Contribution of Japonica Varieties to Greenhouse Gas Reduction

Agriculture had an estimated emission of 5.1 to 6.1 carbon dioxide equivalents per year (Gt CO_2-eq/yr) in 2005. This is 10–12% of the global GHG total. Furthermore, about 60% and 50% of N_2O and CH_4, respectively, of global anthropogenic emissions came from agriculture [82]. Growing rice contributes 25–300 Tg/year, amounting to 10–20% of global methane emissions [83]. In [84,85], South East Asia, where *Indica* rice is massively cultivated, emitted about 10,000 kg of methane per square kilometer, contributing about 90% of the global total methane emissions from rice cultivation. Methane, a potent greenhouse gas, is produced in enormous quantities by bacteria in waterlogged rice fields through anaerobic decomposition of organic material. Nitrous oxide, a powerful greenhouse gas, is also produced by soil microbes in rice fields, but its impact on global warming has been given lesser attention than CH_4 emissions from rice fields. The intensity of N_2O emissions is related to nitrogen (N) fertilizer application rates [86].

Worldwide measurements show that there are large temporal variations in CH_4 fluxes, and these differ distinctly with soil type and texture and the application of organic matter and mineral fertilizer [87]. The reduction in CO_2 with H_2, fatty acids, or alcohols as hydrogen donors, and the transmethylation of acetic acid or methanol by methane-producing bacteria, are the major pathways of CH_4 production in flooded soils. In paddy fields, the kinetics of the reduction processes are strongly affected by the composition and texture of soil and its content of inorganic electron acceptors [88].

There are three main processes of CH_4 release into the atmosphere from rice fields: diffusion loss of CH_4 across the water surface (least significant), as bubbles (ebullition) from paddy soils (important during land clearing and especially if the soil texture is not clayey), and CH_4 transport through rice plants. Plant transport of methane is the most critical loss mechanism, with more than 90% of total CH_4 emitted during the cropping season being released by diffusive or lacunae transport through the aerenchyma system of the rice plants [89,90]. Rice plants develop an intercellular gas space system, the aerenchyma, which provides roots submerged in inundated soils with oxygen (O_2). This gas space system also enables the transport of other gases, including CH_4, N_2O, and carbon dioxide (CO_2), from the soil/sediment to the atmosphere. Consequently, the variation among rice varieties with different growth and developmental progress could result in differences in total GHG emissions among rice varieties [91].

The increasing demand for rice in the future has raised concerns about increasing GHG emissions [92,93]. Considering the importance of rice as a staple food crop which provides more calories to the global population than any other single crop, the cultivation of more climate-friendly varieties would help mitigate the impacts of greenhouse warming that might be associated with increased production.

In [89], substantial differences were observed in the impacts of rice variety type on greenhouse gas emissions, rice yield, and global warming potential (GWP). The CH_4 and N_2O emissions were 6356 and 267 kg CO_2e/ha for *Indica* rice varieties and 4845 and 295 kg CO_2e/ha for *Japonica* rice varieties, respectively. A statistically significant higher yield-scaled GWP occurred in *Indica* rice varieties (1101.72 kg CO_2e/ha) than *Japonica* rice varieties (711.38 kg CO_2e/ha), indicating that the *Japonica* rice varieties released less GHGs with higher yields. The inverse relationship between rice yield increases and GHG emission reductions will aid cropping technique innovation for rice production in achieving higher yield while lowering emissions.

5. Conclusions

Parboiling retained the micronutrient profile of the treated *Japonica* rice, indicating that it could be a potential solution to hidden hunger in rice-consuming countries. Adhesiveness was also reduced in treated *Japonica* rice, thereby providing a quality that most survey respondents found to be desirable. The increasing human population is expected to increase rice demand worldwide. With higher yields, the cultivation of *Japonica* rice varieties will likely have lower GHG emissions compared to *Indica*. With *Japonica* rice being more affected by temperature than geography, parts of Africa, India, and Bangladesh would have conducive weather for growing it. The adoption of the more nutritious higher-yielding *Japonica* rice has the potential to help strengthen global food security and reduce hidden hunger, particularly in the world's five most impoverished countries, while also reducing the GHG emissions of global rice cultivation.

Author Contributions: D.D.U.: Conceptualization, Methodology, Software, Investigation, Formal analysis, Data Curation, Visualization, and Writing—original draft. S.A.-H.: Methodology, Software, Data Analysis, Visualization, and Writing. T.P., Y.H.: Investigation, Validation, Resources, Writing—review and editing, Project administration, Supervision, and Funding acquisition. K.-M.K., A.M., and S.W.: Methodology, Investigation, and Data Curation. All authors have read and agreed to the published version of the manuscript.

Funding: This work was supported by Korea Institute of Planning and Evaluation for Technology in Food, Agriculture and Forestry (IPET) through Agriculture, Food and Rural Affairs Convergence Technologies Program for Educating Creative Global Leader, funded by Ministry of Agriculture, Food and Rural Affairs (MAFRA), Project Nos. (320001-4) and (716001-7), Republic of Korea.

Data Availability Statement: The data presented in this study are available in the article.

Conflicts of Interest: The authors declare no conflict of interest.

Appendix A

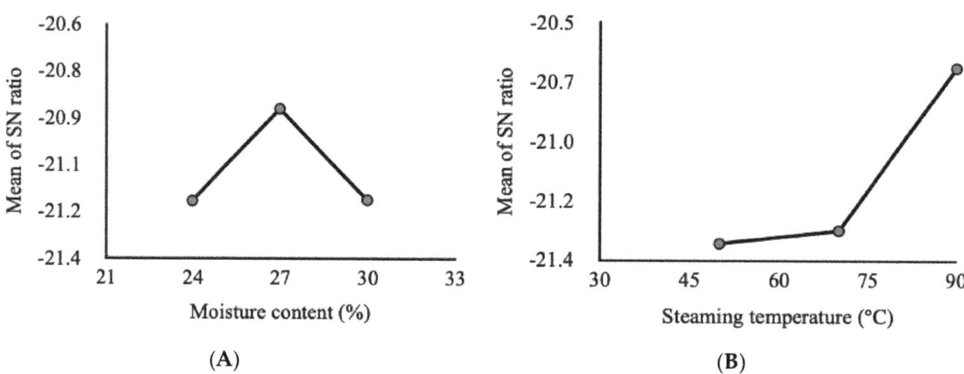

Figure A1. *Cont.*

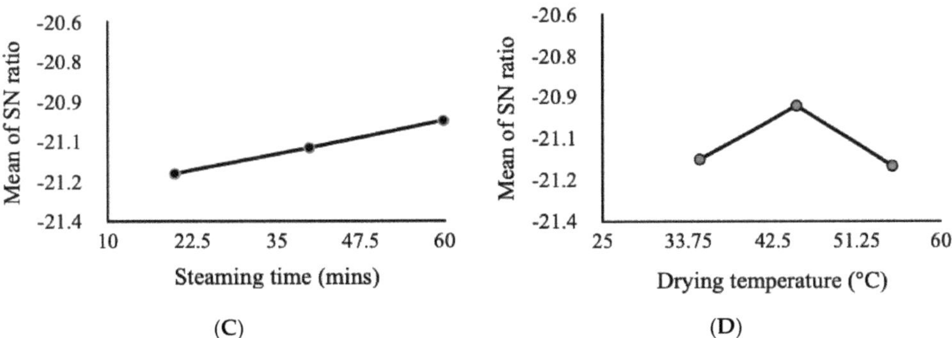

Figure A1. Main effects plot (data means) for SN ratios showing factors with the most impact on adhesiveness of *Japonica* rice: (**A**) moisture content, (**B**) steaming temperature, (**C**) steaming time, (**D**) drying temperature.

Appendix B

- **(Survey Questionnaire)**

  Personal Data

- Age:
- Gender:
- Occupation:
- Country of residence:
- State of residence:

  Survey questions

1. Do you know there are varieties of rice? Yes/No/Not sure
   a. If yes, what type(s):
   b. Which of the above type do you prefer?

2. The length of rice is important to me. (a) Strongly Agree (b) Agree (c) Neither Agree nor Disagree (d) Disagree (e) Strongly Disagree

3. Why is the length of rice important to you?
4. Adhesiveness (stickiness) of rice is important to me. (a) Strongly Agree (b) Agree (c) Neither Agree nor Disagree (d) Disagree (e) Strongly Disagree

5. Why is the adhesiveness of rice important to you?
6. Why is the adhesiveness of rice not important to you?
7. The aroma of rice is important to me. (a) Very Important (b) Important (c) Neither Important nor Unimportant (d) Unimportant (e) Very Unimportant

8. The price of rice is important in your decision for selecting the brand of rice to buy. (a) Very Important (b) Important (c) Neither Important nor Unimportant (d) Unimportant (e) Very Unimportant

9. What do you look out for when selecting rice? (Multiple selections is applicable) (a) Rice brand (b) Length of the grain (c) Adhesiveness after cooking (d) Aroma (e) Price (f) Others (please list)

10. What is the maximum you could pay extra per 10kg bag of rice if you knew it was healthier? (a) Less than ₦1000 (b) ₦1000–2000 (c) ₦3000–4000 (d) More than ₦5000

11. Which source of rice do you prefer? (a) Foreign (b) Local (c) Any

12. Why do you prefer your option in question 11 above?

References

1. Katayama, R.; Wadhwa, D. Half of the World's Poor Live in Just 5 Countries. Available online: https://blogs.worldbank.org/opendata/half-world-s-poor-live-just-5-countries (accessed on 1 August 2021).
2. Index Mondi. Milled Rice Domestic Consumption by Country in 1000 MT—Country Rankings. 2021. Available online: https://www.indexmundi.com/agriculture/?commodity=milled-rice&graph=domestic-consumption (accessed on 1 August 2021).
3. FAOSTAT. Countries by Commodity. Available online: http://www.fao.org/faostat/en/#rankings/countries_by_commodity_imports (accessed on 1 August 2021).
4. World Bank. Nutrition Country Profiles. Available online: https://www.worldbank.org/en/topic/health/publication/nutrition-country-profiles (accessed on 1 August 2021).
5. FAO. The State of Food Security and Nutrition in the World 2019. Available online: http://www.fao.org/3/ca5162en/ca5162en.pdf (accessed on 1 August 2021).
6. FAO. The State of Food Security and Nutrition in the World 2021. Available online: http://www.fao.org/state-of-food-security-nutrition (accessed on 1 August 2021).
7. Von Grebmer, K.; Bernstein, J.; Prasai, N.; Yin, S.; Yohannes, Y.; Towey, O.; Sonntag, A.; Neubauer, L.; de Waal, A. *The Concept of the Global Hunger Index*; IFPRI: Washington, DC, USA, 2015; pp. 6–11.
8. Bhutta, Z.A. Micronutrient needs of malnourished children. *Curr. Opin. Clin. Nutr. Metab. Care* **2008**, *11*, 309–314. [CrossRef]
9. Fraser, D.R. Vitamin D-deficiency in Asia. *J. Steroid Biochem. Mol. Biol.* **2004**, *89*, 491–495. [CrossRef]
10. Chadha, M.; Oluoch, M. Home-based vegetable gardens and other strategies to overcome micronutrient malnutrition in developing countries. *Food Nutr. Agric.* **2003**, *32*, 17–23.
11. Chauhan, B.S.; Jabran, K.; Mahajan, G. *Rice Production Worldwide*; Springer: Berlin/Heidelberg, Germany, 2017; Volume 247.
12. OECD. *OECD-FAO Agricultural Outlook 2019–2028*; OECD: Paris, France, 2019.
13. Kubo, M.; Purevdorj, M. The future of rice production and consumption. *J. Food Distrib. Res.* **2004**, *35*, 128–142.
14. Olivier, J.G.; Schure, K.; Peters, J. *Trends in Global CO_2 and Total Greenhouse Gas Emissions*; PBL Netherlands Environmental Assessment Agency: The Hague, The Netherlands, 2017.
15. World Bank. Land under Cereal Production. Available online: https://data.worldbank.org/indicator/AG.LND.CREL.HA (accessed on 1 August 2021).
16. Statista. Available online: https://www.statista.com/topics/1443/rice/ (accessed on 1 August 2021).
17. Singh, M.; Upadhyaya, H.D. *Genetic and Genomic Resources for Grain Cereals Improvement*; Academic Press: Cambridge, MA, USA; Elsevier: Amsterdam, The Netherlands, 2015.
18. Seck, P.A.; Tollens, E.; Wopereis, M.C.; Diagne, A.; Bamba, I. Rising trends and variability of rice prices: Threats and opportunities for sub-Saharan Africa. *Food Policy* **2010**, *35*, 403–411. [CrossRef]
19. Lançon, F.; Benz, H.D. *Rice Imports in West Africa: Trade Regime and Food Policy Formulation*; Flançon and H.David-Benz: Montpellier, France, 2007.
20. United Nations. Coronavirus: World Risks 'Biblical' Famines Due to Pandemic. Available online: https://www.bbc.com/news/world-52373888 (accessed on 1 August 2021).
21. Cadoni, P.; Angelucci, F. Analysis of incentives and disincentives for Rice in Nigeria. *Gates Open Res.* **2019**, *3*, 643.
22. Linares, O.F. African rice (*Oryza glaberrima*): History and future potential. *Proc. Natl. Acad. Sci. USA* **2002**, *99*, 16360–16365. [CrossRef] [PubMed]
23. NRC. *Lost Crops of Africa: Volume I: Grains*; National Academies Press: Cambridge, MA, USA, 1996.
24. OECD-FAO. Cereals. In *Agricultural Outlook*; OECD: Paris, France, 2018; pp. 109–125. Available online: http://www.fao.org/3/i9166e/i9166e_Chapter3_Cereals.pdf (accessed on 1 August 2021).
25. Champagne, E.T.; Bett-Garber, K.L.; Fitzgerald, M.A.; Grimm, C.C.; Lea, J.; Ohtsubo, K.i.; Jongdee, S.; Xie, L.; Bassinello, P.Z.; Resurreccion, A. Important sensory properties differentiating premium rice varieties. *Rice* **2010**, *3*, 270–281. [CrossRef]
26. Yanjie, X.; Yining, Y.; Shuhong, O.; Xiaoliang, D.; Hui, S.; Shukun, J.; Shichen, S.; Jinsong, B. Factors affecting sensory quality of cooked japonica rice. *Rice Sci.* **2018**, *25*, 330–339. [CrossRef]
27. Odenigbo, A.M.; Ngadi, M.; Ejebe, C.; Woin, N.; Ndindeng, S.A. Physicochemical, cooking characteristics and textural properties of TOX 3145 milled rice. *J. Food Res.* **2014**, *3*, 82. [CrossRef]
28. Chemutai, L.; Musyoki, M.; Kioko, W.; Mwenda, N.; Muriira, K.; Piero, N. Physicochemical characterization of selected Rice (*Oryza Sativa* L.) genotypes based on gel consistency and alkali digestion. *Anal. Biochem.* **2016**, *5*, 285. [CrossRef]
29. Okadome, H. Application of instrument-based multiple texture measurement of cooked milled-rice grains to rice quality evaluation. *Jpn. Agric. Res. Q. JARQ* **2005**, *39*, 261–268. [CrossRef]
30. Oli, P.; Ward, R.; Adhikari, B.; Torley, P. Parboiled rice: Understanding from a materials science approach. *J. Food Eng.* **2014**, *124*, 173–183. [CrossRef]
31. Mir, S.A.; Bosco, S.J.D. Effect of soaking temperature on physical and functional properties of parboiled rice cultivars grown in temperate region of India. *Food Nutr. Sci.* **2013**, *4*, 282. [CrossRef]
32. Bello, M.; Baeza, R.; Tolaba, M. Quality characteristics of milled and cooked rice affected by hydrothermal treatment. *J. Food Eng.* **2006**, *72*, 124–133. [CrossRef]

33. da Fonseca, F.A.; Soares Júnior, M.S.; Caliari, M.; Bassinello, P.Z.; da Costa Eifert, E.; Garcia, D.M. Changes occurring during the parboiling of upland rice and in the maceration water at different temperatures and soaking times. *Int. J. Food Sci. Technol.* **2011**, *46*, 1912–1920. [CrossRef]
34. Miah, M.K.; Haque, A.; Douglass, M.P.; Clarke, B. Parboiling of rice. Part II: Effect of hot soaking time on the degree of starch gelatinization. *Int. J. Food Sci. Technol.* **2002**, *37*, 539–545. [CrossRef]
35. Juliano, B.O. The rice grain and its gross composition. In *Rice: Chemistry and Technology*; American Association of Cereal Chemists: St. Paul, MN, USA, 1985.
36. Pedersen, B.; Eggum, B. The influence of milling on the nutritive value of flour from cereal grains. 4. Rice. *Plant Foods Hum. Nutr.* **1983**, *33*, 267–278. [CrossRef]
37. Adhikari, B.; Howes, T.; Bhandari, B.; Truong, V. Stickiness in foods: A review of mechanisms and test methods. *Int. J. Food Prop.* **2001**, *4*, 1–33. [CrossRef]
38. Uyeh, D.D.; Woo, S.M.; Hong, D.H.; Ha, Y.S. Optimization of parboiling conditions for enhanced Japonica rice milling. *Emir. J. Food Agric.* **2016**, *28*, 764–771. [CrossRef]
39. Ha, Y.; Park, K.; Kim, H.; Hong, D.; Nah, K.; Seo, S. Milled rice recovery rate of paddy with various moisture contents. *J. Korean Soc. Agric. Mach.* **2002**, *27*, 125–132.
40. Tang, J.; Sokhansanj, S. Determination of moisture content of whole kernel lentil by oven method. *Trans. ASAE* **1991**, *34*, 255–0256. [CrossRef]
41. Sonsanguan, N.; Sirisomboon, P. Determination of stickiness characteristic of cooked rice samples in rice industry. In Proceedings of the 14th TSAE National Conference and the 6th TSAE International Conference, Hua Hin, Thailand, 1–4 April 2013; pp. 151–153.
42. Choe, Z.R.; Heu, M.H. Optimum conditions for alkali digestibility test in rice. *J. Korean Crop. Soc.* **1975**, *19*, 7–13.
43. IRRI. *Standard Evaluation System for Rice*; IRRI: Los Baños, Philippines, 1996.
44. AOAC. *Official Methods of Analysis*, 21st ed.; AOAC: Rockville, MD, USA, 2019.
45. Cordero-Lara, K.I. Temperate japonica rice (*Oryza sativa* L.) breeding: History, present and future challenges. *Chil. J. Agric. Res.* **2020**, *80*, 303–314. [CrossRef]
46. Dhanushka Perera, J. Analysis of the Variability of Cardinal Temperatures in Rice; UM2: 2011. Available online: https://agritrop.cirad.fr/570279/1/document_570279.pdf (accessed on 1 April 2021).
47. Hu, Y.-J.; Pei, W.; Zhang, H.-C.; Dai, Q.-G.; Huo, Z.-Y.; Ke, X.; Hui, G.; Wei, H.-Y.; Guo, B.-W.; Cui, P.-Y. Comparison of agronomic performance between inter-sub-specific hybrid and inbred japonica rice under different mechanical transplanting methods. *J. Integr. Agric.* **2018**, *17*, 806–816. [CrossRef]
48. Huang, Y.; Gao, L.; Jin, Z.; Chen, H. Simulating the optimal growing season of rice in the Yangtze River Valley and its adjacent area, China. *Agric. For. Meteorol.* **1998**, *91*, 251–262. [CrossRef]
49. Ueno, K.; Miyoshi, K. Difference of optimum germination temperature of seeds of intact and dehusked japonica rice during seed development. *Euphytica* **2005**, *143*, 271–275. [CrossRef]
50. Heinemann, R.J.B.; Fagundes, P.d.L.; Pinto, E.A.; Penteado, M.d.V.C.; Lanfer-Marquez, U. Comparative study of nutrient composition of commercial brown, parboiled and milled rice from Brazil. *J. Food Compos. Anal.* **2005**, *18*, 287–296. [CrossRef]
51. Adu-Kwarteng, E.; Ellis, W.O.; Oduro, I.; Manful, J.T. Rice grain quality: A comparison of local varieties with new varieties under study in Ghana. *Food Control* **2003**, *14*, 507–514. [CrossRef]
52. Islam, M.R.; Roy, P.; Shimizu, N.; Kimura, T. Effect of Processing Conditions on Physical Properties of Parboiled Rice. *Food Sci. Technol. Res.* **2002**, *8*, 106–112. [CrossRef]
53. Chinnaswamy, R.; Hanna, M.A. Expansion, Color and Shear Strength Properties of Com Starches Extrusion-Cooked with Urea and Salts. *Starch Stärke* **1988**, *40*, 186–190. [CrossRef]
54. Ali, N.; Pandya, A.C. Basic concept of parboiling of paddy. *J. Agric. Eng. Res.* **1974**, *19*, 111–115. [CrossRef]
55. Wimberly, J.E.; International Rice Research, I. *Technical Handbook for the Paddy Rice Postharvest Industry in Developing Countries*; International Rice Research Institute: Los Baños, Philippines, 1983.
56. Gariboldi, F.; Food and Agriculture Organization of the United. *Rice Parboiling*; Food and Agriculture Organization of the United Nations: Rome, Italy, 1984.
57. Adhikaritanayake, T.B.; Noomhorm, A. Effect of continuous steaming on parboiled rice quality. *J. Food Eng.* **1998**, *36*, 135–143. [CrossRef]
58. Sareepuang, K.; Siriamornpun, S.; Wiset, L.; Meeso, N. Effect of soaking temperature on physical, chemical and cooking properties of parboiled fragrant rice. *World J. Agric. Sci.* **2008**, *4*, 409–415.
59. Graham-Acquaah, S.; Manful, J.T.; Ndindeng, S.A.; Tchatcha, D.A. Effects of Soaking and Steaming Regimes on the Quality of Artisanal Parboiled Rice. *J. Food Process. Preserv.* **2015**, *39*, 2286–2296. [CrossRef]
60. Fan, C.; Yu, X.; Xing, Y.; Xu, C.; Luo, L.; Zhang, Q. The main effects, epistatic effects and environmental interactions of QTLs on the cooking and eating quality of rice in a doubled-haploid line population. *Theor. Appl. Genet.* **2005**, *110*, 1445–1452. [CrossRef] [PubMed]
61. Jiang, S.L.; Wu, J.G.; Feng, Y.; Yang, X.E.; Shi, C.H. Correlation Analysis of Mineral Element Contents and Quality Traits in Milled Rice (*Oryza stavia* L.). *J. Agric. Food Chem.* **2007**, *55*, 9608–9613. [CrossRef] [PubMed]
62. Yang, W.; Peng, S.; Dionisio-Sese, M.L.; Laza, R.C.; Visperas, R.M. Grain filling duration, a crucial determinant of genotypic variation of grain yield in field-grown tropical irrigated rice. *Field Crop. Res.* **2008**, *105*, 221–227. [CrossRef]

63. Vaughan, D.A.; Lu, B.-R.; Tomooka, N. The evolving story of rice evolution. *Plant Sci.* **2008**, *174*, 394–408. [CrossRef]

64. Zhang, Q.; Maroof, M.S.; Lu, T.Y.; Shen, B. Genetic diversity and differentiation of indica and japonica rice detected by RFLP analysis. *Theor. Appl. Genet.* **1992**, *83*, 495–499. [CrossRef] [PubMed]

65. Lu, B.-R.; Zheng, K.; Qian, H.; Zhuang, J. Genetic differentiation of wild relatives of rice as assessed by RFLP analysis. *Theor. Appl. Genet.* **2002**, *106*, 101–106. [CrossRef] [PubMed]

66. Zhao, K.; Wright, M.; Kimball, J.; Eizenga, G.; McClung, A.; Kovach, M.; Tyagi, W.; Ali, M.L.; Tung, C.-W.; Reynolds, A. Genomic diversity and introgression in O. sativa reveal the impact of domestication and breeding on the rice genome. *PLoS ONE* **2010**, *5*, e10780. [CrossRef] [PubMed]

67. Vaughan, D.; Balazs, E.; Heslop-Harrison, J. From crop domestication to super-domestication. *Ann. Bot.* **2007**, *100*, 893–901. [CrossRef] [PubMed]

68. Khush, G.S. Origin, dispersal, cultivation and variation of rice. *Plant Mol. Biol.* **1997**, *35*, 25–34. [CrossRef]

69. Huang, X.; Kurata, N.; Wang, Z.-X.; Wang, A.; Zhao, Q.; Zhao, Y.; Liu, K.; Lu, H.; Li, W.; Guo, Y. A map of rice genome variation reveals the origin of cultivated rice. *Nature* **2012**, *490*, 497. [CrossRef]

70. Yu, L.-X.; Nguyen, H.T. Genetic variation detected with RAPD markers among upland and lowland rice cultivars (*Oryza sativa* L.). *Theor. Appl. Genet.* **1994**, *87*, 668–672. [CrossRef] [PubMed]

71. Olsen, K.M.; Caicedo, A.L.; Polato, N.; McClung, A.; McCouch, S.; Purugganan, M.D. Selection under domestication: Evidence for a sweep in the rice waxy genomic region. *Genetics* **2006**, *173*, 975–983. [CrossRef] [PubMed]

72. Xiong, Z.; Zhang, S.; Wang, Y.; Ford-Lloyd, B.V.; Tu, M.; Jin, X.; Wu, Y.; Yan, H.; Yang, X.; Liu, P. Differentiation and distribution of indica and japonica rice varieties along the altitude gradients in Yunnan Province of China as revealed by InDel molecular markers. *Genet. Resour. Crop. Evol.* **2010**, *57*, 891–902. [CrossRef]

73. Lu, B.-R.; Cai, X.; Xin, J. Efficient indica and japonica rice identification based on the InDel molecular method: Its implication in rice breeding and evolutionary research. *Prog. Nat. Sci.* **2009**, *19*, 1241–1252. [CrossRef]

74. Morishima, H.; Oka, H.-I. Phylogenetic differentiation of cultivated rice, XXII. Numerical evaluation of the indica-japonica differentiation. *Jpn. J. Breed.* **1981**, *31*, 402–413. [CrossRef]

75. Peng, S.; Laza, R.C.; Visperas, R.M.; Khush, G.S.; Virk, P.; Zhu, D. Rice: Progress in breaking the yield ceiling. In Proceedings of the 4th International Crop Science Congress, Brisbane, Australia, 26 September–1 October 2004.

76. Xiong, Z.; Zhang, S.; Ford-Lloyd, B.; Jin, X.; Wu, Y.; Yan, H.; Liu, P.; Yang, X.; Lu, B.-R. Latitudinal distribution and differentiation of rice germplasm: Its implications in breeding. *Crop Sci.* **2011**, *51*, 1050–1058. [CrossRef]

77. Dubey, A.; Verma, S.; Goswami, S.; Devedee, A. Effect of Temperature on Different Growth Stages and Physiological Process of Rice crop—A Review. *Bull. Environ. Pharmacol. Life Sci.* **2018**, *7*, 129–136.

78. Kondo, M.; Okamura, T. Growth response of rice plant to water temperature. *Agric. Hortic.* **1931**, *6*, 517–530.

79. Osada, A.; Sasiprapa, V.; Rahong, M.; Dhammanuvong, S.; Chakrabndhu, H. Abnormal occurrence of empty grains of indica rice plants in the dry, hot season in Thailand. *Jpn. J. Crop. Sci.* **1973**, *42*, 103–109. [CrossRef]

80. Ghadirnezhad, R.; Fallah, A. Temperature Effect on Yield and Yield Components of Different Rice Cultivars in Flowering Stage. *Int. J. Agron.* **2014**, *2014*, 846707. [CrossRef]

81. Ikeda, K. Photoperiodic flower induction in rice plants as influenced by light intensity and quality. *JARQ. Jpn. Agric. Res. Q.* **1985**, *18*, 164–170.

82. Smith, P.; Martino, D.; Cai, Z.; Gwary, D.; Janzen, H.; Kumar, P.; McCarl, B.; Ogle, S.; O'Mara, F.; Rice, C.; et al. Agriculture. In *Climate Change 2007: Mitigation. Contribution of Working Group III to the Fourth Assessment Report of the Intergovernmental Panel on Climate Change*; Metz, B., Davidson, O.R., Bosch, P.R., Dave, R., Meyer, L.A., Eds.; Cambridge University Press: Cambridge, UK; New York, NY, USA, 2007. Available online: https://www.ipcc.ch/site/assets/uploads/2018/02/ar4-wg3-chapter8-1.pdf (accessed on 1 August 2021).

83. Geetha Thanuja, K.; Karthikeyan, S. Exploring bio-mitigation strategies to reduce carbon footprint in wetland paddy system. *Bioresour. Technol. Rep.* **2020**, *12*, 100557. [CrossRef]

84. Epule, E.T.; Peng, C.; Mafany, N.M. Methane emissions from paddy rice fields: Strategies towards achieving a win-win sustainability scenario between rice production and methane emission reduction. *J. Sustain. Dev.* **2011**, *4*, 188. [CrossRef]

85. Yan, X.; Akiyama, H.; Yagi, K.; Akimoto, H. Global estimations of the inventory and mitigation potential of methane emissions from rice cultivation conducted using the 2006 Intergovernmental Panel on Climate Change Guidelines. *Glob. Biogeochem. Cycles* **2009**, *23*. [CrossRef]

86. Zou, J.; Huang, Y.; Zheng, X.; Wang, Y. Quantifying direct N_2O emissions in paddy fields during rice growing season in mainland China: Dependence on water regime. *Atmos. Environ.* **2007**, *41*, 8030–8042. [CrossRef]

87. Neue, H.-U.; Sass, R.L. Trace gas emissions from rice fields. In *Global Atmospheric-Biospheric Chemistry*; Springer: Berlin/Heidelberg, Germany, 1994; pp. 119–147.

88. Conrad, R. Control of methane production in terrestrial ecosystems. In *Exchange of Trace Gases between Terrestrial Ecosystems and the Atmosphere*; Andreae, M.O., Schimel, D.S., Eds.; Wiley Chichester: Chichester, UK, 1989; pp. 39–58.

89. Seiler, W.; Holzapfel-Pschorn, A.; Conrad, R.; Scharffe, D. Methane emission from rice paddies. *J. Atmos. Chem.* **1983**, *1*, 241–268. [CrossRef]

90. Schütz, H.; Seiler, W.; Conrad, R. Processes involved in formation and emission of methane in rice paddies. *Biogeochemistry* **1989**, *7*, 33–53. [CrossRef]

91. Zheng, H.; Huang, H.; Yao, L.; Liu, J.; He, H.; Tang, J. Impacts of rice varieties and management on yield-scaled greenhouse gas emissions from rice fields in China: A meta-analysis. *Biogeosci. Discuss.* **2013**, *10*, 19045–19069.
92. van Beek, C.L.; Meerburg, B.G.; Schils, R.L.; Verhagen, J.; Kuikman, P.J. Feeding the world's increasing population while limiting climate change impacts: Linking N_2O and CH_4 emissions from agriculture to population growth. *Environ. Sci. Policy* **2010**, *13*, 89–96. [CrossRef]
93. Van Groenigen, K.J.; Van Kessel, C.; Hungate, B.A. Increased greenhouse-gas intensity of rice production under future atmospheric conditions. *Nat. Clim. Chang.* **2013**, *3*, 288. [CrossRef]

 *foods*

Article

Meat Traceability: Traditional Market Shoppers' Preferences and Willingness-to-Pay for Additional Information in Taiwan

Ardiansyah Azhary Suhandoko [1], Dennis Chia-Bin Chen [2] and Shang-Ho Yang [3,*]

[1] International Master Program of Agriculture, National Chung Hsing University, No. 145 Xingda Rd., South District, Taichung 40227, Taiwan; ianazhary@smail.nchu.edu.tw
[2] Massey College of Business, Belmont University, 1900 Belmont Boulevard, Nashville, TN 37212, USA; dennis.chen@belmont.edu
[3] Graduate Institute of Bio-Industry Management, National Chung Hsing University, No. 145 Xingda Rd., South District, Taichung 40227, Taiwan
* Correspondence: bruce.yang@nchu.edu.tw

Abstract: Due to food scandals that shocked the retailer markets, traceability systems were advocated to regain consumers' confidence and trust. However, while traceability systems can be more easily explored in modern markets, almost no traceability system can be found in traditional markets in Taiwan, especially when buying meat products. This study explored the preference and the willingness-to-pay (WTP) for traceability information of pork products in traditional markets in Taiwan. The random utility theory (RUT) with the contingent valuation method (CVM) was adopted to examine the total of 1420 valid responses in Taiwan. Results show that 80% of traditional market consumers are willing to pay more for traceability information of pork products. Specifically, when consumers (1) know the market price of pork, (2) do not often buy food in the traditional market, (3) live in south or north regions of Taiwan, (4) have a flexible buying schedule, (5) are aware of food safety due to frequently accessing health-related content through media, or (6) think pork grading is very important, they would tend to choose meat products with traceability information. The implication of this study suggests that there is an urgent desire for food safety labeling and traceability information system in traditional markets in Taiwan. Especially, those who usually shop in the higher-price markets are willing to pay the most for this traceability information.

Keywords: traditional market; traceability; willingness-to-pay; pork; Taiwan

Citation: Suhandoko, A.A.; Chen, D.C.-B.; Yang, S.-H. Meat Traceability: Traditional Market Shoppers' Preferences and Willingness-to-Pay for Additional Information in Taiwan. *Foods* **2021**, *10*, 1819. https://doi.org/10.3390/foods10081819

Academic Editors: António Raposo, Renata Puppin Zandonadi and Raquel Braz Assunção Botelho

Received: 4 July 2021
Accepted: 4 August 2021
Published: 6 August 2021

Publisher's Note: MDPI stays neutral with regard to jurisdictional claims in published maps and institutional affiliations.

1. Introduction

In the current industrialized and mass-produced food era, consumers have heightened safety awareness of the food they buy [1–3]. This attention often translates to the purchase behavior of safer products [4–6]. Furthermore, Taiwanese citizens particularly showed interest in food safety over time [7], especially when buying food in modern markets. These safety concerns stem from prior labeling deceptions, illicit ingredients, and food scandals around the globe. Therefore, the implementation of a food safety concept is necessary [8–11] since these concerns can significantly affect consumers' choices [12]. In Taiwan, past scandals, including avian influenza cases (2015) [13], mad cow disease, dioxin defilement, high percentages of pesticides remaining on agriculture products (the 1990s) [14,15], a codling moth larvae outbreak (1990s) [16,17], and the prominence of foot-and-mouth disease (1997) [18,19], lessen citizens' trust in food safety [20]. Therefore, food-safety-oriented consumers must be guaranteed safety and quality as they become more selective in buying.

When shoppers take food safety into account, they automatically consider food quality. These two dimensions often blend when shopping for food [21,22]. These concerns could be addressed through additional product information, certifications, and standards [23], as they can increase shoppers' confidence and their willingness-to-pay (WTP) [7,24]. The presence of these certifications parallels the urgency in applying a traceability system [25]. Therefore, the importance of a traceability system is quite high [26], particularly for producers who aspire towards excellence in food safety and food quality [11,27].

Traceability certification encompasses a system that allows entities along the supply chain to trace forward and back information, including the beginning to final production, multiple stages of processing, and multi-levels of distribution, with the inclusion of the dimensions of breadth, depth, and precision [28–34]. Due to its importance to the food industry, the most advanced traceability systems implement the Internet of Things (IoT) to detect problems from farm to table through the utilization of wearable devices (e.g., mobiles) in real-time [35]. From the consumers' perspective, traceability positively impacts the consumers' confidence level as observed in prior research [10,30,31,36–40]. Moreover, traceability systems saw years of development worldwide since 2002 for the US [31,41], since 2003 for Canada [42], and since 2005 for the EU [43–46]. In Taiwan, traceability has been discussed since 1990 with the Taiwan Good Agricultural Practice (TGAP) and the Good Manufacture Practice (GMP) [47]. However, Taiwan Agriculture and Food Traceability System (TAFTS) was officially established in 2004 [14,47,48]. The history of these systems demonstrates how traceability is favored in this modern era. Considering all these food safety concerns and development, Taiwanese consumers have critical perceptions on this matter, and those perceptions affect their WTP [7,20,49] when purchasing a product in certain markets. These traceability's positive impacts on consumers are found in restaurants, [25,33] modern markets [43,50], and supermarkets [10,49]. As to this day, its impact of traceability on consumers' preference in traditional markets is still limited.

The exigency of further traceability research in traditional markets is situated on the markets' importance to the Taiwanese economy (USD 5–6 billion annually in 2020 and increasing) [51,52]. An estimated 50,000 stalls are employing approximately 100,000 laborers [53]. Moreover, traditional markets have special characteristics attracting the general buyer more than other markets. These attributes include the chance to haggle the price [54–57], the ability to build a connection with vendors (increase social scope), the perceived higher quality level [58–61], and the perceived freshness [62]. Looking at these aspects, meat became the most purchased product in traditional markets, accounting for over 50% of purchases [62]. This might be a result of the cheaper prices, the availability to purchase specific cuts of meat, and the freshly slaughtered selections [55]. Moreover, in the last decade, the Taiwanese increased meat spending in their budget [63]. Especially, pork's popularity in Taiwan increased, which secures Taiwan as the fifth highest in the world in terms of consumption of pork, with 45 kg per capita each year [64].

Although meat traceability applications show promising results [27,43,65], prior research at the consumer stage was focused on consumers' attitudes, certainty, and acceptance of traceability [40,47,66–69]. However, when it comes to consumers' preferences (WTP), most observations of the positive effect of traceability centered around consumers of multiple other kinds of markets than traditional ones [10,25,33,43,49,50]. Lastly, due to the currently limited traceability research on pork products [50,70–75], this study which observes consumers' preference on traceability impact in Taiwanese traditional market for pork products helps fill the gap in research on this topic.

This research was conducted using a contingent valuation method to measure the consumers' fondness (or WTP) and how much more they will pay for pork with traceability information in the Taiwan traditional market [76,77]. The research consisted of questions posed in hypothetical market conditions [78–80]. The questionnaire has an open-ended format, which lets the respondents reply with their own words in assorted dimensions, and it is commonly used for assessing WTP with an abundant amount of information [81–83]. Furthermore, to understand the relationships among consumers' daily habits and their preferences in the market, this study employed RUT [84–87] while utilizing CVM to estimate the WTP of a non-marketed item [88]. In short, this study observed the effect of traceability information for pork on traditional market shoppers' preferences and WTP in Taiwan. Particularly, this research describes the types of consumers sorted by their socio-economic history, shopping habits, and other environmental attributes at both the lower and the higher levels of WTP.

2. Materials and Methods

This traditional market research was conducted across Taiwan in 2015. In an attempt to support the more traditional market in sustaining their economic share and in competing with other types of markets (as the growth of modern markets is to keep mounting up [89–91]), improving the status quo of traditional markets became one of the foundations of the topic selection. To express consumers' WTP and their preferences, the CVM (stated preference method) [82,92] was utilized in this hypothetical market situation [88]. The CVM decision allows a monetary valuation to be applied to examine non-marketed items or attributes from a service product [76,88,93,94]. The WTP calculates both the median and the mean from the collected responses of simulated-market questionnaires reflecting the consumer's budget perception in the real market situation [77,83]. Furthermore, several prior studies stated that the differences are not significant between the real-market WTP valuation and a hypothetical situation [78,79]. A further elaboration of the materials and methods is offered below.

2.1. Theoretical Model Used

When consumers buy a product in a traditional market, they usually have many considerations, starting from their daily shopping habits to the atmosphere of traditional markets. The magnitude of these considerations might be small or large based upon the consumers' utility. This utility can be modeled by adopting the Lancaster consumer theory of RUT [85]. Moreover, this theory can imitate the rational act of consumers going to the market and choosing the option they think most matches their preference (highest utility) [86,87,95]. Since this research applies a non-existent attribute/product (pork belly with traceability information), the application of this theory fits those consumers who pay attention to price and who tend to find utility from an item's characteristics or attributes rather than the item itself [85,96]. The framework of RUT helped us model every shopper's utility as a decision-maker in a given hypothetical market situation [84].

The RUT method models the substitute product as a linear function of several attributes. In the research, the independent variables were categorized into three groups ((A) attributes regarding the consumer's habits in traditional markets, (B) attributes related to pork attributes and their safety, and (C) attributes related to the consumer's socio and economic profile) and formed the mathematical RUT model (1) below:

$$U_{ij} = \beta_k \, X_{ijk} + \varepsilon_{ij} \tag{1}$$

where U_{ij} represents the utilization of $_i$ consumer for pork belly item $_j$ with the application of traceability information, β (utility or preference weights) represents the vectors of coefficients that are homogeneous among traditional market visitors in Taiwan, and X_{ijk} represents the $_k$ attribute for pork belly item $_j$ of the $_i$ consumer. Finally, the ε_{ij} (random residue) represents the random component that encompasses the asymmetric factors and the unobserved variance that might impact the genuine utility located throughout consumers [73,97,98], which are independent and identically distributed (*i.i.d.*), and adheres to the type I extreme value distribution and follows the error term differences of logistic distributions [99]. In short, this theory models the decision-maker's ability to acquire a particular degree of utility from every alternative [66].

2.2. Participants and Survey

The data collection was conducted during July and August of 2015. Based upon previous research, pork belly was specified as the meat product in the questionnaire. Moreover, the determination of the prices and the meat product was conducted during the market research phase. The difference of prices and prominent meat products through the island vary, thus this phase assisted the survey in matching the conditions in Taiwanese perception during the research time. Furthermore, freshly slaughtered swine represented more than 50% of the meat bought from traditional market vendors [100], and pork belly was so favored that the government charged a tariff of up to 40% [64].

The survey's design provided a starting price point. The lower level was 110 TWD per 600 g, the middle level 130 TWD per 600 g, and the upper level 150 TWD per 600 g, as this range was comparable with previous studies' findings and market conditions from the north to the south of Taiwan [101]. Therefore, this design enabled us to model the value of pork belly with traceability information at these price levels.

While the price was charted into three levels, the questionnaire presented the consumers with two choices initially, which included (a) "do know the price" and (b) "do not know or unsure about the price". The option for "do know the price" was further subdivided into (i) 110 TWD (~USD 3.94) per 600 g, (ii) 130 TWD (~USD 4.66) per 600 g, and (iii) 150 TWD (~USD 5.38) per 600 g. The option for (b) "do not know or unsure about the price" was assigned the middle level, which was (iv) 130 TWD (approximately USD 4.66) per 600 g. These prices were selected to keep the monetary values consistent with the particular product and markets [102].

When the survey design was completed, the next phase included randomly distributing and sampling in dual channels. The first one was (1) a direct channel, which was conducted at traditional markets and train stations by providing the survey on an electronic tablet (handily utilized without redundant questions and a situation sheet) with a reward administration (a stainless steel cutlery knife) at the end of survey session. The second channel was (2) an indirect channel, where the survey was administered online (through social media, e.g., Facebook) by completing the survey through a webpage. The survey was designed and deployed through the SurveyMonkey platform.

Once the respondents began the survey, they received background information in the form of a scenario to put their state of mind in a shopping situation within a traditional market for pork belly with traceability information. After reading the information, they completed the screening questions which asked (i) "Have you gone to any Taiwan traditional market in the past year?" and (ii) "Have you purchased any fresh meat items at any Taiwan traditional market in the past year?". Respondents were removed from further analysis whenever they responded "No, I have not gone/have not purchased", "No, I do not believe", or "I am unsure". This phase was critical to mitigate sample bias and to accurately estimate the WTP [103].

The questionnaire involved answering questions surrounding 23 independent variables defined in Table 1.

Further, the dependent variable consisted of both the positive WTP and the WTP of traceability information. In the end, 1420 valid responses were collected. After the data were retrieved, they were cleaned and organized neatly. In summary, the details of all these independent and dependent variables is further spelled out in the sub-section of the theoretical model used and the sample distribution section. The analysis phase is described below.

2.3. Data Analysis

The data analysis centers on the WTP probability for traceability information of traditional market consumers. Because the utility of the consumers might be dependent upon distinctive attributes (such as personal habits or environmental conditions), the effect of traceability information on pork belly in Taiwan's traditional market can be predicted through econometric analysis. The data collected consisted of both continuous and ordinal types, therefore the positive WTP for the observed attribute (pork belly traceability) could be analyzed by the application of the logit model. The notation for its mathematical equation is written on Equation (2) below:

$$p = \mathrm{pr}\,(y_i = 1 \mid x_i) = F(x'\beta) = \frac{e^{x'\beta}}{1 + e^{x'\beta}} = \frac{\exp(x'\beta)}{1 + \exp(x'\beta)} \tag{2}$$

where, in Equation (2), $y_i = 1$ represents the likelihood of a positive preference or WTP, x_i represents independent variables (e.g., (A) market-related habit, (B) safety-and-pork-related information, (C) socio-economy profile). Further, within the logit model, the estimation of marginal effect (M.E.) is shown as $\partial p / \partial xj = F'(x'\beta)\ \beta_j$.

Table 1. The attributes regarding each survey question.

Attributes to their habit in traditional markets	The status as a food purchaser within the family	All the time Sometimes No or never (reference group)
	The duration of traditional market shopping	Less than 30 min (reference group) 30–60 min More than an hour
	The point-in-time of going to traditional markets	Prior to 11:00. 11:00–17:00. In the wake of 17:00. (reference group)
		The recurrence in cooking at the home
Attributes related to pork and its safety	Safety certificate (relevant with pork safety)	Safety certificate is relevant to pork safety
		Safety certificate has no relevance with pork safety (reference group)
	Pork grading information	Very unnecessary (reference group) Unnecessary (reference group) Fair Necessary Very necessary
	Calories and nutrients label	Providing calories and nutrients labels can increase purchase intent
		Providing calories and nutrients labels cannot increase purchase intent (reference group)
	Fat-lean proportion information	Providing fat-lean proportion information can increase purchase intent
		Providing fat-lean proportion information cannot increase purchase intent (reference group)
	The tendency of accessing health-related content from mass media	Accessing health-related content from mass media—never (reference group)
		Accessing health-related content from mass media—seldom (reference group)
		Accessing health-related content from mass media—sometimes (reference group)
		Accessing health-related content from mass media—often
		Accessing health-related content from mass media—almost every time
Attributes to their socio and economic profile	Sex-gender	Female Male (reference group)
	Housewife status	Occupancy as housewife status
		Occupancy not as housewife status (reference group)
	Survey completion location	Northern Taiwan Central Taiwan Southern Taiwan (reference group) Eastern Taiwan (reference group)
	Buyers' origin (rural or urban district)	Urban district Rural district (reference group)
		Age
		Family size
		Range of family earnings every month
		Education level

Source: Assembled by this research.

The next step in data analysis was finding out the extension of WTP. The probability test by the logit model provided the impact of the attributes with the independent variables and whether they manifested positive or negative impacts. Moreover, if the given attributes showed a positive impact, it was possible to run interval regressions to find out how much New Taiwan dollars (TWD) traditional market consumers will pay for the pork belly with traceability information, because interval regression possesses the ability to forecast the WTP within the limits of the 0 to 1 mathematical term of prediction-likelihood [104]. According to the interval barrier of WTP, the application of interval regression in this research was written as notation in Equations (3) and (4) below:

$$y_i^* = x_i'\beta + u_i \tag{3}$$

$$\Pr\left[a_j < y^* \le a_{j+1}\right] = \Pr\left[y^* \le a_{j+1}\right] - \Pr\left[y^* \le a_j\right] = F^*\left(a_{j+1}\right) - F^*\left(a_j\right) \tag{4}$$

In elaborating Equations (3) and (4), y_i^* was examined to be found inside the $(J+1)$ exclusive to one another as intervals $(-\infty, a_1], (a_1, a_2], \ldots, (a_J, \infty)$. Provided the consumers' response from the questionnaire, y^* was located and put in consistent intervals, which were $y^* \le 0, 1 < y^* \le 3, 4 < y^* \le 6, \ldots,$ and $16 \le y^*$. Understanding all these terms and all the 23 independent variables, the empiric specs of the pork belly with traceability information found in Taiwan traditional market are shown as Equation (5):

$$y^* = \beta_0 + \beta_1 X_1 + \beta_2 X_2 + \ldots + \beta_{23} X_{23} + \varepsilon \tag{5}$$

In elaborating Equation (5), the WTP for traceability information on pork belly is defined as y^* and is affected by 23 independent variables, which are defined as X_s. Other terms are β_S, which is defined for the estimation parameters, and ε, which is defined for the unobserved variance.

3. Results

The results of this study are discussed in three sub-sections: (1) sample profile, (2) the probability of positive WTP, and (3) the WTP for traceability information. In general, out of the 1420 respondents, of those people who knew the price, 467 people selected (i) 110 TWD (~USD 3.94) per 600 g, 223 people selected (ii) 130 TWD (~USD 4.66) per 600 g, 102 selected (iii) 150 TWD (~USD 5.38) per 600 g. Further, 628 people selected "do not know or unsure about the price" which was translated to (iv) 130 TWD (~USD 4.66) per 600 g. The further results are pointed out in the tables and the paragraphs below.

3.1. Sample Profile

Based on Table 2, more than three-quarters of the respondents would spend extra money for traceability information, as their positive WTP was positive TWD. These respondents also showed their WTP reached 7 TWD (~USD 0.25) whenever facing the pork belly with traceability information in Taiwan's traditional market. Almost 50% considered themselves as a daily cook at home. Most of them considered themselves to be the major buyer's representative and go to market before 11:00., spending only 30–60 min. Moreover, around 70% of these people believed that traceability qualification is relevant to pork safety. Approximately 40% of them believed that pork grading is necessary to be provided by the sellers and that the fat-lean proportion can increase buyer's WTP; they also often access mass media to find health-related content. Lastly, from the socio-economy profile, the results showed >50% of the respondents were female, with fewer than 20% of them housewives. The average age was 41 years old with at least 15 years of formal education. Most of them lived in an urban district or around northern Taiwan, had a monthly salary of around 65,000 TWD (~USD 2330) every month, and had at least four people living in the house.

Table 2. Description of variables ($n = 1420$).

Variables		Mean	SD	Type	Measurement
Dependent					
Positive WTP (put extra TWD from 0-up)		83%	0.37	BV	Will put = 1 Otherwise = 0
WTP of traceability information		7	5.69	CV	WTP in TWD
Independent					
Market-related habit					
Recurrence of cook-at-home		7	5.17	CV	Frequency in times
Major purchaser status (All the time)		50%	0.50	BV	All the time = 1 Otherwise = 0
Major purchaser status (Sometimes)		32%	0.47	BV	Sometimes = 1 Otherwise = 0
Time-point of shopping (prior to 11:00.)		43%	0.50	BV	11:00 or before = 1 Otherwise = 0
Time-point of shopping (11:00–17:00.)		22%	0.42	BV	11:00–17:00. = 1 Otherwise = 0
Duration of shopping (30–60 min)		50%	0.50	BV	30–60 min = 1 Otherwise = 0
Duration of shopping (more than an hour)		14%	0.34	BV	$\geq 1\,h = 1$ Otherwise = 0
Safety-and-pork-related information					
Safety qualification (relevant with pork safety)		73%	0.44	BV	Relevant = 1 Otherwise = 0
Pork grading (provided by butcher as potential service)	Fair	30%	0.46	BV	Fair = 1 Otherwise = 0
	Necessary	47%	0.50	BV	Necessary = 1 Otherwise = 0
	Very necessary	15%	0.36	BV	Very necessary = 1 Otherwise = 0
Fat-lean proportion information (can increase purchase intent)		36%	0.48	BV	Increase = 1 Otherwise = 0
Calories and nutrients label (can increase purchase intent)		20%	0.40	BV	Increase = 1 Otherwise = 0
Health-related content from mass-media (often and always)		40%	0.49	BV	Often and always = 1 Otherwise = 0
Socio-economy profile					
Sex-gender		66%	0.47	BV	Female = 1 Otherwise = 0
Housewife status		13%	0.33	BV	Housewife = 1 Otherwise = 0
Age		41	10.49	CV	Age in years
Education level		15	2.31	CV	Education in years
Family monthly-earnings		65	31.42	CV	Salary in 1000 TWD
Family size		4	1.51	CV	Amount of people
Urban district		64%	0.48	BV	Urban = 1 Otherwise = 0
Northern Taiwan		49%	0.50	BV	Northern = 1 Otherwise = 0
Central Taiwan		28%	0.45	BV	Central = 1 Otherwise = 0

Source: Assembled by this research. Note: BV = binary variables, CV = continuous variables, SD = standard deviations.

3.2. The Outcome of Positive WTP for Traceable Pork

The logit model result of positive WTP for traceable pork is presented in Table 3. According to Table 3, the log-likelihood and the Wald X^2 values were estimated and valued as -614.55 and 47.82, respectively. This revealed that the overall model specification was valid and significant. Each variable was explained if both coefficients and marginal effect (M.E.) reached a significant level. From the market-related habit, it mattered at what point-of-time consumers shop in a day. Especially, respondents who went to the market in the morning (prior to 11:00) and noon-to-afternoon (11:00–17:00) were significantly different from those who usually shopped during the evening period (after 17:00). These results make sense because traditional market consumers who usually shop after 17:00 are often hurrying to go back home to prepare dinner for the family. Therefore, the traceable pork information may not be as attractive compared to those who shop prior to 17:00. From the safety-and-pork-related side, those who thought a safety certificate is relevant to pork safety were more likely to give positive WTP compared to those who thought it irrelevant. Moreover, respondents who thought that pork grading information is fair to them showed positive WTP for traceability information compared to those who just thought that the pork-grading is either unnecessary or very unnecessary. Lastly, from the socio-economic side, the only variable that showed its significance was age. However, the result showed that the younger Taiwan-traditional-market buyers were more likely to give a positive WTP for traceability information compared to those who were older. In short, the following independent variables which had statistical significance are shown in Table 3: shopping time, safety qualification, pork grading information, and age of the respondents.

3.3. The Estimated WTP of Traceability Information

The WTP elicitation for traceable pork could be evaluated through an interval regression model as shown in Table 4. Because of the effect of price knowledge when eliciting WTP, it was important to distinguish whether respondents knew about the current market price for pork in traditional markets. There are four different interval regression examinations shown in Table 4. The results of log-likelihood and Wald X^2 presented a strong fit by each model examination in each category. Further, most of the consumers in the survey thought that they had knowledge of pork prices in traditional markets ($n = 792$), while others who did not have knowledge about the pork price were slightly lower ($n = 628$).

Table 3. Probability of positive WTP for traceable pork ($n = 1420$).

Independent Variables		Dependent Variables	
		Positive WTP	
		M.E.	*Coef.*
Market-related habit			
Recurrence of cook-at-home		0.00	−0.01
Major-purchaser status (All the time)		−0.05	−0.34
Major-purchaser status (Sometimes)		0.00	−0.01
Time-point of shopping (prior to 11:00)		0.07 ***	0.54 ***
Time-point of shopping (11:00–17:00)		0.07 ***	0.53 ***
Duration of shopping (30–60 min)		0.00	−0.02
Duration of shopping (more than an hour)		−0.02	−0.16
Safety-and-pork-related information			
Safety certificate (relevant with pork safety)		0.05 **	0.37 **
Pork grading (provided by butcher as potential service)	Fair	0.06 **	0.48 *
	Necessary	0.04	0.26
	Very necessary	0.06 *	0.49

Table 3. *Cont.*

	Dependent Variables	
Independent Variables	**Positive WTP**	
	M.E.	*Coef.*
Fat-lean proportion information (can increase purchase intent)	0.02	0.12
Calories and nutrients label (can increase purchase intent)	−0.02	−0.13
Health-related content from mass-media (often and always)	0.03	0.20
Socio-economy profile		
Sex-gender	−0.01	−0.05
Housewife status	−0.01	−0.11
Age	−0.00 ***	−0.02 ***
Education chronicle	0.00	0.02
Family monthly-earnings	0.00	0.00
Family sum	0.00	0.03
Urban district	0.00	0.00
Northern Taiwan	0.01	0.11
Central Taiwan	0.04	0.27
Constant	1.16	
Log-Likelihood	−614.55	
Wald X^2	47.82 ***	
Classification Predication	83.52%	
Goodness-of-fit	1411.47	
Pseudo R^2	0.04	

Source: Assembled by this research. Note: M.E. = marginal effect, Coef. = coefficient, *** = 1%, ** = 5%, * = 10% of statistical significance.

The elicitation of WTP in the interval regression model could be explained as the actual dollar that respondents want to pay extra for the traceable pork. Digging deeper into the results based on respondents' selection of price, the first price outlined showed the people who knew the price and chose 110 TWD (~USD 3.94) per 600 g. It potentially demonstrates that there are still a fair number of consumers who face lower pricing in traditional markets in Taiwan. These Taiwanese with a higher monthly income wanted to pay around 2 cents TWD per 600 g (<USD 0.0007) for traceable pork in traditional markets if compared to those who had a lower monthly income. However, the consumer with lower education also desired to pay more, around 36 cents TWD per 600 g (~USD 0.01), if compared to those who had a higher education level. Lastly, whenever respondents thought that pork grading is very necessary to be provided by a pork-butcher, they wanted to pay a premium extra of about 3.5 TWD per 600 g (~USD 0.13) for traceability information if compared to those who thought the pork grading is unnecessary or very unnecessary.

The middle-price selectors showed a significant difference in WTP on many independent variables. The people who knew the market price at 130 TWD (~USD 4.66) per 600 g cared a great deal about the family, as they showed the premium of WTP at about 1 TWD per 600 g (~USD 0.03) whenever they had more members in the household and resided in an urban district. The Taiwanese with a lower monthly income wanted to pay around 3 cents TWD per 600 g (<USD 0.001) for traceable pork in traditional markets if compared to those who had a higher monthly income. Respondents who typically were not housewives and the prevalence of cooking at home contributed to increased WTP of 2.1 TWD per 600 g (~USD 0.07) if they cooked at home more often. This was related to the next increase in WTP by people who always go traditional-market shopping in the noon–afternoon, and they added 2.76 TWD per 600 g (~USD 0.10) more. To end, these respondents seemed not to be housewives, since these respondents added more WTP (~1.87 to 4.50 TWD per 600 g) for pork grading either necessarily or very necessarily provided by the butcher in traditional markets.

Table 4. The outcome of interval regression for traceability information ($n = 1420$).

Independent Variables	Dependent Variables			Do Not Know the Price
	Do Know the Price			
	WTP 110 TWD (~USD 3.94)	WTP 130 TWD (~USD 4.66)	WTP 150 TWD (~USD 5.38)	WTP 130 TWD (~USD 4.66)
Market-related habit				
Recurrence of cook-at-home	−0.02	0.13 *	−0.28	−0.04
Major purchaser status (All the time)	−0.34	−1.37	3.03	−0.82
Major purchaser status (Sometimes)	0.45	−0.49	6.44 *	0.02
Time-point of shopping (prior to 11:00.)	−1.18	0.58	8.09 ***	0.96
Time-point of shopping (11:00–17:00)	−1.22	2.76 ***	5.59 *	0.84
Duration of shopping (30–60 min)	0.98	0.61	0.73	−0.37
Duration of shopping (more than an hour)	1.08	−0.50	−1.32	−0.67
Safety-and-pork-related information				
Safety certificate (relevant with pork safety)	0.78	0.96	−1.21	0.82
Pork grading (provided by butcher as potential service) — Fair	0.85	1.01	6.20	1.11
Pork grading (provided by butcher as potential service) — Necessary	1.27	1.87 *	7.69	1.04
Pork grading (provided by butcher as potential service) — Very necessary	3.50 **	4.50 ***	9.28 *	3.02 ***
Fat-lean proportion information (can increase purchase intent)	1.04	0.56	−3.70 **	−0.21
Calories and nutrients label (can increase purchase intent)	1.42	−0.37	2.58	−0.76
Health-related content from mass-media (often and always)	0.70	0.67	6.77 ***	1.68 ***
Socio-economy profile				
Sex-gender	−0.03	0.41	−5.57 ***	1.24 **
Housewife status	0.95	−1.97 **	1.08	−1.05
Age	−0.06	0.00	−0.19	−0.06 *
Education level	−0.36 **	0.27	−0.39	0.12
Family monthly-earnings	0.02 *	−0.03 ***	0.00	0.01
Family size	0.38	0.79 ***	−0.23	−0.07
Urban district	−0.11	1.40 *	0.39	0.72
Northern Taiwan	0.92	0.14	−5.50	0.43
Central Taiwan	0.47	0.51	−6.46 *	0.50
Constant	13.69 ***	−6.04	7.73	1.84
Observations (n)	467	223	102	628
Log-Likelihood	−979.48	−450.32	−150.64	−1332.98
Wald X^2	34.73 ***	64.26 ***	41.86 ***	49.07 ***
AIC	2008.97	950.64	351.28	2715.97

Source: Assembled by this research. Note: Coef. = coefficient, M.E. = marginal effect, *** = 1%, ** = 5%, * = 10% of statistical significance.

There were still a large number of respondents who did not know the pork belly price in traditional markets, thus respondents were provided the average market price at 130 TWD (~USD 4.66) as the starting point of WTP elicitation. These people perceived that they would be likely to pay around 1.24 TWD per 600 g (~USD 0.04) more for traceable pork if the respondents were female. Further, younger respondents were likely to pay about 6 cents TWD more per 600 g (~USD 0.002) for traceability information compared to those who were older. If respondents often received health-related content from mass-media, they would have WTP of about 1.68 TWD per 600 g (~USD 0.06) for traceable pork if compared to those who rarely received health-related information. Moreover, the pork grading information had the biggest impact when respondents did not know about the market price. Respondents who thought pork grading information is very necessary would be likely to pay about 3.02 TWD more per 600 g (~USD 0.11) for traceability information compared to those who thought pork grading information is unnecessary or very unnecessary.

Ultimately, the final section of the results elaborated on Taiwanese who knew and chose a price of 150 TWD (~USD 5.38) per 600 g. These results potentially showed that there were still small groups of consumers who faced higher pricing in traditional markets in Taiwan. These respondents likely had the highest range of WTP rise, which started from 4 to 9 TWD per 600 g (~USD 0.14–0.32). However, consumers who were interested in the fat-lean-proportion information may not have been willing to give a positive WTP for traceability information. This may imply that those who were not interested in fat-lean proportion information would be willing to pay for the traceability information. The inclusion of 6 TWD per 600 g (~USD 0.22) to these respondents' WTP was added whenever they were male or considered sometimes-major-shoppers. Moreover, if the respondents in this group were not living in central Taiwan and often listened to health content on media, they had an increase of WTP at around 7 TWD per 600 g (~USD 0.25). Then, the highest supplement to their WTP was found if they thought that pork grading is very necessary to be provided by butchers and had the habit of going to the market prior to 11:00 and 11:00–17:00, as they would pay 6–9 TWD more per 600 g (~USD 0.22–0.32). Since the WTP was estimated in each variable, this study further provides a summation of WTP for each category.

3.4. The Estimated WTP for Each Category

The significant premium of WTP in this section can be explained as the summation of the significant parameters multiplied by the respective sample mean. For example, if the recurrence of cook-at-home was seven times a week on average, then it was multiplied with the estimated parameter of 0.13 for the WTP of 130 TWD (~USD 4.66) per 600 g. Thus, we obtained a calculated WTP of about 0.91 TWD for recurrence of cook-at-home. Repeating the same step, the summation of premium WTP is shown in Table 5.

Respondents who knew the market price at 130 TWD (~USD 4.66) were more influenced by "recurrence of cook-at-home", "time-point of shopping (11:00–17:00)", "necessary and very necessary information on pork grading (provided by butcher as potential service)", "non-housewife", "family monthly earnings", "family size", and urban district", and the summation of WTP was about 10.68 TWD (~USD 0.37). However, we saw several factors affecting WTP estimation when respondents did not know about market price in traditional market. Only the factors of "very necessary information on pork grading (provided by butcher as potential service)", "health-related content from mass-media (often and always)", "female", and "age" influenced the WTP elicitation, and the summation of WTP was about 3.48 TWD (~USD 0.12). Especially, note that respondents who knew the market price possibly had WTP for traceable pork at 10.68 TWD, which was about 7 TWD higher than those who did not know the market price in traditional markets.

Respondents who knew the market price at 110 TWD (~USD 3.94) were influenced more by "very necessary information on pork grading (provided by butcher as potential service)", "educational level", and "family monthly-earnings", and the summation of WTP was about 13.69 TWD (~USD 0.48). However, the respondent who experienced and knew the market price at 150 TWD (~USD 5.38) was more of influenced by "major purchaser status (sometimes)", "time-point of shopping (prior to 11:00)", "time-point of shopping (11:00–17:00)", "very necessary information on pork grading (provided by butcher as potential service)", "fat-lean proportion information (can increase purchase intent)", "health-related content from mass-media (often and always)", "male", and "non-central Taiwan", and the summation of WTP was about 20.44 TWD (~USD 0.72). If we compared the difference of WTP summation between the categories of the highest and the lowest market price, it also showed about 7 TWD difference.

Table 5. The summation of significant premium WTP (*n* = 1420).

Independent Variables		Mean	Do Know the Price			Do Not Know the Price
			110 TWD (~USD 3.94)	130 TWD (~USD 4.66)	150 TWD (~USD 5.38)	130 TWD (~USD 4.66)
Market-related habit						
Recurrence of cook-at-home		7	Ø	0.91	Ø	Ø
Major-purchaser status (Sometimes)		1	Ø	Ø	6.44	Ø
Time-point of shopping (prior to 11:00)		1	Ø	Ø	8.09	Ø
Time-point of shopping (11:00–17:00)		1	Ø	2.76	5.59	Ø
Safety-and-pork-related information						
Pork grading (provided by butcher as potential service)	Necessary	1	Ø	1.87	Ø	Ø
	Very necessary	1	3.50	4.50	9.28	3.02
Fat-lean proportion information (can increase purchase intent)		1	Ø	Ø	−3.70	Ø
Health-related content from mass-media (often and always)		1	Ø	Ø	6.77	1.68
Socio-economy profile						
Sex-gender		1	Ø	Ø	−5.57	1.24
Housewife status		1	Ø	−1.97	Ø	Ø
Age		41	Ø	Ø	Ø	−2.46
Education level		15	−5.40	Ø	Ø	Ø
Family monthly-earnings		65	1.30	−1.95	Ø	Ø
Family size		4	Ø	3.16	Ø	Ø
Urban district		1	Ø	1.40	Ø	Ø
Central Taiwan		1	Ø	Ø	−6.46	Ø
Constant		1	13.69	Ø	Ø	Ø
The Significant Extra-WTP			13.09	10.68	20.44	3.48

Source: Assembled by this research. Note: Ø = denotes null symbol as the values do not come statistically significant.

This final section interprets Figure 1 and shows the significant extra WTP in total. Figure 1 illustrates the total WTP when all attributes were summed based on the price category in traditional markets they visited. This figure shows that every group was significantly different from each other. Furthermore, the largest result of significant WTP extra was from traditional markets that had the highest price from butchers, which represented almost 21 TWD (~USD 0.74). The second largest WTP extra was from traditional markets that had the lowest price from butchers, which revealed about 13 TWD (~USD 0.47). However, note that consumers who knew or did not know about the market price for pork may have had a significant difference in their WTP. Especially, when they did not know about the market price, their WTP for traceable pork was only about 4 TWD (~USD 0.14) for WTP extra. However, if they knew about the market price at 130 TWD (~USD 4.66), their WTP for traceable pork could reach up to 11 TWD (~USD 0.39) for WTP extra.

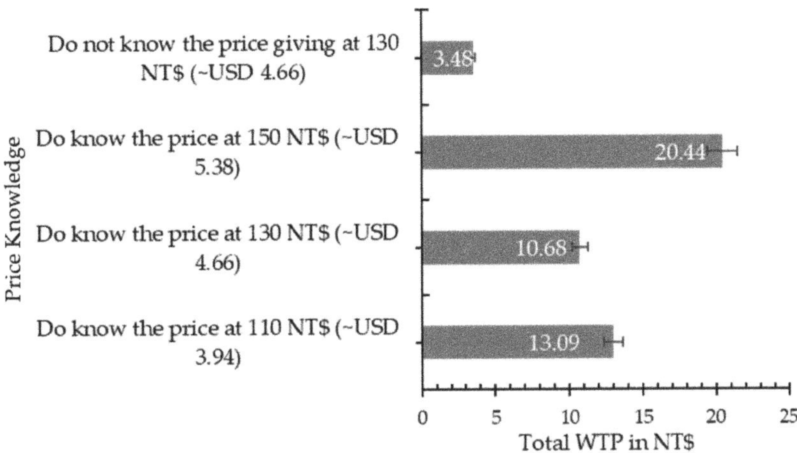

Figure 1. The Total of WTP extra for traceability information ($n = 1420$). Note: The error bars denote 5% of statistical significance.

4. Discussion

This section broadly discusses the results of this study that explored the effect of traceability information on meat products in traditional markets in Taiwan. The first part discusses the sample profile. Based on the reported habits, the respondents in this study are active in cooking based on their frequency of cooking at home. The interpretation could be that respondents cook on average once a day. Alternatively, it can also mean that they cook two or three times a day, then skip one day and cook again the next day. This habit leads to their dependability in shopping in the market since most of them are identified as the major shoppers who are responsible for buying the ingredients. They are also mostly time-efficient buyers who prefer buying products in the morning, seeking certain market benefits such as freshness. Regarding safety-and-pork-related information, most of the respondents in the study seem to care about the safety of food. About 40% of the respondents have an interest in pork grading and fat-lean-proportion, which are both related to safety and pork quality. These results are consistent with a prior study showing 60% of respondents put interest in safety guarantee and meat class for animal-derived items [43]. Further, the food safety comprehension seems to come from the mass media, as fewer than half of respondents access health-related materials. Finally, the last part of the profile is the socio-economy profile. Based on previous research, this study profile has a similar pattern with research done in the market either in the modern or the traditional markets. These profiles are written with a monthly salary around 65,000 TWD per month, four members in the household, the average age of 41, with at least 12 years of education, and most self-identify as housewives [47,105,106].

The next discussion concentrates on the likelihood the respondents show positive impacts. Based on the results in Section 3.2, the goodness-of-fit indicators exhibited the model specification used in the study fit the data well. The discussion begins with the market-related habit, which elicits consumers' positive WTP tendency on the habit related to daytime shopping. This showed that the daytime-preferred respondents give significantly higher positive WTP for traceable pork if compared to the evening shopping schedule. The reason behind this is likely because the evening shoppers usually have work during the daytime and have to rush to quickly buy and return home to prepare dinner for their family [68]. Furthermore, from prior studies, people prefer to buy meat products in the morning for their benefit [107]. From the safety-and-pork-related side, this study finds that people would have positive WTP whenever they were given traceable pork belly in traditional markets. They also think about the safety certification relevance with pork-safety and that pork grading is (moderately or very) necessary and should be included in butcher's stall. These findings show significantly higher positive impacts than people who do not care about these attributes. They are also consistent with preceding studies which declared that knowledge on meat certification and meat classes connects the food safety perception, as it affects consumers' decisions towards traceability products [108]. Lastly, from the socio-economy profile, the respondent's age is discussed. As mentioned before, the results from this study show that WTP is lower as respondents' ages are older. In other words, the inclusion of traceability information in pork belly has a more positive effect on younger people. The average age was 41 in the study. Therefore, this study found that Taiwanese below age 40 would have a higher WTP when viewing traceability of meat products in traditional markets. This finding is relevant to previous research which mentioned that younger consumers below the age of 35 years old would have more likeliness to buy traceable products compared with other age categories [29]. In short, traceability showcases positive WTP generally towards variables.

The subsequent discussion is centered on the WTP for traceability information. From Section 3.3, the respondents in this research are most likely to know the pork belly price. This finding is likely to be supported by the fact that most of these Taiwanese are identified as major purchasers in the household, making their food almost every day in their homes. Knowing that pork is the most purchased and consumed product in traditional markets [78], as the main shopper, they must know the price of pork because they most likely buy or cook pork daily, weekly, or monthly in their main dish.

The next focus from Section 3.3. is the WTP of respondents who selected 110 TWD (~USD 3.94) per 600 g. These people are considered as common pork buyers because they are the majority of those who know about the market price level. Moreover, their attributes have a significant difference in their WTP. The lower education they have, the more WTP they add. The more income they make, the more WTP they supplement. The WTP is about 3.5 TWD when pork grading has the highest necessity to be provided in the market, and they show the second lowest WTP across all groups of the consumer. The finding on common pork buyers' income background is compatible with the previous finding, which stated that, when consumers have a preference for traceability and safer products, they increase their WTP, as they have higher income every month [29,105]. However, for the education attribute, the result creates a new finding on traceability preference, whereas the previous finding on education showed the opposite result [105]. The reason behind this could be the general profile of common pork buyers; they are not always people who have an education of more than twelve years. Common pork buyers can be considered just the normal buyers in traditional markets.

The following discussion is centralized on Taiwanese buyers' WTP who picked the middle price (i.e., 130 TWD (~USD 4.66) per 600 g). These people can be considered career mom pork buyers because their attributes indicate a typical mom who needs to take care of the family and be the family's backbone. This is supported by the rise in their WTP whenever the family size increases (the larger number they have, the higher WTP they put), when they have purchasing budget management (career woman's strategy in planning middle price selection, how often to cook at home, how minimal money should be used in the food expenditures), when the location of living differs (more WTP is found in more municipal districts), when they have a shopping schedule concern (because they are a career woman who works the whole day and thus are only able to buy ingredients in the evening), and when they are thinking about the safety and the quality of the food they buy for the family (e.g., they think pork grading is necessary or very necessary). Career mom pork buyers' preference on evening shopping, number of family members, and pork grading showed (barely cents) an increased WTP. This is supported by a study which stated career woman are found to prefer to buy food in the evening where they can buy food after work hours [109]. Another finding seemed to correspond to prior studies which dictated the more family members there are in the family, the more they care about food quality they serve. In the case of pork grading preferred buyers, they show a higher WTP [110,111].

Although a previous study also mentioned a positive relationship between income and WTP for traceability [24], they were not situated in the price-sensitive class (110 TWD or ~USD 3.94) and still picked middle price (130 TWD or ~USD 4.66). In short, a career woman who takes care of the whole family, including the money flow in purchasing food, stills pick a rational price and still believes in a safe pork product.

The succeeding discussion elaborates upon respondents who did not know the price and who were assigned the middle price. These respondents who did not know the price were younger customers who preferred traceable pork and were curious or showed concern about health-related content from online sources, which affected their pork grading interests. Thus, we labeled these respondents as younger female pork buyers. This curiosity of health and safety materials is relevant with the prevalence of younger consumers' curiosity and concern levels tending to be elevated. Furthermore, females put more positive impact towards traceable products and meat class than males in the previous research [11,111], which supports these discoveries.

However, the difference with people choosing the highest price (150 TWD or ~USD 5.38) is the most positive WTP for traceable pork belly for males rather than females. Because of the big range of WTP from 4 to 9 TWD (~USD 0.14–0.32) among their significant attributes and their pick on the premium price, they can be regarded as the health-enthusiast male buyers. This designation is also supported by their higher WTP regarding low occurrence of being the food shopper (which was not often when they were male), buying food in a flexible time (since males are prone to not having particular times to buy food), caring about health content from mass media and pork grading (which increases their WTP), and caring less about fat-lean-proportion. These findings seem novel to the traceability research since previous research identified that it is primarily females who care about health and quality information [11,111]. However, one of the findings on fat-lean-proportion is supported by prior research which showed increased money preference whenever males were less interested in this information to be provided in regard to food [112].

The final part of the discussion focuses on Section 3.4. This part of the discussion concentrates on the highest significant effect of extra WTP. The highest effect was approximately 21 TWD (~USD 0.74) of extra WTP, which might give more possibility of traceable pork belly to be strategized in premium prices of 171 TWD (~USD 5.99) per 600 g, which is highly impacted by the preference on pork grading. This result was also consistent with the previous result, which suggested that, whenever consumers are given food safety certification, their safety awareness leads them to set aside other attributes to pick the premium price ([92,108,113]), especially when the products include traceability information and consumers are scared based on food scandal events in the past [10,47,92,93]. To summarize, whenever Taiwanese consumers acknowledge food safety causation and know the price, they have a much higher WTP for traceable food.

5. Conclusions

The safety awareness of Taiwanese keeps rising due to food scandals that happened in the past. This leads to the desire to apply food labels on their food with certifications such as traceability. Traceability might help regain consumers' confidence and trust while also helping to increase shoppers' WTP. However, traceability's positive impact is only currently found in modern markets. As traceability information is not often present in Taiwan's traditional market, this research provides novel findings.

In general, the application of traceability information on the pork belly product in traditional markets offers a positive effect related to several given attributes among many kinds of market goers. These positive impacts lean towards consumers who are younger buyers, have flexible shopping schedules, are concerned about safety qualification related to pork safety, and perceive pork grading provisions as (commonly) important.

In conclusion, regarding WTP (in TWD), the most favored attribute by all price electors is the attribute of pork grading, since all groups showed the highest increase of WTP. Additionally, whenever consumers have a high recurrence in watching or listening to health-related content in offline or online media, the mindset of food safety and the addition of WTP increase.

The traceable pork belly also pushes increased WTP whenever traditional market visitors prefer to go food shopping in the evening. Another result is that traceability information impacts are sensitive to gender and income gaps. Connecting with the price selection, whenever these respondents acknowledge the price, there is a chance for them to be situated in price options one level higher. Furthermore, the highest market price venue (150 TWD or ~USD 5.38 per 600 g) also has the highest total of extra WTP that enables producers to establish an even higher new price possibility of about 171 TWD ($\pm$ ~USD 5.99).

Considering these findings, the implication of this research is outlined in two aspects. The first one relates to traditional market vendors. Traditional market vendors may be able to devise their marketing strategy for specific buyers (e.g., ones who choose the premium prices, ones who do not often purchase food in the market, ones who reside in the south or the north area of Taiwan, ones who have a flexible buying schedule, ones who are aware of food safety, ones who frequently access health-related content on online or offline media, and ones who perceive pork grading as very important) and create a new price point of 170 TWD (~USD 5.99) per 600 g. The second direction addresses the government's role, as these results can be applied to encourage food safety issues along with traceability information and pork grading information via mass media online or offline, especially towards particular buyers based on their sex-gender, shopping schedule, and monthly income gaps.

Finally, the limitation of this study may be the hypothetical perception that the market simulation was utilized for this research. Even with the utilization of CVM or RUT and the screening questions asked, the hypothetical condition might have reached the limit of consumers' preference, and real market situations may lead to different results. Thus, in future studies, the introduction of "cheap talk" at the beginning of the questionnaire is needed, as it decreases the hypothetical perception [114,115]. The "cheap talk" method can give insight to the respondents before they start the survey by providing them an explanation or script of possible hypothetical bias that could happen, thus they are asked to disregard the conjecture effect and are recommended to situate themselves in a real market situation.

Author Contributions: Conceptualization, S.-H.Y. and A.A.S.; methodology, S.-H.Y. and A.A.S.; software, S.-H.Y.; validation, S.-H.Y. and A.A.S.; formal analysis, D.C.-B.C., S.-H.Y., and A.A.S.; investigation, S.-H.Y. and A.A.S.; resources, S.-H.Y. and A.A.S.; data curation, S.-H.Y. and A.A.S.; writing—original draft preparation, S.-H.Y. A.A.S., and D.C.-B.C.; writing—review and editing, S.-H.Y., A.A.S. and D.C.-B.C.; visualization, S.-H.Y., A.A.S., and D.C.-B.C.; supervision, D.C.-B.C. and S.-H.Y.; project administration, S.-H.Y.; funding acquisition, S.-H.Y. All authors have read and agreed to the published version of the manuscript.

Funding: This research is supported in part by the funding from the Ministry of Science and Technology (MOST) of Taiwan (funding numbers MOST-103–2410-H-005-005-SSS) and the Council of Agriculture, Executive Yuan, Taiwan (funding numbers 110農科-2.2.3-牧-U2).

Institutional Review Board Statement: The study was conducted according to the guidelines of the Declaration of Helsinki, and approved by the Institutional Review Board (or Ethics Committee) of National Cheng Kung University (protocol code 104-017 and 6 March 2015).

Informed Consent Statement: Not applicable.

Data Availability Statement: The datasets generated for this study are available on request to the corresponding author.

Acknowledgments: The authors acknowledge all the enumerators and the participants who gave their valuable time in order to provide data related to their purchase behavior and preference. The authors are thankful to the Ministry of Science and Technology (MOST) of Taiwan and the Council of the Agriculture, Executive Yuan, Taiwan for their support through the funding for this research. All remaining errors are the authors.

Conflicts of Interest: The authors declare no conflict of interest.

References

1. Wezemael, L.V.; Verbeke, W.; Kügler, J.O.; de Barcellos, M.D.; Grunert, K.G. European consumers and beef safety: Perceptions, expectations and uncertainty reduction strategies. *Food Control* **2010**, *21*, 835–844. [CrossRef]
2. Kondo, N. Automation on fruit and vegetable grading system and food traceability. *Trends Food Sci. Technol.* **2010**, *21*, 145–152. [CrossRef]
3. Bánáti, D. Consumer response to food scandals and scares. *Trends Food Sci. Technol.* **2011**, *22*, 56–60. [CrossRef]
4. Rezai, G.; Mohamed, Z.; Shamsudin, M.N. Non-Muslim consumers' understanding of Halal principles in Malaysia. *J. Islam. Mark.* **2012**, *3*, 35–46. [CrossRef]
5. Golnaz, R.; Zainalabidin, M.; Nasir, S.M.; Chiew, F.C.E. Non-Muslims' awareness of Halal principles and related food products in Malaysia. *Int. Food Res. J.* **2010**, *17*, 667–674. Available online: http://psasir.upm.edu.my/id/eprint/11717/1/Non-Muslims_awareness_of_Halal_principles.pdf (accessed on 23 March 2021).
6. Rahman, M.; Khatun, M.; Rahman, M.H.; Ansary, N.P. Food safety issues in Islam. *Health Saf. Environ.* **2014**, *2*, 132–145. [CrossRef]
7. Huang, C.; Kan, K.; Fu, T. Consumer willingness-to-pay for food safety in Taiwan: A Binary-Ordinal probit model of analysis. *J. Consum. Aff.* **2005**, *33*, 76–91. [CrossRef]
8. Zhao, J.; Chen, A.; You, X.; Xu, Z.; Zhao, Y.; He, W.; Zhao, L.; Yang, S. A panel of SNP markers for meat traceability of Halal beef in the Chinese market. *Food Control* **2018**, *87*, 94–99. [CrossRef]
9. MingQiang, W.; XiaoGang, C. Measures and thoughts on constructing the halal food safety system. *J. Food Saf. Qual.* **2015**, *6*, 2581–2586. Available online: http://www.chinafoodj.com/ch/reader/view_abstract.aspx?flag=1&file_no=20150331004&journal_id=spzljcxb (accessed on 28 March 2021).
10. Loureiro, M.L.; Umberger, W.J. A choice experiment model for beef: What US consumer responses tell us about relative preferences for food safety, country-of-origin labeling, and traceability. *Food Policy* **2007**, *32*, 496–514. [CrossRef]
11. Verbeke, W.; Rutsaert, P.; Bonne, K.; Vermeir, I. Credence quality coordination and consumers' willingness-to-pay for certified halal labeled meat. *Meat Sci.* **2013**, *95*, 790–797. [CrossRef] [PubMed]
12. Saunders, M.; Guenther, J.; Saunders, P.; Dalziel, P.; Rutherford, P. Consumer preferences for attributes in food and beverages in developed and emerging export markets and their impact on the European Union and New Zealand. *Aust. N. Z. J. Eur. Stud.* **2016**, *8*, 53–68. [CrossRef]
13. Liang, W.-S.; He, Y.C.; Wu, H.D.; Li, Y.T.; Shih, T.H.; Kao, G.S.; Guo, H.Y.; Chao, D.Y. Ecological factors associated with persistent circulation of multiple highly pathogenic avian influenza viruses among poultry farms in Taiwan during 2015-17. *PLoS ONE* **2020**, *15*, e0236581. [CrossRef] [PubMed]
14. Liao, P.A.; Chang, H.H.; Chang, C.Y. Why is the food traceability system unsuccessful in Taiwan? Empirical evidence from a national survey of fruit and vegetable farmers. *Food Policy* **2011**, *36*, 686–693. [CrossRef]
15. Wei-han, C. Pesticide residues found in 73% of fruits, vegetables—Taipei Times. *Taipei Times*. 22 January 2016. Available online: https://www.taipeitimes.com/News/taiwan/archives/2016/01/22/2003637796 (accessed on 27 April 2021).
16. Hansen, J.D.; Watkins, M.A.; Heidt, M.L.; Anderson, P.A. Cold storage to control Codling Moth Larvae in fresh apples. *HortTechnology* **2007**, *17*, 195–198. [CrossRef]
17. Fruit Growers News. Codling Moth Could Keep U.S. Apples Out of Taiwan—Fruit Growers News. 15 November 2006. Available online: https://fruitgrowersnews.com/news/codling-moth-could-keep-u-s-apples-out-of-taiwan/ (accessed on 27 April 2021).
18. Lin, T.-Y.; Twu, S.-J.; Ho, M.-S.; Chang, L.-Y.; Lee, C.-Y. Enterovirus 71 outbreaks, Taiwan: Occurrence and recognition. *Emerg. Infect. Dis.* **2003**, *9*, 291–293. [CrossRef]
19. Huang, C.-C.; Jong, M.-H.; Lin, S.-Y. Characteristics of foot and mouth disease virus in Taiwan. *J. Vet. Med. Sci.* **2000**, *62*, 677–679. [CrossRef]
20. Chen, M.-F. Consumer trust in food safety—A multidisciplinary approach and empirical evidence from Taiwan. *Risk Anal.* **2008**, *28*, 1553–1569. [CrossRef]
21. Grunert, K.G. Food quality and safety: Consumer perception and demand. *Eur. Rev. Agric. Econ.* **2005**, *32*, 369–391. [CrossRef]
22. Ambali, A.R.; Bakar, A.N. People's Awareness on Halal foods and products: Potential issues for policy-makers. *Procedia-Soc. Behav. Sci.* **2014**, *121*, 3–25. [CrossRef]
23. International Organization for Standardization. *Food Safety Management*; ISO 22000; International Organization for Standardization: Geneva, Switzerland, 2018; pp. 1–4.
24. Ortega, D.L.; Wang, H.H.; Wu, L.; Olynk, N.J. Modeling heterogeneity in consumer preferences for select food safety attributes in China. *Food Policy* **2011**, *36*, 318–324. [CrossRef]
25. Chiu, J.-Z.; Hsieh, C.-C. Perspectives on food traceability system: A case in chain restaurant franchising. *Int. J. Manag. Econ. Soc. Sci.* **2018**, *7*, 272–282. [CrossRef]
26. Islam, R.; Madkouri, F.E. Assessing and ranking HALMAS parks in Malaysia: An application of importance-performance analysis and AHP. *J. Islam. Mark.* **2018**, *9*, 240–261. [CrossRef]
27. Gellynck, X.; Verbeke, W.; Vermeire, B. Pathways to increase consumer trust in meat as a safe and wholesome food. *Meat Sci.* **2006**, *74*, 161–167. [CrossRef]
28. Thakur, M.; Donnelly, K.A.M. Modeling traceability information in soybean value chains. *J. Food Eng.* **2010**, *99*, 98–105. [CrossRef]

29. Zhang, C.; Bai, J.; Wahl, T.I. Consumers' willingness to pay for traceable pork, milk, and cooking oil in Nanjing, China. *Food Control* **2012**, *27*, 21–28. [CrossRef]
30. Lavelli, V. High-warranty traceability system in the poultry meat supply chain: A medium-sized enterprise case study. *Food Control* **2013**, *33*, 148–156. [CrossRef]
31. Smith, G.C.; Tatum, J.D.; Belk, K.E.; Scanga, J.A.; Grandin, T.; Sofos, J.N. Traceability from a US perspective. *Meat Sci.* **2005**, *71*, 174–193. [CrossRef]
32. Aung, M.M.; Chang, Y.S. Traceability in a food supply chain: Safety and quality perspectives. *Food Control* **2014**, *39*, 172–184. [CrossRef]
33. Wei, Y.-P.; Huang, S.-H. Food traceability system as elevating good corporate social responsibility for fast-food restaurants. *Cogent Bus. Manag.* **2017**, *4*, 1–10. [CrossRef]
34. Golan, E.H.; Krissoff, B.; Kuchler, F.; Calvin, L.; Nelson, K.E.; Price, G.K. Traceability in the U.S. food supply: Economic theory and industry studies. *Agric. Econ. Rep.* **2004**, 33939. [CrossRef]
35. Bader, F.; Jagtap, S. Internet of things-linked wearable devices for managing food safety in the healthcare sector. In *Wearable and Implantable Medical Devices*; Academic Press: Cambridge, MA, USA, 2020; pp. 229–253.
36. Doherty, E.; Campbell, D. Demand for safety and regional certification of food. *Br. Food J.* **2014**, *116*, 676–689. [CrossRef]
37. Kamath, R. Food traceability on blockchain: Walmart's Pork and Mango pilots with IBM. *Med. Klin.* **2018**, *7*, 481–487. [CrossRef]
38. Verbeke, W. The emerging role of traceability and information in demand-oriented livestock production. *Outlook Agric.* **2001**, *30*, 249–255. [CrossRef]
39. Vernede, R.; Verdenius, F.; Broeze, J. Traceability in Food Processing Chains: State of the Art and Future Developments. 2003. Available online: https://www.researchgate.net/publication/40121387_Traceability_in_food_processing_chains_state_of_the_art_and_future_developments (accessed on 27 April 2021).
40. Choe, Y.C.; Park, J.; Chung, M.; Moon, J. Effect of the food traceability system for building trust: Price premium and buying behavior. *Inf. Syst. Front.* **2009**, *11*, 167–179. [CrossRef]
41. Zach, L. 13 -Legal requirements and regulation for food traceability in the United States. In *Advances in Food Traceability Techniques and Technologies Improving Quality throughout the Food Chain*; Espiñeira, M., Santaclara, F.J., Eds.; Woodhead Publishing: Sawston, UK, 2016; pp. 237–260. [CrossRef]
42. GS1 Canada. GS1 Canada—Can-Trace. 2004. Available online: https://gs1ca.org/can-trace/ (accessed on 25 April 2021).
43. Verbeke, W.; Roosen, J. Market differentiation potential of country-of-origin, quality, and traceability labeling. *Estey Cent. J. Int. Law Trade Policy* **2009**, *10*, 20–35. [CrossRef]
44. European Parliament and Council. Regulation (EC) No 178/2002 of 28 January 2002. Available online: http://eur-lex.europa.eu/LexUriServ/LexUriServ.do?uri=OJ:L:2002:031:0001:0024:EN:PDF (accessed on 25 April 2021).
45. Charlier, C.; Valceschini, E. Coordination for traceability in the food chain. A critical appraisal of European regulation. *Eur. J. Law Econ.* **2008**, *25*, 1–15. [CrossRef]
46. Houghton, J.R.; Rowe, G.; Frewer, L.J.; Van, K.E.; Chryssochoidis, G.; Kehagia, O.; Korzen-Bohr, S.; Lassen, J.; Pfenning, U.; Strada, A. The quality of food risk management in Europe: Perspectives and priorities. *Food Policy* **2008**, *33*, 13–26. [CrossRef]
47. Tsai, H.T.; Hong, J.T.; Yeh, S.P.; Wu, T.J. Consumers' acceptance model for Taiwan agriculture and food traceability system. *Anthropologist* **2014**, *17*, 845–856. [CrossRef]
48. TAFTS. Status Quo and Prospect of Taiwan Traceable Agricultural Product. 6 June 2019. Available online: https://taft.coa.gov.tw/ct.asp?xItem=3733&CtNode=321&role=C (accessed on 22 April 2021).
49. Chen, Y.K.; Wang, T.C.; Chen, C.Y.; Huang, Y.C.; Wang, C.Y. Consumer preferences for information on Taiwan's pork traceability system. *Inf. Technol. J.* **2012**, *11*, 1154–1165. [CrossRef]
50. Ubilava, D.; Foster, K. Quality certification vs. product traceability: Consumer preferences for informational attributes of pork in Georgia. *Food Policy* **2009**, *34*, 305–310. [CrossRef]
51. Hsueh, A.; Cohen, S. *Taiwan Retail Food Sector*; USDA: Taipei, Taiwan, 2000; pp. 1–10.
52. Frederick, C.; Chang, C. *Taiwan Retail Foods Annual 2019*; USDA: Washington, DC, USA, 2019.
53. Ford, M.; Chang, C. *Taiwan Retail Foods Annual 2014*; USDA: Washington, DC, USA, 2014. Available online: https://gain.fas.usda.gov/RecentGAINPublications/RetailFoods_TaipeiATO_Taiwan_12-19-2016.pdf (accessed on 29 June 2020).
54. Trappey, C.V. Are Wet Markets Drying Up?—Taiwan Today. *Taiwan Review, Taiwan Today*. 1 March 1997. Available online: https://taiwantoday.tw/news.php?post=12589&unit=8,29,32,45 (accessed on 2 May 2020).
55. Goldman, A.; Ramaswami, S.; Krider, R.E. Barriers to the advancement of modern food retail formats: Theory and measurement. *J. Retail.* **2002**, *78*, 281–295. [CrossRef]
56. Varela, O.; Huang, W.; Sanyang, S.E. Consumer behavior and preference in the fruit markets of Taiwan. *Agri. Econ. Mark. J.* **2009**, *2*, 19–28. Available online: https://www.researchgate.net/publication/275521129_Consumer_Behavior_and_Preference_in_the_Fruit_Markets_of_Taiwan (accessed on 2 July 2020).
57. Ackerman, D.; Tellis, G. Can culture affect prices? A cross-cultural study of shopping and retail prices. *J. Retail.* **2001**, *77*, 57–82. [CrossRef]
58. Kniazeva, M.; Belk, R.W. Supermarkets as libraries of postmodern mythology. *J. Bus. Res.* **2010**, *63*, 748–753. [CrossRef]
59. Paswan, A.; Pineda, M.D.L.D.S.; Ramirez, F.C.S. Small versus large retail stores in an emerging market-Mexico. *J. Bus. Res.* **2010**, *63*, 667–672. [CrossRef]

60. Jones, M.A.; Reynolds, K.E.; Mothersbaugh, D.L.; Beatty, S.E. The positive and negative effects of switching costs on relational outcomes. *J. Serv. Res.* **2007**, *9*, 335–355. [CrossRef]
61. Malhotra, N.K.; Ulgado, F.M.; Wu, L.; Agarwal, J.; Shainesh, G. Dimensions of service quality in developed and developing economies: Multi-country cross-cultural comparisons. *Int. Mark. Rev.* **2005**, *22*, 256–278. [CrossRef]
62. Gorton, M.; Sauer, J.; Supatpongkul, P. Wet markets, supermarkets and the 'Big Middle' for food retailing in developing countries: Evidence from Thailand. *World Dev.* **2011**, *39*, 1624–1637. [CrossRef]
63. Faostat. Meat Consumption Per Capita in Taiwan | Helgi Library. 18 October 2018. Available online: https://www.helgilibrary.com/indicators/meat-consumption-per-capita/taiwan/ (accessed on 29 April 2021).
64. Wu, D. *The Pork Meat Market in Taiwan*; Flanders Investment & Trade: Taipei, Taiwan, 2017; pp. 1–22.
65. Dalvit, C.; Marchi, M.D.; Cassandro, M. Genetic traceability of livestock products: A review. *Meat Sci.* **2007**, *77*, 437–449. [CrossRef]
66. Lilavanichakul, A.; Boecker, A. Consumer acceptance of a new traceability technology: A discrete choice application to Ontario ginseng. *Int. Food Agribus. Manag. Rev.* **2013**, *16*, 25–50. [CrossRef]
67. Song, M.; Liu, L.; Wang, Z.; Nanseki, T. Consumers' attitudes to food traceability system in China: Evidence from the Pork market in Beijing. *J. Fac. Agric. Kyushu Univ.* **2008**, *53*, 569–574. [CrossRef]
68. Rijswijk, W.V.; Frewer, L.J.; Menozzi, D.; Faioli, G. Consumer perceptions of traceability: A cross-national comparison of the associated benefits. *Food Qual. Prefer.* **2008**, *19*, 452–464. [CrossRef]
69. Verbeke, W.; Ward, R.W. Consumer interest in information cues denoting quality, traceability, and origin: An application of ordered probit models to beef labels. *Food Qual. Prefer.* **2006**, *17*, 453–467. [CrossRef]
70. Dickinson, D.L.; Bailey, D. Meat traceability: Are U.S. consumers willing to pay for it? *J. Agric. Resour. Econ.* **2002**, *27*, 348–364. Available online: http://www.jstor.org/stable/40987840 (accessed on 27 April 2021).
71. Dickinson, D.; Hobbs, J.; Bailey, D. A Comparison of US and Canadian Consumers' Willingness to Pay for Red-Meat Traceability. 2003. Available online: https://www.researchgate.net/publication/23696640_A_Comparison_of_US_and_Canadian_Consumers\T1\textquoteright_Willingness_To_Pay_for_Red-Meat_Traceability (accessed on 27 April 2021).
72. Dickinson, D.; Bailey, D. *Willingness-to-Pay for Information: Experimental Evidence on Product Traceability from the U.S.A., Canada, the U.K., and Japan*; Working Papers; Department of Economics, Utah State University: Logan, UT, USA, 2003; Available online: https://ideas.repec.org/p/usu/wpaper/2003-12.html (accessed on 27 April 2021).
73. Enneking, U. Willingness-to-pay for safety improvements in the German meat sector: The case of the Q&S label. *Eur. Rev. Agric. Econ.* **2004**, *31*, 205–223. [CrossRef]
74. Loureiro, M.L.; Umberger, W.J. Assessing consumer preferences for country-of-origin labeling. *J. Agric. Appl. Econ.* **2005**, *37*, 49–63. [CrossRef]
75. Meuwissen, M.P.M.; Lans, I.A.V.D.; Huirne, R.B.M. Consumer preferences for pork supply chain attributes. *NJAS-Wagening. J. Life Sci.* **2007**, *54*, 293–312. [CrossRef]
76. Hanemann, W.M. Welfare evaluations in contingent valuation experiments with discrete responses. *Am. J. Agric. Econ.* **1984**, *66*, 332–341. [CrossRef]
77. Syafiq, A.; Lukman, B.; Suprehatin. Consumer awareness and willingness to pay for halal-certified of Beef in Bogor area. *J. Halal Prod. Res.* **2019**, *2*, 51–59. [CrossRef]
78. Carlsson, F.; Martinsson, P. Do hypothetical and actual marginal willingness to pay differ in choice experiments?: Application to the valuation of the environment. *J. Environ. Econ. Manag.* **2001**, *41*, 179–192. [CrossRef]
79. Lusk, J.L.; Schroeder, T.C. Are choice experiments incentive compatible? A test with quality differentiated Beef Steaks. *Am. J. Agric. Econ.* **2004**, *86*, 467–482. [CrossRef]
80. Owusu, V.; Anifori, M.O. Consumer willingness to pay a premium for organic fruit and vegetable in Ghana. *Int. Food Agribus. Manag. Rev.* **2013**, *16*, 67–86. [CrossRef]
81. Hanemann, M.; Loomis, J.; Kanninen, B. Statistical efficiency of double-bounded dichotomous choice contingent valuation. *Am. J. Agric. Econ.* **1991**, *73*, 1255–1263. [CrossRef]
82. Hanemann, W.M. Valuing the environment through contingent valuation. *J. Econ. Perspect.* **1994**, *8*, 19–43. Available online: http://www.jstor.org/stable/2138337 (accessed on 30 June 2020). [CrossRef]
83. Carmona-Torres, C.; Calatrava-Requena, J. *Bid Design and its Influence on the Stated Willingness to Pay in a Contingent Valuation Study*; International Association of Agricultural Economists: Queensland, Australia, 2006; Available online: https://ideas.repec.org/p/ags/iaae06/25367.html (accessed on 18 April 2021).
84. Cascetta, E. *Transportation Systems Engineering: Theory and Methods*; Kluwer Academic Publishers: Napoli, Italy, 2001; Volume 49.
85. Lancaster, K.J. A new approach to consumer theory author. *J. Polit. Econ.* **1966**, *74*, 132–157. Available online: http://www.jstor.com/stable/1828835 (accessed on 30 June 2020). [CrossRef]
86. Hanemann, W.M.; Kanninen, B.J. The Statistical Analysis of Discrete-Response CV Data. *Oxf. Scholarsh. Online* **1996**, *123*. [CrossRef]
87. McFadden, D. Econometric models of probabilistic choice. In *Structural Analysis of Discrete Data with Econometric Applications*; University Berkeley: Berkeley, CA, USA, 1974; pp. 198–227. Available online: https://eml.berkeley.edu/discrete/ch5.pdf (accessed on 14 December 2020).

88. Perman, R.; Ma, Y.; McGilvray, J.; Common, M. *Natural Resource and Environmental Economics*, 3rd ed.; Pearson Education Limited-Addison Wesley: Harlow, UK, 2003; Volume 6.

89. Jang, Y.J.; Kim, W.G.; Bonn, M.A. Generation Y consumers' selection attributes and behavioral intentions concerning green restaurants. *Int. J. Hosp. Manag.* **2011**, *30*, 803–811. [CrossRef]

90. Kueh, K.; Voon, B.H. Culture and service quality expectations: Evidence from Generation Y consumers in Malaysia. *Manag. Serv. Qual.* **2007**, *17*, 656–680. [CrossRef]

91. Department of Household Registration Affairs. Dept. of Household Registration. Ministry of the Interior. Republic of China (Taiwan)—HISTORY. 2018. Available online: https://www.ris.gov.tw/app/en/3911 (accessed on 5 May 2020).

92. Cicia, G.; Colantuoni, F. Willingness to pay for traceable meat attributes a meta-analysis. *Int. J. Food Syst. Dyn* **2010**, *1*, 252–263. [CrossRef]

93. Dickinson, D.L.; Bailey, D.V. Experimental evidence on willingness to pay for red meat traceability in the United States, Canada, the United Kingdom, and Japan. *J. Agric. Appl. Econ.* **2005**, *37*, 537–548. [CrossRef]

94. Mccluskey, J.; Grimsrud, K.; Ouchi, H.; Wahl, T. Bovine Spongiform Encephalopathy in Japan: Consumers' food safety perceptions and willingness to pay for tested Beef. *Aust. J. Agric. Resour. Econ.* **2005**, *49*, 197–209. [CrossRef]

95. Miller, S.; Tait, P.; Saunders, C.; Dalziel, P.; Rutherford, P.; Abell, W. Estimation of consumer willingness-to-pay for social responsibility in fruit and vegetable products: A cross-country comparison using a choice experiment. *J. Consum. Behav.* **2017**, *16*, 2017. [CrossRef]

96. Viegas, I.; Nunes, L.C.; Madureira, L.; Fontes, M.A.; Santos, J.L. Beef credence attributes: Implications of substitution effects on consumers' WTP. *J. Agric. Econ.* **2014**, *65*, 600–615. [CrossRef]

97. Louviere, J.; Hensher, D.; Swait, J.; Adamowicz, W. Combining sources of preference data. In *Stated Choice Methods: Analysis and Applications*; Hensher, D.A., Swait, J.D., Louviere, J.J., Eds.; Cambridge University Press: Cambridge, UK, 2000; pp. 227–251.

98. Spencer-Cotton, A. Applied choice analysis 2nd ed. *Aust. J. Agric. Resour. Econ.* **2016**, *60*, 17–18. [CrossRef]

99. Maddala, G.S. *Limited-Dependent and Qualitative Variables in Econometrics*; Cambridge University Press: Cambridge, UK, 1983.

100. Council of Agriculture. *Agriculture Statistics Yearbook 2012*; Council of Agriculture: Taipei, Taiwan, 2012.

101. Liu, Y.; Sheu, Y. Study on Pork consumption and CAS sign recognize for different urban degree (in traditional Chinese). *Chin. J. Agribus. Manag.* **2002**, *8*, 223–245. Available online: http://www.airitilibrary.com/Publication/alDetailedMesh?DocID=a0000144-200212-x-8-223-245-a (accessed on 3 July 2020).

102. Tait, P.; Saunders, C.; Guenther, M.; Rutherford, P. Emerging versus developed economy consumer willingness to pay for environmentally sustainable food production: A choice experiment approach comparing Indian, Chinese and United Kingdom lamb consumers. *J. Clean. Prod.* **2016**, *124*, 65–72. [CrossRef]

103. Arrow, K.; Solow, R.; Portney, P.; Leamer, E.; Radner, R.; Schuman, H. *Report of the NOAA Panel on Contingent Valuation*; National Oceanic and Atmospheric Administration: Washington, DC, USA, 1993; Volume 58, Available online: https://www.researchgate.net/publication/235737401_Report_of_the_NOAA_panel_on_Contingent_Valuation (accessed on 18 April 2021).

104. Prathiraja, P.; Ariyawardana, A. Impact of nutritional labeling on consumer buying behavior. *Sri Lankan J. Agric. Econ.* **2003**, *5*, 35–46. [CrossRef]

105. Wu, L.; Wang, S.; Zhu, D.; Hu, W.; Wang, H. Chinese consumers' preferences and willingness to pay for traceable food quality and safety attributes: The case of pork. *China Econ. Rev.* **2015**, *35*, 121–136. [CrossRef]

106. Hsu, J.L.; Chang, W.H. Market segmentation of fresh meat shoppers in Taiwan. *Int. Rev. Retail. Distrib. Consum. Res.* **2002**, *12*, 423–436. [CrossRef]

107. Chamhuri, N.; Batt, P.J. Factors influencing the consumer's choice of retail food store. *Stewart Postharvest Rev.* **2009**, *5*, 1–7. [CrossRef]

108. Angulo, A.M.; Gil, J.M. Risk perception and consumer willingness to pay for certified beef in Spain. *Food Qual. Prefer.* **2007**, *18*, 1106–1117. [CrossRef]

109. Maruyama, M.; Wu, L. Quantifying barriers impeding the diffusion of supermarkets in China: The role of shopping habits. *J. Retail. Consum. Serv.* **2014**, *21*, 383–393. [CrossRef]

110. Putri, W.R.; Samsudin, M.; Rianto, E.; Susilowati, I. Consumers' willingness to pay for Halal labelled Chicken Meat. *J. Din. Manaj.* **2017**, *8*, 122–133. [CrossRef]

111. Makweya, F.L.; Oluwatayo, I.B. Consumers preference and willingness to pay for graded beef in Polokwane municipality, South Africa. *Ital. J. Food Saf.* **2019**, *8*, 46–54. [CrossRef]

112. Banović, M.; Chrysochou, P.; Grunert, K.G.; Rosa, P.J.; Gamito, P. The effect of fat content on visual attention and choice of red meat and differences across gender. *Food Qual. Prefer.* **2016**, *52*, 42–51. [CrossRef]

113. Lee, J.; Han, D.B.; Nayga, R.; Lim, S.S. Valuing traceability of imported beef in Korea: An experimental auction approach. *Aust. J. Agric. Resour. Econ.* **2011**, *55*, 360–373. [CrossRef]

114. Lusk, J.L. Effects of cheap talk on consumer willingness-to-pay for Golden Rice. *Am. J. Agric. Econ.* **2003**, *85*, 840–856. Available online: http://www.jstor.org/stable/1244768 (accessed on 21 April 2021). [CrossRef]

115. Tonsor, G.T.; Shupp, R.S. Cheap talk scripts and online choice experiments: 'Looking Beyond the Mean'. *Am. J. Agric. Econ.* **2011**, *93*, 1015–1031. Available online: http://www.jstor.org/stable/41240379 (accessed on 21 April 2021). [CrossRef]

Article

Government Policy for the Procurement of Food from Local Family Farming in Brazilian Public Institutions

Panmela Soares [1], Suellen Secchi Martinelli [2], Mari Carmen Davó-Blanes [3,*], Rafaela Karen Fabri [4], Vicente Clemente-Gómez [5] and Suzi Barletto Cavalli [1]

[1] Nutrition Post-Graduation Programme, Nutrition Department, Universidade Federal de Santa Catarina, Florianópolis 88040-900, Brazil; panmela.soares@gmail.com (P.S.); sbcavalli@gmail.com (S.B.C.)
[2] Nutrition Department, Universidade Federal de Santa Catarina, Florianópolis 88040-900, Brazil; suellen.smartinelli@gmail.com
[3] Department of Community Nursing, Preventive Medicine and Public Health and History of Science, University of Alicante, 03690 Alicante, Spain
[4] Observatory of Studies on Healthy and Sustainable Food, Florianópolis 88040-900, Brazil; rafa.kf@gmail.com
[5] Public Health Research Group, University of Alicante, 03690 Alicante, Spain; vicente.clemente@ua.es
* Correspondence: mdavo@ua.es

Abstract: This study aims to explore and compare Brazilian public institutional food services' characteristics concerning the implementation of the government policy for the procurement of food from family farming (FF) and the opinions of food service managers on the benefits and difficulties of its implementation. We conducted a cross-sectional study employing an online questionnaire. The results were stratified by purchase. The Chi-square and Fisher's Exact tests were applied. Five hundred forty-one food services' managers participated in the study. Most claimed to buy food from FF, and this acquisition was more frequent among those working in institutions of municipalities <50,000 inhabitants, and educational and self-managed institutions. Those buying from FF developed more actions to promote healthy and sustainable food. Most recognized that the purchase could boost local farming and the economy and improve the institution's food. However, the managers believe that the productive capacity of FF, the lack of technical assistance to farmers, production seasonality, and the bureaucratic procurement process hinder this type of purchase. The self-management of food services and the small size of the municipality might be associated with implementing the direct purchase policy from FF, which can contribute to building healthier and more sustainable food systems. However, the lack of public management support and the weak productive fabric may pose an obstacle to its maintenance or dissemination. The strengthening and consolidation of these policies require more significant government investments in productive infrastructure for family farming.

Keywords: public policies; family farming; health promotion; sustainability

Citation: Soares, P.; Martinelli, S.S.; Davó-Blanes, M.C.; Fabri, R.K.; Clemente-Gómez, V.; Cavalli, S.B. Government Policy for the Procurement of Food from Local Family Farming in Brazilian Public Institutions. *Foods* **2021**, *10*, 1604. https://doi.org/10.3390/foods10071604

Academic Editors: António Raposo, Renata Puppin Zandonadi and Raquel Braz Assunção Botelho

Received: 16 June 2021
Accepted: 7 July 2021
Published: 10 July 2021

1. Introduction

We face a global syndemic setting, where obesity, malnutrition, and climate change coexist and represent the main current challenge to human health, the environment, and the planet [1]. Higher availability and consumption of ultra-processed foods with high amounts of sugar, sodium, and saturated and trans-fatty acids, and reduced consumption of fruits, vegetables, legumes, and whole grains in countries with different socioeconomic contexts are observed [2,3]. At the same time, an increase in adverse impacts on the environment and society results from modern agricultural activities and the increasing distance between production and consumption [4,5].

Building healthier and more sustainable food systems is one of the objectives of the international political agenda [6]. The World Health Organization (WHO) and the Food and Agriculture Organization of the United Nations (FAO), within the framework of the

Second International Conference on Nutrition (2014), urge countries to develop strategies for building more sustainable and healthier food systems. The main recommendations are strengthening food production and processing by small farmers and family farmers and promoting the availability of healthy diets in public facilities, such as educational institutions, hospitals, and prisons [7,8].

Moreover, there is a concern with the increased number of people with chronic non-communicable diseases [9], and the higher negative environmental and social impact from the current food system [10] can be seen in the United Nations 2030 Agenda for Sustainable Development [6].

Among the global objectives to achieve the Agenda's goals are developing strategies that promote a healthier diet, integrated with more sustainable food systems, that allow satisfying food needs in an inclusive, fair, and respectful way with the environment. For this reason, countries from different political, economic, and social contexts are developing policies to promote a healthy diet among the population, including, at the same time, sustainability principles [11–13].

Some countries are establishing nutritional criteria for food supply to these institutions to improve nutrition in public institutions, such as restricting the supply of foods with high amounts of sugar, sodium, and saturated and trans fats [11,14,15]. In the same vein, governments are using public institutions' purchasing power to encourage production and consumption ways that are more environmentally, economically, and socially sustainable. Countries such as Brazil, the United States, Paraguay [16], Italy [17], Sweden, Denmark, Austria, and Scotland [13,18] have incorporated proximity criteria (purchase of food from farmers in the region) for the acquisition of food in public institutions.

Brazil has implemented food services in several public institutions to strengthen the population's food security, such as schools, through the School Food Program (with free meals for the entire school population in the public education network) [19]; university restaurants (with low-cost meals for students from public universities or free meals for socially vulnerable students) [20,21]; popular restaurants (with low-cost meals for the socially vulnerable population) [22,23]. Moreover, other public institutions in the country, such as prisons and hospitals, provide meals for the population served. The planning, management, and implementation of the food service of public institutions is the responsibility of a public agency called the Managing Unit (MU). In school meals, the MUs are the state or municipal education secretariats; for university restaurants, the MU is the university itself; in popular restaurants, MUs are the municipal or state social development secretariats. The MU can define the type of food service management, which can be self-managed or outsourced. In all cases, the MU must ensure that the regulations in force regarding the purchase and quality of the food offered comply [19,22,24]. Among the current regulations, the mandatory purchase of food from family farmers stands out [24].

The direct purchase policy of foods from local family farming, small-scale agriculture with predominantly family labor and an intrinsic relationship with property [25], could positively affect producers and consumers [26–28].

In Brazil, family farming represents 76.8% of establishments and occupies 23% of the area of all agricultural establishments in the country [29]. However, the lack of public policies aimed at these farmers and the difficulty of accessing increasingly globalized markets threaten these farmers' permanence in the countryside.

In 2003, several public policies were developed in an unprecedented way to strengthen the family farming productive sector. Public food purchases made through a bidding process based on economic criteria started to adopt other priority criteria with implementing the Food Acquisition Program (Programa de Aquisição de Alimentos—PAA). Thus, food produced by family farmers was now included in the purchase of public institutions (hospitals, universities, and schools), prioritizing regional and less structured (socially vulnerable) producers, such as settlers of the agrarian reform and indigenous peoples [30]. In 2009, the federal government made public purchasing of food from family farmers extensive and mandatory for all public schools. For this purpose and through specific regulations, a minimum expenditure of program resources was stipulated to purchase food from family farmers directly [19,31]. The positive experience of the school food program contributed to establishing the PAA-Institutional Purchase modality, expanding the possibility of direct acquisition of food from family farming for all public institutional food services (university restaurants, popular restaurants, and hospitals), which became mandatory in 2016 [24].

There is evidence on the environmental, social, and health benefits of implementing the purchase of food from local producers in public institutions [11,13,32]. However, many public institutions are not committed to its implementation, even in the face of the benefits and the existence of a normative framework that encourages the purchase of food from family farmers [12,33,34]. Previous studies have shown that the implementation of food purchases from family farmers is more frequent in municipalities in rural areas, which may be due to the presence of a greater number of family farmers and with more significant support from the local government [33]. In addition, the non-implementation of the procurement policy may be related to implementation constraints. Previous studies suggest that the region's productive fabric may affect the implementation of the purchase [35].

Knowing the benefits and difficulties of implementing the food procurement policy from the viewpoint of managers involved with the planning, managing, and implementing of food services in public institutions can be helpful to develop strategies to strengthen this initiative. Moreover, knowing the characteristics of institutions buying food from family farmers is a starting point for planning strategies that facilitate their implementation in other public institutions.

Thus, this work aims to explore and compare Brazilian public institutional food services' characteristics concerning implementing the government policy for the procurement of food from family farming and the opinions of food service managers on the benefits and difficulties of its implementation.

2. Materials and Methods

This descriptive and analytical cross-sectional study was conducted from 2019 to 2020 by sending an electronic questionnaire (Supplementary Materials) to the MUs of the public institutional food services throughout the Brazilian territory (University Restaurants, School Food Programs, and Popular Restaurants).

The questionnaire was addressed to the manager responsible for the food service of public institutions. Usually, the manager is in charge of planning the menu, preparing the purchase list, selecting suppliers, and supporting the food purchase process.

A structured and self-completed online questionnaire was used and elaborated on the Google Forms platform to collect data. The questionnaire was built from the updated literature on the subject and from questionnaires previously elaborated by the research team in other studies with the same theme [33,36–38]. The questionnaire was reviewed by food and farming experts and tested before its application.

The questionnaire contained questions on the following topics: (a) identification of participants—Position; (b) characterization of the institution—Country Region (North/ Northeast/Midwest/Southeast/South); State; Institution type (School, University, Others (the other category included popular restaurants and other institutional restaurants, such as hospitals and prisons)); Administration (Municipal/State, Regional, or Federal/Federal District); Food services management type (Self-management/Others); Number of people served by the institution (n), Cost of lunch per person per day (Reais); Development of actions to promote healthy food (Yes/No); Development of actions to promote sustainable food (Yes/No); (c) identification of the food purchase process from local family farming—direct purchase of food from local family farming (Yes/No); Year of onset of direct food purchase from local family farming; Purchased food groups (Yes/No); (d) opinion on the benefits and difficulties of direct food purchase from local family farming in public institutions. Closed-ended questions were used for item "d" (yes/no/don't know), containing the main benefits and difficulties of local food purchase identified in the literature. The questionnaire contained a list with statements about the direct purchase from family farming, and respondents should answer whether these were difficulties and benefits observed in their workplace.

The questionnaire was sent by e-mail from April 2019 to January 2020, with the help of an e-mail manager. We used the e-mail addresses available on the web pages of the state education secretariats (MU of school food program in the state school system); Federal Universities and Institutes (MU of university restaurants); and the Ministry of Social Development (Organization linked to popular restaurants). Complementarily and to increase the response rate, a telephone contact was made with the State Education Secretariats, university restaurants, and popular restaurants in the 26 Brazilian states and the Federal District.

The research objectives were explained during the telephone call, and professionals were invited to join the study. If participants accepted the invitation, additional information about the survey and the link to access the questionnaire were sent by e-mail. This procedure was repeated several times until reaching the maximum number of participants from the country's different regions within ten months, stipulated for this research stage. In total, 232 food service MU managers from public institutions in the country were contacted by e-mail or telephone (School food program MU (n = 53); University restaurants MU (n = 93); and MU of popular restaurants (n = 86)).

When contacting the MUs, their collaboration was requested to disseminate the research to other MUs responsible for public institutions' food services in their region to increase the response rate. Moreover, we requested the collaboration of the ten regional councils of nutritionists and the cooperating centers on school food and nutrition to disseminate the research.

Data were analyzed with the SPSS software (IBM Corp., Armonk, NY, USA). A descriptive analysis was performed to explore the characteristics of the participating institutional food services, stratifying data by region, type of institution, and food purchase from family farming. The Chi-square test or Fisher's exact test was applied to identify the association between the characteristics of the institutional food services concerning the implementation of the government policy for the procurement of food from family farming, comparing the characteristics of the institutional food services that bought from family farming with those that did not. The same procedure was used to analyze the study participants' opinions concerning the benefits and difficulties of buying food from family farming, comparing the opinion of those who bought from family farming with those who did not. We decided to merge the answers "no" and "don't know" to analyze the opinions on the purchase's benefits due to the low percentage of these options.

The Human Research Ethics Committee of the Federal University of Santa Catarina approved the study under Opinion N° 3.344.854.

3. Results

Five hundred forty-one institutional food services' managers from different locations in the country participated in the study. Managers of school food and university restaurants from almost every state in the country participated in the study (except for Sergipe, Amapá, and Acre in the case of MUs of school food and Amazonas and Rondônia in the case of the MUs of university restaurants).

Most participants were nutritionists (74%), and 26% were categorized as other professionals (technicians, administrative staff, and social workers). Most of them claimed to carry out management activities (81.7% planning menus, 78.2% preparing shopping lists, and 68% buying food). Figure 1 shows the distribution of participants by country region, state, and institution type in which they worked. The region with the largest number of participants was the Southeast (n = 181), followed by the South (n = 163) and Northeast (n = 118). The ones with the lowest number were the Midwest and the North, with 49 and 39 participants. The states with the largest number of participants were Santa Catarina (n = 92), Minas Gerais (n = 87), São Paulo (n = 42), and Rio Grande do Sul (n = 41).

Regarding the type of institution, most participants reported working for school food MUs ($n = 292$), followed by MUs of university restaurants ($n = 135$).

Institution type

		Amount		
■ School		· 2		60
■ University		◎ 20		80
▦ Others		◉ 40		92

Region	School	University	Others	Total
Southeast	98	38	45	181
South	83	34	46	163
Northeast	71	38	9	118
Midwest	20	13	7	40
North	20	12	7	39
Total	292	135	114	541

State	
Santa Catarina	92
Minas Gerais	87
São Paulo	42
Rio Grande do Sul	41
Paraná	30
Rio de Janeiro	29
Bahia	27
Ceará	23
Espírito Santo	23
Alagoas	15
Rio Grande do Norte	15
Pará	13
Pernambuco	13
Goiás	12
Mato Grosso	10
Piauí	10
Distrito Federal	9
Mato Grosso do Sul	9
Tocantins	9
Amazonas	8
Maranhão	8
Paraíba	4
Acre	3
Sergipe	3
Amapá	2
Rondônia	2
Roraima	2

Total institutions by state

Figure 1. Distribution of study participants by region, State, and type of institution in which they operate.

Regarding the purchase of food to supply the institution, 69.9% of the participants ($n = 378$) stated that the units where they worked bought part of the products directly from family farmers to supply food services. Figure 2 shows the cumulative number of participants by year of onset of direct food purchase from local family farmers, stratified by type of institution and region of the country. An increase in the number of managers from schools and other institutions buying from local farming has been observed since 2010 and from universities starting in 2016.

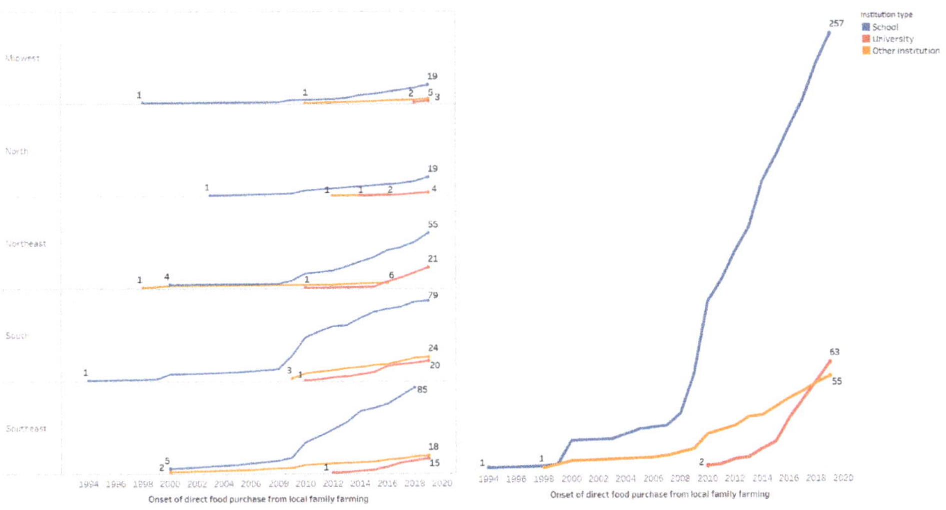

Figure 2. Accumulated number of managing units of public institutional food services per year of onset of direct food purchase from local family farming by type of institution and region of the country.

Figure 3 shows the food groups purchased from family farming by the MUs that implement the government's food procurement policy, stratified by region of the country. In all regions, the typical food group purchased was vegetables, followed by fruits and processed foods. The South had the highest percentage of institutions that bought food processed by family agribusiness (76%), followed by the Northeast (62%). Restaurants in the Midwest bought legumes (11%) and cereals (7%) the least, a value well below the national average (44% and 28%, respectively).

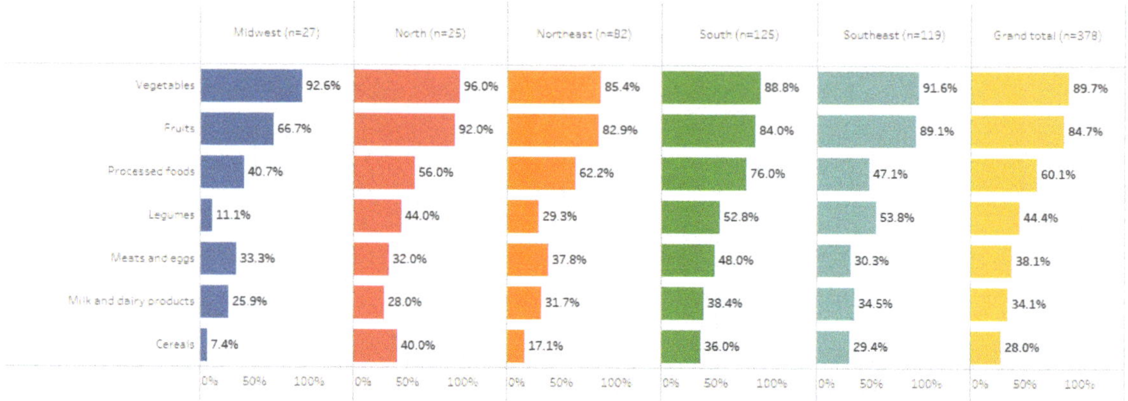

Figure 3. Food groups purchased from family farming by the food service management units of public institutions that implement the government's food procurement policy, stratified by region of the country.

Table 1 shows the characteristics of the managing units of the food services of public institutions, stratified by direct purchase of food from local family farming. Most institutional food services were administered by the Municipality (49%) and bought more food from local producers (81%). An association was also found between food purchase from family farming and self-managed food services, located in municipalities with fewer than 50,000 inhabitants, and public schools (National School Food Program), which served more than 3000 people and with a unit meal costing less than three Brazilian reais ($p < 0.05$). Moreover, participants who bought food from family farming developed more actions to promote healthy and sustainable food (74.3% and 78.3%) ($p < 0.05$). No association was observed between the region and food purchases from local family farming.

Table 2 shows the participants' opinions regarding the benefits of buying food from local family farming, stratified by the purchase made by the managing unit of the food service of public institutions. Most participants in the two groups recognize that the purchase of local food from family farming can contribute to the food system's sustainability, boosting local farming and the economy while improving the institution's food. A statistically significant association was identified between the purchase from family farming and the positive opinion about the increased supply of vegetables and fruits on the menu ($p < 0.05$). However, no associations were noted between purchasing from local farmers and opinions on other benefits.

Table 1. Characteristics of public institutional food services in which the study participants operate, stratified according to the local family farmers' food purchase.

	N (%) 541 (100)	Family Farming Food Purchase	
		Yes, *n* (%) 378 (69.9)	No, *n* (%) 163 (30.1)
Administration [a, c]			
State, Regional, and Federal District	136 (25.9)	87 (64)	49 (36)
Municipal	257 (48.9)	209 (81.3)	48 (18.7)
Federal	133 (25.2)	74 (55.6)	59 (44.4)
Region			
North	39 (7.2)	25 (64.1)	14 (35.9)
Northeast	118 (21.8)	82 (69.5)	36 (30.5)
Midwest	40 (7.4)	27 (67.5)	13 (32.5)
Southeast	181 (33.5)	119 (65.7)	62 (34.3)
South	163 (30.1)	125 (76.7)	38 (23.3)
Self-managed restaurant [a]			
Yes	302 (55.8)	249 (82.5)	53 (17.5)
No	239 (44.2)	129 (54.0)	110 (46.0)
The institution develops actions to promote healthy food [a]			
Yes	428 (79.1)	318 (74.3)	110 (25.7)
No	113 (20.9)	60 (53.1)	53 (46.9)
The institution develops actions to promote sustainable food [a]			
Yes	318 (58.8)	249 (78.3)	69 (21.7)
No	223 (41.2)	129 (57.8)	94 (42.2)
Municipality's size [a]			
<50,000 inhabitants	183 (33.8)	161 (88)	22 (12)
50,000–310,000 inhabitants	174 (32.2)	116 (66.7)	58 (33.3)
>310,000 inhabitants	184 (34)	101 (54.9)	83 (45.1)
Institution type [a]			
School	292 (54)	257 (88)	35 (12)
University	135 (25)	63 (46.7)	72 (53.3)
Other institutions	114 (21)	58 (50.9)	56 (49.1)
N° people serviced by the institution [b, c]			
≤500	187 (34.8)	114 (61)	73 (39)
501–3000	171 (31.8)	124 (72.5)	47 (27.5)
3001+	179 (33.4)	138 (77.1)	41 (22.9)
Cost of lunch/person/day [a,c]			
≤R$3.00	145 (33.9)	122 (84.1)	23 (15.9)
R$3.01–R$8.00	153 (35.7)	97 (63.4)	56 (36.6)
R$8.01+	130 (30.4)	71 (54.6)	59 (45.4)

[a] $p < 0.001$; [b] $p < 0.05$; [c] Total < 541—Lost data.

Table 2. Opinion of study participants concerning the benefits of buying food from local family farming in public institutional food service, stratified according to the family farmers' food purchase.

Benefits **		Total	Food Purchase from Family Farming		*p*-Value
		N (%)	Yes, *n* (%)	No, *n* (%)	
		541 (100)	378 (69.9)	163 (30.1)	
Stimulates the local economy *	Yes	532 (98.3)	372 (98.4)	160 (98.2)	1
	No/Don't know	9 (1.7)	6 (1.6)	3 (1.8)	
Increases the amount of food produced in the region *	Yes	525 (97.1)	369 (97.6)	156 (95.7)	0.269
	No/Don't know	16 (2.9)	9 (2.4)	7 (4.3)	
Increases the variety of food produced in the region	Yes	510 (94.3)	357 (94.4)	153 (93.9)	0.790
	No/Don't know	31 (5.7)	21 (5.6)	10 (6.1)	
Increases food processing in the region	Yes	446 (82.4)	314 (83.1)	132 (81)	0.558
	No/Don't know	95 (17.6)	64 (16.9)	31 (19)	
Increases the supply of fresh food in the institution	Yes	512 (94.6)	362 (95.8)	150 (92)	0.076
	No/Don't know	29 (5.4)	16 (4.2)	13 (8)	
Increases the supply of vegetables and fruits on the institution's menu	Yes	476 (88)	342 (90.5)	134 (82.2)	0.007
	No/Don't know	65 (12)	36 (9.5)	29 (17.8)	
Contributes to the revival of food traditions	Yes	509 (94.1)	354 (93.7)	155 (95.1)	0.514
	No/Don't know	32 (5.9)	24 (6.3)	8 (4.9)	
Improves the quality of the food offered by the institution	Yes	519 (95.9)	365 (96.6)	154 (94.5)	0.261
	No/Don't know	22 (4.1)	13 (3.4)	9 (5.5)	
Contributes to the sustainability of the food system *	Yes	533 (98.5)	373 (98.7)	160 (98.2)	0.702
	No/Don't know	8 (1.5)	5 (1.3)	3 (1.8)	
Increases the farmer's income *	Yes	530 (98)	372 (98.4)	158 (96.9)	0.320
	No/Don't know	11 (2)	6 (1.6)	5 (3.1)	
Ensures market for food produced by family farmers *	Yes	530 (98)	371 (98.1)	159 (97.5)	0.741
	No/Don't know	11 (2)	7 (1.9)	4 (2.5)	

* Fisher's exact test. ** Categories "No" and "Don't know" were grouped for analysis due to the small number of responses.

Table 3 presents the participants' opinions regarding the difficulties in buying food from local family farming, stratified according to the purchase made by the managing unit of the food service of public institutions. Most respondents (> 50%) acknowledged the difficulty in making the purchase as the limited production capacity to meet the institution's food demand (52.5%); production seasonality (62.8%); the bureaucratic purchase process (53.2%); the lack of information from farmers' organizations on the possibility of selling (54.2%); the lack of technical assistance to farmers (62.8%); a limited number of farmers' food-selling organizations (52.1%); and the limited food processing infrastructure (53%). On the other hand, most participants did not consider the following hardships as difficulties: consumers' product acceptance (76%); restaurants' product storage infrastructure (50.6%); the institution's access to information on purchasing possibilities (54.5%); and the amounts paid for products (62.1%). The participants' different opinions have been observed regarding some difficulties: family farming food cost; health surveillance's food sale criteria; lack of support from public management; the number of existing farmers in the region; and product delivery logistics. In these cases, the majority (more than 50%) was not reached in any answer options (yes/no/don't know).

Table 3. Opinion of the study participants concerning the difficulties of buying food from family farming in public institutional food service, stratified according to the family farmers' food purchase.

Difficulties		Total	Food Purchase from Family Farming	
		N (%) 541 (100)	Yes, *n* (%) 378 (69.9)	No, *n* (%) 163 (30.1)
Demand for food greater than family farming production capacity [a]	Yes	284 (52.5)	201 (53.2)	83 (50.9)
	No	184 (34)	145 (38.4)	39 (23.9)
	Don't know	73 (13.5)	32 (8.5)	41 (25.2)
The seasonality of local production does not satisfy the demand for food required by the institution [a]	Yes	340 (62.8)	245 (64.8)	95 (58.3)
	No	143 (26.4)	112 (29.6)	31 (19)
	Don't know	58 (10.7)	21 (5.6)	37 (22.7)
The institutional purchase of food is a very bureaucratic process [a]	Yes	288 (53.2)	185 (48.9)	103 (63.2)
	No	189 (34.9)	164 (43.4)	25 (15.3)
	Don't know	64 (11.8)	29 (7.7)	35 (21.5)
Food sold by family farmers is more expensive than other foods [a]	Yes	224 (41.4)	169 (44.7)	55 (33.7)
	No	232 (42.9)	181 (47.9)	51 (31.3)
	Don't know	85 (15.7)	28 (7.4)	57 (35)
Family farm foods are not well accepted by consumers [a]	Yes	74 (13.7)	59 (15.6)	15 (9.2)
	No	411 (76)	307 (81.2)	104 (63.8)
	Don't know	56 (10.4)	12 (3.2)	44 (27)
The food sale criteria established by health surveillance [a]	Yes	209 (38.6)	154 (40.7)	55 (33.7)
	No	240 (44.4)	181 (47.9)	59 (36.2)
	Don't know	92 (17)	43 (11.4)	49 (30.1)
Lack of institutional restaurant infrastructure for food storage [a]	Yes	218 (40.3)	161 (42.6)	57 (35)
	No	274 (50.6)	196 (51.9)	78 (47.9)
	Don't know	49 (9.1)	21 (5.6)	28 (17.2)
Lack of support from public management [a]	Yes	219 (40.5)	125 (33.1)	94 (57.7)
	No	258 (47.7)	223 (59)	35 (21.5)
	Don't know	64 (11.8)	30 (7.9)	34 (20.9)
The institution lacks information on the possibility of buying food from family farming [a]	Yes	190 (35.1)	98 (25.9)	92 (56.4)
	No	295 (54.5)	251 (66.4)	44 (27)
	Don't know	56 (10.4)	29 (7.7)	27 (16.6)
Farmers lack information on the possibility of selling food to public institutions [a]	Yes	293 (54.2)	199 (52.6)	94 (57.7)
	No	165 (30.5)	141 (37.3)	24 (14.7)
	Don't know	83 (15.3)	38 (10.1)	45 (27.6)
Lack of technical assistance for farmers [a]	Yes	340 (62.8)	245 (64.8)	95 (58.3)
	No	104 (19.2)	88 (23.3)	16 (9.8)
	Don't know	97 (17.9)	45 (11.9)	52 (31.9)
Low amounts paid by institutions for family farming products [a]	Yes	73 (13.5)	40 (10.6)	33 (20.2)
	No	336 (62.1)	289 (76.5)	47 (28.8)
	Don't know	132 (24.4)	49 (13)	83 (50.9)

Table 3. *Cont.*

Difficulties		Total	Food Purchase from Family Farming	
		N (%) **541 (100)**	**Yes, *n* (%)** **378 (69.9)**	**No, *n* (%)** **163 (30.1)**
There are few family farmers in the region [a]	Yes	202 (37.3)	154 (40.7)	48 (29.4)
	No	231 (42.7)	188 (49.7)	43 (26.4)
	Don't know	108 (20)	36 (9.5)	72 (44.2)
Few family farming organizations sell food in the region [a]	Yes	282 (52.1)	222 (58.7)	60 (36.8)
	No	146 (27)	123 (32.5)	23 (14.1)
	Don't know	113 (20.9)	33 (8.7)	80 (49.1)
Farmers' organizations lack the necessary infrastructure for food processing [a]	Yes	287 (53)	223 (59)	64 (39.3)
	No	122 (22.6)	105 (27.8)	17 (10.4)
	Don't know	132 (24.4)	50 (13.2)	82 (50.3)
Product delivery logistics is very costly for family farmers and does not make the sale worth the while [a]	Yes	204 (37.7)	150 (39.7)	54 (33.1)
	No	183 (33.8)	163 (43.1)	20 (12.3)
	Don't know	154 (28.5)	65 (17.2)	89 (54.6)

[a] ($p < 0.001$).

Statistically significant differences were observed in all variables studied ($p < 0.001$) (Table 3) when comparing the opinion of the participants working in the MU of the food service of public institutions that bought food from local family farmers with those that did not. Most of the participants working in managing units of public food services that purchased food from family farming did not consider as purchase implementation difficulties: food cost (47.9%); health surveillance's food sale criteria (47.9%); the number of farmers (49.7%); and product delivery logistics (43.1%). Meanwhile, many of those who did not buy could not express an opinion on these aspects (35%, 30%, 44.2%, and 54.6%, respectively). The results also show that almost 60% of participants working in food services buying from family farming considered the small number of farmers' food-selling organizations (58.7%) and the lack of farmers organizations' food processing infrastructure (59%) as factors hindering the implementation of food purchase from local family farming. On the other hand, approximately 50% of the participants in the group that did not buy from local family farming were unable to express their opinion on these aspects (small number of farmers' food selling organizations (49.1%) and the lack of farmers organizations' food processing infrastructure (50.3%)). Moreover, participants in institutional food services that did not buy from family farming thought that bureaucracy (63.2%), lack of support from public management (57.7%), and the institution's information about the possibility of buying food from local family farming (56.4%) hinder its implementation. These aspects were not considered a difficulty for 49%, 59%, and 66% of those participating in the MU of the food service of public institutions buying food from family farming.

4. Discussion

This study explored and compared Brazilian public institutional food services' characteristics concerning the implementation of the government policy for the procurement of food from family farming and the opinions of food service managers on the benefits and difficulties of its implementation. The implementation of the purchase of food from local family farmers was more frequent in the MUs of school food programs, self-managed institutions, municipalities with a smaller number of inhabitants, and food services serving a larger number of people. Moreover, those who implemented the family farming purchase policy developed more actions to promote healthy and sustainable food. However, the country's region did not influence the implementation of the purchase from local family farmers. According to the study participants, the direct purchase from local family farming can contribute to the food system's sustainability. However, aspects related to the region's productive fabric (such as low production capacity, seasonality, lack of technical assistance, low number of farmers' organizations) can hinder the implementation of local purchasing. In addition, operational aspects, such as bureaucracy, support from public management, and access to information about the purchasing process, were identified as difficulties by most participants that had not implemented the government policy for purchasing food from the local family farming.

According to our study, the MUs of the school food program implemented the government purchasing policy more than other institutions. Moreover, purchases from family farming increased from 2010 on. These results may be related to the longer time elapsed since the mandatory food purchase from local family farms for schools. The purchase became mandatory for public schools in the country as of 2010 [19,31], while other public institutional food services were mandated only in 2016 [24]. This outcome is similar to that of a previous study that suggests the importance of regulatory frameworks to drive the implementation of food purchase policies from local producers [39].

Our results also agree with previous studies that identified that the implementation of the direct purchase policy of food from local farmers is more frequent in self-managed food service institutions [33,40] and smaller, rural municipalities [33]. This situation may be related to a more structured, productive fabric in these regions [41].

On the other hand, our study did not identify differences in the implementation of the direct food purchase from local family farming by region of the country, which differs from a previous study carried out with the Brazilian School Food Program [40]. However, we should consider that our study includes institutions other than schools and that the time elapsed since the mandatory purchase from farmers in the region was implemented may have allowed implementation in a more significant number of institutions.

Educational environments such as schools and universities are a strategic space for developing health promotion actions [42]. The food supply in these institutions is an opportunity to promote healthy and more sustainable food. As in studies by Soares and collaborators [33], our results indicate that purchasing from local producers can favor the development of these actions in public institutional food services. This may be because professionals involved in the direct purchase of food from local farmers may be more sensitive to food, health, and sustainability issues and highlights the potential of local food purchase in institutional food services as a health promotion tool. In this sense, the opinion of this study's participants corroborates the evidence in previous studies [11,12], indicating that food purchase from farmers in the region has a positive impact on local production and supply of fruits and vegetables in public institutional food service, contributing to reviving traditional foods and improving food quality. The most purchased food from local family farming in all regions of the country was fruits and vegetables. On the other hand, our results show differences in the purchase of some food groups, such as legumes and cereals, which may be related to the farming characteristics of each region or their culinary traditions [43,44].

Study participants also stated that local purchasing could positively influence the local economy and farmers' income. Institutional food purchase is an essential mechanism for establishing stable markets for local farming [45]. Considering that the economic impact generated may be modest, product processing is an alternative to add value to the products [46]. In this sense, and similarly to a previous study [11], foods processed by family agribusinesses were among the most purchased foods. On the other hand, our study's participants affirmed that farmers' food processing organizations' infrastructure is the difficulty of purchasing from local family farmers. The investment in infrastructure for food processing can strengthen the direct purchase policies from family farming, generating higher added value to the products and positive impacts on farmers' income and the local economy. In this sense, it is worth highlighting among our results the regional differences in the purchase of processed food by family agro-industries, which suggests inequality in the infrastructure available for food processing in different regions of the country. Knowing these differences is an essential step for directing efforts to build more sustainable food systems.

One of the family farming food procurement policy objectives is strengthening less structured productive sectors, such as family farming, through the establishment of a stable market [47]. However, the participants' opinion shows that aspects related to the region's productive fabric hinder purchases from local family farmers. Therefore, the feasible implementation of the governmental policy to purchase food from family farming requires the region to have organized family farmers to meet the institution's demand. In this sense, it is worth noting that institutional restaurants demand large amounts of food daily. As an example, the university restaurant at the Federal University of Santa Catarina demands an average of 34 tons of fruit and vegetables per month [38].

However, family farmers are not always prepared to meet this demand, which can be a problem for food service MUs that need food periodically. This situation is evidenced in our results, where we identified that the limited production capacity, the lack of technical assistance to farmers for food production, and the limited food processing infrastructure hamper the purchase of food from family farming.

Technical support and technological assistance can provide mechanisms for less structured farmers to produce food competitively [48]. However, farmers with advanced age, low education, and low per capita income, and those who sell directly to the consumer use this service less [49]. Therefore, strengthening public technical assistance agencies in the country can be an essential government tool to help small producers plan production and enter the institutional market.

Depending on food availability in the region [41], the implementation requires the adaptation/adequacy of the institution's food supply to local food production. Besides nutritional aspects, menus and food shopping lists must be constructed considering the productive capacity and the production's seasonality [50]. The implementation of direct purchase policies from farmers in the region requires training efforts and new institutional food services' work routines, which can translate into positive impacts for the provision of healthy food in the institution, the local economy, and the environment. A study carried out in the EU indicates that a regional supply could lead to a 5–8% decrease in greenhouse gas emissions [51]. The authors emphasize that local purchases should be associated with organic products to curb environmental impacts.

The bureaucratic procurement process was also identified as a difficulty in implementing food purchase from family farming. The purchase has several technical and administrative requirements that can, in some cases, take months to become effective. Furthermore, unlike a purchase from a retailer, farmers need planning and time to produce food according to the delivery schedule (quantities per delivery). In this sense, the delay in resolving the purchase process can lead to farmers' lost production or a shortage of food services should farmers not have the food for delivery when the purchase is made official. A previous study identified that farmers' guarantee of food delivery is one of the concerns of agents involved with the purchase of local food in food services [52].

The difficulties perceived by the participants in implementing the purchase from local family farming differed according to their experiences (buying/not buying). The bureaucratic purchasing process, the lack of support from public management, and the lack of information from the institution on the possibility of acquiring food from family farmers in the region were identified as a difficulty by most participants working in institutions not buying from family farming. This result may be because those who worked in public institutional food services that started the purchase did not face or have already overcome these implementation difficulties. Knowing these differences can help develop intervention strategies more appropriate when implementing the direct purchase of food from family farming in each institution. Our results suggest that training and dissemination strategies can boost the implementation of direct purchasing policies. However, strengthening and consolidating these policies requires more significant effort in productive infrastructure (increasing food production/storage/transport/processing capacity), as was pointed out by the participants of institutional food services buying food from farmers in the region.

The implementation of the family farm food purchase policy can positively affect the quality of food offered in public institutions, the local economy, and the environment. Knowing the benefits and difficulties of its implementation can be helpful for planning and implementing food and nutrition public policies. Our results suggest that strengthening and consolidating the government's food procurement policy requires more significant government investment in productive infrastructure for family farming.

When interpreting the results, we should consider that they stem from a convenience sample. Participation in the study was voluntary, which may interfere with the results. The opinions of the participants may be influenced by their professional experience. Moreover, those involved in buying food from local family farming could be more motivated to participate in the study. In addition, the data collection technique used did not allow us to know the actual response rate. However, the sample revealed a very different picture of the implementation of policies for the direct purchase from family farmers in Brazilian public institutional food services, with participants from different regions and diverse experiences. This study achieved a first approximation of the characteristics of institutional restaurants that implement the direct purchase policy of food from local farmers and grasped the opinion of the participants involved in the institutional food purchase process on the benefits and difficulties of buying.

5. Conclusions

The existence of a normative framework favors the implementation of direct purchase policies from local family farming. Nevertheless, self-management of food services and the municipality's small size also seem to be associated with its implementation. This type of food supply in institutions can contribute to building healthier and more sustainable food systems. However, the lack of information, the lack of public management support, and the weak regional productive fabric may pose an obstacle to its maintenance or dissemination to a higher number of institutions.

Supplementary Materials: The following are available online at https://www.mdpi.com/article/10.3390/foods10071604/s1, Compra de alimentos em Restaurante institucional.

Author Contributions: Conceptualization, P.S., S.S.M., and S.B.C.; methodology, P.S. and S.S.M.; formal analysis, P.S. and V.C.-G.; investigation, P.S. and S.S.M.; data curation, P.S.; writing—original draft preparation, P.S.; writing—review and editing, P.S., S.S.M., S.B.C., V.C.-G., R.K.F., and M.C.D.-B.; visualization, V.C.-G.; supervision, S.B.C. and M.C.D.-B.; funding acquisition, S.B.C. All authors have read and agreed to the published version of the manuscript.

Funding: This research was funded by the National Council for Scientific and Technological Development, Ministry of Science, Technology, and Innovation of Brazil: Call CNPq/MCTIC N° 016/2016.

Data Availability Statement: The data presented in this study are available on request from the corresponding author. The data are not publicly available due to privacy.

Acknowledgments: We are grateful to the Graduate Program in Nutrition, Federal University of Santa Catarina (UFSC), for the postdoctoral scholarship of the National Postdoctoral Program of the Coordination for the Improvement of Higher Education Personnel (PNPD/CAPES), Brazil, granted to Panmela Soares.

Conflicts of Interest: The authors declare no conflict of interest. The funders had no role in the study's design; in the collection, analyses, or interpretation of data; in the writing of the manuscript; or in the decision to publish the results.

References

1. Swinburn, B.A.; Kraak, V.I.; Allender, S.; Atkins, V.J.; Baker, P.I.; Bogard, J.R.; Brinsden, H.; Calvillo, A.; De Schutter, O.; Devarajan, R.; et al. The Global Syndemic of Obesity, Undernutrition, and Climate Change: The Lancet Commission report. *Lancet* **2019**, *393*, 791–846. [CrossRef]
2. Monteiro, C.A.; Moubarac, J.C.; Cannon, G.; Ng, S.W.; Popkin, B. Ultra-processed products are becoming dominant in the global food system. *Obes. Rev. Off. J. Int. Assoc. Study Obes.* **2013**, *14* (Suppl. 2), 21–28. [CrossRef] [PubMed]
3. Maciejewski, G. Food consumption in the Visegrad Group Countries—Towards a healthy diet model. *Stud. Ekon.* **2018**, *361*, 20–32.
4. Kastner, T.; Erb, K.-H.; Haberl, H. Rapid growth in agricultural trade: Effects on global area efficiency and the role of management. *Environ. Res. Lett.* **2014**, *9*, 034015. [CrossRef]
5. IPES-Food. *From Uniformity to Diversity: A Paradigm Shift from Industrial Agriculture to Diversified Agroecological Systems*; International Panel of Experts on Sustainable Food Systems (IPES-Food): Brussels, Belgium, 2016.
6. United Nations Development Programme. *Sustainable Development Goals*; United Nations: New York, NY, USA, 2015.
7. Food and Agriculture Organization of the United Nations; World Health Organization. *Second International Conference on Nutrition (ICN2)*; FAO: Rome, Italy, 2014.
8. World Health Organization. *Action Framework for Developing and Implementing Public Food Procurement and Service Policies for a Healthy Diet*; WHO: Geneva, Switzerland, 2021.
9. World Health Organization. *Global Status Report on Noncommunicable Diseases 2014*; World Health Organization: Geneva, Switzerland, 2014.
10. Willett, W.; Rockström, J.; Loken, B.; Springmann, M.; Lang, T.; Vermeulen, S.; Garnett, T.; Tilman, D.; DeClerck, F.; Wood, A.; et al. Food in the Anthropocene: The EAT Lancet Commission on healthy diets from sustainable food systems. *Lancet* **2019**, *393*, 447–492. [CrossRef]
11. Soares, P.; Davó-Blanes, M.C.; Martinelli, S.S.; Melgarejo, L.; Cavalli, S.B. The effect of new purchase criteria on food procurement for the Brazilian school feeding program. *Appetite* **2017**, *108*, 288–294. [CrossRef]
12. Nicholson, L.; Turner, L.; Schneider, L.; Chriqui, J.; Chaloupka, F. State Farm-to-School Laws Influence the Availability of Fruits and Vegetables in School Lunches at US Public Elementary Schools. *J. Sch. Health* **2014**, *84*, 310–316. [CrossRef]
13. Barling, D.; Andersson, G.; Bock, B.; Canjels, A.; Galli, F.; Gourlay, R.; Hoekstra, F.; De Iacovo, F.; Karner, S.; Mikkelsen, B.E.; et al. *Revaluing Public Sector Food Procurement in Europe: An Action Plan for Sustainability*; Foodlinks: Wageningen, The Netherlands, 2013.
14. Rosettie, K.L.; Micha, R.; Cudhea, F.; Peñalvo, J.L.; O'Flaherty, M.; Pearson-Stuttard, J.; Economos, C.D.; Whitsel, L.P.; Mozaffarian, D. Comparative risk assessment of school food environment policies and childhood diets, childhood obesity, and future cardiometabolic mortality in the United States. *PLoS ONE* **2018**, *13*, e0200378. [CrossRef]
15. Piekarz-Porter, E.; Leider, J.; Turner, L.; Chriqui, J.F. District Wellness Policy Nutrition Standards Are Associated with Healthier District Food Procurement Practices in the United States. *Nutrients* **2020**, *12*, 3417. [CrossRef]
16. Swensson, L.F.J.; Tartanac, F. Public food procurement for sustainable diets and food systems: The role of the regulatory framework. *Glob. Food Secur.* **2020**, *25*, 100366. [CrossRef]
17. Sonnino, R. Quality food, public procurement, and sustainable development: The school meal revolution in Rome. *Environ. Plan. A* **2009**, *41*, 425. [CrossRef]
18. Smith, J.; Andersson, G.; Gourlay, R.; Karner, S.; Mikkelsen, B.E.; Sonnino, R.; Barling, D. Balancing competing policy demands: The case of sustainable public sector food procurement. *J. Clean. Prod.* **2016**, *112 Pt 1*, 249–256. [CrossRef]
19. Brazil. Lei n° 11.947 de 16 de Junho de 2009. Dispõe Sobre o Atendimento da Alimentação Escolar e do Programa Dinheiro Direto na Escola aos Alunos da Educação Básica; Altera as Leis n° 10.880, de 9 de Junho de 2004, 11.273, de 6 de Fevereiro de 2006, 11.507, de 20 de Julho de 2007; Revoga Dispositivos da Medida Provisória n° 2.178-36, de 24 de Agosto de 2001, e a Lei n° 8.913, de 12 de Julho de 1994; E dá Outras Providências; 2009. Available online: http://www.planalto.gov.br/ccivil_03/_ato2007-2010/2009/lei/l11947.htm (accessed on 15 August 2020).
20. Mussio, B.R. A Alimentação no Âmbito da Assistência Estudantil para o Ensino Superior: Uma Análise das Universidades Federais Brasileiras. Master's Thesis, Universidade Federal da Fronteira do Sul, Chapecó, Brazil, 2015.
21. Brazil. Decreto n° 7.234, de 19 de Julho de 2010: Dispõe Sobre o Programa Nacional de Assistência Estudantil—PNAES. 2010. Available online: https://rd.uffs.edu.br/handle/prefix/721 (accessed on 10 June 2021).
22. Padrão, S.M.; Aguiar, O.B. Restaurante popular: A política social em questão. *Phys. Rev. Saúde Colet.* **2018**, *28*, e280319. [CrossRef]
23. Brazil Ministério da Cidadania. Programa Restaurante Popular. Available online: https://www.gov.br/cidadania/pt-br/acesso-a-informacao/carta-de-servicos/desenvolvimento-social/inclusao-social-e-produtiva-rural/programa-restaurante-popular# (accessed on 10 June 2021).
24. Brazil. Decreto n° 8.473, de 22 de Junho de 2015. Estabelece, No Âmbito da Administração Pública Federal, o Percentual Mínimo Destinado à Aquisição de Gêneros Alimentícios de Agricultores Familiares e Suas Organizações, Empreendedores Familiares Rurais e Demais Beneficiários da Lei n° 11.326, de 24 de Julho de 2006, e dá Outras Providências.; 2015. Available online: http://www.planalto.gov.br/ccivil_03/_ato2015-2018/2015/decreto/D8473.htm (accessed on 17 January 2021).
25. Garner, E.; Campos, A.P. *Identifying the "Family Farm": An Informal Discussion of the Concepts and Definitions*; Food and Agriculture Organization of the United Nations: Rome, Italy, 2014.

26. World Health Organization. *Global Strategy on Diet, Physical Activity and Health*; 57th World Health Assembly; WHO: Geneva, Switzerland, 2004.
27. Food and Agriculture Organization of the United Nations. *Alimentación Escolar Directa de la Agricultura Familiar y las Posibilidades de Compra: Estudio de Caso de Ocho Países (Versión Preliminar)*; Fortalecimiento de Programas de Alimentación Escolar en el Marco de la Iniciativa América Latina y Caribe Sin Hambre 2025; FAO: Rome, Italy, 2013; p. 107.
28. World Food Programme. *State of School Feeding Worldwide*; World Food Programme: Rome, Italy, 2013.
29. Brazil. Instituto Brasileiro de Geografia e Estatística. Censo Agropecuário. Available online: https://censoagro2017.ibge.gov.br/templates/censo_agro/resultadosagro/pdf/agricultura_familiar.pdf (accessed on 12 April 2020).
30. Brazil. Lei n° 10.696 de 02 de Julho de 2003. Dispões Sobre a Repactuação e o Alongamento de Dívidas Oriundas de Operações de Crédito Rural, e dá Outras Providências; 2003. Available online: http://www.planalto.gov.br/ccivil_03/leis/2003/l10.696.htm (accessed on 15 January 2021).
31. Brazil. Fundo Nacional de Desenvolvimento da Educação. Resolução n° 38, 16 de Julho de 2009. Dispõe Sobre o Atendimento da Alimentação Escolar aos Alunos da Educação Básica no Programa Nacional de Alimentação Escolar—PNAE; 2009. Available online: https://www.fnde.gov.br/index.php/acesso-a-informacao/institucional/legislacao/item/3341-resolu%C3%A7%C3%A3o-cd-fnde-n%C2%BA-38-de-16-de-julho-de-2009 (accessed on 23 August 2020).
32. Harris, D.; Lott, M.; Lakins, V.; Bowden, B.; Kimmons, J. Farm to institution: Creating access to healthy local and regional foods. *Adv. Nutr.* **2012**, *3*, 343–349. [CrossRef] [PubMed]
33. Soares, P.; Caballero, P.; Davó-Blanes, M.C. Compra de alimentos de proximidad en los comedores escolares de Andalucía, Canarias y Principado de Asturias. *Gac. Sanit.* **2017**, *31*, 446–452. [CrossRef] [PubMed]
34. Soares, P.; Martínez-Milan, M.A.; Caballero, P.; Vives-Cases, C.; Davó-Blanes, M.C. Alimentos de producción local en los comedores escolares de España. *Gac. Sanit.* **2017**, *31*, 466–471. [CrossRef] [PubMed]
35. Soares, P.; Suárez-Mercader, S.; Comino, I.; Martínez-Milán, M.A.; Cavalli, S.B.; Davó-Blanes, M.C. Facilitating Factors and Opportunities for Local Food Purchases in School Meals in Spain. *Int. J. Environ. Res. Public Health* **2021**, *18*, 2009. [CrossRef]
36. Cavalli, S.B.; Melgarejo, L.; Soares, P.; Martinelli, S.S.; Fabri, R.K.; Ebone, M.V.; Rodrigues, V.M. Planejamento e operacionalização do fornecimento de vegetais e frutas pelo Programa de Aquisição de Alimentos para a alimentação escolar. In *Avaliação de Políticas Públicas: Reflexões Acadêmicas Sobre o Desenvolvimento Social e o Combate à Fome*; Cunha, J.V.Q., Pinto, A.R., Bichir, R.M., Paula, R.F.S., Eds.; Ministério do Desenvolvimento Social e Combate à Fome. Secretaria de Avaliação e Gestão da Informação: Brasília, Brazil, 2014; Volume 4 Segurança alimentar e nutricional.
37. Soares, P.; Martinelli, S.S.; Melgarejo, L.; Davó-Blanes, M.C.; Cavalli, S.B. Potencialidades e dificuldades para o abastecimento da alimentação escolar mediante a aquisição de alimentos da agricultura familiar em um município brasileiro. *Ciênc. Saúde Colet.* **2015**, *20*, 1891–1900. [CrossRef]
38. Martinelli, S.S.; Soares, P.; Fabri, R.K.; Campanella, G.R.A.; Rover, O.J.; Cavalli, S.B. Potencialidades da compra institucional na promoção de sistemas agroalimentares locais e sustentáveis: O caso de um restaurante universitário. *Segur. Aliment. Nutr.* **2015**, *22*, 558–573. [CrossRef]
39. Schneider, L.; Chriqui, J.; Nicholson, L.; Turner, L.; Gourdet, C.; Chaloupka, F. Are Farm to School Programs More Common in States with Farm to School Related Laws? *J. Sch. Health* **2012**, *82*, 210–216. [CrossRef]
40. De Oliveira Machado, P.M.; de Abreu Soares Schmitz, B.; González-Chica, D.A.; Tittoni Corso, A.C.; de Assis Guedes de Vasconcelos, F.; Garcia Gabriel, C. Compra de alimentos da agricultura familiar pelo Programa Nacional de Alimentação Escolar (PNAE): Estudo transversal com o universo de municípios brasileiros. *Ciênc. Saúde Colet.* **2018**, *23*, 4153–4164. [CrossRef] [PubMed]
41. Botkins, E.R.; Roe, B.E. Understanding participation in farm to school programs: Results integrating school and supply-side factors. *Food Policy* **2018**, *74*, 126–137. [CrossRef]
42. Ottawa Charter. Ottawa Charter for health promotion. In Proceedings of the First International Conference on Health Promotion, Ottawa, ON, Canada, 21 November 1986; pp. 17–21.
43. Brazil. *Guia Alimentar Para a População Brasileira*; Ministério da Saúde, Secretaria de Atenção à Saúde, Departamento de Atenção Básica: Brasília, Brazil, 2014.
44. Instituto Brasileiro de Geografia e Estatística (Brazil). *Censo Agropecuário 2017: Resultados Definitivos IBGE*; Instituto Brasileiro de Geografia e Estatística: Rio de Janeiro, Brazil, 2017.
45. Wittman, H.; Blesh, J. Food Sovereignty and Fome Zero: Connecting Public Food Procurement Programmes to Sustainable Rural Development in Brazil. *J. Agrar. Chang.* **2015**, *17*, 81–105. [CrossRef]
46. Hughes, D.W.; Isengildina-Massa, O. The economic impact of farmers' markets and a state level locally grown campaign. *Food Policy* **2015**, *54*, 78–84. [CrossRef]
47. Food and Agriculture Organization of the United Nations. *Strengthening Sector Policies for Better Food Security and Nutrition Results. Public Food Procurement*; FAO: Rome, Italy, 2018.
48. Berchin, I.I.; Nunes, N.A.; de Amorim, W.S.; Alves Zimmer, G.A.; da Silva, F.R.; Fornasari, V.H.; Sima, M.; de Andrade Guerra, J.B.S.O. The contributions of public policies for strengthening family farming and increasing food security: The case of Brazil. *Land Use Policy* **2019**, *82*, 573–584. [CrossRef]
49. Rocha Junior, A.B.; de Freitas, J.A.; da Cunha Cassuce, F.C.; Almeida Lima Costa, S.M. Análise dos determinantes da utilização de assistência técnica por agricultores familiares do Brasil em 2014. *Rev. Econ. Sociol. Rural* **2019**, *57*, 181–197. [CrossRef]

50. Bianchini, V.U.; Martinelli, S.S.; Soares, P.; Fabri, R.K.; Cavalli, S.B. Criteria adopted for school menu planning within the framework of the Brazilian School Feeding Program. *Rev. Nutr.* **2020**, 33. [CrossRef]
51. Cerutti, A.K.; Contu, S.; Ardente, F.; Donno, D.; Beccaro, G.L. Carbon footprint in green public procurement: Policy evaluation from a case study in the food sector. *Food Policy* **2016**, *58*, 82–93. [CrossRef]
52. Pinard, C.A.; Smith, T.M.; Carpenter, L.R.; Chapman, M.; Balluff, M.; Yaroch, A.L. Stakeholders' Interest in and Challenges to Implementing Farm-to-School Programs, Douglas County, Nebraska, 2010–2011. *Prev. Chronic Dis.* **2013**, *10*, E210. [CrossRef]

Article

Slaughter Conditions and Slaughtering of Pregnant Cows in Southeast Nigeria: Implications to Meat Quality, Food Safety and Security

Ugochinyere J. Njoga [1], Emmanuel O. Njoga [2,*], Obichukwu C. Nwobi [2], Festus O. Abonyi [3], Henry O. Edeh [4], Festus E. Ajibo [5], Nichodemus Azor [5], Abubakar Bello [6,*], Anjani K. Upadhyay [7], Charles Odilichukwu R. Okpala [8,*], Małgorzata Korzeniowska [8] and Raquel P. F. Guiné [9,*]

[1] Department of Veterinary Obstetrics and Reproductive Diseases, Faculty of Veterinary Medicine, University of Nigeria, Nsukka 410001, Nigeria; ugochinyere.njoga@unn.edu.ng

[2] Department of Veterinary Public Health and Preventive Medicine, Faculty of Veterinary Medicine, University of Nigeria, Nsukka 410001, Nigeria; obichukwu.nwobi@unn.edu.ng

[3] Department of Animal Health and Production, Faculty of Veterinary Medicine, University of Nigeria, Nsukka 410001, Nigeria; festus.abonyi@unn.edu.ng

[4] Department of Animal Science, Faculty of Agriculture, University of Nigeria, Nsukka 410001, Nigeria; henry.edeh@unn.edu.ng

[5] Department of Animal Health and Production, Enugu State Polytechnic, Iwollo 401139, Nigeria; ejikeajibo5@gmail.com (F.E.A.); azornichodemus@gmail.com (N.A.)

[6] Faculty of Veterinary Medicine, Wroclaw University of Environmental and Life Sciences, 50-375 Wrocław, Poland

[7] Heredity Healthcare & Lifesciences, 206-KIIT TBI, Patia, Bhubaneswar, Odisha 751024, India; upadhyayanjanikumar6@gmail.com

[8] Faculty of Biotechnology and Food Sciences, Wroclaw University of Environmental and Life Sciences, 50-375 Wroclaw, Poland; malgorzata.korzeniowska@upwr.edu.pl

[9] CERNAS Research Centre, Polytechnic Institute of Viseu, 3504-510 Viseu, Portugal

* Correspondence: njoga.emmanuel@unn.edu.ng (E.O.N.); abubakar.bello@upwr.edu.pl (A.B.); charlesokpala@gmail.com (C.O.R.O.); raquelguine@esav.ipv.pt (R.P.F.G.)

Citation: Njoga, U.J.; Njoga, E.O.; Nwobi, O.C.; Abonyi, F.O.; Edeh, H.O.; Ajibo, F.E.; Azor, N.; Bello, A.; Upadhyay, A.K.; Okpala, C.O.R.; et al. Slaughter Conditions and Slaughtering of Pregnant Cows in Southeast Nigeria: Implications to Meat Quality, Food Safety and Security. *Foods* **2021**, *10*, 1298. https://doi.org/10.3390/foods10061298

Academic Editors: António Raposo, Renata Puppin Zandonadi and Raquel Braz Assunção Botelho

Received: 28 April 2021
Accepted: 3 June 2021
Published: 5 June 2021

Publisher's Note: MDPI stays neutral with regard to jurisdictional claims in published maps and institutional affiliations.

Abstract: The increase in the slaughter of pregnant cows (SPCs) for meat (except as may be approved by veterinarians on health grounds to salvage the animal) is unethical. SPCs for meat is also counterproductive, detrimental to food security, and may enhance zoonotic disease transmission. In this context, therefore, this current study examined slaughter conditions and the slaughtering of pregnant cows, and the implications for meat quality, food safety, and food security in Southeast Nigeria. The direct observational method was employed to examine the slaughterhouse activities, from when the cattle arrived at the lairage to the post-slaughter stage. A pre-tested and validated closed-ended-questionnaire was used to elicit information on causes of the SPCs and the method of disposal of eviscerated foetuses. Pregnancy status of cows slaughtered was determined by palpation followed by visual examination of the eviscerated and longitudinal incised uteri. The study lasted for six months during which 851 cows out of 1931 slaughtered cattle were surveyed. Assessment/decision-making protocol of slaughterhouse conditions, welfare conditions of slaughter-cattle, reasons for sale or slaughter of pregnant cows, distribution of pregnant cows slaughtered, method of disposal of eviscerated foetuses, and estimated economic losses of SPCs were delineated. Of the 851 cows examined, 17.4% (148/851) were pregnant while 43.2% (64/148) of the total foetuses recovered were in their third trimester. Major reasons adduced for SPCs by proportion of involved respondents were: ignorance of the animals' pregnancy status (69.7%, 83/119), high demand for beef (61.3%, 73/119), preference for large-sized cattle (47.9%, 57/119), economic hardship (52.1%, 62/119) and diseases conditions (42.9%. 51/119). The conduct of SPCs for meat would not be profitable. This is because within six months, an estimated loss of about 44,000 kg of beef, equivalent to ₦ 70.1 million or $186,400 would be associated with SPCs and the consequential foetal wastages. If losses were to be replicated nationwide across slaughterhouses, 4.3 tons of beef estimated at ₦ 8.6 billion or $23 million would be wasted. Improving slaughter conditions and the welfare of slaughter-cattle in Nigerian slaughterhouses through advocacy, training of slaughterhouse workers, and strict implementation of laws promoting humane slaughter practices is imperative. Preventing SPCs for meat and inhumane slaughter practices at the slaughterhouse would enhance the welfare needs of slaughter cattle, grow the national herd size, and improve meat safety as well as food security.

Keywords: animal welfare; bovine foetal wastage; food safety and security; meat quality; national herd size; slaughter of pregnant cows

1. Introduction

Globally, achieving high-level beef/meat quality from any given cattle slaughterhouse requires that optimum levels of good practices must be upheld. Regardless of location, cattle pre-slaughter and slaughter operations remain very critical to meat quality and food security [1–3]. There are pre- and post-slaughter conditions/factors in slaughterhouses that influence meat quality and food security. For instance, pre-slaughter factors include high animal density during transport, inadequate lairage facilities at the slaughterhouse, long-distance travel, (poor) handling practices during loading and unloading and during transport, as well as unskilled drivers [4–7]. On the other hand, post-slaughter factors include humidity, temperature, meat pH, microbial contamination and storage time, which would affect the keeping and nutritional qualities of the meat [8]. Moreover, the welfare of animals at slaughter indeed affects meat quality, which undermines food safety, public health, and economic viability in meat production or processing enterprises [2,9]. Additionally, the stress inflicted during slaughter can change the meat characteristics [10–12]. Further, bad slaughter management operations can decrease carcass quality, which could arise because of cross-contamination [2,13]. Some carcass/meat defects caused by pre-slaughter stress or poor welfare conditions are Pale, Soft, Exudative (PSE) and Dark Firm and Dry (DFD) meats [14,15]. Humane handling of slaughter-cattle prohibits cruel practices such as dragging, dropping, throwing or hoisting during the slaughter process [16]. Factors that are used to evaluate compliance to humane slaughter practices include the proportion of the slaughter animals that: (a) were successfully stunned at the first attempt; (b) were rendered unconscious post first stunning; (c) vocalised during stunning and bleeding; (d) fall during handling; and (e) moved with an electric goad [16,17].

Animal welfare (AW) refers to the overall well-being of non-human animals, especially domesticated/farmed animals reared for companionship and food (meat) production [18]. Typically, an animal is considered to be in a good state of welfare when it is healthy and well-nourished, under zero pain, distress or fear, and able to easily express its instinctive behaviour [18]. Good AW, among other things, requires the provision of appropriate feed and shelter, humane handling, and humane slaughter [19]. The principles of AW permit responsible use of animals for human benefits (companionship, food, recreation and work), but such animals must be protected and cared for in a manner that minimises fear, pain, stress, and suffering [20]. In Nigeria, the Department of Veterinary and Pest Control Services (DVPCS) is responsible for protection and promotion of AW. Actually, the DVPCS in Nigeria is under the supervision of the Chief Veterinary Officer (CVO), who also functions as the World Organisation for Animal Health delegate. The CVO oversees animal health and welfare issues, well known to involve: (a) animal disease prevention and control; (b) prevention of the slaughter of gravid animals (unless approved by an accredited veterinary doctor); and (c) humane animal handling/slaughter practices.

Adherence to AW standards, other than an ethical issue, may enhance the microbiological safety of beef and beef products, and therefore protect the consumers from infections with zoonotic pathogens. Stress factors and poor AW can increase susceptibility of slaughter-cattle, especially the pregnant ones, to diseases which may be zoonotic and hence transmissible to humans via the food chain [21]. Globally, approximately 600 million foodborne illnesses and 420,000 deaths have annually been attributed to microbial pathogens, largely owed to poor food (as well as meat) processing practices [9]. Through the provisions of the Meat Edict of 1968 in Nigeria, the slaughter of pregnant female animals (PFAs) is prohibited, so as to improve AW and conserve livestock resources. This specific edict equally forbids the slaughter of gravid animals, with the exception of emergency slaughter. The purpose for this is to relieve animal suffering, and this has to be recommended by a veterinarian. Despite the obsolete nature of this edict, maternal slaughter and the resultant foetal wastage has continued unabated [22–24]. The reason for this could also be attributed to the poor implementation of this edict. Moreover, the immediate cause(s) of this menace (maternal slaughter) still remains elusive. This may be due to the upsurge in the demand for edible animal products in Nigeria [25], occasioned by the rapidly increasing human population, currently estimated at 210 million, which is based on a 3.5% annual growth rate [26].

Besides animal cruelty, the slaughter of pregnant female animals (PFAs) remains counterproductive, threatens food security, and exerts immense loss in livestock revenue/resources [27–30]. Slaughter of pregnant cows (SPCs) would not only depopulate productive female animals, but also depletes future herds via consequential foetal wastages. As such, this would jeopardise efforts towards achieving self-sufficiency in provision of enough edible animal protein, especially in developing countries [31]. SPCs could lead to the introduction of exotic zoonoses (bovine spongiform encephalopathy and variant Creutzfeldt–Jakob disease) through meat importation due to deficits in animal protein occasioned by excessive off-takes from the national herd [31]. In developing countries, the diminution of animal protein is a major public health problem associated with the slaughter of PFAs [32], which may ensue through the unwarranted off-takes from the national herd without commensurate replacement, as epitomised in SPCs and the consequential bovine foetal wastages. Despite the fact that SPCs and associated AW and economic implications are preventable, the practice regrettably persists in Nigeria [24,33–37], where demand for edible animal products incidentally supersedes the supply [38,39].

SPCs appears to increasingly be the status quo, despite the fact that unauthorised slaughter of gravid females for meat connotes animal cruelty and counteracts the growth of the national herd. Therefore, determining the current status and its underpinning drivers are warranted considering that the last available report on SPCs in South-East Nigeria is over three decades old [40]. Additionally, it appears there is no published report/pathway to determine the conditions/welfare of both slaughtered cattle/slaughterhouses in Nigeria. Additionally, knowing the number of pregnant cattle slaughtered for meat may provide useful clues on slaughterhouses' compliance to AW needs, especially in Nigeria. Further, determining the root-causes of SPCs will guide informed decision making regarding improved AW, growing the national cattle herd size and boosting the livestock economy in Nigeria. To supplement existing literature, therefore, this current study examined the slaughter conditions and slaughtering of pregnant cows, and the implications for meat quality, food safety, and food security in South-East Nigeria.

2. Materials and Methods

2.1. Area/Location of Study

The demographics, climatic and orographic conditions of South-East Nigeria, the study location, have been previously described [41,42]. The South-East is one of the six geopolitical zones of Nigeria that consists of five states—Abia, Anambra, Ebonyi, Enugu and Imo (Figure 1). The zone is located on latitude $5°45'00''$ N and longitude $8°30'00''$ E. The South-East zone has a total land area of about 40,000 km^2, and borders Cameroon at the Eastern boundary. The region had an estimated population of about 40 million people in 2019 [43], but realistically the population is now approximately 60 million. The population density of the South-East ranges from 140 to 390 persons per km^2 (depending on the state or town), as against Nigeria's average population density of 226 per km^2; making the zone one of the most densely populated regions in Africa.

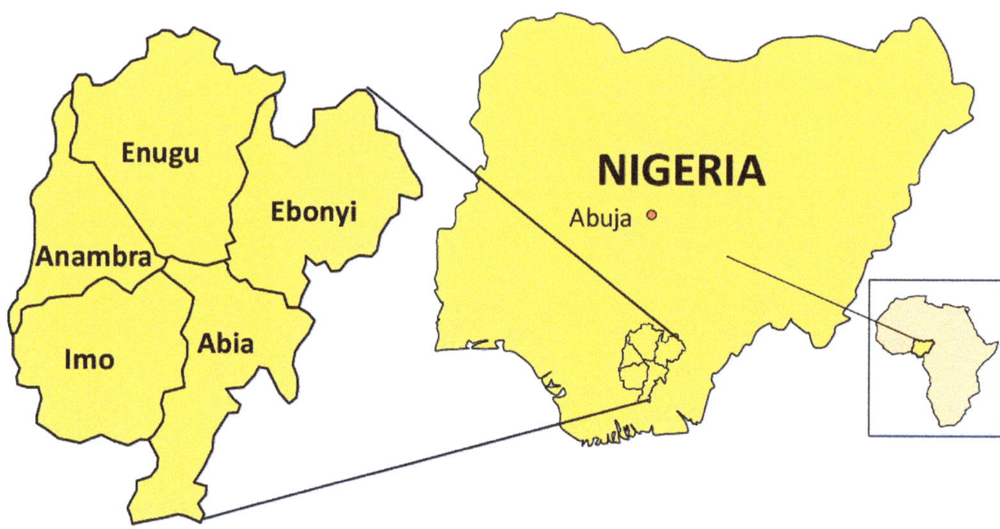

Figure 1. Map of the study area, the South-East geopolitical zone of Nigeria, showing the positions of the zone in African and Nigerian maps, and the constituent five Nigerian states that comprise the South-Eastern part.

2.2. Schematic Overview of the Experimental Program

The schematic overview of the experimental program, from the study design to the age estimations and disposal of eviscerated bovine foetuses from major slaughter facilities in South-East Nigeria, is shown in Figure 2. For emphasis, this current study was designed to examine how slaughter conditions and slaughtering of pregnant cows impact on the meat quality, food safety, and food security, with specific reference to the South-Eastern part of Nigeria. The study adopted a cross-sectional study design which consisted of two phases. The first was the pathway determination of the status of slaughter-cattle, from the process of off-loading, to the period of waiting at the lairage, and during slaughter. This was conducted by direct observational methods, as described by Holmes [44]. The second was determination of the prevalence and causes of SPCs in the study area. Three major slaughter facilities in the study area, Nsukka, Kwata and Akwata slaughterhouses, being the major slaughterhouses with a combined slaughter capacity of about 70% of all cattle consumed in the zone [45], were purposively selected for the study.

2.3. Ethical Approval

Institutional ethics approval was deemed not necessary for this survey activity. This is because the researchers did not slaughter the cattle surveyed, but only examined slaughtered pregnant cows during routine meat/carcass inspection activities performed at the studied slaughterhouses. Pre-slaughter conditions of the animals were determined by direct observational study, and there was no physical restraint/handling of the live animals by the researchers. However, the slaughterhouses' associations approved the use of the research instrument (i.e., questionnaire) prior to the survey. Importantly, given that the survey involved interviews of slaughterhouse workers, informed consent was orally obtained prior to their participation. Additionally, this study adhered to the code of ethics of the World Medical Association Declaration of Helsinki [46], and the participation of every respondent included in this survey was entirely voluntary.

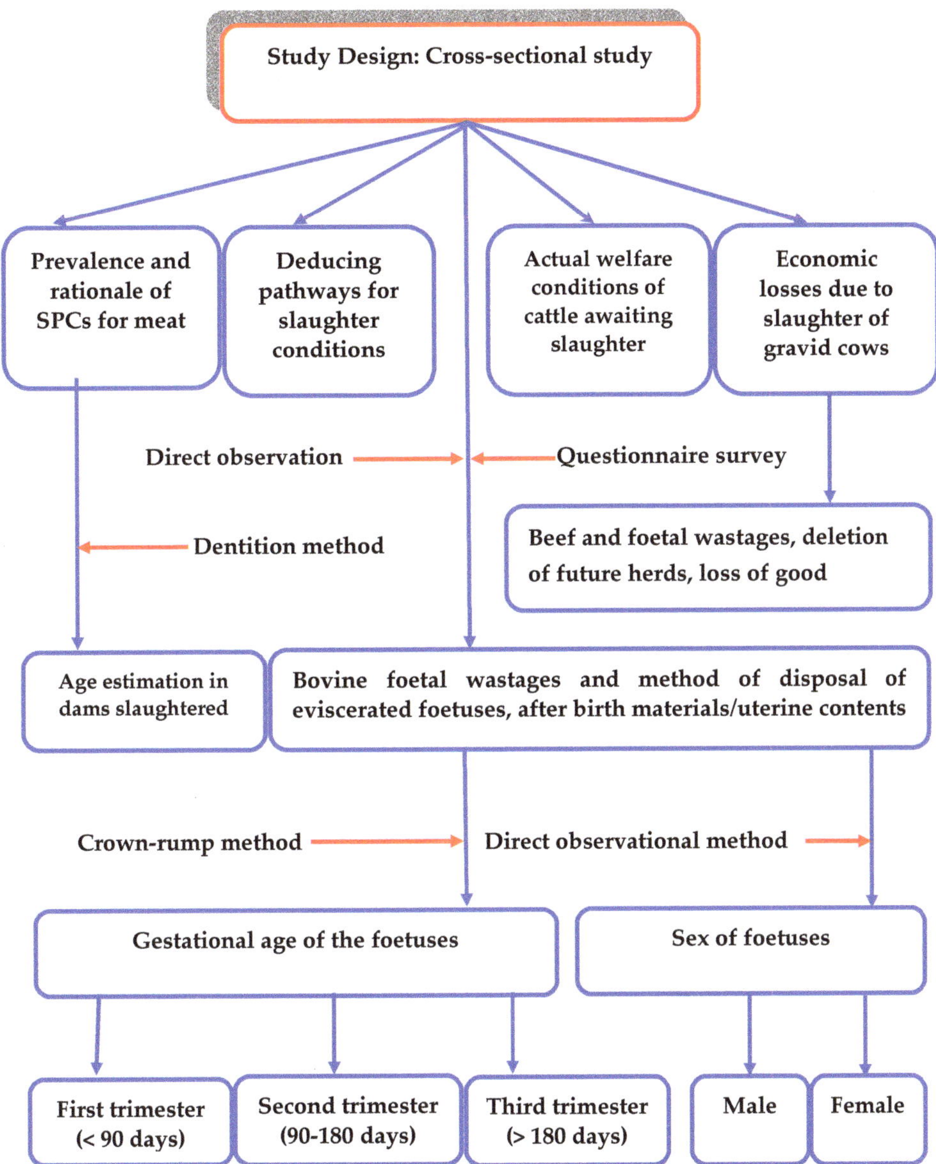

Figure 2. The schematic overview of the experimental program, from the study design, to the age estimations, and the disposal of eviscerated bovine foetuses from major slaughter facilities in South-East Nigeria.

2.4. Research Visits and Sample Size Determination

The selected slaughterhouses were visited weekly, precisely on Saturdays (as more cattle are usually slaughtered on weekends) for six months, (three months during the dry/hot season and another three months during the wet/rainy season) for data collection. A minimum sample size of 385 cows was calculated for the study on pregnancy status of slaughter-cattle, using the formula: $n = Z^2 P (1 - P)/d^2$; where n = required sample size, Z = normal deviate (1.96) at 5% significance level, and P = assumed prevalence of bovine foetal wastages (BFW). The sample size calculation was based on a 50% estimated prevalence, as described by Pourhoseingholi and other workers [47], since there is no published report, to the best of knowledge, on BFW in South-East Nigeria. However, for accuracy and buoyancy of data, a total of 851 cows out of 1931 slaughtered cattle surveyed during the six month period, were selected by simple random sampling technique (toss of coin) in this study.

2.5. Development and Validation of Research Instrument

The research instrument took the form of a structured and pre-tested closed-ended questionnaire. This was used to elicit information underscoring the SPCs and its implications on the meat quality, food safety and security. Specifically, the questionnaire structure had two parts. One part sought to ascertain the pathways to determine the slaughter conditions/welfare. This involved questions that aimed to reveal the major steps followed prior to decision-making about welfare conditions within the studied slaughterhouses. The other part sought to ascertain the possible causes of SPCs and method of disposal of eviscerated foetuses or gravid uterine contents, from which implications of SPCs on the meat quality, food safety and security could be deciphered. This involved questions that aimed to reveal the reasons for sale or slaughter of pregnant cows, distribution of pregnant cows slaughtered, the method of disposal of eviscerated foetuses, and the estimated economic losses of SPCs.

The initial draft questionnaire was subjected to face and content validations consistent with the method described by Bolarinwa [48]. Thereafter, it was pilot tested on 20 respondents in order to detect and correct possible errors that may arise during the actual survey process. Additionally, Cronbach's alpha test was performed to ascertain the reliability of the data/questionnaire in determining the parameters of interest, using IBM® SPSS statistics version 25 (SPSS Inc., Chicago, IL, USA). The test yielded a reliability coefficient (alpha value) of 0.841 (which was ≥ 0.7), and therefore ascertained the questionnaire reliable and valid, and ready for this current work.

2.6. Administration of Research Instrument

A total of 119 slaughterhouse workers, selected from a sampling frame of 356 workers, across the three selected slaughterhouses, participated in the study. The slaughterhouse workers surveyed included butchers, cattle dealers/marketers, animal attendants (herders) and meat sellers. These workers performed various roles, one complementing the other in some cases. For instance, a butcher could be herding the cattle after abattoir operations. In another instance, the cattle marketers may butcher the slaughtered animals and also sell the meat themselves. A simple random sampling technique (toss of coin) was used to select the 119 respondent from over 200 slaughterhouse workers, who willingly volunteered to partake in the survey.

The questionnaire survey was conducted during each visit. The survey was conducted in the form of an interview, and the responses recorded accordingly. The veterinarian and an assistant would ask questions from the questionnaire, and the slaughterhouse worker would provide answers. In some cases where respondents were unable to fully understand the question posed in the English language, the native language was used without change of content or context, but to enhance participation, ensure least pressure (to the respondent), and elicit appropriate response. Utmost care was taken to ensure that each respondent was not surveyed more than once.

2.7. Elaboration of Questionnaire Parameters

2.7.1. Pathways to Determine the Slaughter Conditions

The task here was to determine what would eventually be called "assessment/decision-making protocol of slaughterhouse conditions/welfare". This would help identify and subsequently establish in logical sequence the steps that would lead to decision-making as regards conditions/welfare of slaughterhouses. This involved questions aimed to delineate major factors that could serve as checks, as well as help establish right decisions regards the condition/welfare of the slaughterhouse. Brainstorming sessions with (slaughterhouse) workers, together with the veterinarians that conducted the study, would enable deductions/reflections on the positive and negative welfare conditions, to proffer potential way-outs.

2.7.2. Pregnancy Status of the Selected Slaughtered Cows

The pregnancy status of the selected slaughtered cows was ascertained by visual inspection and palpation of the uteri and the uterine horns for gross/macroscopic evidence(s) of pregnancy. The uteri and the horns were eviscerated post slaughter, longitudinally incised, and then systematically and thoroughly inspected (post-mortem) for evidence of pregnancy (presence of membrane slip, foetus/foetuses, placentation, placentome-foetal cotyledon and maternal caruncle or other afterbirth materials). The ages of the dams were estimated by the dentition method as described by Pace and Wakeman [49]. For the foetuses, estimation of the gestational age was performed by crown-rump length measurement using the formula: $X = 2.5 (Y + 21)$; where X = age of the foetus in days and Y = crown-rump length in cm as described by Kouamo and other researchers [50]. Thereafter, the gestational age of the recovered foetuses were categorised as first (<90 days), second (90–180 days) or third (>180 days) trimester. Similarly, sex of the foetuses was determined by visual examination of the external genitalia at the inguinal region or below the base of the tail.

2.7.3. Estimated Loses of Beef from Slaughtered Pregnant Cows (SPCs)

Losses in beef production associated with SPCs was estimated based on a 63% carcass yield [51], and an average maturity live weight of 460 kg per cattle, as described by Ndi and co-workers [52]. The estimation assumed that the 148 foetuses would be born alive and raised to maturity, but allowance for 5% pre-maturity mortality was factored in as recommended by Ndi and co-workers [52]. Additionally, the associated economic losses were determined by mark-up pricing methods, as described by Crowson [53], using the prevailing price of beef in the local markets. The monetary value was estimated in Naira (₦) and converted to US Dollars ($) based on the current market price of ~₦ 2000 (roughly $5.3 at the Central Bank of Nigeria (CBN) official exchange rate of ₦ 380 per US Dollar as of 01 April 2021) per kilo of beef.

2.8. Statistical Analysis

The resultant data were collated, analysed and presented in tables and figures. Fisher's exact test was used to determine whether there is significant association ($p \leq 0.05$) between SPCs and seasons, age, breed, months of the year and slaughterhouse locations. The statistical significance was set at $p < 0,05$ (95% confidence interval). IBM® SPSS statistics version 25 (SPSS Inc., Chicago, IL, USA) was used to run the statistical analysis.

3. Results

3.1. Establishing the Pathway of Slaughter Conditions/Welfare

Brainstorming sessions with (slaughterhouse) workers, together with the veterinarians that conducted the study, enabled the deduction of the feasible pathway in regard to the slaughterhouse conditions/welfare decision-making activity. The schematic diagram of assessment/decision-making protocols of slaughterhouse conditions/welfare is shown in Figure 3. It shows the major steps, from the assembly of the animal welfare team, to the actual decision-making actions. Table 1 shows the considerations of cattle slaughter conditions/welfare and corresponding way-outs, from environment at the lairage, to during the slaughter of cattle. The considerations of cattle slaughter conditions/welfare were either positive or negative. The way-outs were aimed to alleviate the negative (and strengthen the positive) decision-making.

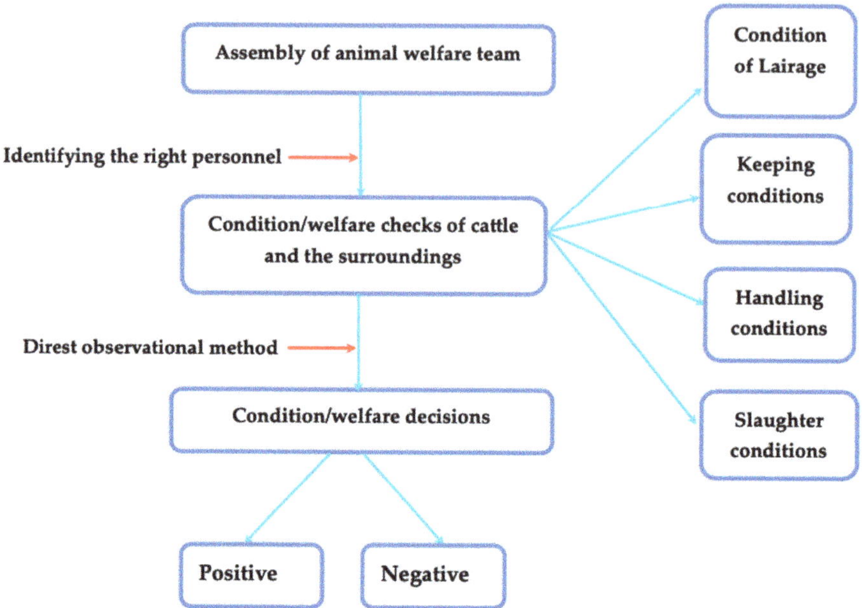

Figure 3. A schematic flow of assessment/decision-making protocol of slaughterhouse conditions/welfare.

Table 1. Positive and negative cattle welfare conditions and corresponding way-outs, from environment conditions at the lairage, up to the actual slaughter process/practices.

Positive Cattle Welfare Condition	Negative Cattle Welfare Condition	Way-Outs/Remarks
CATTLE ENVIRONMENT AT THE LAIRAGE		
* Adequate shelter at the lairage ** Moderate stocking density and hence adequate floor space ** Absence of wounds, lacerations or fight injury ** Ability to express innate abilities like mooing, jumping, etc	* No shelter, marshy or water-logged floor ** Overcrowded or high stocking density ** Presence of wounds, lacerations and injuries/fractures due to fighting caused by overstocking or starvation/thirst ** Depressed, afraid, in a state of pain and hence unable to express innate abilities	Provision of shelter, adequate floor space, feed/water and conducive environment at the holding pen/lairage; to limit overcrowding, stress, fighting and hence injuries and lacerations
PRE-SLAUGHTER CATTLE CONDITION		
** No slipping or falling ** No hitting or mal-handling ** No dragging to the kill floor ** No vocalisation/groaning during handling	** Dragging, hitting, falling, groaning, etc during pre-slaughter handling	Training and retraining on pre-slaughter human handling. Mechanisation of slaughter and processing processed to limit excessive use of force or manual labour. Enactment and or strict implementation of human animal handling before and during slaughter.
DURING SLAUGHTER OF CATTLE		
** Stunning prior to bleed ** Stunning to bleeding interval of less than 5 s ** No dropping, throwing or hoisting during the slaughter ** All animals successfully stunned (were rendered unconscious) at the first attempt. ** No vocalisation/groaning during stunning or bleeding	** No stunning prior to bleeding ** Stunning to bleeding interval of $\geq$ 5 s ** Dropping, throwing or hoisting during the slaughter ** No or few animals successfully stunned (were rendered unconscious) at the first attempt ** Vocalisation/groaning during stunning or bleeding	Training and retraining on modern and human slaughter practices. Mechanisation of slaughter and processing processed to limit excessive use of force or manual labour. Provision of modern slaughter facilities to automate slaughter process and limit meat contamination. Enactment and or strict implementation of human animal handling before and during slaughter.

* Veterinarians' field experience; ** Sourced from published works [2,16,54–56].

3.2. Reasons for Sale or Slaughter of Pregnant Cows

The reasons for sale or slaughter of pregnant cows for meat among herders and abattoir workers in South-Eastern Nigeria are shown in Table 2. Some of the reasons for SPCs and a proportion of the respondents were: ignorance of the animals' pregnancy status (69.7%, 83/119), high demand for beef (61.3%, 73/148), buyers' preference for large-sized animals (47.9%, 57/148), feed scarcity during drought (26.9%, 32/148), economic hardship (52.1%, 62/148) and disease conditions (42.9%, 51/148).

Table 2. Reasons for sale or slaughter of pregnant cows for meat among herders and abattoir workers (*n* = 119 *) in South-Eastern Nigeria.

Reasons/Causes	Number (%) of Respondents		
	Yes	No	No Response
High demand for beef	73 (61.3)	21 (17.6)	25 (21.0)
Economic hardship	62 (52.1)	57 (47.9)	0
Ignorance of the pregnancy status of the animal	83 (69.7)	32 (26.9)	4 (3.4)
Preference for pregnant cows because of size	57 (47.9)	46 (38.7)	16 (13.4)
Feed scarcity during dry seasons	32 (26.9)	87 (73.1)	0
Disease conditions	51 (42.9)	42 (35.3)	26 (21.8)

* Total number of herders and abattoir workers in the 3 slaughterhouses surveyed.

3.3. Distribution of Pregnant Cows Slaughtered

The age, season and breed distribution of pregnant cows (*n* = 148) slaughtered in South-East Nigeria is shown in Table 3. The spatial and temporal distribution of pregnant cows (*n* = 148) slaughtered in major slaughterhouses in South-East Nigeria is shown in Table 4. The number of cows slaughtered statistically differed across the age groups (*p* = 0.029) and season (*p* = 0.038), but not breed (*p* > 0.05). The majority of the slaughtered pregnant cows (60.1% = prevalence) were in their active reproductive age (4–8 years). Approximately 59% (87/148) of the pregnant animals were slaughtered during the dry/hot season. Most of the pregnant cows (80.4%, 119/148) were of the White Fulani breed. The number of cows slaughtered statistically differed across the month of slaughter (*p* = 0.021), but not slaughter location (*p* > 0.05). Similarly, greater proportions (34.5%) of the pregnant cows were slaughtered in December, while only 38.5% (57/148) were killed at the Akwata slaughterhouse, located in Enugu State, Nigeria.

Table 3. Age, season and breed distribution of pregnant cows (*n* = 148) slaughtered in South-East Nigeria.

Variables	Number (%) of Cows Slaughtered	Number (%) of Pregnant Cows Slaughtered	Prevalence	*p*-Value
AGE				
<4 years	126 (14.8)	30 (23.8)	20.3	0.029 *
4–8 years	498 (58.5)	89 (17.8)	60.1	
>8 years	227 (26.7)	29 (12.8)	19.6	
SEASON				
Wet	418 (49.1)	61 (14.6)	41.2	0.038 *
Dry	433 (50.9)	87 (20.1)	58.8	
BREED				
White Fulani	621 (72.9)	119 (19.2)	80.4	0.149
Sokoto gudali	130 (15.3)	18 (13.8)	12.2	
Red bororo	63 (7.4)	7 (11.1)	4.7	
Mixed breeds	37 (4.3)	4 (10.8)	2.7	

* Significant statistical association (*p* ≤ 0.05), Fisher's exact test.

Table 4. Spatial and temporal distribution of pregnant cows (*n* = 148) slaughtered in major slaughterhouses in South-East Nigeria.

Variables	Number (%) of Cows Slaughtered	Number (%) of Pregnant Cows Slaughtered	Prevalence	*p*-Values
MONTHS				
December	197 (23.1)	51 (25.9)	34.5	0.021 *
January	157 (18.4)	24 (15.3)	16.2	
February	129 (15.2)	21 (16.3)	14.2	
July	118 (13.9)	17 (14.4)	11.5	
August	121 (14.2)	16 (13.2)	10.8	
September	129 (15.2)	19 (14.7)	12.8	
SLAUGHTER LOCATIONS				
Enugu	327 (38.4)	57 (17.4)	38.5	0.406
Nsukka	315 (37.0)	49 (15.6)	33.1	
Awka	209 (24.6)	42 (20.1)	28.4	

* Significant statistical association (*p* ≤ 0.05), Fisher's exact test.

The age and sex distribution of foetuses (n = 148) recovered from pregnant cows slaughtered in South-East Nigeria is shown in Table 5. Furthermore, 55.4% (82/148) of foetuses recovered were male, while 22.3% (33/148), 34.5% (51/148) and 43.2% (64/148) were in their first, second and third trimester of gestation, respectively. Significant association ($p \leq 0.05$) existed between SPCs and age and season at p = 0.029 and 0.038, respectively. In the same vein, there was significant association ($p \leq 0.05$) between SPCs and months of the year, but none was found between SPCs and slaughterhouse location at p = 0.406.

Table 5. Age and sex distribution of foetuses (n = 148) recovered from pregnant cows slaughtered in Southeast, Nigeria.

Variables	Number (%) of Foetuses
AGE	
First trimester	33 (22.3)
Second trimester	51 (34.5)
Third trimester	64 (43.2)
SEX	
Male	82 (55.4)
Female	66 (44.5)

3.4. Method of Disposal of Eviscerated Foetuses

The method of disposal of eviscerated foetuses among abattoir workers (n = 119) surveyed in South-East Nigeria is shown in Table 6. Eviscerated foetuses or uterine contents were sold for human consumption, 17.6% (21/119); preparation of dog food, 27.7% (33/119) or disposed by open refuse dump methods, 54.6% (65/119). Likewise, 23.5% (28/119) of the slaughterhouse workers surveyed sold foetuses or afterbirth materials for the feeding of pigs or fish (Table 6). Only 5.9% (7/119), and 3.4% (4/119) of the respondents disposed the foetuses or afterbirth materials by burial and incineration, respectively.

Table 6. Method of disposal of eviscerated foetuses among abattoir workers (n = 119) surveyed in South-East Nigeria.

Method of Foetus Disposal *	Number (%) of YES Respondents
Sold foetuses for human consumption	21 (17.6)
Sold foetuses for preparation of dog food	33 (27.7)
Sold eviscerated foetus for feeding of fishes or pigs	28 (23.5)
Dumped unsold foetus or gravid uterine contents in municipal refuse dump	65 (54.6)
Incinerated unsold foetus or gravid uterine tissues	4 (3.4)
Buried unsold foetus or gravid uterine contents	7 (5.9)

* Many respondents disposed foetuses by more than one method.

3.5. Estimated Economic Losses of SPCs

As regards the economic losses inherent in SPCs and the consequential bovine foetal wastages, if the 148 foetuses were born alive and reared to maturity, this would have yielded 42,890 kg of beef (based on 63% carcass yield and average maturity live weight of 460 kg). If the same numbers of foetuses were wasted in all of the slaughterhouses nationwide, 4.3 tons of beef would have been lost. In monetary terms (based on current market price of ₦ 2000 or $5.3 per kg of beef), this would amount to a loss of ₦ 70.1 million ($186,400), which corresponds to a loss of ₦ 8.6 billion ($23 million) nationwide (data not shown). To estimate the annual economic losses associated in SPCs and the resultant foetal wastages, the aforementioned figures would have to be doubled, since the estimated losses were computed for just a period of six months in which the study lasted.

4. Discussion

Identifying the right personnel, particularly with the appropriate experience, as shown in Figure 3, is very crucial in ensuring that the conditions/welfare checks of the cattle and its surroundings are adequately assessed. Additionally, from Figures 2 and 3, the direct observation method is adapted to help ensure that the final decision-making arrived at is evidence-based. Conditions of the lairage, keeping, handling and slaughter conditions are among the key factors deemed needful in ensuring that optimal welfare checks are adequate. On the other hand, Table 1 has elaborated positive and negative conditions foreseeable in various cattle slaughter situations. Particularly, the way-outs further substantiate how to ensure the authenticity, consistency and standardisation of decision-making. Implementing the decision-making strategy herein (refer to Figure 3 and Table 1) can help maintain humane slaughter practices of cattle thriving in slaughterhouses. To see that humane slaughter practices apply to the slaughter animals, some factors have to be considered, for example the number of cattle that: (1) were successfully stunned at the first attempt, (2) were rendered unconscious post first stunning, (3) vocalised during stunning and bleeding, (4) fall during handling, and (5) moved with an electric goad (where applicable) [16,17]. At the post-slaughter of cattle, there could arise carcass/meat defects such as PSE and DFD, which can be associated with an abnormal meat pH. This generally happens when food-producing animals, prior to slaughter, are subject to poor welfare conditions [14]. Such meat defects pose detrimental effects to the quality indices, which impedes profitability and sustainability of any meat production enterprise [15]. Therefore, cruel practices like dragging, dropping, throwing or hoisting during the slaughter process should be prohibited, as it is not part of the humane handling practices [16]. Notably, poor AW together with stress factors can increase disease susceptibility of slaughter-cattle, especially the pregnant ones. Such disease conditions may be zoonotic, and hence transmissible to humans via the food chain [21].

The findings herein, that 17.4% of the cows slaughtered for meat were pregnant, is unacceptably high from AW and animal production perspectives (Table 3). This portends flagrant disregard and insensitivity to the welfare needs of the pregnant cows. To protect the welfare of pregnant animals, the Nigerian Meat Edict of 1968, the Animal Disease (Control) Act of 2004 (as amended), as well as the Council Regulation (No 1/2005) of the EU [57], prohibits the transportation of gravid animals, especially at the late stage of their gestation. The laws/regulations likely foresee that the welfare needs of the dam and that of their foetuses may not be met en route, due to the delicate nature of pregnant animals. Therefore, the SPCs only climaxed the abuse of the animals' welfare, which was violated *ab initio* when the gravid dams were transported to the abattoirs for slaughter. Although Nigeria does not have a stand-alone legislation on the welfare of animals, legislations guiding the use and ownership of animals, whether in the form of Acts, Code, and Laws, of which some are considered obsolete, have been amended to promote AW. The development of the Nigerian Animal Welfare Strategy Framework in 2016 is one of such amendments. This welfare strategy, which applies to all domestic and captive animals, proscribes all forms of cruelty, deliberate pain or suffering to animals, through negligence or failure to act by the animal owners/custodians/care giver. Additionally, the Animal Disease (Control) Act of 2004 (as amended) provides some additional protections for farm animals, including limiting stocking density during transportation to ensure adequate ventilation. This specific Act also provides for feeding en route, for slaughter/trade animals transported over very long distances. However, there appears to be no significant progress, particularly to ensure the strict implementation of these legal statutes. The operationalisation of standard AW practices in Nigeria remains very challenging, because the nation's cattle production sub-sector is almost entirely in the hands of "Fulani" herdsmen, who engage in the nomadic system of cattle-rearing that involves (calves or pregnant cows) trekking over very long distances daily in search of food or water [58–60].

To actualise the effective implementation of humane slaughter practices in Nigerian slaughterhouses remains very challenging, largely due to the limited mechanised slaughter facilities [61]. Moreover, stunning prior to bleeding during the slaughter process can greatly reduce pain perception. However, the 'halal' slaughter widely practiced in Nigeria appears yet to fully embrace stunning, particularly penetrative percussive stunning, even though some workers [62,63] have reported this stunning type able to knock out the consciousness in slaughter animals with ease. Stress due to decreased pre-slaughter welfare conditions (no housing and harsh climatic conditions) observed in this study can lower the immune status of slaughter cattle. This situation can aid the acquisition and dissemination of economically important livestock diseases, thereby worsening the already precarious welfare condition of the animals [64–66]. This is where and why the age of the cattle and season of slaughter remains very crucial. In addition, the humane pre-slaughter handling is key to preventing an abrupt increase in the rate of pH decline at the early phases of post-mortem, because this situation can greatly affect both physiochemical and sensory qualities of the meat [7,56,67,68]. For instance, the water-holding capacity (WHC) is a key index of meat that demonstrates the ability of post-mortem muscle to retain moisture despite external pressures (cooking/heating, gravity). Moreover, pre-slaughter stress or exhaustion increases heart and respiratory rates, including muscle metabolic activities. This situation will deplete the muscle glycogen available for enzymatic conversion to lactic acid in the meat tissue, which is required to optimise the post-mortem pH (5.5–5.7) [69–72], prolong the shelf life, and limit microbial spoilage. Humane pre-slaughter practices would improve the sensory qualities of meat such as tenderness (or texture), juiciness and flavour [73,74].

A number of poor pre-slaughter practices were noted in this study. These included dragging, strangulation and excessive beating, which can increase the levels of creatine kinase (CK) and aspartate aminotransferase (AST) in meat. These enzyme markers are indicative of muscle damage and low meat quality, due to their adverse effects on meat colour (appearance), taste, and other quality indicators [75]. Agbeniga and Webb [76] reported that reducing both anxiety/stress and pre-slaughter stunning enhanced bleed-out, decreased blood retention in the trachea, as well as blood splash in the lungs, which prolonged the meat shelf-life quality and limited the proliferation of spoilage bacteria. Enhanced bleed-out during slaughter is cardinal to food safety, as it limits microbial meat spoilage. Inefficient bleeding during slaughter owing to shock or fright-induced vasocon-striction can withhold blood in edible tissues [77], which facilitates meat decomposition and spoilage bacteria proliferation [78]. Further, high tissue cortisol level caused by pre-slaughter stress also facilitates lipid oxidation and activities of autolysis enzymes, which would enhance putrefaction or formation of PSE meat, even under storage (refrigeration or frozen) conditions [1,12,78,79].

As per the welfare of the foetuses, the scientific debate has been on-going, for and against the bridge of AW inherent in foetal wastages as a result of the SPCs. The contentions focus on whether a foetus can perceive pain during maternal slaughter, and if so, at what stage of gestation does the pain perception begin. While the EFSA [80] affirmed that foetuses do not feel pain during the first and second trimesters, as the relevant physical and neurological structures only develop during the third trimester, there are expert opinions that suggest there could be between a 1%–33% likelihood that foetuses may experience pain during the last stage of gestation (third trimester). Corroborating this, recent expert opinions from the field of developmental neuroscience suggest that pain perception in foetuses may even start earlier than might have been previously thought. Derbyshire and Bockmann [81] reported that the brain cortex and the associated tracts, which are responsible for pain perception, could develop and emerge functional during the latter phase of the second trimester. Despite this, the "foetal pain" may be sub-cortically based, even though the relatively undifferentiated experiences of discomfort may be due to neural processing at levels below the cerebral cortex [82]. For actual pain perception to occur, the foetus needs to be both sentient and conscious [83]. During the gestational life, foetuses are maintained in a sleep-like unconscious mode through the activities of neuro-inhibitors [83]. Consciousness only appears after birth following a substantial withdrawal of the neuro-inhibitors particularly adenosine [83,84]. This unconscious state makes pain perception in the foetus most unlikely. This implies that on event of unforeseen death of the dam, the foetus simply passes from the unconscious state to death with no or minimal pain. This argument, however, does not justify the unapproved slaughter of pregnant animals.

The high rate (17.4%) at which pregnant cows were slaughtered and foetuses recovered (Tables 4 and 5) shown in this current study is rather surprising. This suggests that AW awareness in the study area appears to be low. With the exception of the SPCs, on the grounds of animal health and other welfare considerations, as may be approved by a veterinarian on ethical grounds, maternal slaughter on the basis of increased meat demand connotes cruelty and is hence unacceptable. It appears that the slaughter of gravid cows and the associated welfare issues is not restricted to Nigeria, as SPCs have been reported in more advanced countries such as the UK [85], Germany [86] and more recently in Denmark [87]. However, the magnitude of the practice and the reasons for the slaughter may vary from place to place. While the EFSA [80] reported that an average of 3% of dairy cows, and 1.5 % of beef cattle slaughtered in the EU were in their last stage of gestation, our findings shows that 43.2% of pregnant cows slaughtered were at their third trimester. The slaughter of terminally gravid cows is worrisome, as pregnancies at the third trimester of gestation should be detected with ease, even by visual assessment. The slaughter of pregnant cows/heifers at the latter stages of pregnancy, as found in this work (Table 5), agrees with the findings of Wosu [40], in which 74% of pregnant cows slaughtered were in their second or third trimesters. This destructive practice does not only indicate animal cruelty but also threatens the development or growth of the livestock industry through excessive off-takes, depletion of the future herd, as well as loss of good genetic traits [88].

In the current study, SPCs within six months would lead to an estimated beef loss of about 44,000 kg, which would be equivalent to about ₦ 70.1 million or $186,400, with consequential foetal wastage. If this were to be replicated nationwide across slaughterhouses, there could be an estimated loss of about 4.3 tons of beef, equivalent to about ₦ 8.6 billion or $23 million. Further, if SPCs are allowed to persist unabated, the already precarious situation of acute animal protein shortfall in the country may worsen. This is because the 21 million cattle in the country, with an average annual growth rate of 1.85% [89], may not provide all the beef and dairy needs of approximately 210 million Nigerians with a population growth rate of 3.5% per annum [26]. Given that most developing countries are yet to achieve self-sufficiency in beef production [90], it is mind blowing to imagine the colossal loss of 4.3 tons of beef or ₦ 8.6 billion ($23 million) due to SPCs and the resultant BFW. These avoidable losses are major setbacks to food security and economic development, in a country where meat supply grossly lags behind the demand [91].

Apart from the losses and the inherent economic wastages, SPCs may facilitate the spread of zoonotic pathogens inhabiting the reproductive tract of cows. *Brucella* and *Leptospira* organisms inhabit the reproductive and urinary tracts. Zoonotic transmission of brucellosis in abattoirs is possible via direct contact with placenta, eviscerated foetuses, uterine secretions or afterbirth materials from infected animals [92–95]. *Leptospira* infection can result from direct or indirect exposure to body fluids (urine) from the reproductive or urinary tracts of an infected animal [96,97], especially during abattoir meat processing. During the dressing or gutting of an infected animal, the contamination of the processed meat as well as the infection of the slaughterhouse workers would most likely be inevitable, if slaughterhouse workers do not wear personal protective equipment during routine operations [98]. It is also possible for the abattoir environment to be contaminated with the zoonotic organisms from food animals grazing around infected areas, which become onward transmissions of the pathogens to humans. This can apply to eviscerated foetuses discarded by unconventional methods (Table 6), which can aid environmental contamination and disease spread.

In all ramifications, the conduct of SPCs for meat would not be profitable. Herders, who may have erroneously culled their cows for infertility reasons, probably due to poor proficiency in pregnancy detection, would be at a loss. The butchers and beef sellers are equally at loss, as pregnant animals, especially those in their late stage of pregnancy, yield less meat than non-pregnant ones [87,99,100]. Further, the quality of meat from pregnant animals is doubtful. Meats sourced from pregnant animals are watery, have high pH and peak shear force values, poor eye appeal and also smell and tastes abnormal, due to high progesterone tissue levels [100,101]. In order to remedy the problem of slaughtering pregnant cows, decisive steps should be taken to provide trained personnel who are proficient in pregnancy diagnosis at the slaughterhouses. One major step in this regard is the engagement of more veterinarians and other trained animal health workers as meat inspectors and animal attendants, respectively. In Nigeria, the local and state governments have the responsibility to oversee and regulate the operations of slaughterhouses domiciled in their locality; however, some governments have contracted the oversight and regulatory functions to privately owned organisations. As a result, ante-mortem inspections, including pregnancy diagnosis in female animals intended for slaughter, have been reduced to mere collection of revenue per animal slaughtered [102,103]. To worsen the problem, veterinarians, who by training are competent on pregnancy and disease diagnosis in animals, are hardly employed by the contractors, probably to save costs. This development may have resulted in the high rate of SPCs being reported, and should be reversed by engagement of trained personnel at the abattoirs to improve the welfare conditions of slaughter-cattle and halt the SPCs.

Seasonality in the SPCs demonstrated herein (Table 3) could be a useful clue exploitable for effective control of AW challenges during the dry/hot seasons. A huge quantity of animal feed could be stocked and sold off to livestock farmers during draught and/or festive periods, at a subsidised rate, as an incentive to prevent sale or slaughter of gravid cows for meat. Significant loss of body conditions, as a result of drought-related food and water scarcities during the dry/hot season, usually compels livestock owners to salvage their animals by selling them off, irrespective of their pregnancy status [89,104]. Additionally, diversification of meat production through intensification of poultry production and other food animals could lessen the increased meat demand during festive periods, which usually prompts livestock farmers to sell off their animals, including pregnant ones, as buyers tend to offer higher prices. On the part of livestock farmers, there is a need for supervised breeding (especially in outdoor/extensive farming systems), proper recording of all breeding activities and routine on-farm pregnancy testing. This will help to limit farmers' ignorance on the pregnancy status of their animals, which was one of the major factors underpinning the sale of SPCs for meat (Table 2). In this regard, farmers' training and retraining of abattoir workers on some on-farm and rapid pregnancy diagnostic methods will be a step in the right direction. Also, massive public enlightenment campaigns on AW and the need to curtail the SPCs for meat is recommended, since the welfare of animals, particularly slaughter-cattle, is not prioritised in most developing countries.

5. Concluding Remarks

The slaughter of pregnant cows (SPCs) and its implications on the meat quality and food safety/security has been examined in this current study. Of the 851 cows, 17.4% (148/851) were pregnant while 43.2% (64/148) of the total foetuses recovered were in their third trimester. The finding that 17.4% of the cows slaughtered for meat were pregnant is unacceptably high from AW and animal production perspectives. The high rate (17.4%) at which pregnant cows was slaughtered is rather surprising, and suggests that AW awareness in the study area appears low. In this study, SPCs within six months would estimate beef loss amounting to 44,000 kg, which would be equivalent to ₦ 70.1 million or \$186,400, with consequential foetal wastages. Despite this, the SPCs for meat is unprofitable in all ramifications, and its seasonality demonstrated herein could be useful clue for effective control of AW challenges in slaughterhouses in South-East Nigeria.

Strict implementation of the provisions of the Meat Edict of 1968 and Animal Disease (Control) Act of 2004, including humane slaughter and the prohibition of slaughter of PFAs, remains imperative. In fact, an amendment of the laws or enactment of the Nigerian Animal Welfare Strategy Framework drafted in 2016 in line with the present day AW issues, particularly at the slaughterhouses, is long overdue. Incorporating compulsory pregnancy diagnosis of all slaughter female animals in the ante-mortem inspection protocol is warranted. Proper labelling of meats obtained from pregnant animals, to guide buyers' discretion, should be mandatory. Slaughter facilities and persons defaulting on the issue of humane slaughter practices should be publicly reprimanded to deter others from doing the same. Overall, SPCs for meat and inhumane slaughter practices at the slaughterhouse should be prevented, to enhance the welfare needs of slaughter cattle, grow the national herd size, and improve meat safety as well as food security in Nigeria.

Author Contributions: Conceptualisation, U.J.N., E.O.N., O.C.N., H.O.E.; Data curation, U.J.N., O.C.N., F.O.A., F.E.A., N.A., A.B., A.K.U., C.O.R.O., M.K., R.P.F.G.; Literature search/synthesis, U.J.N., E.O.N., O.C.N., F.O.A., H.O.E., F.E.A., C.O.R.O.; Software, O.C.N., F.O.A., F.E.A., N.A., A.B., A.K.U.; Formal analysis, U.J.N., E.O.N., H.O.E., F.E.A., N.A., A.K.U.; Creation of figures, E.O.N., N.A., C.O.R.O.; Project administration, E.O.N., F.O.A., C.O.R.O., M.K., R.P.F.G.; Methodology, U.J.N., E.O.N., F.E.A., N.A., C.O.R.O.; Funding acquisition, C.O.R.O., M.K. and R.P.F.G.; Supervision, E.O.N., C.O.R.O., M.K., R.P.F.G.; Validation, O.C.N., F.O.A., H.O.E., F.E.A., A.K.U., A.B., C.O.R.O., M.K., R.F.P.G.; Visualisation, O.C.N., N.A., A.B., A.K.U., C.O.R.O., M.K., R.P.F.G.; Writing—original draft, U.J.N., E.O.N., O.C.N., F.O.A., N.A., H.O.E.; Writing—review and editing, A.B., A.K.U., C.O.R.O., M.K., R.P.F.G. All authors have read and agreed to the submitted version of the manuscript.

Funding: Publication financed by the project UPWR 2.0: international and interdisciplinary programme of development of Wrocław University of Environmental and Life Sciences, co-financed by the European Social Fund under the Operational Program Knowledge Education Development, under contract no. POWR.03.05.00-00-Z062/18 of 4 June 2019.

Institutional Review Board Statement: Not applicable.

Informed Consent Statement: All slaughterhouse workers provided their informed consent orally and individually prior to participation in this current survey.

Data Availability Statement: Data sharing not applicable.

Acknowledgments: The authors A.B., C.O.R.O. and M.K. acknowledge financial support from the Wrocław University of Environmental and Life Sciences, Poland. The author R.P.F.G. acknowledges financial support from the Polytechnic Institute of Viseu, Portugal.

Conflicts of Interest: The authors declare no conflict of interest.

References

1. Xing, T.; Gao, F.; Tume, R.K.; Zhou, G.; Xu, X. Stress Effects on Meat Quality: A Mechanistic Perspective. *Compr. Rev. Food Sci. Food Saf.* **2019**, *18*, 380–401. [CrossRef] [PubMed]
2. Cevallos-Almeida, M.; Burgos-Mayorga, A.; Gómez, C.A.; Lema-Hurtado, J.L.; Lema, L.; Calvache, I.; Jaramillo, C.; Ruilova, I.C.; Martínez, E.P.; Estupiñán, P. Association between animal welfare indicators and microbiological quality of beef carcasses, including Salmonella spp., from a slaughterhouse in Ecuador. *Veter. World* **2021**, *14*, 918–925. [CrossRef]
3. Okpala, C.; Nwobi, O.; Korzeniowska, M. Assessing Nigerian Butchers' Knowledge and Perception of Good Hygiene and Storage Practices: A Cattle Slaughterhouse Case Analysis. *Foods* **2021**, *10*, 1165. [CrossRef] [PubMed]
4. McCleery, D.R.; Stirling, J.M.E.; McIvor, K.; Patterson, M.F. Effect of ante-and postmortem hide clipping on the microbiological quality and safety and ultimate pH value of beef carcasses in an EC-approved abattoir. *J. Appl. Microbiol.* **2008**, *104*, 1471–1479. [CrossRef]
5. Węglarz, A. Meat quality defined based on pH and colour depending on cattle category and slaughter season. colour and pH as determinants of meat quality dependent on cattle category and slaughter season. *Czech. J. Anim. Sci.* **2010**, *55*, 548–556. [CrossRef]
6. Węglarz, A. Effect of pre-slaughter housing of different cattle categories on beef quality. *Anim. Sci. Pap. Rep.* **2011**, *29*, 43–52.
7. Abubakar, A.; Zulkifli, I.; Goh, Y.; Kaka, U.; Sabow, A.; Imlan, J.; Awad, E.; Othman, A.; Raghazli, R.; Mitin, H.; et al. Effects of Stocking and Transport Conditions on Physicochemical Properties of Meat and Acute-Phase Proteins in Cattle. *Foods* **2021**, *10*, 252. [CrossRef]
8. Delgado, H.; Cedeño, C.; de Ocaii, N.M.; Villoch, A. Hygienic quality of the meat obtained at slaughterhouses in Manabí- Ecuador. *J. Anim. Health* **2015**, *37*, 1–9.
9. Yenealem, D.G.; Yallew, W.W.; Abdulmajid, S. Food Safety Practice and Associated Factors among Meat Handlers in Gondar Town: A Cross-Sectional Study. *J. Environ. Public Health* **2020**, *2020*, 7421745. [CrossRef]
10. Bautista, J.H.; De Oaxaca, M.U.A.B.J.; López, J.L.A.; Rincón, F.G.R. Pre-mortem handling effect on the meat quality. *Nacameh* **2013**, *7*, 41–64. [CrossRef]

11. Sinclair, M.; Fryer, C.; Phillips, C.J.C. The Benefits of Improving Animal Welfare from the Perspective of Livestock Stakeholders across Asia. *Animal* **2019**, *9*, 123. [CrossRef] [PubMed]

12. Food and Agricultural Organisation of the United Nations (FAO). Effects of Stress and Injury on Meat and by-Product Quality. Available online: http://www.fao.org/3/x6909e/x6909e04.htm (accessed on 10 April 2020).

13. Sheridan, J. Sources of contamination during slaughter and measures for control. *J. Food Saf.* **1998**, *18*, 321–339. [CrossRef]

14. Njisane, Y.Z.; Muchenje, V. Farm to abattoir conditions, animal factors and their subsequent effects on cattle behavioural responses and beef quality—A review. *Asian Australas. J. Anim. Sci.* **2016**, *30*, 755–764. [CrossRef]

15. Fernandes, J.; Hemsworth, P.; Coleman, G.; Tilbrook, A. Costs and Benefits of Improving Farm Animal Welfare. *Agriculture* **2021**, *11*, 104. [CrossRef]

16. Grandin, T. Auditing animal welfare at slaughter plants. *Meat Sci.* **2010**, *86*, 56–65. [CrossRef]

17. Food and Agricultural Organisation of the United Nations (FAO). Slaughter of Livestock. Available online: http://www.fao.org/3/x6909e/x6909e09.htm (accessed on 24 January 2021).

18. American Veterinary Medical Association (AVMA). Animal Welfare: What Is It? Available online: https://www.avma.org/resources/animal-health-welfare/animal-welfare-what-it (accessed on 5 April 2021).

19. OIE (World Organization for Animal Health). Chapter 7.1. Introduction to the Recommendations for Animal Welfare. Terrestrial Animal Health Code 2010. Available online: http://www.oie.int/index.php?id=169&L=0&htmfile=chapitre_1.7.1.htm (accessed on 21 February 2021).

20. Brown, M.J.; Winnicker, J.; Animal Welfare. *American College of Laboratory Animal Medicine, Laboratory Animal Medicine*, 3rd ed.; Fox, J.G., Anderson, L.C., Otto, G., Pritchett-Corning, K.R., Whary, M.T., Eds.; Academic Press: Cambridge, MA, USA, 2015; pp. 1653–1672. [CrossRef]

21. Iannetti, L.; Neri, D.; Santarelli, G.A.; Cotturone, G.; Vulpiani, M.P.; Salini, R.; Antoci, S.; Di Serafino, G.; Di Giannatale, E.; Pomilio, F.; et al. Animal welfare and microbiological safety of poultry meat: Impact of different at-farm animal welfare levels on at-slaughterhouse Campylobacter and Salmonella contamination. *Food Control.* **2020**, *109*, 106921. [CrossRef]

22. Mshelia, G.; Maina, V.; Aminu, M. Foetometrics and Economic Impact Analysis of Reproductive Wastages in Ruminant Species Slaughtered in North-Eastern Nigeria. *J. Anim. Prod. Adv.* **2015**, *5*, 1. [CrossRef]

23. Odeh, S.; Dawuda, P.M.; Oyedipe, E.O.; Obande, G.E. Incidence of foetal wastage in slaughtered cattle at Wurukum abattoir, Makurdi, Benue State. *Vom. J. Vet. Sci.* **2015**, *10*, 41–50.

24. Adebowale, O.O.; Ekundayo, O.; Awoseyi, A. Female cattle slaughter and foetal wastage: A case study of the Lafenwa abattoir, Ogun state, Nigeria. *Cogent Food Agric.* **2020**, *6*, 6. [CrossRef]

25. Abonyi, F.O.; Njoga, E.O. Prevalence and determinants of gastrointestinal parasite infection in intensively managed pigs in Nsukka agricultural zone, Southeast, Nigeria. *J. Parasit. Dis.* **2020**, *44*, 31–39. [CrossRef]

26. Worldometer, Nigerian Population. Available online: https://www.worldometers.info/world-population/nigeria-population/ (accessed on 4 April 2021).

27. Algers, B. Animal welfare-recent developments in the field. *CAB Rev. Perspect. Agric. Veter. Sci. Nutr. Nat. Resour.* **2011**, *6*, 1–10. [CrossRef]

28. Burroughs, T.; Knobler, S.; Lederberg, J. (Eds.) The Importance of Zoonotic Diseases. In *The Emergence of Zoonotic Diseases: Understanding the Impact on Animal and Human Health: Workshop Summary*; National Academies Press: Washington, DC, USA, 2002. Available online: https://www.ncbi.nlm.nih.gov/books/NBK98094 (accessed on 14 March 2021).

29. Njoga, E.O.; Ariyo, O.E.; Nwanta, J.A. Ethics in veterinary practice in Nigeria: Challenges and the way-forward. *Niger. Vet. J.* **2019**, *40*, 85–93. [CrossRef]

30. Baer-Nawrocka, A.; Sadowski, A. Food security and food self-sufficiency around the world: A typology of countries. *PLoS ONE* **2019**, *14*, e0213448. [CrossRef] [PubMed]

31. Brown, P.; Will, R.G.; Bradley, R.; Asher, D.M.; Detwiler, L. Bovine Spongiform Encephalopathy and Variant Creutzfeldt-Jakob Disease: Background, Evolution, and Current Concerns. *Emerg. Infect. Dis.* **2001**, *7*, 6–16. [CrossRef]

32. Henchion, M.; Hayes, M.; Mullen, A.M.; Fenelon, M.; Tiwari, B. Future Protein Supply and Demand: Strategies and Factors Influencing a Sustainable Equilibrium. *Foods* **2017**, *6*, 53. [CrossRef] [PubMed]

33. Cadmus, S.I.; Adesokan, H.K. Bovine fetal wastage in Southwestern Nigeria: A survey of some abattoirs. *Trop. Anim. Health Prod.* **2010**, *42*, 617–621. [CrossRef]

34. Salami, S.O.; Raji, M.A.; Ameh, J.A. Foetal wastages through slaughtering of pregnant cows in Zaria, Nigeria. *Sahel J. Vet. Sci.* **2010**, *9*, 21–24.

35. Iliyasu, D.; Ogwu, D.; Yelwa, H.; Yaroro, I.; Ayinla, A.; Usman, A.; Sharadi, B.; Grace, K. Incidence of Bovine pregnancy wastage in Maiduguri abattoir, Borno State Nigeria. *Int. J. Livest. Res.* **2015**, *5*, 59. [CrossRef]

36. Amuta, P.O.; Tordue, K.A.; Kudi, C.A.; Mhomga, L.I. Economic Implication of Foetal Wastages through Slaughter of Pregnant Pigs: A Case Study of the Makurdi Municipal Abattoir in Benue State, Nigeria. *Asian J. Res. Anim. Vet. Sci.* **2018**, *1*, 1–8.

37. Anyaku, C.E.; Ajagun, E.J.; Oshagbemi, A.O. Foetal Wastage at the Oja-tuntun Slaughterhouse in Ogbomoso Town, South Western Nigeria: A Three Year Retrospective Study. *Asian J. Res. Zoo* **2019**, *2*, 1–7. [CrossRef]

38. Odoemena, K.G.; Walters, J.P.; Kleemann, H.M.A. System Dynamics Model of Supply-Side Issues Influencing Beef Consumption in Nigeria. *Sustainability* **2020**, *12*, 3241. [CrossRef]

39. Vries-Ten Have, J.; Owolabi, A.; Steijns, J.; Kudla, U.; Melse-Boonstra, A. Protein intake adequacy among Nigerian infants, children, adolescents and women and protein quality of commonly consumed foods. *Nutr. Res. Rev.* **2020**, *33*, 102–120. [CrossRef] [PubMed]
40. Wosu, L. Calf wastage through slaughtering of pregnant cows in Enugu abattoir, Nigeria. *Rev. Elev. Med. Vet. Pays Trop. Fr.* **1988**, *41*, 97–98.
41. Nwanta, J.A.; Shoyinka, S.V.O.; Chah, K.F.; Onunkwo, J.I.; Onyenwe, I.W.; Eze, J.I.; Iheagwam, C.N.; Njoga, E.O.; Onyema, I.; Ogbu, K.I.; et al. Production characteristics, disease prevalence, and herd health management of pigs in Southeast Nigeria. *J. Swine Health Prod.* **2011**, *19*, 331–339.
42. Anonymous. Climate of Nigeria. Available online: https://www.britannica.com/place/Nigeria/Climate (accessed on 11 April 2021).
43. Worldometers. Nigeria Population in 2019. Available online: http://www.worldometers.info/world-population/nigeria-population/ (accessed on 22 January 2021).
44. Holmes, A. Direct Observation. In *Encyclopedia of Autism Spectrum Disorders*; Volkmar, F.R., Ed.; Springer: New York, NY, USA, 2013. [CrossRef]
45. Nwanta, J.A.; Onunkwo, J.I.; Ezenduka, E.V.; Phil-Eze, P.O.; Egege, S.C. Abattoir operations and waste management in Nigeria: A review of challenges and prospects. *Sokoto J. Vet. Sci.* **2008**, *7*, 61–67.
46. World Medical Association (WMA). World Medical Association Declaration of Helsinki: Ethical principles for medical research involving human subjects. *JAMA* **2013**, *310*, 2191–2194. [CrossRef]
47. Pourhoseingholi, M.A.; Vahedi, M.; Rahimzadeh, M. Sample size calculation in medical studies. *Gastroenterol. Hepatol. Bed Bench* **2013**, *6*, 14–17.
48. Bolarinwa, O.A. Principles and methods of validity and reliability testing of questionnaires used in social and health science researches. *Niger. Postgrad. Med. J.* **2015**, *22*, 195–201. [CrossRef] [PubMed]
49. Pace, J.E.; Wakeman, D.L. Determining the Age of Cattle by Their Teeth, University of Florida, IFAS Extension. 2003. Available online: http://www.edis.ifas.ufl.edu (accessed on 11 April 2020).
50. Kouamo, J.; Saague, A.M.N.; Zoli, A.P. Determination of age and weight of bovine foetus (*Bos indicus*) by biometry. *J. Livest. Sci.* **2018**, *9*, 9–15.
51. PennState Extension. Understanding Beef Carcass Yields and Losses During Processing. Available online: https://extension.psu.edu/understanding-beef-carcass-yields-and-losses-during-processing (accessed on 21 February 2021).
52. Ndi, C.; Tambi, N.E.; Agharih, N.W. Reducing Calf Wastage from Slaughtering of Pregnant Cows in Cameroon. Available online: http://www.fao.org/documents/art_dett.asp?lang=en&art_id=54949 (accessed on 21 May 2020).
53. Crowson, P. Price Determination. In *Economics for Managers*; Palgrave Macmillan: London, UK, 1985. [CrossRef]
54. Frimpong, S.; Gebresenbet, G.; Bobobee, E.; Aklaku, E.D.; Hamdu, I. Effect of Transportation and Pre-Slaughter Handling on Welfare and Meat Quality of Cattle: Case Study of Kumasi Abattoir, Ghana. *Vet. Sci.* **2014**, *1*, 174–191. [CrossRef]
55. Huertas, S.M.; van Eerdenburg, F.; Gil, A.; Piaggio, J. Prevalence of carcass bruises as an indicator of welfare in beef cattle and the relation to the economic impact. *Vet. Med. Sci.* **2015**, *1*, 9–15. [CrossRef]
56. Faucitano, L. Preslaughter handling practices and their effects on animal welfare and pork quality. *J. Anim. Sci.* **2018**, *96*, 728–738. [CrossRef] [PubMed]
57. European Commission (EC). *Council Regulation No 1/2005 of 22 December 2004 on the Protection of Animals during Transport. and Related Operations and Amending Directives 64/432/EEC and 93/119/EC and Regulation (EC) No 1255/97*; European Commission (EC): Brussels, Belgium, 2005.
58. Okaeme, A.; Ayeni, J.; Oyatogun, M.; Wari, M.; Haliru, M.; Okeyoyin, A. Cattle movement and its ecological implications in the Middle Niger Valley area of Nigeria. *Environ. Conser.* **1988**, *15*, 311–316. [CrossRef]
59. Lawal-Adebowale, O.A. Dynamics of ruminant livestock management in the context of the Nigerian agricultural system. In *Livestock Production*; IntechOpen: London, UK, 2012. [CrossRef]
60. FAO. *The Future of Livestock in Nigeria: Opportunities and Challenges in the Face of Uncertainty*; Food and Agricultural Organization of the United Nations: Rome, Italy, 2019. Available online: http://www.fao.org/3/ca5464en/ca5464en.pdf (accessed on 28 January 2021).
61. Omotosho, O.O.; Emikpe, B.O.; Lasisi, O.T.; Oladunjoye, O.V. Pig slaughtering in Southwestern Nigeria: Peculiarities, animal welfare concerns and public health implications. *Afr. J. Infect. Dis.* **2016**, *10*, 146–155. [CrossRef]
62. Nakyinsige, K.; Man, Y.B.C.; Aghwan, Z.A.; Zulkifli, I.; Goh, Y.M.; Bakar, F.A.; Al-Kahtani, H.A.; Sazili, A.Q. Stunning and animal welfare from Islamic and scientific perspectives. *Meat Sci.* **2013**, *95*, 352–361. [CrossRef]
63. Abdullah, F.; Borilova, G.; Steinhauserova, I. Halal criteria versus conventional slaughter technology. *Animals* **2019**, *9*, 530. [CrossRef] [PubMed]
64. Oloruntoba, E.O.; Adebayo, A.M.; Omokhodion, F.O. Sanitary conditions of abattoirs in Ibadan, Southwest Nigeria. *Afr. J. Med. Med. Sci.* **2014**, *43*, 231–237. [PubMed]
65. Biobaku, K.T.; Amid, S.A. Predisposing factors associated with diseases in animals in Nigeria and possible botanical immunos-timulants and immunomodulators: A review. *BJVM* **2018**, *16*, 87–101. [CrossRef]
66. Dahl, G.E.; Tao, S.; Laporta, J. Heat stress impacts immune status in cows across the life cycle. *Front. Vet. Sci.* **2020**, *7*, 116. [CrossRef] [PubMed]

67. Huff-Lonergan, E.; Lonergan, S.M. Mechanisms of water-holding capacity of meat: The role of postmortem biochemical and structural changes. *Meat Sci.* **2005**, *71*, 194–204. [CrossRef] [PubMed]
68. Huff-Lonergan, E. Fresh meat water-holding capacity. In *Woodhead Publishing Series in Food Science, Technology and Nutrition, Improving the Sensory and Nutritional Quality of Fresh Meat*; Kerry, J.P., Ledward, D., Eds.; Woodhead Publishing: Cambridge, UK, 2009; pp. 147–160. [CrossRef]
69. Chulayo, A.Y.; Tada, O.; Muchenje, V. Research on pre-slaughter stress and meat quality: A review of challenges faced under practical conditions. *Appl. Anim. Husb. Rural Develop.* **2012**, *5*, 1–6.
70. Brewer, M.S. *Chemical and Physical Characteristics of Meat | Water-Holding Capacity. Encyclopedia of Meat Sciences*, 2nd ed.; Dikeman, M., Devine, C., Eds.; Academic Press: Cambridge, MA, USA, 2014; pp. 274–282. [CrossRef]
71. Birhanu, A.F. Review on Beef Cattle Pre-slaughter Stress and Its Effect on Meat and Carcass Quality. *Adv. Life Sci. Technol.* **2020**, *78*. [CrossRef]
72. Ponnampalam, E.N.; Hopkins, D.L.; Bruce, H.; Li, D.; Baldi, G.; Bekhit, A.E.-D. Causes and Contributing Factors to "Dark Cutting" Meat: Current Trends and Future Directions: A Review. *Compr. Rev. Food Sci. Food Saf.* **2017**, *16*, 400–430. [CrossRef] [PubMed]
73. López-Pedrouso, M.; Rodriguez, J.M.L.; Purriños, L.; Oliván, M.; García-Torres, S.; Sentandreu, M.Á.; Lorenzo, J.M.; Zapata, C.; Franco, D. Sensory and Physicochemical Analysis of Meat from Bovine Breeds in Different Livestock Production Systems, Pre-Slaughter Handling Conditions, and Ageing Time. *Foods* **2020**, *9*, 176. [CrossRef]
74. Terlouw, E.M.C.; Picard, B.; Deiss, V.; Berri, C.; Hocquette, J.-F.; Lebret, B.; Lefèvre, F.; Hamill, R.; Gagaoua, M. Understanding the Determination of Meat Quality Using Biochemical Characteristics of the Muscle: Stress at Slaughter and Other Missing Keys. *Foods* **2021**, *10*, 84. [CrossRef]
75. Loudon, K.M.; Tarr, G.; Lean, I.J.; Polkinghorne, R.; McGilchrist, P.; Dunshea, F.R.; Gardner, G.E.; Pethick, D.W. The Impact of Pre-Slaughter Stress on Beef Eating Quality. *Animal* **2019**, *9*, 612. [CrossRef]
76. Agbeniga, B.; Webb, E. Effect of slaughter technique on bleed-out, blood in the trachea and blood splash in the lungs of cattle. *S. Afr. J. Anim. Sci.* **2012**, *42*, 524–529. [CrossRef]
77. Addis, M. Major Causes Of Meat Spoilage and Preservation Techniques: A Review. *Food Sci. Qual. Manag.* **2015**, *41*, 101–114.
78. Nakyinsige, K.; Fatimah, A.B.; Aghwan, Z.A.; Zulkifli, I.; Goh, Y.M.; Sazili, A.Q. Bleeding Efficiency and Meat Oxidative Stability and Microbiological Quality of New Zealand White Rabbits Subjected to Halal Slaughter without Stunning and Gas Stun-killing. *Asian Australas. J. Anim. Sci.* **2014**, *27*, 406–413. [CrossRef] [PubMed]
79. Luong, N.-D.M.; Coroller, L.; Zagorec, M.; Membré, J.-M.; Guillou, S. Spoilage of Chilled Fresh Meat Products during Storage: A Quantitative Analysis of Literature Data. *Microorganisms* **2020**, *8*, 1198. [CrossRef]
80. EFSA. Scientific opinion on the animal welfare aspects in respect of the slaughter or killing of pregnant livestock animals (cattle, pigs, sheep, goats, horses). *EFSA J.* **2017**, *15*, 4782. [CrossRef]
81. Derbyshire, S.W.; Bockmann, J.C. Reconsidering fetal pain. *J. Med. Ethics.* **2020**, *46*, 3–6. [CrossRef] [PubMed]
82. Campbell, M.L.; Mellor, D.J.; Sandøe, P. How should the welfare of fetal and neurologically immature postnatal animals be protected? *Anim. Welf.* **2014**, *23*, 369–379. [CrossRef] [PubMed]
83. Mellor, D.J.; Diesch, T. Onset of sentience: The potential for suffering in fetal and newborn farm animals. *Appl. Anim. Behav. Sci.* **2006**, *100*, 48–57. [CrossRef]
84. Mellor, D.J. Galloping colts, fetal feelings, and reassuring regulations: Putting animal-welfare science into practice. *J. Vet. Med. Educ.* **2010**, *37*, 94–100. [CrossRef]
85. Singleton, G.H.; Dobson, H. A survey for the reasons for culling pregnant cows. *Vet. Rec.* **1995**, *136*, 162–165. [CrossRef]
86. Maurer, P.; Lücker, E.; Riehn, K. Slaughter of pregnant cattle in German abattoirs—Current situation and prevalence: A cross-sectional study. *BMC Vet. Res.* **2016**, *12*, 91. [CrossRef] [PubMed]
87. Nielsen, S.S.; Sandøe, P.; Kjølsted, S.U.; Agerholm, J.S. Slaughter of pregnant cattle in Denmark: Prevalence, gestational age, and reasons. *Animals* **2019**, *9*, 392. [CrossRef] [PubMed]
88. Fayemi, P.O.; Muchenje, V. Maternal slaughter at abattoirs: History, causes, cases and the meat industry. *SpringerPlus* **2013**, *2*, 125. [CrossRef] [PubMed]
89. Dunka, H.I.; Buba, D.M.; Gurumyen, Y.G.; Oragwa, A.O.; Oziegbe, S.D.; Patrobas, M.N. Economic losses associated with the slaughter of pregnant animals in Jos abattoir. *Int. J. Adv. Res.* **2017**, *5*, 1047–1052.
90. FAO. Current Worldwide Annual Meat Consumption, Per Capita, Livestock and Fish Primary Equivalent. Available online: http://faostat.fao.org (accessed on 14 December 2018).
91. Njoga, E.O.; Onunkwo, J.I.; Chinwe, C.E.; Ugwuoke, W.; Nwanta, J.A.; Chah, K.F. Assessment of antimicrobial drug administration and antimicrobial residues in food animals in Enugu State, Nigeria. *Trop. Anim. Health Prod.* **2018**, *50*, 897–902. [CrossRef]
92. Aworh, M.K.; Okolocha, E.; Kwaga, J.; Fasina, F.; Lazarus, D.; Suleman, I.; Poggensee, G.; Nguku, P.; Nsubuga, P. Human brucellosis: Seroprevalence and associated exposure factors among abattoir workers in Abuja, Nigeria—2011. *Pan Afr. Med. J.* **2013**, *16*, 103. [CrossRef]
93. Godfroid, J. Brucellosis in livestock and wildlife: Zoonotic diseases without pandemic potential in need of innovative one health approaches. *Arch. Public Health* **2017**, *75*, 34. [CrossRef]
94. Bamaiyi, P.H. Prevalence and risk factors of brucellosis in man and domestic animals: A review. *Int. J. One Health* **2016**, *2*, 29–34. [CrossRef]

95. Awah-Ndukum, J.; Mouiche, M.M.M.; Kouonmo-Ngnoyum, L.; Bayang, H.N.; Manchang, T.K.; Poueme, R.S.N.; Kouamo, J.; Ngu-Ngwa, V.; Assana, E.; Feussom, K.J.M.; et al. Seroprevalence and risk factors of brucellosis among slaughtered indigenous cattle, abattoir personnel and pregnant women in Ngaoundéré, Cameroon. *BMC Inf. Dis.* **2018**, *18*. [CrossRef]
96. Swai, E.S.; Schoonman, L. A survey of zoonotic diseases in trade cattle slaughtered at Tanga city abattoir: A cause of public health concern. *Asian Pac. J. Trop. Biomed.* **2012**, *2*, 55–60. [CrossRef]
97. Guernier, V.; Goarant, C.; Benschop, J.; Lau, C.L. A systematic review of human and animal leptospirosis in the Pacific Islands reveals pathogen and reservoir diversity. *PLoS Negl. Trop. Dis.* **2018**, *12*, e000650. [CrossRef] [PubMed]
98. Ekere, S.O.; Njoga, E.O.; Onunkwo, J.I.; Njoga, U.J. Serosurveillance of *Brucella* antibody in food animals and role of slaughterhouse workers in spread of *Brucella* infection in Southeast Nigeria. *Vet. World* **2018**, *11*, 1171–1178. [CrossRef]
99. Nonga, H.E. A review on cattle foetal wastage during slaughter and its impacts to the future cattle herds in Tanzania. *Livest. Res. Rural. Dev.* **2015**, *27*. Available online: http://www.lrrd.org/lrrd27/12/nong27251.html (accessed on 21 February 2021).
100. Okorie-Kanu, O.J.; Ezenduka, E.V.; Okorie-Kanu, C.O.; Anyaoha, C.O.; Attah, C.A.; Ejiofor, T.E.; Onwumere-Idoloh, S.O. Slaughter of pregnant goats for meat at Nsukka slaughterhouse and its economic implications: A public health concern. *Vet. World* **2018**, *11*, 1139–1144. [CrossRef] [PubMed]
101. Wythes, J.R.; Shorthose, W.R.; Fordyce, G.; Underwood, D.W. Pregnancy effects on carcass and meat quality attributes of cows. *Anim. Prod. Sci.* **1990**, *51*, 461–468. [CrossRef]
102. Kalio, G.A.; Ali-Uchechukwu, A. Assessment of abattoirs operations and hygiene practices in Obio-Akpor Local Government Area, Rivers State, Nigeria. *Niger. J. Anim. Prod.* **2019**, *46*, 73–81. [CrossRef]
103. Gali, A.U.; Abdullahi, H.A.; Umaru, G.A.; Zailani, S.A.; Adamu, S.G.; Hamza, I.M.; Jibrin, M.S. Assessment of operational facilities and sanitary practices in Zangon Shanu abattoir, Sabon Gari Local Government Area, Kaduna State, Nigeria. *J. Vet. Med. Anim. Health* **2020**, *12*, 36–47. [CrossRef]
104. Madzingira, O. Animal welfare considerations in food-producing animals. In *Animal Welfare*; Abubakar, M., Manzoor, S., Eds.; IntechOpen: London, UK, 2018. [CrossRef]

Article

Understanding Food Waste, Food Insecurity, and the Gap between the Two: A Nationwide Cross-Sectional Study in Saudi Arabia

Nora A. Althumiri [1,*], Mada H. Basyouni [1,2], Ali F. Duhaim [3], Norah AlMousa [1,4], Mohammed F. AlJuwaysim [1,5] and Nasser F. BinDhim [1,3,6]

[1] Sharik Association for Health Research, Riyadh 13326, Saudi Arabia; mada.basyouni@sharikhealth.net (M.H.B.); 2170001797@iau.edu.sa (N.A.); 216006361@student.kfu.edu.sa (M.F.A.); nasser.bindhim@sharikhealth.net (N.F.B.)
[2] Ministry of Health, Riyadh 11176, Saudi Arabia
[3] Saudi Food and Drug Authority, Riyadh 13513, Saudi Arabia; afduhaim@sfda.gov.sa
[4] Public Health Department, Imam Abdulrahman Bin Faisal University, Dammam 31441, Saudi Arabia
[5] Pharmacy College, King Faisal University, AlAhsa 31982, Saudi Arabia
[6] College of Medicine, Alfaisal University, Riyadh 11533, Saudi Arabia
* Correspondence: nora@althumiri.net

Citation: Althumiri, N.A.; Basyouni, M.H.; Duhaim, A.F.; AlMousa, N.; AlJuwaysim, M.F.; BinDhim, N.F. Understanding Food Waste, Food Insecurity, and the Gap between the Two: A Nationwide Cross-Sectional Study in Saudi Arabia. *Foods* **2021**, *10*, 681. https://doi.org/10.3390/foods10030681

Academic Editors: Bahar Aliakbarian, António Raposo, Renata Puppin Zandonadi and Raquel Braz Assunção Botelho

Received: 2 March 2021
Accepted: 20 March 2021
Published: 23 March 2021

Publisher's Note: MDPI stays neutral with regard to jurisdictional claims in published maps and institutional affiliations.

Abstract: Background: Food waste and food insecurity may co-exist in various balances in developing and developed countries. This study aimed to explore the levels of food waste and food insecurity, the factors associated with them, and their relationships at the household and individual levels in Saudi Arabia. Methods: This study was a nationwide cross-sectional survey conducted via computer-assisted phone interviews in January 2021. Quota sampling was utilized to generate balanced distributions of participants by gender across all the administrative regions of Saudi Arabia. Data collection included household demographics, food waste and disposal, the Food Insecurity Experience Scale (FIES), and the Household Food Insecurity Access Scale (HFIAS). Results: Out of the 2807 potential participants contacted, 2454 (87.4%) completed the interview. The mean age was 31.4 (SD = 11.7; range = 18–99) and 50.1% were female. The weighted prevalence of uncooked food waste in the last four weeks was 63.6% and the cooked food waste was 74.4%. However, the food insecurity weighted prevalence at the individual level (FIES) was 6.8%. In terms of food insecurity at the household level (HFIAS), 13.3% were in the "severely food insecure" category. Moreover, this study found that "moderately food insecure" households were associated with an increased likelihood to waste uncooked food (relative risk (RR) = 1.25), and the "mildly food insecure" (RR = 1.21) and "moderately food insecure" (RR = 1.17) households were associated with an increased likelihood to waste cooked food. However, "food secure" households were associated with a decreased likelihood to waste cooked food (RR = 0.56). Finally, this study identified four household factors associated with food waste and three household factors that were associated with "severe food insecurity." Conclusions: This first national coverage study to explore food waste and food insecurity at the individual level and household level, identified household factors associated with food waste and food insecurity and identified new associations between food waste and food insecurity in Saudi Arabia. The associations found between food waste and food insecurity are potential areas of intervention to reduce both food waste and food insecurity at the same time, toward achieving the Sustainable Development Goal (SDG) targets related to food waste and food security.

Keywords: food security; food waste; household; food insecurity; Saudi Arabia

1. Background

As Saudi Arabia imports around 80% of its food needs and this trend is expected to increase, the importance of food waste and food insecurity as concepts are expected to grow exponentially in Saudi Arabia in the near future [1–3]. This importance is stimulated by the global coronavirus disease 2019 (COVID-19) pandemic, which has affected the food supply chain and is expected to have a long-term effect on food production and the supply chain globally [4–6]. Although Saudi Arabia mainly relies on food imported from all around the globe, due to limited agricultural production, high levels of food waste are still occurring [7,8].

Despite this local context, food waste is a global issue. Food waste reduction is one of the global Sustainable Development Goal (SDG) targets [9]. Target 12.3 of the SDGs aims to halve the global food waste per capita at the retail and consumer levels and to reduce food losses along the production and supply chains, including post-harvest losses, by 2030 [9]. Food waste and food insecurity may coexist in various balances in developing and developed countries [10–12]. Surplus food management and the reduction of food waste are increasingly acknowledged as being levers for the mitigation of food insecurity, as both surplus food reduction at the source and its recovery for human consumption are critical elements in the local and global food security effort [11].

In terms of the definitions of food waste and food security, food waste—which occurs at the end of the food chain (retail and final consumption)—is acknowledged to be a huge problem worldwide, even though the definition of various terms and information collection processes are not yet well harmonized [1,11]. However, according to the Food and Agriculture Organization (FAO), food security exists when "all people, at all times, have physical and economic access to sufficient, safe and nutritious food to meet their dietary needs and food preferences for an active and healthy life" [13].

Food waste at the consumption (household and individual) level can be categorized into fresh or cooked food waste and unprepared or uncooked food, such as canned or frozen food waste. It also can be categorized based on food type. In Saudi Arabia, a national study conducted in 2018 led by the Saudi Grains Organization to estimate food losses and food waste found that waste was estimated at 2.33 million tons, which represents 18.9% of the volume of the targeted food groups in the study [8]. Most importantly, 57% of the food waste was found to happen at the consumption level [8]. The most wasted types of food in the study were rice at 31%, bread at 25%, meat at 19%, and fruits and vegetables at 16% [8].

Nevertheless, food security rests on three pillars: Food availability (existence of sufficient quantities), food access (households are able to obtain the quantities required), and food utilization (appropriate nutrition and hygiene) [14]. In terms of food insecurity in Saudi Arabia, the FAO report Near East and North Africa, Regional Overview of Food Security and Nutrition used the Food Insecurity Experience Scale (FIES) and stated that severe food insecurity in Saudi Arabia between 2015 and 2017 was 8.1%. However, one study looked at food insecurity during the COVID-19 pandemic in Saudi Arabia using an online convenience sample recruited via social media platforms and used the Household Food Insecurity Access Scale (HFIAS) and found that the overall prevalence of food insecurity was 37.7% in the study sample, with 17.0% of participants experiencing moderate and severe food insecurity [15]. As this is the only study exploring food security on a household level in Saudi Arabia, the prevalence is very high and alarming.

Although few previous efforts have explored food waste at the consumption level and food insecurity separately, it is important to explore them on a large national level together to bring more insight into the relationship between food waste and food insecurity. Thus, this study aimed to explore the levels of food waste and food insecurity, the factors associated with them, and their relationships at the household and individual levels in Saudi Arabia. The remainder of this paper is structured as follows. The main section presents the method used to conduct this study, the survey question, and the outcomes measured. Section 3 then presents the results based on the objectives of this study. Section 4 provides a summary of the results and a discussion of the important points.

2. Method

2.1. Study Design

For this study, we conducted a nationwide cross-sectional survey in Saudi Arabia via computer-assisted phone interviews in January 2021.

2.2. Sampling and Sample Size

A proportionate quota-sampling procedure was used to obtain an approximately balanced allocation of participants based on gender across the 13 administrative regions of Saudi Arabia (Ar Riyadh, Macca Al Mukaramah, Eastern Province, Asir, Baha, Jazan, Najran, Al Madinah, Al Qassim, Hail, Tabuk, Northern Borders, and Al Jouf), leading to a total of 26 quotas. The sample size was generated using a medium effect size at 0.30, with an 80% sampling power and a 95% confidence interval (CI), to compare regions at the gender level [16]. Therefore, 94 participants were needed in each quota, and a total sample size of 2444 participants was required in this research project [16]. The total sample size in each administrative region was 190 participants at an approximately 0.20 effect size [16]. This study used "QPlatform", an electronic data collection platform, which has eligibility and sampling functions, to manage the sample distribution process and to prevent human bias in the sampling procedure [17]. Two factors, gender and region, were used to determine the adherence to the sampling quota.

2.3. Participant Recruitment

In this study, we recruited Saudi residents, Arabic-speaking adults ($\geq$ 18 years old). We used a random phone number list that was obtained from the Sharik Association for Health Research (Sharik research participants' database) to identify prospective participants [18]. The Sharik research participants' database includes individuals who provided consent and are willing to participate in future studies. The database includes more than 75,000 potential research participants distributed across the 13 administrative regions of Saudi Arabia and continues to expand [18]. Potential participants were called up to three times. If there was no response, the number of a new participant with similar demographical characteristics was generated from the Sharik database. After explaining the study to the potential participant, consent to participate in the study was obtained, and the interviewer assessed the eligibility of each participant on the data collection platform, based on the gender and region quota completion criteria. Once the sampling quota was complete, recruitment ceased automatically [17]. For the household-related questions, the participant responded for the entire household.

2.4. Survey Design and Outcome Measures

The survey was divided into five sections.

1. The demographics section. This section included age, gender, region, educational level, household's net income, number of people living in the household, number of children living in the household, elderly people living in the household, and social support status.
2. The household food waste of uncooked items (such as canned food or fresh vegetables). In this section, we asked the participants if they had wasted any uncooked food items within the last four weeks and the frequency of such behavior. If the household had wasted any uncooked items, then they were directed to answer three more questions about the reason for the food waste, the type of food, and how/where the items were wasted.

Food waste for uncooked food items was calculated for the overall sample and categorized based on the reason of food waste, type of food, and how/where the items were wasted.

3. The household food waste of cooked items. In this section, we asked the participants if they had wasted any cooked food items within the last four weeks and the frequency

of such behavior. If the household had wasted any cooked food items, then they were directed to answer four more questions about the reason for the food waste, the source of the cooked food, the type of food, and how/where the items were wasted.

Food waste for cooked food items was calculated for the overall sample and categorized based on the reason for the food waste, the source of the cooked food, the type of food, and how/where the items were wasted.

4. The individual food insecurity experience measured via the FIES [19,20]. The FIES was the first tool to be used to measure food insecurity at the individual level globally and was validated by the Food and Agriculture Organization (FAO) in 151 countries, including in Saudi Arabia [21]. The FIES has eight questions, each of which is asked with a recall period of 12 months [19,20]. Respondents answer yes/no to the eight questions and the responses are aggregated to provide raw scores ranging from 0 to 8. Food insecurity is then classified into three categories: (1) Food secure (FS) with raw scores = 0–3; (2) moderate food insecurity (MFI) with raw scores = 4–6; (3) severe food insecurity (SFI) with raw scores = 7–8 [19,20].

5. Household food insecurity was measured via the HFIAS for measurement of food access [22]. The HFIAS is an adaptation of the approach used to estimate the prevalence of food insecurity in the United States annually [22]. The scale was validated in the Arabic language and has been widely used by United Nations countries to measure household food insecurity [22–24]. The HFIAS has nine questions, each of which is asked with a recall period of four weeks [22]. If the respondent answers "yes" to an occurrence question, a frequency question is asked to determine whether the condition happened 1 = rarely (once or twice), 2 = sometimes (three to ten times), or 3 = often (more than ten times) in the past four weeks [22]. We used the interviewer instructions and questionnaire administration guide recommended by the Food and Nutrition Technical Assistance III Project [22]. Four indicators were used to determine the characteristics of and changes in household food insecurity (access) in the surveyed population, including (1) household food insecurity access-related conditions, (2) household food insecurity access-related domains, (3) household food insecurity access scale scores, and (4) household food insecurity access prevalence. The calculation method for each indicator is explained in detail in the Food and Nutrition Technical Assistance III Project guide, version 3 [22].

We conducted several rounds of linguistic validation to ensure clarity and understanding of the survey questions via focus groups, where group members were asked to review and discuss survey questions and answers. Based on the outcomes of linguistic validation procedures, the survey was further edited, and the final version was approved.

2.5. Data Analysis

Descriptive analysis including (proportions, mean, median, etc.) were used to describe the variables. The prevalence of all variables including food waste, reasons, types, and disposal method by the HFIAS and FIES were weighted to equal the adult population within regions in Saudi Arabia, according to the General Authority of Statistics 2017 Census Report [25]. Bivariate analysis was used to compare the categorical variables. To identify the household factors associated with food waste and severe food insecurity, a logistic regression model, including household characteristic variables including monthly household income, total people living in the household, number of children living in the household, elderly family members living in the household, and receipt of social benefits or aids, was used.

3. Results

Out of the 2807 potential participants contacted, 2454 (87.4%) completed the interview. The mean age was 31.4 (SD = 11.7; range = 18–99) and 50.1% were female. Table 1 shows the participants' demographics and household characteristics.

Table 1. Participants' demographics and household characteristics in the sample.

Variable	*n* (%)
Sex	
Female	1230 (50.1)
Male	1224 (49.9)
Education level	
Less than bachelor's degree	1077 (43.9)
Bachelor's degree and above	1377 (56.1)
Regions	
Asir	189 (7.7)
Baha	190 (7.7)
Eastern region	189 (7.7)
Hail	191 (7.8)
Jazan	185 (7.5)
Al Jouf	183 (7.5)
Madinah	192 (7.8)
Makkah	191 (7.8)
Najran	190 (7.7)
Northern border	182 (7.4)
Qassim	191 (7.8)
Riyadh	191 (7.8)
Tabuk	190 (7.7)
Monthly household income *	
Less than SAR 5000	385 (15.7)
SAR 5001 to 8000	489 (19.9)
SAR 8001 to 11,000	394 (16.1)
SAR 11,001 to 13,000	290 (11.8)
SAR 13,001 to 16,000	304 (12.4)
SAR 16,001 to 20,000	250 (10.2)
More than SAR 20,000	342 (13.9)
Total people living in the household	
0 to 2	277 (11.3)
3 to 5	779 (31.7)
6 to 8	779 (37.5)
9 and above	478 (19.5)
Number of children living in the household	
0	702 (28.6)
1 to 4	1619 (66.0)
5 and above	133 (5.4)
Elderly family members living in the household	
Yes	637 (26.0)
No	1817 (74.0)
Receiving social benefits or aids	
Yes	364 (14.8)
No	2090 (85.2)

* SAR = USD 3.75.

3.1. Food Waste

The weighted national prevalence of uncooked food waste in the last four weeks was 63.6%, while for the cooked food waste, it was 74.4%. Table 2 shows food waste, reasons, types, and disposal method in the weighted sample for uncooked and cooked food.

Table 2. Food waste, reasons, types, and disposal method in the weighted sample.

Variable	*n* (%)
Wasted any uncooked food in the last 4 weeks	
Never	892 (36.4)
1 day a week maximum	1041 (42.5)
2 to 3 days a week	411 (16.8)
4 days or more per week	103 (4.2)
Reasons for wasting uncooked food in the last 4 weeks *	
Expired food	1081 (44.2)
Old but not expired food	248 (10.1)
Recalled by authority	74 (3.0)
Caused extreme allergic reaction	87 (3.6)
Caused mild allergic reaction	66 (2.7)
Caused digestive issues	119 (4.9)
We don't need it anymore	388 (15.9)
Not used for a long time (clearing storage)	663 (27.1)
Other	128 (5.2)
Types of wasted uncooked food in the last 4 weeks *	
Fresh dairy products	864 (35.3)
Long-life dairy products	206 (8.4)
Fruits and vegetables	637 (26.0)
Egg	107 (4.4)
Baby food	94 (3.8)
Meat	93 (3.8)
Canned food	265 (10.8)
Poultry	109 (4.4)
Nuts	66 (2.7)
Fish	51 (2.1)
Long-life juice	129 (5.3)
Soft drinks	114 (4.7)
Eastern sweets	182 (7.5)
Chocolates	131 (5.3)
Uncooked rice	63 (2.6)
Other	227 (9.3)
Disposal methods of uncooked food in the last 4 weeks *	
Trash bin	878 (35.9)
Feeding stray animals	990 (40.5)

Table 2. *Cont.*

Variable	*n* (%)
Feeding pets	279 (11.4)
Used as compost	57 (2.3)
Food donation to individuals	339 (13.9)
Food recycling	41 (1.7)
Food donation to nonprofit organizations	109 (4.5)
Other	67 (2.7)
Wasted any cooked food in the last 4 weeks	
Never	626 (25.6)
1 day a week maximum	984 (40.2)
2 to 3 days a week	617 (25.2)
4 days or more per week	220 (9.0)
Reasons for wasting cooked food in the last 4 weeks *	
It was more than what we could eat	1168 (47.8)
We do not store cooked food or leftovers	474 (19.4)
We don't have enough space to store it	135 (5.5)
Spoiled	963 (39.3)
Burned or was not cooked properly	262 (10.7)
Other	145 (5.9)
Types of wasted cooked food in the last 4 weeks *	
Bread and bakeries	916 (37.4)
Pies and pastries	679 (27.7)
Cooked vegetables and seasoned salads	657 (26.8)
Cooked rice	1293 (52.8)
Meat	492 (20.1)
Poultry	763 (31.2)
Fish	184 (7.5)
Egg	167 (6.8)
Grains and legumes	350 (14.3)
Other	196 (8.0)
Disposal methods of cooked food in the last 4 weeks *	
Trash bin	732 (29.9)
Feeding stray animals	1183 (48.3)
Feeding pets	371 (15.2)
Used as compost	64 (2.6)
Food donation to individuals	502 (20.5)
Food recycling	60 (2.5)
Food donation to nonprofit organizations	137 (5.6)
Other	80 (3.3)

Table 2. *Cont.*

Variable	*n* (%)
Sources of wasted cooked food in the last 4 weeks *	
Restaurant	1166 (47.7)
Home	1633 (66.8)
Local family food business **	257 (10.5)
Neighbors or relatives	205 (8.4)

* Participant can provide more than one answer; ** cooked in the home of a family business entity and delivered by order.

3.2. Individual Food Insecurity Experience

The individual food insecurity experience analysis showed that 2281 (93.2%) were in the FS category, 122 (5.0%) were in the MFI category, and 44 (1.8%) were in the SFI category.

3.3. Household Food Insecurity Access

Table 3 shows the results of two HFIAS indicators: (1) Household food insecurity access-related conditions and (2) household food insecurity access-related domains. The weighted average Household Food Insecurity Access Scale score was 1.6 (SD = 3.9; range = 0–27).

In terms of the household food insecurity access prevalence in the weighted sample, 1815 (74.2%) were in the "food secure" category, 226 (9.2%) were in the "mildly food insecure" category, 104 (4.2%) were in the "moderately food insecure" category, and 325 (13.3%) were in the "severely food insecure" category.

3.3.1. Factors Associated with Food Waste

The regression model showed that only two household factors were associated with uncooked food waste. Receiving social benefits or aids (odds ratio (OR) = 1.35; 95% CI = 1.04–1.75; p = 0.025) was significantly associated with increasing the likelihood of uncooked food waste compared to those not receiving social benefits in the model. A household income of less than SAR 5000 (OR = 0.62; 95% CI = 0.46–0.85; p = 0.003) and SAR 5001 to 8000 (OR = 0.61; 95% CI = 0.46–0.81; p = 0.001) was significantly associated with decreasing the likelihood of uncooked food waste compared to the highest income group in the model.

In terms of cooked food waste, the regression model showed three associated factors. A household income of SAR 11,001 to 13,000 (OR = 1.48; 95% CI = 1.01–2.16; p = 0.041) and SAR 13,001 to 16,000 (OR = 1.68; 95% CI = 1.17–2.43; p = 0.005) was significantly associated with increasing the likelihood of cooked food waste compared to the highest income group in the model. Elderly family members living in the household (OR = 1.41; 95% CI = 1.12–1.78; p = 0.005) was significantly associated with an increased likelihood of cooked food waste compared to no elderly family members living at home in the model. A total of nine people or more living in the household (OR = 1.91; 95% CI = 1.30–2.85; p = 0.001) was significantly associated with increasing the likelihood of cooked food waste compared to those living with zero to two people in the model.

3.3.2. Factors Associated with Severe Food Insecurity

The regression model showed that only three household factors were associated with severe food insecurity. A household income of less than SAR 5000 (OR = 12.49; 95% CI = 6.47–24.11; p < 0.001), SAR 5001 to 8000 (OR = 5.67; 95% CI = 2.93–10.95; p < 0.001), SAR 8001 to 11,000 (OR = 3.391; 95% CI = 1.70–6.78; p = 0.001), SAR 11,001 to 13,000 (OR = 7.37; 95% CI = 3.71–14.65; p < 0.001), and SAR 13,001 to 16,000 (OR = 4.10; 95% CI = 2.01–8.34; p < 0.001) was significantly associated with increasing the likelihood of severe food insecurity compared to the highest income group in the model. Elderly family members living in the household (OR = 2.43; 95% CI = 1.70–3.17; p < 0.001) was significantly associated with increasing the likelihood of severe food insecurity compared to no elderly family members living at home in the model. One to four (OR = 1.49; 95% CI = 1.09–8.34; p = 0.012) and five or more children living in the household (OR = 3.97; 95% CI = 2.32–6.80; p < 0.001) were significantly associated with increasing the likelihood of severe food insecurity compared to no children living at home in the model.

Table 3. The weighted prevalence and frequency of each item in the Household Food Insecurity Access Scale (HFIAS).

HFIAS Domains		HFIAS Item	No n (%)	Yes (Total) n (%)	Frequency of Experience n (%)		
					Yes (Rarely)	Yes (Sometimes)	Yes (Often)
Anxiety and uncertainty	1.	In the past four weeks, did you worry that your household would not have enough food?	2048 (83.7)	399 (16.3)	289 (11.8)	67 (2.7)	43 (1.7)
Insufficient quality	2.	In the past four weeks, were you or any household member not able to eat the kinds of foods you preferred because of a lack of resources?	2027 (82.8)	419 (17.2)	253 (10.3)	122 (5.0)	45 (1.8)
	3.	In the past four weeks, did you or any household member have to eat a limited variety of foods due to a lack of resources?	2039 (83.4)	408 (16.6)	279 (11.4)	81 (3.3)	48 (2.0)
	4.	In the past four weeks, did you or any household member have to eat some foods that you really did not want to eat because of a lack of resources to obtain other types of food?	2144 (87.6)	302 (12.4)	192 (7.8)	83 (3.4)	27 (1.1)
Insufficient food intake	5.	In the past four weeks, did you or any household member have to eat a smaller meal than you felt you needed because there was not enough food?	2209 (90.3)	237 (9.7)	148 (6.1)	57 (2.3)	31 (1.3)
	6.	In the past four weeks, did you or any household member have to eat fewer meals in a day because there was not enough food?	2186 (89.4)	260 (10.6)	169 (6.9)	63 (2.6)	28 (1.1)
	7.	In the past four weeks, was there ever no food to eat of any kind in your household because of lack of resources to get food?	2207 (90.2)	239 (9.8)	148 (6.1)	56 (2.3)	35 (1.4)
	8.	In the past four weeks, did you or any household member go to sleep at night hungry because there was not enough food?	2189 (89.5)	257 (10.5)	179 (7.3)	49 (2.0)	29 (1.2)
	9.	In the past four weeks, did you or any household member go a whole day and night without eating anything because there was not enough food?	2265 (92.6)	181 (7.4)	123 (5.0)	34 (1.4)	24 (1.0)

3.3.3. Associations between Food Waste and Food Insecurity

In terms of uncooked food waste, the chi-square analysis showed no significant differences between uncooked food waste and individual food insecurity in the FIES. In addition, there were no significant differences between uncooked food waste and household food secure, mildly food insecure, and severely food insecure categories of the HFIAS. However, there were significant differences between uncooked food and the household moderately food insecure category of the HFIAS X^2 (1, n = 2447) = 10.5, p = 0.001, in which the moderately food insecure participants were associated with increased likelihood of wasting uncooked food (relative risk (RR) = 1.25; 95% CI = 1.25–1.39).

In terms of cooked food waste, chi-square analysis showed no significant differences between cooked food waste and individual food insecurity in the FIES. In addition, there were no significant differences between cooked food waste and the household severely food insecure category of the HFIAS. However, there were significant differences between cooked food and the household mildly food insecure category of the HFIAS X^2 (1, n = 2447) = 25.9, p < 0.001, in which mildly food insecure households were associated with an increased likelihood of wasting cooked food (RR = 1.21; 95% CI = 1.15–1.28). In addition, there were significant differences between cooked food and the household moderately food insecure category of the HFIAS X^2 (1, n = 2447) = 8.3, p = 0.004, in which moderately food insecure households were associated with an increased likelihood of wasting cooked food (RR = 1.17; 95% CI = 1.08–1.27). Finally, there were significant differences between cooked food and the household food secure category of the HFIAS X^2 (1, n = 2447) = 8.3, p = 0.004, in which food secure households were associated with decreased likelihood of wasting cooked food (RR = 0.56; 95% CI = 0.44–0.70).

4. Discussion

This study aimed to explore the levels of food waste and food insecurity, the factors associated with them, and their relationships at the household and individual levels in a nationwide sample of adults from Saudi Arabia.

The results showed that the weighted prevalence of uncooked food waste in the previous four weeks was 63.6%, and the cooked food waste prevalence was 74.4%. However, the food insecurity weighted prevalence at the individual level was 6.8%. In terms of food insecurity at the household level, 13.3% of participants were in the severely food insecure category. Moreover, this study identified four household factors associated with food waste: Receiving social benefits or aids, household income, elderly family members in the household, and total number of people in the household. In addition, the household fac-tors that were associated with severe food insecurity were household income, elderly family members in the household, and the number of children in the household. Finally, this study explored the association between food insecurity and food waste and found that moderately food insecure households had an increased likelihood of wasting uncooked food. Moreover, the mildly and moderately food insecure households had an increased likelihood of wasting cooked food; however, food secure households had a decreased likelihood of wasting cooked food.

Our study showed a high level of food waste at the household level in the previous four weeks, with prevalence levels of 63.6% and 74.4%, respectively. These values are higher than the prevalence of general food waste at the household level in the previous four weeks identified in a national survey in 2018 that found a weighted prevalence at 60% [26]. This confirms the previous finding that 57% of food waste happens at the consumption level [8]. Nevertheless, food waste at the consumption level is a global challenge, accounting for 61% of all food loss and food waste in North America and Oceania and 52% in Europe. Moreover, a recent study in South Korea found that approximately 63% of households discharge food waste of less than 500 g a day [27]. Thus, a global effort to combat food waste is urgently needed now more than ever before to protect global food security and food sustainability.

In addition to the high prevalence of food waste found in this study, the methods of food waste disposal are also a challenge, as the two major methods of disposal found in this study were trash bins and feeding stray animals. As well as being unstainable methods of food waste management, they are also harmful to the environment. However, although there is growing interest in the relegalization of using food waste as animal feed, the practice of feeding stray animals could pose environmental and health threats, especially in residential areas [28,29]. One study identified some major challenges for food waste management in Saudi Arabia, including solid waste segregation, inadequate legislations, well-accepted traditional landfill disposal practices, public attitudes, a lack of awareness, and uncertainty of the acceptability of food waste byproducts [28]. These challenges need to be prioritized and addressed at the national level in the near future.

Moving on to the prevalence of food insecurity, the prevalence of food insecurity at the individual level in the previous 12 months was slightly lower than the figure generated by the FAO in 2018, which was 8.1% in Saudi Arabia, and was also lower than the average prevalence of food insecurity at the individual level in the Gulf cooperation region, which was found to be 7.6% [29]. However, in terms of the prevalence of food insecurity at the household level in the previous four weeks, the values found in this study were relatively high; however, unfortunately, no previous national-level studies have used the same measurement tool in Saudi Arabia or within the Gulf cooperation region to allow a meaningful comparison. Despite this, it is important to mention that the 4 weeks measured in this study fell in the period of COVID-19 restrictions, including limiting gatherings to 20 people and closing all dine-in restaurants and food facilities in all regions of Saudi Arabia, which may have contributed to the food insecurity recall by the participants. In addition, cooked food banks rely on the use of food left over from restaurants and large events, and thus they could not operate efficiently during this period of restrictions.

In terms of the factors associated with food waste and food insecurity at the household level, these factors seem logical in a local culture context. One study highlighted that "Food waste patterns are shaped by cultural approaches to the special events and every-day shopping, cooking, and eating. Saudis place high value on generous hospitality; providing ample food is a gesture of welcome" [3]. However, no previous studies con-ducted in Saudi Arabia have explored the associations of household characteristics with food waste or food insecurity to allow a meaningful comparison to be conducted.

However, the associations between food waste and food insecurity are interesting. Although there was no association between food waste and food insecurity at the individual level, there were significant associations between food waste and food insecurity at the household level. One simple explanation might be related to the fact that food waste was measured on a household level, not an individual level. However, the findings showed that food waste is likely to be greater among mildly and moderately food insecure households, while it is lower among food secure households. These findings show that food waste and food insecurity may co-exist in the same household, which represents both a challenge in understanding why this phenomenon is happening and an opportunity to address the food waste issue and to improve food security levels.

Nevertheless, this study has some limitations. It could be criticized for using quota sampling, which has an associated risk of selection bias, rather than using a random probability sampling approach. However, the costs of probabilistic sampling are significantly greater, and for this project intention and the type of variables, the risk of some (generally low-level) [17,18] bias was considered acceptable [30,31]. In addition, using a proportional large sample with large number of quotas plays an important role in reducing the selection bias [30,31]. Currently, in Saudi Arabia, the only way to generate a random national-level sample is via a household survey, which also has some significant limitations due to sociocultural factors and was not possible during periods of COVID-19 restrictions. However, the recruitment and sampling methods used in this research project have been used successfully in many national projects in Saudi Arabia [32–34]. Another point that may represent a limitation in relation to the HFIAS use in this study is that some of the participants may not be the head of the household, however, according to the national authority of statistics in Saudi Arabia, the method nationally adopted of direct contact with the household specifies that it can be through direct contact with the head of the household or any adult member of the household who is familiar with the household affairs [35]. In addition, the HFIAS administration guide followed in this study did not specify which member of the household must be interviewed [22].

5. Conclusions

This first national coverage study to explore food waste and food insecurity at the household level identified household factors associated with food waste and food insecurity and identified new associations between food waste and food insecurity in Saudi Arabia. The associations found between food waste and food insecurity are potential areas of intervention for simultaneously reducing food waste and food insecurity, which could aid in achieving the SDG targets related to food waste and food security. Some of the interventions that may help include public awareness campaigns of food waste, food security, and food waste management, providing food management educational programs to households receiving social benefits, encouraging investment in the food recycling industry, and reexamining regulations related to the utilization of food waste as animal feed.

Author Contributions: N.A.A., M.H.B., A.F.D., and N.F.B. contributed to the conceptual design and design of the survey tool. All authors contributed to the writing and review of the manuscript. N.A.A., N.A., and M.F.A. supervised and managed the data collection process. N.A.A. and N.F.B. cleaned and analyzed the data. All authors have read and agreed to the published version of the manuscript.

Funding: The publication related cost was covered by Al-Dawaa Medical Services Co. (DMSCO).

Institutional Review Board Statement: The institutional review board of the Sharik Association for Health Research approved this research project (application no. 01-2021), in accordance with national research ethics regulations in Saudi Arabia.

Informed Consent Statement: Informed consent was obtained from all the participants in this study verbally during the phone call.

Data Availability Statement: The dataset can be obtained from the Sharik Association for Health Research upon request.

Conflicts of Interest: The authors declare that this project was conducted in the absence of any commercial or financial relationships that could be interpreted as a potential conflict of interest.

References

1. Food Nations AOotU. *Global Food Losses and Food Waste–Extent, Causes and Prevention*; Food and Agricultural Organization of the United Nations: Düsseldorf, Germany, 2011; Available online: https://www.futuredirections.org.au/wp-content/uploads/2015/07/Food_and_Water_Security_in_the_Kingdom_of_Saudi_Arabia.pdf (accessed on 22 January 2021).
2. Canadian Trade Commissioner Service. Agri-Food Sector Profile. 2013. Available online: https://www.futuredirections.org.au/wp-content/uploads/2015/07/Agri-Food-Sector-Profile-Saudi-Arabia.pdf (accessed on 22 January 2021).
3. Baig, M.B.; Gorski, I.; Neff, R.A. Understanding and addressing waste of food in the Kingdom of Saudi Arabia. *Saudi J. Biol. Sci.* **2019**, *26*, 1633–1648. [CrossRef] [PubMed]
4. Stephenson, J. COVID-19 outbreaks among food production workers may intensify pandemic's disproportionate effects on people of color. In *JAMA Health Forum 2020*; American Medical Association: Chicago, IL, USA, 2020.
5. Laborde, D.; Martin, W.; Swinnen, J.; Vos, R. COVID-19 risks to global food security. *Science* **2020**, *369*, 500–502. [CrossRef] [PubMed]
6. Pereira, M.; Oliveira, A.M. Poverty and food insecurity may increase as the threat of COVID-19 spreads. *Public Health Nutr.* **2020**, *23*, 3236–3240. [CrossRef]
7. World Bank. Saudi Arabia Food Products Imports. 2018. Available online: https://wits.worldbank.org/CountryProfile/en/Country/SAU/Year/2018/TradeFlow/Import/Partner/all/Product/16-24_FoodProd (accessed on 22 January 2021).
8. Saudi Grains Organization. National Food Losses and Food Waste Indicators. 2019. Available online: https://www.sago.gov.sa/Content/Files/Baseline_180419.pdf (accessed on 23 January 2021).
9. United Nations. Ensure Sustainable Consumption and Production Patterns. Available online: https://sdgs.un.org/goals/goal12 (accessed on 23 January 2021).

10. Kantor, L.S.; Lipton, K.; Manchester, A.; Oliveira, V. Estimating and addressing America's food losses. *Natl. Food Rev.* **1997**, *20*, 2–12.
11. Garrone, P.; Melacini, M.; Perego, A. Opening the black box of food waste reduction. *Food Policy* **2014**, *46*, 129–139. [CrossRef]
12. Thi, N.B.D.; Kumar, G.; Lin, C.-Y. An overview of food waste management in developing countries: Current status and future perspective. *J. Environ. Manag.* **2015**, *157*, 220–229. [CrossRef] [PubMed]
13. United Nations Food and Agriculture Organization. Rome Declaration on World Food Security, World Food Summit Plan of Action. 1996. Available online: http://www.fao.org/3/w3613e/w3613e00.htm (accessed on 23 January 2021).
14. Warr, P. Food insecurity and its determinants. *Aust. J. Agric. Resour. Econ.* **2014**, *58*, 519–537. [CrossRef]
15. Mumena, W.A. Impact of COVID-19 curfew on eating habits, food Intake, and weight According to food security status in Saudi arabia: A retrospective study. *Prog. Nutr.* **2020**, *22*. [CrossRef]
16. Cohen, J. *Statistical Power Analysis for the Behavioral Sciences*; Academic Press: Cambridge, MA, USA, 2013.
17. BinDhim, N.F. Smart Health Project. 2012. Available online: https://shproject.net/ (accessed on 22 January 2021).
18. Sharik Association for Health Research. Available online: https://sharikhealth.com/ (accessed on 22 January 2021).
19. Nord, M.; Cafiero, C.; Viviani, S. Methods for estimating comparable prevalence rates of food insecurity experienced by adults in 147 countries and areas. *J. Phys. Conf. Ser.* **2016**, *772*, 012060. [CrossRef]
20. Ballard, T.J.; Kepple, A.W.; Cafiero, C.; Schmidhuber, J. Better Measurement of Food Insecurity in the Context of Enhancing Nutrition. 2014. Available online: https://www.ernaehrungs-umschau.de/fileadmin/Ernaehrungs-Umschau/pdfs/pdf_2014/02_14/EU02_2014_M094_M097_-_38e_41_engl.pdf (accessed on 23 January 2021).
21. Smith, M.D.; Rabbitt, M.P.; Coleman-Jensen, A. Who are the world's food insecure? New evidence from the Food and Agriculture Organization's food insecurity experience scale. *World Dev.* **2017**, *93*, 402–412. [CrossRef]
22. Coates, J.; Swindale, A.; Bilinsky, P. Household Food Insecurity Access Scale (HFIAS) for Measurement of Food Access: Indicator Guide: Version 3. 2007. Available online: http://www.fao.org/fileadmin/user_upload/eufao-fsi4dm/doc-training/hfias.pdf (accessed on 23 January 2021).
23. Naja, F.; Hwalla, N.; Fossian, T.; Zebian, D.; Nasreddine, L. Validity and reliability of the Arabic version of the Household Food Insecurity Access Scale in rural Lebanon. *Public Health Nutr.* **2015**, *18*, 251–258. [CrossRef]
24. Castell, G.S.; Rodrigo, C.P.; de la Cruz, J.N.; Bartrina, J.A. Household food insecurity access scale (HFIAS). *Nutr. Hosp.* **2015**, *31*, 272–278.
25. General Authority for Statistics. 2017 Census Report. Available online: https://www.stats.gov.sa/en/857-0 (accessed on 23 January 2021).
26. Althumiri, N.A.; Alammari, N.S.; Almubark, R.A.; Alnofal, F.A.; Alkhamis, D.J.; Alharbi, L.S.; Algabbani, A.M.; BinDhim, N.F.; Alqahtani, A.S. The National Survey of Health, Diet, Physical Activity and Supplements among Adults in Saudi Arabia. *Food Drug Regul. Sci. J.* **2018**, *1*, 1–17. [CrossRef]
27. Kim, S.; Lee, S.H. Examining Household Food Waste Behaviors and the Determinants in Korea Using New Questions in a National Household Survey. *Sustainability* **2020**, *12*, 8484. [CrossRef]
28. Mu'azu, N.D.; Blaisi, N.I.; Naji, A.A.; Abdel-Magid, I.M.; AlQahtany, A. Food waste management current practices and sustainable future approaches: A Saudi Arabian perspectives. *J. Mater. Cycles Waste Manag.* **2019**, *21*, 678–690. [CrossRef]
29. Food and Agriculture Organization. Near East and North Africa, Regional Overview of Food Security and Nutrition. 2019. Available online: https://reliefweb.int/sites/reliefweb.int/files/resources/ca3817en.pdf (accessed on 22 January 2021).
30. Groves, R.M.; Fowler, F.J., Jr.; Couper, M.P.; Lepkowski, J.M.; Singer, E.; Tourangeau, R. *Survey Methodology*; John Wiley & Sons: Hoboken, NJ, USA, 2011; Volume 561.
31. Marsh, C.; Scarbrough, E. Testing 9 Hypotheses About Quota Sampling. *J. Mark. Res. Soc.* **1990**, *32*, 485–506.
32. BinDhim, N.F.; Althumiri, N.A.; Basyouni, M.H.; Alageel, A.A.; Alghnam, S.; Al-Qunaibet, A.M.; Almubarak, R.A.; Aldhukair, S.; Ad-Dab'bagh, Y. Saudi Arabia Mental Health Surveillance System (MHSS): Mental health trends amid COVID-19 and comparison with pre-COVID-19 trends. *Eur. J. Psychotraumatol.* **2021**, *12*, 1875642. [CrossRef]
33. Althumiri, N.A.; Basyouni, M.H.; AlMousa, N.; AlJuwaysim, M.F.; BinDhim, N.F.; Alqahtani, S.A. Prevalence of Self-Reported Food Allergies and Their Association with Other Health Conditions among Adults in Saudi Arabia. *Int. J. Environ. Res. Public Health.* **2021**, *18*, 347. [CrossRef] [PubMed]
34. Algabbani, A.M.; Althumiri, N.A.; Almarshad, A.M.; BinDhim, N.F. National prevalence, perceptions, and determinants of tobacco consumption in Saudi Arabia. *Food Drug Regul. Sci. J.* **2019**, *2*, 1–17. [CrossRef]
35. General Authority for Statistics. Methodology of the Housing Statistics. Available online: https://www.stats.gov.sa/en/%D8%A7%D9%84%D9%85%D9%86%D9%87%D8%AC%D9%8A%D8%A7%D8%AA/methodology-housing-survey (accessed on 17 March 2021).

Article

Food Traceability: A Consumer-Centric Supply Chain Approach on Sustainable Tomato

Foivos Anastasiadis *, Ioanna Apostolidou and Anastasios Michailidis

Deptartment of Agricultural Economics, School of Agriculture, Aristotle University of Thessaloniki, 541-24 Thessaloniki, Greece; ioanapost3@yahoo.gr (I.A.); tassosm@auth.gr (A.M.)
* Correspondence: anastasiadis.f@gmail.com

Abstract: Technological advances result in new traceability configurations that, however, cannot always secure transparency and food safety. Even in cases where a system guarantees transparency, the actual consumer involvement and a real consumer-based perspective cannot always be ensured. The importance of such consumer centricity is vital, since it is strongly associated with effective supply chains that properly fulfil their end-users' needs and requests. Thus, the objective of this paper was to explore the level of consumer centricity in food supply chains under a traceability system. The methodological approach employed a framework of two studies validating subsequently a similar set of variables, using initially consumers data and then supply chain actors data. The supply chain of sustainable tomato was selected to design the studies. The level of agreement between datasets suggested the level of the supply chain consumer centricity. Findings showed health, trust, quality, nutrition, and safety-related values to be significant for the consumers towards accepting a traceability system. The supply chain actors also accepted a traceability system based on the fact that their customers' needs rely on the exact same beliefs, indicating a high level of consumer centricity. The current work underlines the magnitude of consumer centricity in food supply chains and provides an easy and straightforward framework for its exploration. Key implications suggest the design of more effective supply chain and consumer-based strategies for the food industry. Policymakers could also adopt the concept of consumer centricity to further improve the food industry.

Keywords: food supply chains; food safety; tomato; Greece; end-to-end approach; sustainability

Citation: Anastasiadis, F.; Apostolidou, I.; Michailidis, A. Food Traceability: A Consumer-Centric Supply Chain Approach on Sustainable Tomato. *Foods* **2021**, *10*, 543. https://doi.org/10.3390/foods10030543

Academic Editors: António Raposo, Renata Puppin Zandonadi and Raquel Braz Assunção Botelho

Received: 28 January 2021
Accepted: 3 March 2021
Published: 5 March 2021

1. Introduction

Consumers by default are a vital part of any supply chain (SC). According to several supply chain definitions [1,2] and, most importantly, based on the supply chain management concept [3,4], the entire supply chain should serve the end-user of the product/service produced and fulfilled by the supply chain, i.e., the consumer. Nonetheless, theoretical and academic approaches usually are quite far from reality. The continuous pressure on the economic viability of businesses—and eventually their supply chains—results in an overall focus on high revenues and even higher margins [5]. Consequently, in practice (versus theory), there are quite a few cases were supply chains choose to operate against their end-users needs. This could be due to the absence of cooperation among different supply chain actors, challenging the integration of key supply chain processes, and to a myopic view focused on specific levels rather than on the entire chain [6]. The repeated food scandals in food supply chains (for instance, horse meat [7], eggs–fipronil [8], pork–hepatitis [9]) perfectly illustrate the above point.

Nowadays, consumers pay close attention to food safety, requesting high-quality food with information transparency and safety assurances [10]. Food traceability systems can reduce information asymmetry, implement safety governance across geographical boundaries, and help increase consumer faith in food safety to achieve a sustainable agricultural development [11]. From a consumer perspective, traceability helps to build trust, provide peace of mind, and increase confidence in the food system. For the growers, traceability is part of an overall cost-effective quality management system that can also assist in continuous improvement and minimization of the impact of safety hazards. It also facilitates the rapid and effective recall of products and the determination and settlement of liabilities [12]. The implementation and benefits of a traceability system depend on consumers' awareness of food safety, thus an increasing need for such systems that provide transparent information on the quality and safety of food supply chains. The greatest benefit of food traceability could be realized when it is implemented at the entire supply chain level and engages all the participants involved [13]. Therefore, there is a conceptual link between food traceability and consumer-centric supply chains.

The current study aims to further explore the level of consumer centricity in food supply chains implementing traceability systems. Such an investigation shall reveal any association between these concepts. Greek sustainable tomato supply chain was chosen as the research area, due to the high production and consumption of tomaotes in Greece. Two different surveys took place, one based on the consumers and the other on the supply chain actors. The analysis resulted in tangible evidence regarding the relationship between traceability and consumer centricity. The analysis also provides further insights into the nature and importance of consumer-centric food supply chains.

2. Materials and Methods

In this section, initially, we separately present the reviews of key studies on consumer-centric food supply chains and food traceability. Afterwards, the relationship between these concepts is further explored in the literature, revealing a specific research gap. Finally, we describe the methodology employed.

2.1. Consumer-Centric Supply Chains

Every company claims a customer-driven strategy. However, the diverse interpretation of their customers' needs by the numerous value-chain entities, either external, i.e., suppliers and retailers, or internal like R&D managers, results in loss of focus. Thus, a need for frameworks and further research that can help companies develop strategies that are actually customer-centric [14]. This sub-section presents and discusses several examples of consumer-centric approaches in food supply chains. These cases employ diverse methods, have different specific objectives, and use various terminologies (e.g., consumer-driven, customer-oriented, demand-driven, consumer co-creation in new product development), yet all are relevant to the same concept: consumer-centric food supply chains.

A study on the organic tomato supply chain in The Netherlands recommends shifting from the Supply Chain Management concept to Value Chain Management and ultimately to move towards more customer-based strategies. It suggests considering consumers' sustainability concerns in the direction of "greening" the supply chains end to end, gaining competitive advantage through a better position of the entire firm's "greening" [15]. Another study on the blueberry supply chain in Italy highlights the impact of a demand-driven supply chain towards the activation of innovation processes throughout the fresh fruit supply chains. This case exemplifies the integration between a business firm and a research centre, resulting in a new consumer-centred dialogue with the market. The main output was a novel packaging system enabling the maintenance of the quality of the fruits for a longer period, improving the exports, increasing the turnover of the associated group, and improving the remuneration of the fruit growers [16].

Research on pork supply chains towards delivering superior eating quality to consumers explored why there is more emphasis on mass production—which usually is combined with traceability systems and high food safety procedures—than on essential quality cues. A more consumer-centric approach suggests combining a hedonic assessment of meat attributes end to end in the value chain, revealing the complexity regarding to the formation, communication, and provision of attribute-related information to the final consumers [17]. Product modularity and customization to consumer's requirements is another form of consumer-centric supply chain approach. A study investigating the relationships between sustainability, operations, and marketing suggests that such configurations may generate higher demand due to greater satisfaction provision. They also reveal an association between modularity and the development of sustainable products, which, in turn, may enhance sustainable consumption [18].

New product development and co-creation with the active involvement of the consumers is a key element in the consumer-centric supply chain configuration. A paper exemplifies a novel methodology for the creation and design of new beverages, employing a digital crowdsourcing tool to reveal consumer preferences. Afterwards, these preferences were transformed into actionable directions, that allowed a rapid production of the beverage for tasting and evaluation. Hence, a closed-loop beverage design method based on consumers co-creation was applied [19]. Another new product development framework highlighting the importance of consumer-centric supply chains suggests the development, integration, and triangulation of personalized food profiles for customized healthier products [20]. Such reconfigurations suggest digital manufacturing and modularity principles for securing an authentic production–consumption alignment while producing personalized healthier food products [6].

2.2. Food Traceability

The food industry is becoming more customer-oriented and needs faster response times to deal with food scandals and incidents. An effective traceability system helps to minimize the production and distribution of unsafe or poor-quality products, thereby minimizing the potential for bad publicity, liability, and recalls. The current food labelling system cannot guarantee that the food is authentic, safe, and of good quality. Therefore, traceability is applied as a tool to assist in the assurance of food safety and quality as well as to achieve consumer confidence [10]. Food traceability has gained considerable importance, particularly following several food safety incidents during which traceability systems were weak or absent [21]. Traceability appears as a tool to comply with legislation and to meet food safety and quality requirements. It is considered to be an effective safety- and quality-monitoring system with the potential to improve safety within food chains, as well as to increase consumer confidence and to connect producers and consumers [11].

To supply top-quality, safe, and nutritious foods, as well as rebuild public confidence in the food chain, the design and implementation of whole chain traceability from farm to end-user have become an important part of the overall food quality assurance system [12]. FAO stated that managing food safety and quality should be a shared responsibility of all actors in the food chain including governments, industry and consumers [22]. Because of globalization in the food trade, effective food control systems are essential to protect the health and safety of consumers. The foremost responsibility of food control is to enforce the food law(s) protecting the consumer against unsafe, impure, and fraudulently presented food [23]. Food traceability systems can provide a reliable and continuous information flow in supply chains, identify root causes of problems, and recall high-risk products from the market [24].

Therefore, food traceability systems can reduce consumer information asymmetry and food safety risks [25–27]. Given that traceability is mainly a quality assurance tool, its implementation depends on many factors linked to the supply chain [28,29] and trade-related issues [30]. The use of traceability as a tool to increase consumer confidence in food safety is also mainly linked to distrust in the food system that should be addressed by the government [30–32].

2.3. Research Gap

The increasing consumer interest on short food supply chains highlights the importance of more information and more direct contact with the producers [33], revealing the need for innovative traceability systems and further active consumer involvement in the food supply system; hence, more attnetion on consumer-centric supply chains. Recent technological advances have significantly contributed to traceability in food supply chains.

For instance, blockchain technology increases trust, authenticity, and transparency [34]. However, there are several challenges such as the need for Internet-of-Things devices throughout the food supply chain to capture all the necessary data from several parties (supply chain actors) and transfer the data to the blockchain [35]. Moreover, blockchain technology is quite new, and long-term impacts of its implementation in the food chain are not yet proven [34]. Besides, with this technology, there is no actual consumer involvement, thus no consumer centricity in the supply chain.

The paradigm of blockchain technology reveals an issue regarding the role of traceability and transparency per se in the food supply chains. The approaches about consumer-oriented traceability are limited in the literature (as illustrated above in Sections 2.1 and 2.2). However, some studies highlighted the fact that consumers value a bi-directional connection to the supply chain through the ability to give feedback via a consumer-oriented food traceability app, supporting in this way sustainable food consumption [36]. In the same direction, another study revealed the importance of customer-delivered value of a food traceability system, that contributes to enhancing their intention to purchase traceable food [37]. Despite the above examples, the majority of extant research on food traceability has taken either a firm-centric approach or a reflexive and non-thorough consumer-based orientation. Whether and how food supply chain actors value their consumers' motivation for a traceability system—including their relevant needs and requests concerning such system—remain to be explored in significant detail. It seems appropriate and necessary, then, to unpack the food supply chain consumer centricity concept for the case of traceability.

2.4. Methodology

The review of previous studies on consumer-centric food supply chains and food traceability revealed a gap in the literature concerning the way the entire supply chain responds to consumers' claims and requests. The case of traceability is the ideal scenario to test the level of consumer centricity in food supply chains due to the, by default, involvement of all the stakeholders through the entire supply chain. Given the certain contribution of the retailers, wholesalers, logistics, manufactures, and producers at the respective supply chain level, a traceability system requires strong collaboration among all its participants to properly function. The question, however, is whether the supply chain actors collaborate under their personal agendas and respond to their customers' needs and requests. The literature review shows a strong possibility of "accidental" response to consumers requests, e.g., no consumer involvement (ergo, no consumer-centricity), yet a satisfactory reaction to transparency claims. Thus, the objective of the study is to explore in depth several aspects of consumer centricity to conclude whether its level is satisfactory throughout the supply chain. An overall research question of the current study could be formulated as follows:

What is the level of consumer centricity in food supply chains applying a traceability system?

To explore such a research question, we designed two studies. The first study (study A) is a consumer survey to identify consumers' perceptions and concerns regarding a traceability system. Similarly, the second study (study B) is a survey on supply chain stakeholders (i.e., retailers, wholesalers, manufactures, and producers), exploring whether they accept a traceability system based on their consumers' perceptions for such a system. As illustrated in the methodological framework (Figure 1), the level of agreement between consumers and supply chain actors on the significant values associated with a traceability system indicates the supply chain's level of consumer centricity.

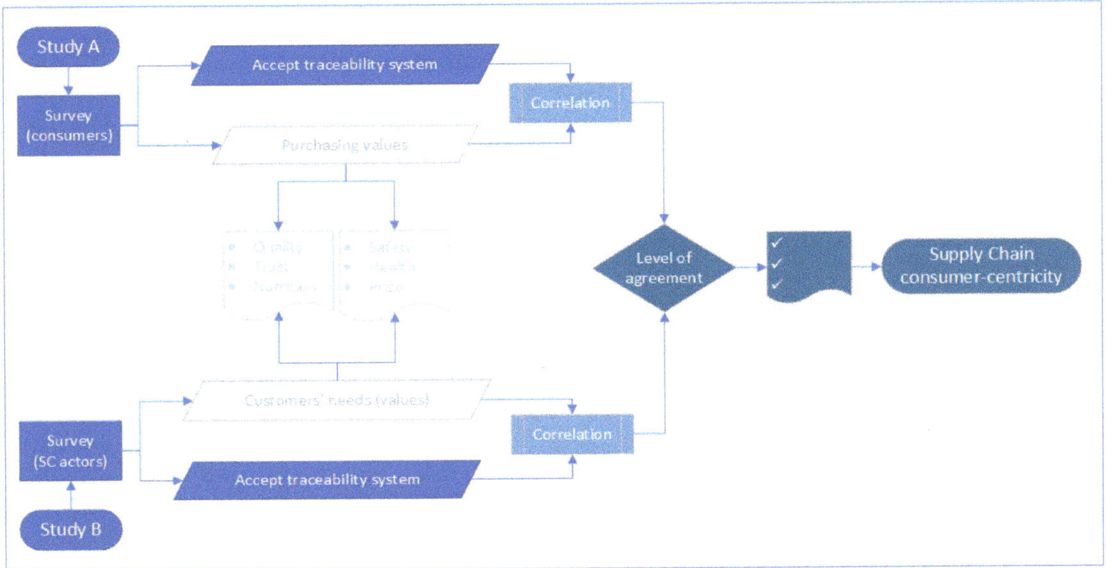

Figure 1. Methodological framework (source: authors).

The supply chain of sustainable tomato was selected to design the studies. That includes all supply actors from the production to the consumption of the tomatoes produced, processed, and sold with sustainable practices, for example, under organic or integrated management. This is essential, so we can present a consistent scenario, with a specific product, to the consumers during the survey in study A and more importantly to target all the stakeholders through the entire supply chain in study B. Regarding data collection, we selected the Metropolitan area of Thessaloniki in Greece as a research area for both studies, due to the high consumption of tomatoes and the availability in the region of the necessary supply chain stakeholders [38].

For study A, we designed and disseminated a questionnaire using SurveyMonkey, exploring consumers' acceptance of a traceability system for sustainable tomato. In addition, we asked consumers to rank the importance of certain purchasing values focused on issues related to: (i) health, (ii) quality, (iii) price, (iv) trust, (v) nutritional value, and (vi) food safety. Most of these questions were framed in five-point Likert-scale intervals, so as to encourage participation and minimize the cognitive burden on the respondents. As a result, the key factors revealed were Quality, Trust, Nutrition, Health, Safety, and Price, consistently with previous studies [39–43]. The sampling method involved an online random sample [44]. The questionnaire was circulated for 4 months and resulted in 729 valid responses.

For study B, we also stuctures and circulated a similar questionnaire using SurveyMonkey, exploring supply chain stakeholders' acceptance of a traceability system for sustainable tomato. Moreover, the participants had to rank the importance of their customers' specific values (using a five-point Likert scale) about their decision to accept a traceability system. The key values used were the same six values from study A, resulting in the same factors (Quality, Trust, Nutrition, Health, Safety, and Price) plus three more business-oriented values: Reliability, Better Market Access, and Stakeholders' Sustainability [45,46]. In this study, the sample was selected following the expert sampling technique [47]. Participants were experts in the area of study, and this sampling method also involved sample assembling of a group of people that could use their experience. The reason for using expert sampling was to have a better way of constructing the views of individuals that were expert in a definite area. The sampling objective was to include representatives from the entire tomato supply chain, i.e., producers, middlemen, wholesalers, manufactures, logistics, and retailers. Given the significantly smaller number of supply chain stakeholders compared to consumers and the objective to achieve—an analogy among the available stakeholders in every category and those in our sample—we concluded with a sample of 69 valid responses.

3. Results

Initially, in this section, we separately present, in their respective subsections, the analysis and the results of the two studies. Then, we discuss these results by reflecting on the methodological framework and the status of the supply chain consumer centricity.

3.1. Study A (Consumers)

Descriptive statistics of the demographic characteristics (Table 1) suggested a representative sample based on other related studies [48–50]. Reliability analysis confirmed that the scale was reliable, Cronbach's alpha coefficients ranged from 0.712 to 0.809, exceeding the minimum standard of 0.60 suggested by Malhotra [51].

Table 1. Descriptive statistics of consumers' demographic characteristics.

	Education		
		Frequency	**Percent**
	High school	71	9.7
	University	329	45.1
Valid	Masters	178	24.4
	PhD	153	20.7
	Total	729	100.0
	Gender		
	Female	474	65.0
Valid	Male	255	35.0
	Total	729	100.0
	Age		
	18–24	135	18.5
	25–34	138	18.9
	35–44	173	23.7
Valid	45–54	164	22.5
	55–64	96	13.2
	65+	23	3.2
	Total	729	100.0
	Annual Household Income		
	<€7999	126	17.3
	€8000–€14,999	202	27.7
	€15,000–€24,999	190	26.1
	€25,000–€39,000	154	21.1
Valid	€40,000–€59,999	44	6.0
	€60,000–€74,999	7	1.0
	€75,999–€99,999	6	8
	Total	729	100.0
	Family Status		
	Married	354	48.6
Valid	Single	375	51.4
	Total	729	100.0

Bivariate Spearman correlation was employed, since Pearson's correlation assumptions were violated [52], to explore the relation between Accept traceability system and the six purchasing factors (Quality, Trust, Nutrition, Health, Safety, and Price). Table 2 presents the output of the statistical analysis using SPSS. Quality, Trust, Nutrition, Safety, and Health were positively related to Accept traceability system, with a correlation coefficients rho = 0.129, rho = 0.144, rho = 0.156, rho = 0.218, and rho = 0.156, respectively, which were also significant at $p < 0.01$ probability. Only Price was not statistically significantly related to Accept traceability system.

Table 2. Correlations (consumers).

Accept Traceability System	Spearman Correlation Coefficient (*rho*)
Quality	0.129 **
Trust	0.144 **
Nutrition	0.156 **
Safety	0.218 **
Heath	0.156 **
Price	0.029

** Indicates that correlation is significant at the 0.01 level (two-tailed).

3.2. Study B (SC Stakeholders)

The sampling procedure (as justified above in Section 3) suggested a representative sample, including a sufficient number of every stakeholder through the entire supply chain. Reliability analysis confirmed that the scale was reliable, with Cronbach's alpha coefficients rangein from 0.805 to 0.913, exceeding the minimum standard of 0.60 suggested by Malhotra [51].

Bivariate Spearman correlation was employed, since Pearson's correlation assumptions were violated [52], to explore the relation between *Accept traceability system* and the six customers' values (Quality, Trust, Nutrition, Health, Safety, and Price) plus the three business-related variables (Reliability, Better Market Access, and Stakeholders' Sustainability). Table 3 presents the output of the statistical analysis using SPSS. Trust, Health, Safety, Quality, Nutrition, Price, and Reliability awee positively related to Accept traceability system, with correlation coefficients rho = 0.612, rho = 0.539, rho = 0.584, rho = 0.422, rho = 0.564, rho = 0.341, rho = 0.442, respectively, which were also significant at $p < 0.01$ probability. Stakeholders' Sustainability, and Better Market Access were positively related to Accept traceability system, with correlation coefficients of rho = 0.277 and rho = 0.343, significant at $p < 0.05$ and $p < 0.01$, respectively.

Table 3. Correlations, supply chain (SC) stakeholders.

Accept Traceability System	Spearman Correlation Coefficient (rho)
Increase consumers' trust	0.612 **
Protect consumers' health	0.539 **
Increase food safety	0.584 **
Better food quality	0.422 **
Positive on products nutritional value	0.564 **
Improve stakeholders' sustainability	0.277 *
Product price	0.341 **
Stakeholders' reliability	0.442 **
Better market access	0.343 **

*, ** Indicates correlation is significant at the 0.05 level (two-tailed) and 0.01 level (two-tailed), respectively.

4. Discussion

This section discusses the results, reflecting on the supply chain consumer centricity dimention. Study A resulted in five (out of six) significant factors related to the acceptance of a traceability system. That means consumers have a positive perception of a traceability system. Specifically, applying such a system to the sustainable tomato supply chain enhances consumers' trust in the entire supply chain. Also, it improves their perception of a product's quality and its nutritional value, while increasing their sense of product's safety. Overall, applying a traceability system to food products makes consumers believe that these products are more beneficial to their health. These findings are consistent with previous studies [24,53].

Study B resulted in seven (out of nine) significant factors related to the acceptance of a traceability system. That means sustainable tomato supply chain actors have a positive perception about a traceability system. Particularly, the supply chain stakeholders believe that the acceptance of a traceability system would increase their customers' trust. Moreover, they believe that it would improve their customers' perception of the nutritional value and both the safety and the quality of their products. Overall, they believe that such a system makes their customers believe that traceable products protect their health. On top of these customer-related values, supply chain stakeholders believe that a traceability system will have a positive effect on their reliability and on products' prices—thus strengthening their reputation—which is consistent with their previous positive perceptions. Past studies also suggest a similar supply chain stakeholder's behavior [54], though not necessarily related to their customers' perceptions.

Comparing the output of both studies, according to the methodological framework, we should highlight the absolute matching among the significant purchasing values of the consumers and the supply chain actors' perceptions on their customers' needs and requests (values). Reviewing such an outcome from a supply chain perspective results in prominent conclusions concerning the consumer centricity of a sustainable tomato sector that has adopted a traceability system. The tomato supply chain end-users (the consumers) appreciate the positive effect of a traceability system on the product's quality, safety, and nutritional value, on their trust for these products, and overall, on their health. Identically, the rest of the supply chain actors (the retailers, wholesalers, logistic operators, manufactures, and producers) value in a similar manner the exact same beliefs, considering the positive effect of a traceability system on their customers. In other words, the supply chain responds to the requests of the consumers and considers adopting a traceability system following a uniform end-to-end logic. This a priori test of traceability-related variables among consumers and subsequently multiple supply chain actors is the main conceptual contribution of the current study.

The set of variables (purchasing-related values) tested in study A and next tested in study B (as customers' needs according to supply chain actors) corresponds to the depth of a consumer centricity exploration: the larger the set of variables, the more thorough the exploration. Similarly, the higher the agreement on these variables among consumers and supply chain stakeholders, the higher the level of consumer centricity. This means that adopting a traceability system for such a supply chain shall be very well appreciated by its consumers.

5. Conclusions

Findings from study A suggested consumer's acceptance of a traceability system for a sustainable tomato supply chain based on the significance of quality, trust, nutrition, safety, and health-related values. Similarly, study B verified that the supply chain actors accept a traceability system based on the fact that their customers' needs rely on the exact same beliefs, indicating a high level of consumer centricity. Reaching such a conclusion before investing the implementation of a food traceability system is extremely valuable. It is not just a positive signal regarding the supply chain effectiveness but also predefines a significant consumer engagement in the presence of a traceability system. In line with previous studies [36], such level of supply chain consumer centricity indicates a bi-directional connection that could result in interactive communication between consumers and multiple supply chain actors.

Underling the magnitude of supply chains' consumer centricity is the key contribution of the current study. The framework employed to achieve this is also significant. Despite the availability of a few studies on consumer-oriented supply chains [14,16,17,55,56], the extant literature fails to provide a straightforward approach to explore the actual consumer centricity at a supply chain level.

The obvious implication of the current research findings is a recommendation to apply the framework to the existing traceability systems and further improve these systems by adopting consumer involvement and engagement principles. Moreover, there are wider applications that go beyond just a positive impact on traceability systems, with several managerial and policy-making implications. Concerning the former, a consumer centricity assessment shall be part of any management toolbox, especially among C-level executives and decision-makers. A firm's strategy and the design of future investments must be viewed from a consumer's perspective. Regarding the latter, the food industry, and generally, the agricultural sector could also benefit from shifting their policy towards a more consumer-centric direction. Policymakers should realize that the entire food system has to work towards fulfilling consumers' needs. Therefore, the current study establishes a necessity of better understanding the importance of supply chains consumer centricity and adopting its principles in the policy-making process.

Limitations of the study and, simultaneously, recommendations for further research regard the generalization of the findings due to the fact that this work was based on a single case and a national sampling. Similar studies should be designed in several countries for multiple food supply chains, covering many types of products, i.e., vegetables, meat, and dairy. Moreover, additional variables and multivariate statistical analyses could enhance the current framework. Nevertheless, caution is necessary about adding variables, which would increase the complexity of the system and its analysis. In fact, one of the objectives (and also key advantages) of the framework is the fact that it is easy and straightforward.

Author Contributions: Conceptualization, F.A.; methodology, F.A.; formal analysis, F.A.; investigation, F.A.; resources, F.A.; data curation, F.A.; writing—original draft preparation, F.A.; writing—review and editing, F.A.; visualization, F.A.; review, I.A.; project administration, I.A.; funding acquisition, F.A. and I.A.; supervision, A.M. All authors have read and agreed to the published version of the manuscript.

Funding: This project has received funding from the Hellenic Foundation for Research and Innovation (HFRI) and the General Secretariat for Research and Innovation (GSRI), grant agreement No [1786].

Institutional Review Board Statement: Not applicable.

Informed Consent Statement: Not applicable.

Data Availability Statement: Not applicable.

Conflicts of Interest: The authors declare no conflict of interest.

References

1. Christopher, M. Logistics and Supply Chain Management: Strategies for Reducing Cost and Improving Service (Second Edition). *Int. J. Logist. Res. Appl.* **1999**, *2*, 103–104. [CrossRef]
2. Chopra, S.; Meindl, P. *Supply Chain Management. Strategy, Planning & Operation*, 3rd ed.; Prentice Hall: Upper Saddle River, NJ, USA, 2007; pp. 265–275.
3. Burgess, K.; Singh, P.J.; Koroglu, R. Supply chain management: A structured literature review and implications for future research. *Int. J. Oper. Prod. Manag.* **2006**, *26*, 703–729. [CrossRef]
4. Storey, J.; Emberson, C.; Godsell, J.; Harrison, A. Supply chain management: Theory, practice and future challenges. *Int. J. Oper. Prod. Manag.* **2006**, *26*, 754–774. [CrossRef]
5. Anderson, D.; Britt, F.; Favre, D.J. The 7 principles of supply chain management. *Supply Chain Manag. Rev.* **2007**, *11*, 41–46.
6. Anastasiadis, F. Consumer-centric food supply chains for healthy nutrition. In Proceedings of the 21st Cambridge International Manufacturing Symposium "GLOBALISATION 2.0 Rethinking supply chains in the new technological and political landscape", Cambridge, UK, 28–29 September 2017.
7. Levitt, T. Three Years on from the Horsemeat Scandal: 3 Lessons We Have Learned. The Guardian. 2017. Available online: https://www.theguardian.com/sustainable-business/2016/jan/07/horsemeat-scandal-food-safety-uk-criminal-networks-supermarkets (accessed on 23 January 2021).
8. Roberts, R. Eggs Recall: 700,000 Distributed in UK Contaminated with Fipronil Pesticide. The Independent. 2017. Available online: http://www.independent.co.uk/news/uk/home-news/eggs-recall-uk-pesticide-fipronil-contaminated-germany-lidl-aldi-netherlands-a7886041.html (accessed on 23 January 2021).

9. Gander, K. UK Supermarket' May Have Infected Thousands with Hepatitis E from Sauseges and Ham. The Independent. 2017. Available online: http://www.independent.co.uk/life-style/food-and-drink/sausages-ham-pork-hepatitis-virus-hev-supermarket-uk-tesco-public-health-england-a7903431.html (accessed on 23 January 2021).
10. Aung, M.M.; Chang, Y.S. Traceability in a food supply chain: Safety and quality perspectives. *Food Control.* **2014**, *39*, 172–184. [CrossRef]
11. Kher, S.V.; Frewer, L.J.; De Jonge, J.; Wentholt, M.; Davies, O.H.; Luijckx, N.B.L.; Cnossen, H.J. Experts ' perspectives on the implementation of traceability in Europe. *Br. Food J.* **2010**, *112*, 261–274. [CrossRef]
12. Opara, L.U. Traceability in agriculture and food supply chain: A review of basic concepts, technological implications, and future prospects. *Food Agric. Environ.* **2003**, *1*, 101–106.
13. Zhu, L. Economic Analysis of a Traceability System for a Two-Level Perishable Food Supply Chain. *Sustainability* **2017**, *9*, 682. [CrossRef]
14. Simons, R. Choosing the Right Customer. *Harv. Bus. Rev.* **2014**, *92*, 48.
15. Anastasiadis, F.; van Dam, Y.K. Consumer driven supply chains: The case of Dutch organic tomato. *Agric. Eng. Int. CIGR J.* 2014; Special issue 2014: Agri-food and biomass supply chains, 11–20.
16. Peano, C.; Girgenti, V.; Baudino, C.; Giuggioli, N.R. Blueberry Supply Chain in Italy: Management, Innovation and Sustainability. *Sustainability* **2017**, *9*, 261. [CrossRef]
17. Schulze-Ehlers, B.; Anders, S. Towards consumer-driven meat supply chains: Opportunities and challenges for differentiation by taste. *Renew. Agric. Food Syst.* **2017**, *33*, 73–85. [CrossRef]
18. Ülkü, M.A.; Hsuan, J. Towards sustainable consumption and production: Competitive pricing of modular products for green consumers. *J. Clean. Prod.* **2017**, *142*, 4230–4242. [CrossRef]
19. Jreissat, M.; Isaev, S.; Moreno, M.; Makatsoris, C. Consumer Driven New Product Development in Future Re-Distributed Models of Sustainable Production and Consumption. *Procedia CIRP* **2017**, *63*, 698–703. [CrossRef]
20. Anastasiadis, F.; Aranda-Jan, C.; Tsolakis, N.; Srai, J.S. Consumer-centric New Food Product Development: A Conceptual Framework. In Proceedings of the 47th EMAC conference: "People Make Marketing", Glasow, UK, 29 May–1 June 2018.
21. Heyder, M.; Theuvsen, L.; Hollmann-Hespos, T. Investments in tracking and tracing systems in the food industry: A PLS analysis. *Food Policy* **2012**, *37*, 102–113. [CrossRef]
22. FAO. FAO's Strategy for a Food Chain Approach to Food Safety and Quality: A Framework Document for the Development of Future Strategic Direction. Available online: http://www.fao.org/DOCREP/MEETING/006/y8350e.HTM (accessed on 23 January 2021).
23. FAO/WHO. Assuring Food Safety and Quality Guidelines for Strengthening National Food Control Systems. Available online: https://www.who.int/foodsafety/publications/guidelines-food-control/en/ (accessed on 25 January 2021).
24. Van Rijswijk, W.; Frewer, L.J.; Menozzi, D.; Faioli, G. Consumer perceptions of traceability: A cross-national comparison of the associated benefits. *Food Qual. Prefer.* **2008**, *19*, 452–464. [CrossRef]
25. Dandage, K.; Badia-Melis, R.; Ruiz-García, L. Indian perspective in food traceability: A review. *Food Control.* **2017**, *71*, 217–227. [CrossRef]
26. Jin, S.; Zhou, L. Consumer interest in information provided by food traceability systems in Japan. *Food Qual. Prefer.* **2014**, *36*, 144–152. [CrossRef]
27. Lee, J.Y.; Han, D.B.; Nayga, R.M.; Lim, S.S. Valuing traceability of imported beef in Korea: An experimental auction approach*. *Aust. J. Agric. Resour. Econ.* **2011**, *55*, 360–373. [CrossRef]
28. Pekkirbizli, T.; Almadani, M.I.; Theuvsen, L. Food safety and quality assurance systems in Turkish agribusiness: An empirical analysis of determinants of adoption. *Econ. Agro Aliment.* **2015**, 31–55. [CrossRef]
29. Hofstede, G.J.; Fritz, M.; Canavari, M.; Oosterkamp, E.; Van Sprundel, G. Towards a cross-cultural typology of trust in B2B food trade. *Br. Food J.* **2010**, *112*, 671–687. [CrossRef]
30. Rimpeekool, W.; Seubsman, S.-A.; Banwell, C.; Kirk, M.; Yiengprugsawan, V.; Sleigh, A. Food and nutrition labelling in Thailand: A long march from subsistence producers to international traders. *Food Policy* **2015**, *56*, 59–66. [CrossRef]
31. Canavari, M.; Centonze, R.; Hingley, M.; Spadoni, R. Traceability as part of competitive strategy in the fruit supply chain. *Br. Food J.* **2010**, *112*, 171–186. [CrossRef]
32. Chang, A.; Tseng, C.-H.; Chu, M.-Y. Value creation from a food traceability system based on a hierarchical model of consumer personality traits. *Br. Food J.* **2013**, *115*, 1361–1380. [CrossRef]
33. Elghannam, A.; Mesias, F.J.; Escribano, M.; Fouad, L.; Horrillo, A.; Escribano, A.J. Consumers' Perspectives on Alternative Short Food Supply Chains Based on Social Media: A Focus Group Study in Spain. *Foods* **2019**, *9*, 22. [CrossRef]
34. Köhler, S.; Pizzol, M. Technology assessment of blockchain-based technologies in the food supply chain. *J. Clean. Prod.* **2020**, *269*, 122193. [CrossRef]
35. Kayikci, Y.; Subramanian, N.; Dora, M.; Bhatia, M.S. Food supply chain in the era of Industry 4.0: Blockchain technology implementation opportunities and impediments from the perspective of people, process, performance, and technology. *Prod. Plan. Control.* **2020**, 1–21. [CrossRef]
36. Lawo, D.; Neifer, T.; Esau, M.; Vonholdt, S.; Stevens, G. WITHDRAWN: From Farms to Fridges: A Consumer-Oriented Design Approach to Sustainable Food Traceability. *Sustain. Prod. Consum.* **2021**, *27*, 282–297. [CrossRef]

37. Yuan, C.; Wang, S.; Yu, X. The impact of food traceability system on consumer perceived value and purchase intention in China. *Ind. Manag. Data Syst.* **2020**, *120*, 810–824. [CrossRef]
38. Anastasiadis, F.; Apostolidou, I.; Michailidis, A. Mapping Sustainable Tomato Supply Chain in Greece: A Framework for Research. *Foods* **2020**, *9*, 539. [CrossRef]
39. Chamhuri, N.; Batt, P.J. Consumer perceptions of food quality in Malaysia. *Br. Food J.* **2015**, *117*, 1168–1187. [CrossRef]
40. Cope, S.; Frewer, L.J.; Houghton, J.; Rowe, G.; Fischer, A.R.H.; De Jonge, J. Consumer perceptions of best practice in food risk communication and management: Implications for risk analysis policy. *Food Policy* **2010**, *35*, 349–357. [CrossRef]
41. Nikolova, H.D.; Inman, J.J. Healthy Choice: The Effect of Simplified Point-of-Sale Nutritional Information on Consumer Food Choice Behavior. *J. Mark. Res.* **2015**, *52*, 817–835. [CrossRef]
42. Nuttavuthisit, K.; Thøgersen, J. The Importance of Consumer Trust for the Emergence of a Market for Green Products: The Case of Organic Food. *J. Bus. Ethic.* **2017**, *140*, 323–337. [CrossRef]
43. Paul, J.; Rana, J. Consumer behavior and purchase intention for organic food. *J. Consum. Mark.* **2012**, *29*, 412–422. [CrossRef]
44. Ilieva, J.; Baron, S.; Healey, N.M. Online Surveys in Marketing Research. *Int. J. Mark. Res.* **2002**, *44*, 1–14. [CrossRef]
45. Sun, S.; Wang, X. Promoting traceability for food supply chain with certification. *J. Clean. Prod.* **2019**, *217*, 658–665. [CrossRef]
46. Sheu, Y.C.; Yen, H.R.; Chae, B. Determinants of supplier-retailer collaboration: Evidence from an international study. *Int. J. Oper. Prod. Manag.* **2006**, *26*, 24–49. [CrossRef]
47. Singh, K. *Quantitative Social Research Methods*; Sage Publications: Thousand Oaks, CA, USA; London, UK, 2007.
48. Dimara, E.; Skuras, D. Consumer evaluations of product certification, geographic association and traceability in Greece. *Eur. J. Mark.* **2003**, *37*, 690–705. [CrossRef]
49. Kehagia, O.C.; Colmer, C.; Chryssochoidis, M.G. Consumer valuation of traceability labels: A cross-cultural study in Germany and Greece. *Br. Food J.* **2017**, *119*, 803–816. [CrossRef]
50. Kehagia, O.; Chrysochou, P.; Chryssochoidis, G.; Krystallis, A.; Linardakis, M. European Consumers' Perceptions, Definitions and Expectations of Traceability and the Importance of Labels, and the Differences in These Perceptions by Product Type. *Sociol. Rural.* **2007**, *47*, 400–416. [CrossRef]
51. Malhotra, N.K. *Marketing Research: An Applied Orientation*; Pearson Education International: Cranbury, NJ, USA, 2004.
52. Field, A.P. *Discovering Statistics Using IBM SPSS Statistics*, 5th ed.; SAGE Publications: London, UK; Thousand Oaks, CA, USA, 2018; ISBN 9781526419514.
53. Dani, D.; Maria, C.C.R.; Zdenka, J.; Hana, B.; Simona, J. Consumers' response to different shelf life food labelling. *Qual. Assur. Saf. Crop. Foods* **2020**, *12*, 24–34. [CrossRef]
54. Pappa, I.C.; Iliopoulos, C.; Massouras, T. What determines the acceptance and use of electronic traceability systems in agri-food supply chains? *J. Rural. Stud.* **2018**, *58*, 123–135. [CrossRef]
55. Buurma, J.; Lamine, C.; Haynes, I. Transition to consumer-driven value chains in The Netherlands. *Acta Hortic.* **2012**, 69–75. [CrossRef]
56. Mishra, N.; Singh, A.; Rana, N.P.; Dwivedi, Y.K. Interpretive structural modelling and fuzzy MICMAC approaches for customer centric beef supply chain: Application of a big data technique. *Prod. Plan. Control.* **2017**, *28*, 945–963. [CrossRef]

MDPI
St. Alban-Anlage 66
4052 Basel
Switzerland
Tel. +41 61 683 77 34
Fax +41 61 302 89 18
www.mdpi.com

Foods Editorial Office
E-mail: foods@mdpi.com
www.mdpi.com/journal/foods